Contents

Preface

Electromagnetics is a discipline developed over centuries, demonstrated by innumerable literatures and versatile applications. Extensive courses and reading assignments may be required to cover all worthy subjects if students or professional engineers try to master this field.

Creative ideas are often inspired by extending concepts over different fields. Comprehensive study over various subjects definitely helps if time allows and efficiency is not of concern. On the other hand, getting a global picture over relevant subjects seems more realistic. In the era of knowledge economy, accumulation of information is accelerating at an amazing rate. Access to literatures has never been so convenient, yet judgment and organization become more critical skills and capabilities. Bearing relevant background knowledge definitely helps to acquire desired information, and hopefully may inspire new ideas more fluently.

This book provides well-organized materials for students or professional engineers to master the field of electromagnetic applications more effectively. This book may also provide them with a sketch to prepare for advanced studies or pursue higher degrees in this field.

Michael Hackett from the Kluwer Academic Publishers is deeply appreciated for inviting me to edit this book. Editing task should become relatively easy as long as a competitive team of contributors can be recruited. Having served as Secretariat of the 2001 Asia-Pacific Microwave Conference, my impression on the participants and their contributions was still fresh at the moment of receiving this invitation. After carefully reviewing the conference proceedings, a candidate list was prepared, which more or less reflected my familiarity with the subject matters. I am very glad that most of the invitees shared my vision and agreed to take part in this book project.

After collecting the first drafts, style issue was immediately brought to my attention. For readers not very familiar with the subject matters, changing style in different Chapters may spoil their appetite and discourage their further exploration. Hence, many efforts were spent to ensure consistent style in all Chapters. Although this endeavor turns out to be more time consuming than I expected, benefits to readers make it worthwhile. Meanwhile, a few Chapters were carefully drafted to bridge some gaps between Chapters. Some of my graduate students were summoned to help collecting and reviewing relevant materials. Occasionally, I attended some short courses or workshops to update knowledge.

Chapters in this book are aligned from concepts to practices, from component design to system architecture, and from small scale to large scale. Each Chapter focuses on specific topic and is organized to be self-sufficient. Each chapter covers concise description of relevant background information, major issues, current trend, future challenges and useful references for further reading. Meanwhile, the terminology is carefully revised to maintain consistency over the whole volume. With these efforts, this book becomes suitable as textbook for senior or graduate courses in microwave engineering because many important and interesting materials are well organized for the disposal of lecturers and students. Many Chapters have been delivered in a graduate course "Applied Electromagnetics" at National Taiwan University by myself to fine-tune the alignment of Chapters.

I would like to express my sincerest appreciation to all the peers who generously provide their knowledge to this book project. I would also like to thank my graduate students who participate in preparing materials for some Chapters, they are Tze-Hsuan Chang, Yuan-Shun Cheng, Ching-I Cho, Hsuan-Sheng Chou, Hsiao-Lun Hsu, Monsen Leu, Chi-Yu Peng, Chun-Wei Wu and Tsung-Sang Yang. My graduate students, Tze-Hsuan Chang, Chih Feng Chou, Jing Je Gau, Cheng Wei Lan and Wen-Zhou Wu, are appreciated for their help to transform files to uniform format, enhance quality of some figures and prepare lecture materials.

Above all, I appreciate the invitation of Kluwer Academic Publishers and Michael Hackett as well as their patience to allow schedule delay for polishing the manuscript. I would also like to thank Deborah Doherty of KAP for her quick response whenever editing assistance is needed.

My wife, Ming-Whei, is appreciated for taking over additional errands beyond her own tight schedule. I also appreciate the good company of my little boy, Bill, whenever I worked on this book at home.

JEAN-FU KIANG, PH.D., PROFESSOR
Department of Electrical Engineering and
Graduate Institute of Communication Engineering
National Taiwan University
Taipei, Taiwan, ROC
June 2003

Chapter 1

REVIEW OF ELECTROMAGNETIC THEORY

Jean-Fu Kiang
Department of Electrical Engineering and
Graduate Institute of Communication Engineering
National Taiwan University
Taipei, Taiwan, ROC

Abstract In this Chapter, relevant electromagnetic theories are reviewed to facilitate the understanding of various mechanisms presented in this book.

Keywords: Maxwell's equations, boundary conditions, wave equations, time-harmonic, potential function, plane wave, layered medium, microstrip line, coplanar waveguide, spectral domain, integral equations, method of moments, resonator, perturbation, periodic structures, electrostatic, magnetostatic, far field, stationary phase approximation, dyadic Green's function.

1. Maxwell's Equations

1.1 Integral Form

Observations on electric and magnetic fields in all possible environments can be summarized by the following four Maxwell's equations in their integral form

$$\oint_C \bar{E}\cdot d\bar{\ell} = -\frac{d}{dt}\iint_S \bar{B}\cdot d\bar{a} \qquad \text{Faraday's law}$$

$$\oint_C \bar{H}\cdot d\bar{\ell} = \frac{d}{dt}\iint_S \bar{D}\cdot d\bar{a} + \iint_S \bar{J}\cdot d\bar{a} \qquad \text{Ampere's law}$$

$$\oiint_S \bar{D}\cdot d\bar{a} = \iiint_V \rho\, dv \qquad \text{electric Gauss' law}$$

$$\oiint_S \bar{B} \cdot d\bar{a} = 0 \qquad \text{magnetic Gauss' law}$$

where $\bar{E}(\bar{r},t)$, $\bar{H}(\bar{r},t)$, $\bar{D}(\bar{r},t)$ and $\bar{B}(\bar{r},t)$ are the electric field intensity (V/m), magnetic field intensity(A/m), electric flux density (As/m^2) and magnetic flux density(Vs/m^2), respectively; $\bar{J}(\bar{r},t)$ is the electric current density(A/m^2), and $\rho(\bar{r},t)$ is the electric charge density(C/m^3). Fig.1.1 shows examples of closed surface and closed contour. The Faraday's law states that the integral of tangential electric field along a closed contour is equal to the negative temporal rate of change of the total magnetic flux flowing through that closed contour. The Ampere's law states that the integral of tangential magnetic field along a closed contour is equal to the total current plus the temporal rate of change of the total electric flux flowing through that closed contour. Hence, the temporal rate of change of the electric flux density can be viewed as an equivalent electric current density. The electric Gauss' law states that the total electric flux flowing out of a closed surface is equal to the total charge enclosed by that surface. The magnetic Gauss' law states that the total magnetic flux flowing out of a closed surface is equal to zero, which implies that there is no net magnetic charge everywhere. Since the Maxwell's equations are induced from observations, their validity can be challenged only by new evidences of observation.

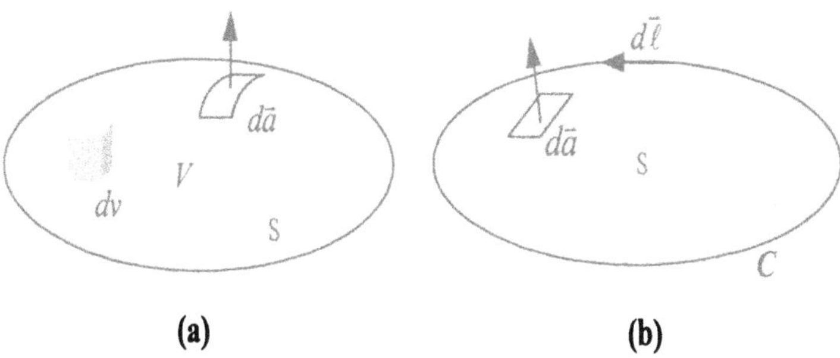

Figure 1.1. (a) Space V enclosed by a closed surface S, (b) open surface S encircled by a closed contour C.

The following physical law of charge conservation is indispensable when applying the Maxwell's equations to solve practical problems.

$$-\frac{d}{dt}\iiint_V \rho dv = \oiint_S \bar{J} \cdot d\bar{a}$$

which states that the total current flowing out of a closed surface is equal to the negative temporal rate of change of the total charge enclosed by that surface.

1.2 Differential Form

The Stoke's theorem relates a closed contour integral of a vector field to the curl of that vector field over the surface encricled by that closed contour. The divergence theorem relates a closed surface integral of a vector field to the divergence of that vector field over the space enclosed by that closed surface, namely

$$\oint_C \bar{V} \cdot d\bar{\ell} = \iint_S \nabla \times \bar{V} \cdot d\bar{a}$$

$$\oiint_S \bar{V} \cdot d\bar{a} = \iiint_V \nabla \cdot \bar{V} dv$$

These two theorems can be applied to transform the integral form of Maxwell's equations and charge conservation law into their differential counterparts as

$$\nabla \times \bar{E} = -\frac{\partial \bar{B}}{\partial t} \qquad \text{Faraday's law}$$

$$\nabla \times \bar{H} = \frac{\partial \bar{D}}{\partial t} + \bar{J} \qquad \text{Ampere's law}$$

$$\nabla \cdot \bar{D} = \rho \qquad \text{electric Gauss' law}$$

$$\nabla \cdot \bar{B} = 0 \qquad \text{magnetic Gauss' law}$$

$$\nabla \cdot \bar{J} = -\frac{\partial \rho}{\partial t} \qquad \text{charge conservation law}$$

By using the vector identity $\nabla \cdot \nabla \times \bar{V} = 0$, the two Gauss' laws can be derived by taking the divergence of the Faraday's law and Ampere's law, respectively. Hence, the Faraday's law, Ampere's law and charge conservation law are the fundamental laws from which other theories and equations can be deduced.

1.3 Boundary Conditions

Proper form of the Maxwell's equations need to be derived around the interfaces between two different media which are encountered in many practical problems. Consider a rectangular contour embedded in two media as shown in Fig.1.2(a), the Faraday's law in its integral form is applied along this contour to have $(\bar{E}_1 - \bar{E}_2) \cdot \hat{\ell}\ell = -d(\ell h \bar{B} \cdot \hat{t})/dt$,

assuming $h \ll \ell$. By letting ℓ approach 0, we have $\hat{n} \times (\bar{E}_1 - \bar{E}_2) = 0$. Similarly, the Ampere's law can be reduced to $\hat{n} \times (\bar{H}_1 - \bar{H}_2) = \bar{J}_s$ where $\bar{J}_s$ is the surface current density(A/m) on the interface, usually resides on a conductor.

Next, conside a pillbox with circular cross section as shown in Fig.1.2(b), the electric Gauss' law is reduced to $\bar{D}_1 \cdot \hat{n}A - \bar{D}_2 \cdot \hat{n}A = \rho_s A$, assuming $2\pi Rh \ll A$. By letting A approach 0, we have $\hat{n} \cdot (\bar{D}_1 - \bar{D}_2) = \rho_s$ where ρ_s is the surface charge density(C/m^2) on the interface, usually resides on a conductor. Similarly, the magnetic Gauss' law and charge conservation law can be reduced to $\hat{n} \cdot (\hat{B}_1 - \hat{B}_2) = 0$ and $\hat{n} \cdot (\bar{J}_1 - \bar{J}_2) = -d\rho_s/dt$, respectively.

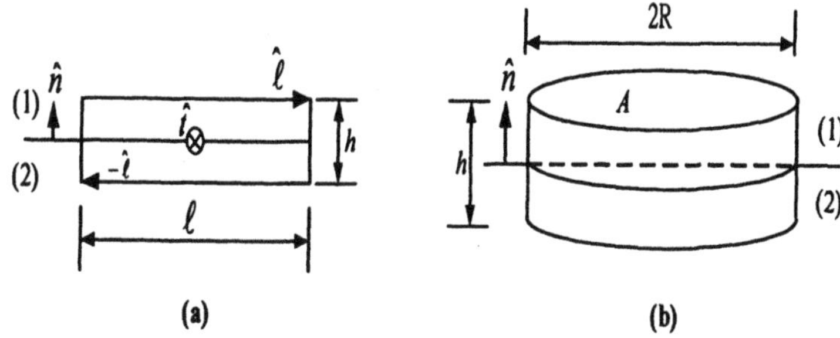

Figure 1.2. (a) A rectangular contour embedded in two media, (b) a pillbox with circular cross section embedded in two media.

In summary, the boundary conditions are

$$\hat{n} \times (\bar{E}_1 - \bar{E}_2) = 0 \qquad \text{Faraday's law}$$
$$\hat{n} \times (\bar{H}_1 - \bar{H}_2) = \bar{J}_s \qquad \text{Ampere's law}$$
$$\hat{n} \cdot (\bar{D}_1 - \bar{D}_2) = \rho_s \qquad \text{electric Gauss' law}$$
$$\hat{n} \cdot (\hat{B}_1 - \hat{B}_2) = 0 \qquad \text{magnetic Gauss' law}$$
$$\hat{n} \cdot (\bar{J}_1 - \bar{J}_2) = -\frac{d\rho_s}{dt} \qquad \text{charge conservation law}$$

2. Wave Equations

To solve for the 12 scalar unknowns in $\bar{E}$, $\bar{H}$, $\bar{D}$ and $\bar{B}$ form the Faraday's law and Ampere's law, assuming that $\bar{J}$ and ρ are known, we need six more independent scalar equations which are provided by the constitutive relations

$$\bar{D} = \bar{\bar{\epsilon}} \cdot \bar{E} + \bar{\bar{\zeta}} \cdot \bar{H}$$
$$\bar{B} = \bar{\bar{\mu}} \cdot \bar{H} + \bar{\bar{\chi}} \cdot \bar{E}$$

For a general bianisotropic medium, $\bar{\bar{\epsilon}}$, $\bar{\bar{\zeta}}$, $\bar{\bar{\mu}}$ and $\bar{\bar{\chi}}$ are dyadics relating flux densities to field intensities. For a linear isotropic medium, the above constitutive relations reduce to

$$\bar{D} = \epsilon\bar{E} = \epsilon_0\bar{E} + \bar{P}$$
$$\bar{B} = \mu\bar{H} = \mu_0\bar{H} + \bar{M}$$

where $\bar{P}$ is the polarization density and $\bar{M}$ is the magnetization density.

Consider a source-free region in free space where $\bar{J} = 0$, $\rho = 0$, $\bar{D} = \epsilon_0\bar{E}$ and $\bar{B} = \mu_0\bar{H}$, the Maxwell's equations are reduced to

$$\nabla \times \bar{E} = -\mu_0\frac{\partial\bar{H}}{\partial t}$$
$$\nabla \times \bar{H} = \epsilon_0\frac{\partial\bar{E}}{\partial t}$$
$$\nabla \cdot \bar{E} = 0$$
$$\nabla \cdot \bar{H} = 0$$

Taking curl of the first equation, and using the vector identity $\nabla \times \nabla \times \bar{V} = \nabla\nabla \cdot \bar{V} - \nabla^2\bar{V}$, we have

$$\nabla^2\bar{E} - \mu_0\epsilon_0\frac{\partial^2\bar{E}}{\partial t^2} = 0$$

which is the Helmholtz equation or wave equation for the electric field. The wave equation for magnetic field can be derived in the same manner.

3. Time-Harmonic Fields

Any field component in the time domain can be transformed into its frequency domain counterpart by using Fourier transform. Consider a time-harmonic or signle-tone field component with angular frequency ω, its spatial-temporal wave form can be expressed as the real part of a space-dependent complex variable multiplied by a time harmonic as

$$A(\bar{r}, t) = \text{Re}[\underline{A}(\bar{r})e^{j\omega t}]$$

where A can be $\bar{E}$, $\bar{H}$, $\bar{D}$, $\bar{B}$, $\bar{J}$ or ρ.

The Faraday's law can be tansformed to

$$\text{Re}[\nabla \times \underline{\bar{E}}(\bar{r})e^{j\omega t}] = -\text{Re}[\underline{\bar{B}}(\bar{r})\frac{\partial}{\partial t}e^{j\omega t}], \qquad \forall t$$

which renders the following equation

$$\nabla \times \underline{\bar{E}}(\bar{r}) = -j\omega\underline{\bar{B}}(\bar{r}) \tag{1.1}$$

Notice that the complex field variables are function of space only, and the time derivative is replaced by a product with $j\omega$. The other three Maxwell's equations and the charge conservation law can be transformed in a similar mannar to have

$$\nabla \times \underline{\bar{H}}(\bar{r}) = j\omega \underline{\bar{D}}(\bar{r}) + \underline{\bar{J}}(\bar{r}) \tag{1.2}$$

$$\nabla \cdot \underline{\bar{D}}(\bar{r}) = \underline{\rho}(\bar{r}) \tag{1.3}$$

$$\nabla \cdot \underline{\bar{B}}(\bar{r}) = 0 \tag{1.4}$$

$$\nabla \cdot \underline{\bar{J}}(\bar{r}) = -j\omega \underline{\rho}(\bar{r})$$

The underline can be omitted hereafter without incurring ambiguity. When ω appears in an expression, the field variables are complex function of space only. When t appears in an expression, the field variables are real function of both space and time. In a source-free region in free space, the following wave equation can be derived by taking curl of (1.1)

$$\nabla^2 \bar{E} + k_0^2 \bar{E} = 0 \tag{1.5}$$

where $k_0^2 = \omega^2 \mu_0 \epsilon_0$. Either electrostatic or magnetostatic problem can be considered as a limiting case of time-harmonic with $\omega = 0$, in which (1.5) reduces to the Laplace equation

$$\nabla^2 \bar{E} = 0$$

4. Potential Functions

As auxiliary variables are often introduced to facilitate the solution of differential equations, potential functions are introduced to solve the Maxwell's equations. Define a vector potential function, $\bar{A}$, such that

$$\bar{B} = \nabla \times \bar{A} \tag{1.6}$$

which satisfies (1.4). Substituting (1.6) into (1.1), we have

$$\nabla \times (\bar{E} + j\omega \bar{A}) = 0$$

By observing the vector identity $\nabla \times \nabla\Phi = 0$, the above equation can be solved to have

$$\bar{E} = -j\omega \bar{A} - \nabla\Phi \tag{1.7}$$

Substituting (1.6) and (1.7) into (1.2), we have

$$\nabla\nabla \cdot \bar{A} - \nabla^2 \bar{A} = k_0^2 \bar{A} - j\omega\mu_0\epsilon_0 \nabla\Phi + \mu_0 \bar{J} \tag{1.8}$$

A vector is uniquely determined if both its divergence and curl are given. The curl of vector $\bar{A}$ has been defined in (1.6), hence $\bar{A}$ can be uniquely determined if the divergence of $\bar{A}$ follows the Lorentz gauge that

$$\nabla \cdot \bar{A} = -j\omega\mu_0\epsilon_0\Phi$$

which is proposed to simplify (1.8) to

$$(\nabla^2 + k_0^2)\bar{A} = -\mu_0\bar{J} \tag{1.9}$$

As a consequence, (1.3) is reduced to

$$(\nabla^2 + k_0^2)\Phi = -\frac{\rho}{\epsilon_0} \tag{1.10}$$

Both (1.9) and (1.10) have the form of an inhomogeneous wave equation. In the static limit with $\omega = 0$, (1.9) and (1.10) reduce to the Poisson equations

$$\nabla^2\bar{A} = -\mu_0\bar{J}, \qquad \nabla^2\Phi = -\frac{\rho}{\epsilon_0}$$

Consider a Hertzian dipole located at the origin with current distribution of $\bar{J} = \hat{z}I\ell\delta(\bar{r})$, the associated vector potential, $\bar{A}$, is conveniently chosen as $\bar{A} = \hat{z}A$. Then, (1.9) is reduced to a scalar equation

$$(\nabla^2 + k_0^2)A = -\mu_0 I\ell\delta(\bar{r}) \tag{1.11}$$

At $\bar{r} \neq 0$, (1.11) can be expressed in the spherical coordinates as

$$\frac{d^2}{dr^2}(rA) + k_0^2(rA) = 0$$

with a homogeneous solution of $A = ce^{-jk_0 r}/r$, where the other homogeneous solution, $c'e^{jk_0 r}/r$, is discarded because it violates the principle of causality. To determine the unknown coefficient, c, first take a volume integral of (1.11) over a small spheroidal space with radius r and centered at the origin, then let the radius approach zero to have $c = \mu_0 I\ell/(4\pi)$. Thus, we have

$$\bar{A}(\bar{r}) = \hat{z}\frac{\mu_0 I\ell}{4\pi r}e^{-jk_0 r}$$

Linear superposition can be applied to calculate the vector potential, $\bar{A}(\bar{r})$, generated by an arbitrary current distribution, $\bar{J}(\bar{r}')$, over space V as

$$\bar{A}(\bar{r}) = \iiint_V d\bar{r}' \frac{e^{-jk_0|\bar{r}-\bar{r}'|}}{4\pi|\bar{r}-\bar{r}'|}\mu_0\bar{J}(\bar{r}') \tag{1.12}$$

Similarly, the scalar potential generated by an arbitrary distribution of charge, $\rho(\bar{r}')$, over space V can be expressed as

$$\Phi(\bar{r}) = \iiint_V d\bar{r}' \frac{e^{-jk_0|\bar{r}-\bar{r}'|}}{4\pi|\bar{r}-\bar{r}'|} \frac{\rho(\bar{r}')}{\epsilon_0} \tag{1.13}$$

In the static limit with $\omega = 0$, (1.12) and (1.13) reduce to

$$\bar{A}(\bar{r}) = \iiint_V d\bar{r}' \frac{\mu_0 \bar{J}(\bar{r}')}{4\pi|\bar{r}-\bar{r}'|}$$

$$\Phi(\bar{r}) = \iiint_V d\bar{r}' \frac{\rho(\bar{r}')}{4\pi\epsilon_0|\bar{r}-\bar{r}'|}$$

which are the magnetostatic and electrostatic solution, respectively.

5. Construction of Homogeneous Solutions

5.1 Vector Potential Approach

Consider a source-free region in free space with $\nabla \cdot \bar{D} = 0$. One may start by defining

$$\bar{D} = -\nabla \times \bar{F} \tag{1.14}$$

Substituting (1.14) into the Ampere's law, we have

$$\bar{H} = -j\omega\bar{F} - \nabla\Phi^f \tag{1.15}$$

where Φ^f is an arbitrary scalar function. Substituting (1.14) and (1.15) into the Faraday's law, we have

$$-\nabla\nabla \cdot \bar{F} + \nabla^2\bar{F} = -k_0^2\bar{F} + j\omega\mu_0\epsilon_0\nabla\Phi^f \tag{1.16}$$

Next, apply another Lorentz gauge, $\nabla \cdot \bar{F} = -j\omega\mu_0\epsilon_0\Phi^f$, to simplify (1.16) to a homogeneous wave equation as

$$(\nabla^2 + k_0^2)\bar{F} = 0$$

As a consequence, the scalar potential, Φ^f, also satisfies the homogeneous wave equation

$$(\nabla^2 + k_0^2)\Phi^f = 0$$

By applying superposition technique to (1.6), (1.7), (1.14) and (1.15), fields can generally be expressed in terms of vector potentials $\bar{A}$ and $\bar{F}$

as

$$\bar{E} = -\nabla \times \frac{\bar{F}}{\epsilon_0} + \frac{1}{j\omega\epsilon_0}\nabla \times \nabla \times \frac{\bar{A}}{\mu_0}$$
$$\bar{H} = \nabla \times \frac{\bar{A}}{\mu_0} + \frac{1}{j\omega\mu_0}\nabla \times \nabla \times \frac{\bar{F}}{\epsilon_0} \tag{1.17}$$

Proper vector potentials can be chosen contingent upon the problems to be solved. For example, fields of TM_z modes can be derived by choosing $\bar{F} = 0$ and $\bar{A}/\mu_0 = \hat{z}\psi$, fields of TE_z modes can be derived by choosing $\bar{A} = 0$ and $\bar{F}/\epsilon_0 = \hat{z}\psi'$.

5.2 Field Decomposition Approach

For certain problems, it is convenient to decompose the fields into longitudinal and transversal components with respect to a specific direction. Take $\hat{z}$ direction for example, the del operator, ∇, and fields can be decomposed as

$$\nabla = \nabla_s + \hat{z}\frac{\partial}{\partial z}, \qquad \bar{E} = \bar{E}_s + \bar{E}_z, \qquad \bar{H} = \bar{H}_s + \bar{H}_z$$

The Faraday's law and Ampere's law are then factorized as

$$\nabla_s \times \bar{E}_s = -j\omega\mu\bar{H}_z \tag{1.18}$$
$$\nabla_s \times \bar{H}_s = j\omega\epsilon\bar{E}_z \tag{1.19}$$
$$\nabla_s \times \bar{E}_z + \frac{\partial}{\partial z}\hat{z} \times \bar{E}_s = -j\omega\mu\bar{H}_s \tag{1.20}$$
$$\nabla_s \times \bar{H}_z + \frac{\partial}{\partial z}\hat{z} \times \bar{H}_s = j\omega\epsilon\bar{E}_s \tag{1.21}$$

Substituting (1.21) into (1.20), we have

$$j\omega\epsilon\nabla_s \times \bar{E}_z + \frac{\partial}{\partial z}\nabla_s H_z - \frac{\partial^2}{\partial z^2}\bar{H}_s = k^2\bar{H}_s \tag{1.22}$$

where $k^2 = \omega^2\mu\epsilon$. Assume that $\bar{H}_s$ is a space-harmonic function in the $\hat{z}$ direction, namely $\partial^2\bar{H}_s/\partial z^2 = -k_z^2\bar{H}_s$. Then, (1.22) can be reduced to

$$\bar{H}_s = \frac{1}{k^2 - k_z^2}\left[\frac{\partial}{\partial z}\nabla_s H_z + j\omega\epsilon\nabla_s \times \bar{E}_z\right] \tag{1.23}$$

Similarly, we have

$$\bar{E}_s = \frac{1}{k^2 - k_z^2}\left[\frac{\partial}{\partial z}\nabla_s E_z - j\omega\epsilon\nabla_s \times \bar{H}_z\right] \tag{1.24}$$

By substituting (1.23) into (1.19), we obtain $(\nabla_s^2 - k_z^2 + k^2)\bar{E}_z = 0$. Similarly, by substituting (1.24) into (1.18), we have $(\nabla_s^2 - k_z^2 + k^2)\bar{H}_z = 0$. Once proper forms of H_z and E_z are chosen, the other field components can be derived by using (1.23) and (1.24). Notice that $H_z = 0$ and $E_z \neq 0$ for the TM$_z$ modes, $H_z \neq 0$ and $E_z = 0$ for the TE$_z$ modes.

6. Plane Wave Functions

The scalar wave equation, $(\nabla^2 + k^2)\psi(\bar{r}) = 0$, can be expressed in the Cartesian coordinates as

$$\frac{\partial^2\psi}{\partial x^2} + \frac{\partial^2\psi}{\partial y^2} + \frac{\partial^2\psi}{\partial z^2} + k^2\psi = 0 \tag{1.25}$$

By using the separation-of-variables technique, let $\psi(\bar{r}) = X(x)Y(y)Z(z)$, and (1.25) is reduced to the following three harmonic equations

$$\frac{d^2X}{dx^2} + k_x^2 X = 0, \quad \frac{d^2Y}{dy^2} + k_y^2 Y = 0, \quad \frac{d^2Z}{dz^2} + k_z^2 Z = 0 \tag{1.26}$$

where the separation constants k_x, k_y and k_z satisfy the dispersion relation $k_x^2 + k_y^2 + k_z^2 = k^2$. The solutions to (1.26) are harmonic functions which can be sinusoidals to represent standing waves or exponentials to represent traveling waves, contingent upon the problems encountered.

A plane wave function is constituted if we choose $X(x) = e^{-jk_x x}$, $Y(y) = e^{-jk_y y}$ and $Z(z) = e^{-jk_z z}$, thus $\psi(\bar{r}) = e^{-j\bar{k}\cdot\bar{r}}$ with $\bar{k} = \hat{x}k_x + \hat{y}k_y + \hat{z}k_z$. In general, $\bar{k}$ is a complex vector with $\bar{k} = \bar{\beta} - j\bar{\alpha}$. Hence, $\psi(\bar{r})$ represents a decaying or growing plane wave with equiphase surfaces on $\bar{\beta}\cdot\bar{r}$ = constant and equiamplitude surfaces on $\bar{\alpha}\cdot\bar{r}$ = constant. In a lossy medium, the wave number, k, becomes complex, namely $k = k' - jk''$. Thus, we have $k^2 = k'^2 - k''^2 - 2jk'k''$. On the other hand, $k^2 = \bar{k}\cdot\bar{k} = \beta^2 - \alpha^2 - j2\bar{\alpha}\cdot\bar{\beta}$. In a lossless medium with $k'' = 0$, the above equation implies that $\bar{\alpha}\cdot\bar{\beta} = 0$, which further implies that either $\alpha = 0$(uniform plane wave) or $\bar{\alpha}$ is perpendicular to $\bar{\beta}$ (evanescent or surface wave).

7. Fields in Layered Media

Consider a guided wave problem in a layered medium as shown in Fig.1.3. The fields are generally hybrid, consisting of both TM$_z$ and TE$_z$ components. The vector potential approach can be used to construct the solutions by choosing $\bar{A}/\mu_0 = \hat{z}\psi$ and $\bar{F}/\epsilon_0 = \hat{z}\psi'$ in each layer. Assume that a wave mode is guided along the $\hat{y}$ direction with a propagation constant k_y. Mathematically, the x dependence of both vector potentials can be expressed as Fourier integrals in the $\hat{x}$ direction.

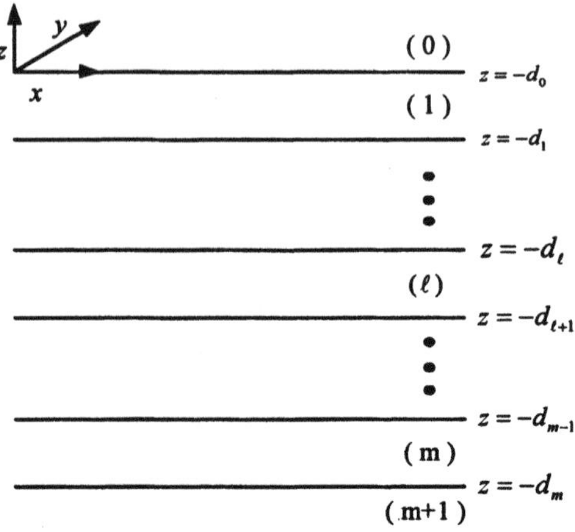

Figure 1.3. Homogeneous layers stacked in $\hat{z}$ direction.

For each component in the Fourier integral with fixed k_y and k_x, the functional form in z is chosen to be traveling waves in the $\pm\hat{z}$ directions. Thus, we have

$$\psi_0(\bar{r}) = e^{-jk_y y} \int_{-\infty}^{\infty} dk_x e^{-jk_x x} e_0^U(\bar{k}_s) e^{-jk_{0z} z_0}$$

$$\psi_0'(\bar{r}) = e^{-jk_y y} \int_{-\infty}^{\infty} dk_x e^{-jk_x x} h_0^U(\bar{k}_s) e^{-jk_{0z} z_0}$$

$$\psi_\ell(\bar{r}) = e^{-jk_y y} \int_{-\infty}^{\infty} dk_x e^{-jk_x x} \left[e_\ell^U(\bar{k}_s) e^{-jk_{\ell z} z_\ell} + e_\ell^D(\bar{k}_s) e^{jk_{\ell z} z_\ell} \right]$$

$$\psi_\ell'(\bar{r}) = e^{-jk_y y} \int_{-\infty}^{\infty} dk_x e^{-jk_x x} \left[h_\ell^U(\bar{k}_s) e^{-jk_{\ell z} z_\ell} + h_\ell^D(\bar{k}_s) e^{jk_{\ell z} z_\ell} \right]$$

$$1 \le \ell \le m$$

$$\psi_{m+1}(\bar{r}) = e^{-jk_y y} \int_{-\infty}^{\infty} dk_x e^{-jk_x x} e_{m+1}^D(\bar{k}_s) e^{jk_{(m+1)z} z_{m+1}}$$

$$\psi_{m+1}'(\bar{r}) = e^{-jk_y y} \int_{-\infty}^{\infty} dk_x e^{-jk_x x} h_{m+1}^D(\bar{k}_s) e^{jk_{(m+1)z} z_{m+1}} \qquad (1.27)$$

where $\bar{k}_s = \hat{x}k_x + \hat{y}k_y$, $k_s^2 = k_x^2 + k_y^2$, $k_{\ell z} = \sqrt{k_\ell^2 - k_s^2}$ and $k_\ell^2 = \omega^2\mu_0\epsilon_\ell$ with $0 \le \ell \le m+1$, $e_0^U(\bar{k}_s)$, $h_0^U(\bar{k}_s)$, $e_\ell^U(\bar{k}_s)$, $e_\ell^D(\bar{k}_s)$, $h_\ell^U(\bar{k}_s)$, $h_\ell^D(\bar{k}_s)$, $e_{m+1}^D(\bar{k}_s)$ and $h_{m+1}^D(\bar{k}_s)$ are unknown coefficients to be determined. Notice that the spatial distribution associated with a specific k_x in the integral represents a plane wave propagating in the direction of $\bar{k} = \hat{x}k_x + \hat{y}k_y + \hat{z}k_z$. The integral representation is based on the conjecture that any wave function can be expressed as a superposition of plane waves.

All the field components can be derived from the potentials in (1.27) using (1.17). To facilitate the derivations, reflection coefficients are defined as $e_\ell^U(\bar{k}_s) = R_{\cap\ell}^{\rm TM} e_\ell^D(\bar{k}_s)$ and $h_\ell^U(\bar{k}_s) = R_{\cap\ell}^{\rm TE} h_\ell^D(\bar{k}_s)$. By imposing the continuity conditions at $z = -d_\ell$ that $E_{\ell x} = E_{(\ell+1)x}$, $E_{\ell y} = E_{(\ell+1)y}$, $H_{\ell x} = H_{(\ell+1)x}$ and $H_{\ell y} = H_{(\ell+1)y}$ with $1 \le \ell \le m$, recursive formulas for the reflection coefficients are obtained as

$$R_{\cap\ell}^{\rm TM} = \frac{R_{\ell(\ell+1)}^{\rm TM} + R_{\cap(\ell+1)}^{\rm TM} e^{-2jk_{(\ell+1)z}h_{\ell+1}}}{1 + R_{\ell(\ell+1)}^{\rm TM} R_{\cap(\ell+1)}^{\rm TM} e^{-2jk_{(\ell+1)z}h_{\ell+1}}}, \qquad 1 \le \ell \le m-1$$

$$R_{\cap\ell}^{\rm TE} = \frac{R_{\ell(\ell+1)}^{\rm TE} + R_{\cap(\ell+1)}^{\rm TE} e^{-2jk_{(\ell+1)z}h_{\ell+1}}}{1 + R_{\ell(\ell+1)}^{\rm TE} R_{\cap(\ell+1)}^{\rm TE} e^{-2jk_{(\ell+1)z}h_{\ell+1}}}, \qquad 1 \le \ell \le m-1$$

$$R_{\cap m}^{\rm TM} = \frac{k_{mz}\epsilon_{m+1} - k_{(m+1)z}\epsilon_m}{k_{mz}\epsilon_{m+1} + k_{(m+1)z}\epsilon_m}, \qquad R_{\cap m}^{\rm TE} = \frac{k_{mz} - k_{(m+1)z}}{k_{mz} + k_{(m+1)z}}$$

where h_ℓ is the thickness of layer ℓ.

7.1 Microtrip Line

Fig.1.4 shows the cross section of a uniform microstrip line. Assume that the wave modes are guided along the $\hat{y}$ direction with a propagation constant k_y. The field components in each layer can be derived from the vector potentials associated with TM_z and TE_z mode, respectively. Surface current on the strip is accounted for by the discontinuity of tangential magnetic fields at $z = -d_0$ by $\hat{n} \times (\bar{H}_0 - \bar{H}_1) = \bar{J}_s$. The surface current can be expressed as a Fourier integral in the $\hat{x}$ direction as

$$J_{s\alpha}(\bar{r}_s) = e^{-jk_y y} \int_{-\infty}^{\infty} dk_x e^{-jk_x x} \tilde{J}_{s\alpha}(\bar{k}_s), \qquad \alpha = x, y$$

Spectral coefficients of the surface current are related to the field coefficients via the discontinuity condition, $\hat{n} \times (\bar{H}_0 - \bar{H}_1) = \bar{J}_s$. The other boundary condition requires that the tangential electric fields vanish on the strip surface. Thus, a set of integral equations with the surface

currents as unknowns are obtained as

$$\int_{-\infty}^{\infty} dk_x e^{-jk_x x} \left[g_{xx}(\bar{k}_s)\tilde{J}_{sx}(\bar{k}_s) + g_{xy}(\bar{k}_s)\tilde{J}_{sy}(\bar{k}_s) \right] = 0, \qquad \text{on strip}$$

$$\int_{-\infty}^{\infty} dk_x e^{-jk_x x} \left[g_{yx}(\bar{k}_s)\tilde{J}_{sx}(\bar{k}_s) + g_{yy}(\bar{k}_s)\tilde{J}_{sy}(\bar{k}_s) \right] = 0, \qquad \text{on strip}$$

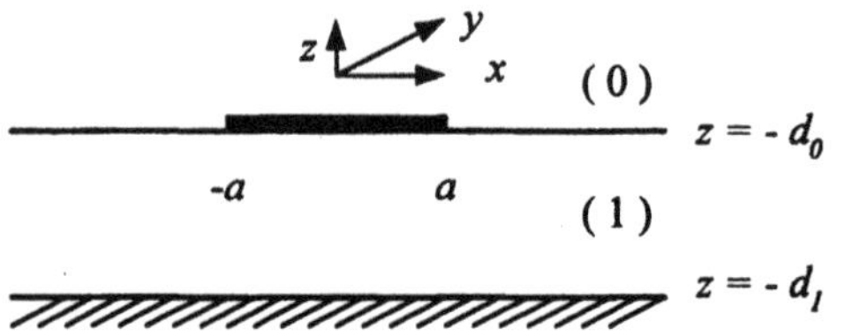

Figure 1.4. Cross section of a uniform microstrip line.

Method of moments is generally used to solve the integral equations numerically. A set of basis functions are empirically chosen to expand the continuous unknown surface currents, resulting in a finite number of expansion coefficients. Then, substitute the current expansions into the integral equations, choose another set of weighting functions which can be of the same form as the basis functions. Take the inner product, with proper definition, of the weighting functions with the resulting integral equations to convert them into a matrix equation. Numerical methods like Muller's method can be applied to search for the propagation constants which are zeros of the matrix determinant.

7.2 Patch Resonator

To calculate the resonant frequencies of a patch resonator, first express the vector potentials associated with TM_z and TE_z fields as two-dimensional Fourier integrals in the $\hat{x}$ and $\hat{y}$ directions as

$$\psi_1(\bar{r}) = \iint_{-\infty}^{\infty} d\bar{k}_s e^{-j\bar{k}_s\cdot\bar{r}_s} \left[e_1^U(\bar{k}_s)e^{-jk_{1z}z_1} + e_1^D(\bar{k}_s)e^{jk_{1z}z_1} \right]$$

$$\psi_1'(\bar{r}) = \iint_{-\infty}^{\infty} d\bar{k}_s e^{-j\bar{k}_s\cdot\bar{r}_s} \left[h_1^U(\bar{k}_s)e^{-jk_{1z}z_1} + h_1^D(\bar{k}_s)e^{jk_{1z}z_1} \right]$$

where $\bar{r}_s = \hat{x}x + \hat{y}y$. All the field components can be derived in the same manner as in the micorstrip line problem. Coupled integral equa-

tions are derived by imposing proper continuity conditions at the layer interfaces and discontinuity conditions at the interfaces embedding the patches. Method of moments is then applied to convert the coupled integral equations into a matrix equation from which the resonant frequencies can be obtained.

7.3 Coplanar Waveguide

Fig.1.5 shows the cross section of a coplanar waveguide. Since perfect electric conductor occupies most of the interface at $z = -d_0$, it is more convenient to solve for the unknown tangential fields over the gaps than the unknown surface current on the conductor. The equivalence principle is applied to replace the gaps by equivalent magnetic surface currents backed by a perfect electric conductor, and the magnetic surface currents are related to the tangential electric fields across the gaps by $\bar{M} = \bar{E}_0 \times \hat{n} = \bar{E}_1 \times \hat{n}$. Fields in the two regions separated by the perfect electric conductor can be expressed in terms of the same magnetic surface currents. Continuity of tangential magnetic fields across the gaps relates the fields in these two regions to render a set of coupled integral equations with the magnetic surface current as unknowns. Method of moments is then applied to solve for the propagation constant k_y and relevant parameters.

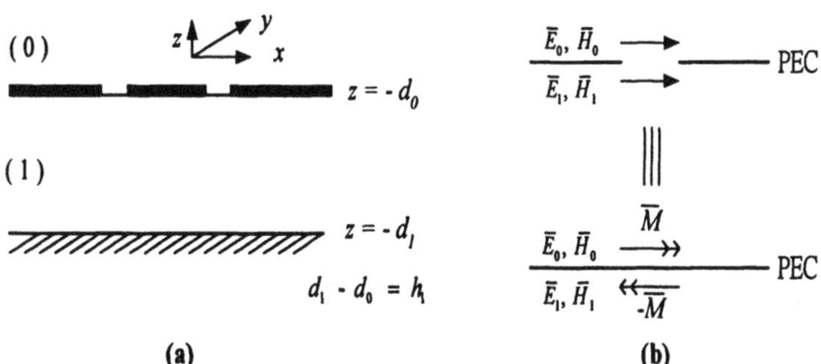

Figure 1.5. (a) Cross section of a uniform coplanar waveguide, (b) equivalent principle applied over the gaps.

8. Perturbation Technique

The perturbation technique is based on the assumption that the physical parameters of interest change smoothly with slight change of the

problem domain. Hence, the solution in a known problem domain can be modified to estimate the new solution in another problem domain which is slightly different from the known domain. Fig.1.6 shows a cavity resonator with problem domain V_0 which is enclosed by a perfect electric conductor and is filled with medium (μ, ϵ). The internal field distributions, $(\bar{E}_0, \bar{H}_0)$, and the resonant frequency, ω_0, satisfy the following equations

$$\nabla \times \bar{E}_0 = -j\omega_0 \mu \bar{H}_0 \tag{1.28}$$

$$\nabla \times \bar{H}_0 = j\omega_0 \epsilon \bar{E}_0 \tag{1.29}$$

When the problem domain is slightly changed to V, the internal field distributions, $(\bar{E}, \bar{H})$, and the resonant frequency, ω, are related as

$$\nabla \times \bar{E} = -j\omega \mu \bar{H} \tag{1.30}$$

$$\nabla \times \bar{H} = j\omega \epsilon \bar{E} \tag{1.31}$$

By calculating $\bar{H} \cdot (1.28)^* - \bar{E}_0^* \cdot (1.31) + \bar{H}_0^* \cdot (1.30) - \bar{E} \cdot (1.29)^*$, we have $\nabla \cdot (\bar{E}_0^* \times \bar{H} + \bar{E} \times \bar{H}_0^*) = j(\omega_0 - \omega)\mu \bar{H}_0^* \cdot \bar{H} + j(\omega_0 - \omega)\epsilon \bar{E}_0^* \cdot \bar{E}$. Integrate this expression over V_0 and apply proper boundary conditions

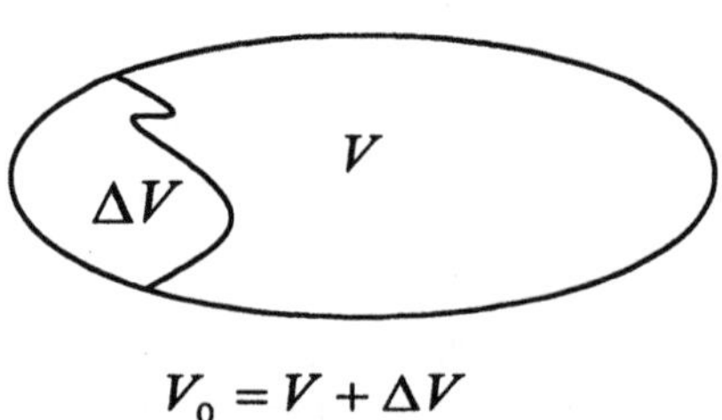

Figure 1.6. A cavity resonator enclosed by perfect electric conductor.

on the field components, we have

$$\omega - \omega_0 = \frac{j \iint_{\Delta S} d\bar{s} \cdot (\bar{E} \times \bar{H}_0^*)}{\iiint_{V_0} dv(\mu \bar{H}_0^* \cdot \bar{H} + \epsilon \bar{E}_0^* \cdot \bar{E})} \cong \frac{j \iiint_{\Delta V} dv \nabla \cdot (\bar{E}_0 \times \bar{H}_0^*)}{\iiint_{V_0} dv(\mu |\bar{H}_0|^2 + \epsilon |\bar{E}_0|^2)}$$

where $\bar{E}$ and $\bar{H}$ are approximated by $\bar{E}_0$ and $\bar{H}_0$, respectively, in the last step. By using (1.28) and (1.29), the above equation is further reduced to

$$\frac{\omega - \omega_0}{\omega_0} \cong \frac{\iiint_{\Delta V} dv(\mu |\bar{H}_0|^2 - \epsilon |\bar{E}_0|^2)}{\iiint_{V_0} dv(\mu |\bar{H}_0|^2 + \epsilon |\bar{E}_0|^2)}$$

where the numerator is the difference between the average magnetic and electric energies in ΔV, and the denominator is the sum of the average magnetic and electric energies in V_0. Note that only the fields in the known problem domain are required to estimate the resonant frequency shift.

9. Periodic Structures

One-dimensional periodic structures in layered media can be constructed by periodically modulating the medium parameters like permittivity or permeability along the guiding direction, or by inserting conducting objects periodically along the guiding direction. Assume that the structure has a period P, and the modes are guided along the $\hat{y}$ direction, all the field components or potential functions will satisfy the homogeneous wave equation

$$\left(\frac{\partial^2}{\partial y^2} + \nabla_t^2 + k^2\right)\phi(y, \bar{r}_t) = 0 \tag{1.32}$$

where $\bar{r}_t = \hat{x}x + \hat{z}z$. If the medium parameters are periodically modulated, k^2 will become a periodic function of y. If conducting objects are periodically inserted in the medium, k^2 will be a constant, but the boundary conditions will become periodic functions in the $\hat{y}$ direction. $\phi(y + P, \bar{r}_t)$ will be a solution to (1.32) too if $\phi(y, \bar{r}_t)$ is a solution. Let $\phi(y, \bar{r}_t) = \psi(y)\eta(\bar{r}_t)$ by using the separation-of-variables technique, (1.32) can be reduced to

$$\begin{aligned} &\eta'' + k_t^2\eta = 0 \\ &\psi'' + (k^2 - k_t^2)\psi = 0 \end{aligned} \tag{1.33}$$

Assume that $\psi_1(y)$ and $\psi_2(y)$ are two independent solutions to (1.33), namely

$$\psi_1'' + (k^2 - k_t^2)\psi_1 = 0 \tag{1.34}$$

$$\psi_2'' + (k^2 - k_t^2)\psi_2 = 0 \tag{1.35}$$

The combination of $(1.34)\psi_2 - (1.35)\psi_1$ gives $\psi_2\psi_1'' - \psi_1\psi_2'' = 0$. Integrating this expression over $y_1 \le y \le y_2$ with arbitrary y_1 and y_2, we have $\psi_2(y_2)\psi_1'(y_2) - \psi_1(y_2)\psi_2'(y_2) = \psi_2(y_1)\psi_1'(y_1) - \psi_1(y_1)\psi_2'(y_1)$, which implies that $\psi_2(y)\psi_1'(y) - \psi_1(y)\psi_2'(y)$ is a constant. Since $\psi_1(y+P)$ and $\psi_2(y + P)$ are also solutions to (1.33), they can be expressed as linear combinations of $\psi_1(y)$ and $\psi_2(y)$ as

$$\begin{aligned} \psi_1(y + P) &= \alpha_{11}\psi_1(y) + \alpha_{12}\psi_2(y) \\ \psi_2(y + P) &= \alpha_{21}\psi_1(y) + \alpha_{22}\psi_2(y) \end{aligned} \tag{1.36}$$

After arithmetic manipulations, we have

$$\begin{aligned}&\psi_1(y+P)\psi_2'(y+P)-\psi_1'(y+P)\psi_2(y+P)\\&=(\alpha_{11}\alpha_{22}-\alpha_{12}\alpha_{22})[\psi_1(y)\psi_2'(y)-\psi_1'(y)\psi_2(y)]\end{aligned}$$

which implies $\alpha_{11}\alpha_{22}-\alpha_{12}\alpha_{21}=1$.

Any solution to (1.33), $F(y)$, can be expressed as a linear combination of $\psi_1(y)$ and $\psi_2(y)$ as $F(y)=A\psi_1(y)+B\psi_2(y)$. By changing the variable y to $y+P$, and using (1.36), we have

$$\begin{aligned}F(y+P)&=A\psi_1(y+P)+B\psi_2(y+P)\\&=(A\alpha_{11}+B\alpha_{21})\psi_1(y)+(A\alpha_{12}+B\alpha_{22})\psi_2(y)\end{aligned} \quad (1.37)$$

If $F(y)$ represents a mode guided in the $\hat{y}$ direction with propagation constant k_y, we have

$$F(y+P)=e^{-jk_yP}F(y) \quad (1.38)$$

The phase term can be factored out of $F(y)$ as $F(y)=e^{-jk_yy}\Phi(y)$. Applying (1.38), we have $\Phi(y+P)=e^{jk_y(y+P)}F(y+P)=e^{jk_yy}F(y)=\Phi(y)$, which implies that $\Phi(y)$ is a periodic function of y with period P.

Substituting (1.38) into (1.37), we obtain $A(\alpha_{11}-e^{-jk_yP})+B\alpha_{21}=0$ and $A\alpha_{12}+B(\alpha_{22}-e^{-jk_yP})=0$. To have nontrivial A and B, k_y must satisfy $e^{-2jk_yP}-e^{-jk_yP}(\alpha_{11}+\alpha_{22})+(\alpha_{11}\alpha_{22}-\alpha_{12}\alpha_{21})=0$, of which the solution is

$$e^{-jk_yP}=\frac{\alpha_{11}+\alpha_{22}}{2}\pm\sqrt{\left(\frac{\alpha_{11}+\alpha_{22}}{2}\right)^2-1} \quad (1.39)$$

Define $\cosh\theta=(\alpha_{11}+\alpha_{22})/2$, then (1.39) is reduced to $e^{-jk_yP}=e^{\pm\theta}$. From this functional form, it is observed that if k_y is a solution, $\pm(k_y+2nP/\pi)$ with n an arbitrary integer are also solutions. The modes associated with propagation constant of $\pm(k_y+2nP/\pi)$ are called the space-harmonics or Floquet modes of order n.

10. Static Formulations

Full-wave approaches usually fail at low frequencies while the static solution renders reasonable accuracy. Hence, a quasistatic solution derived from the static solution can be used as an approximate dynamic solution. The deviation between quasistatic solution and dynamic solution usually grows with increasing frequency. For example, in coaxial cables or wire pair with homogeneous background medium, the dominant mode is a pure TEM mode. In the cross section, the electric field distribution is exactly the same as the electrostatic field, and the magnetic field distribution is exactly the same as the magnetostatic field. Spatial variation along the guidance direction and temporal variation of the

TEM mode definitely indicate dynamic behaviors. As the background medium becomes inhomogeneous or lossy, or the conductors becomes imperfect, field distributions of the dominant mode will deviate from that of a pure TEM mode, more obvious when the frequency increases. Hence, this mode is called a quasi-TEM mode.

10.1 Two-Dimensional Electrostatic Formulation

In the dc limit, the electric field can be derived from a scalar potential as $\bar{E}(\bar{r}) = -\nabla\Phi(\bar{r})$. By applying the electric Gauss' law, $\nabla \cdot \bar{D}(\bar{r}) = 0$, in a homogeneous medium, we have the Laplace equation, $\nabla^2\Phi(\bar{r}) = 0$. By assuming that the field is uniform along the $\hat{y}$ direction, the Laplace eqution is reduced to

$$\left(\frac{\partial^2}{\partial x^2} + \frac{\partial^2}{\partial z^2}\right)\Phi(x, z) = 0$$

The potential function is then expanded in the $\hat{x}$ direction as a Fourier integral

$$\Phi(x, z) = \int_{-\infty}^{\infty} dk_x e^{-jk_x x}\tilde{\Phi}(k_x)e^{-jk_z z}$$

where the separation relation, $k_x^2 + k_z^2 = 0$, can be viewed as the dispersion relation in the dc limit with $\omega = 0$.

For the microstrip line shown in Fig.1.4, potential distributions in each layer can be expressed as

$$\Phi_0(x, z) = \int_{-\infty}^{\infty} dk_x e^{-jk_x x} e_0^U(k_x)e^{-|k_x|z_0}$$

$$\Phi_1(x, z) = \int_{-\infty}^{\infty} dk_x e^{-jk_x x}\left[e_1^U(k_x)e^{-|k_x|z_1} + e_1^D(k_x)e^{|k_x|z_1}\right]$$

where the upward(downward) decaying term in $\hat{z}$ direction can be viewed as the dc limit of the upward(downward) propagating term in the dynamic field expressions. The boundary conditions require that $\Phi_1(x, z) = 0$ at $z = 0$, $\Phi_0(x, z) = \Phi_1(x, z)$, $D_{0z}(x, z) - D_{1z}(x, z) = \rho_s(x)$ at $z = h$, and $\Phi_0(x, z) = V$ on the strip surface. Thus, we obtain the following integral equation

$$\int_{-\infty}^{\infty} dk_x e^{-jk_x x}G(k_x)\tilde{\rho}_s(k_x) = V, \qquad \text{on strip} \qquad (1.40)$$

where the Green's function is

$$G(k_x) = \frac{1}{|k_x|} \frac{\tanh |k_x| h}{\epsilon_0 \tanh |k_x| h + \epsilon_1}$$

Method of moments is then applied to solve the integral equation for the surface charge density on strip. The per-unit-length charge, Q, can be obtained by integrating the surface charge density over the strip width, and the per-unit-length capacitance can be defined as $C = Q/V$.

The static approach can also be extended to study the problem with patches embedded in a layered medium. The potential function in each layer now becomes a two-dimensional integral over k_x and k_y, and the the separation relation becomes $k_x^2 + k_y^2 + k_z^2 = 0$. For a multilayered problem, coefficients relating upward decaying and downward decaying terms can be defined, similar to the reflection coefficients defined in the dynamic problem, to facilitate the solution of boundary-value problems.

10.2 Two-Dimensional Magnetostatic Formulation

For the same microstrip line as shown in Fig.1.4, a two-dimensional magnetostatic problem can also be formulated. The magnetic Gauss' law, $\nabla \cdot \bar{B} = 0$, implies that the magnetic flux density, $\bar{B}$, can be expressed in terms of a vector potential function as $\bar{B} = \nabla \times \bar{A}$. The two-dimensional structure suggests that $\bar{A}(\bar{r}) = \hat{y}A_y(x, z)$. The Ampere's law, $\nabla \times \bar{H} = 0$, implies that $\nabla^2 A_y = 0$. The magnetic flux per unit length flowing between the strip and ground plane can be calculated as

$$\Psi = \iint\limits_S \bar{B} \cdot da = \iint\limits_S \nabla \times \bar{A} \cdot d\bar{a} = \oint \bar{A} \cdot d\bar{\ell} = A_y(S) - A_y(G)$$

where $A_y(S)$ and $A_y(G)$ are the potentials on the strip surface and ground, respectively. The potential functions in each layer can be expressed as the following Fourier integrals

$$A_{0y}(x, z) = \int_{-\infty}^{\infty} dk_x e^{-jk_x x} h_0^U(k_x) e^{-|k_x| z_0}$$

$$A_{1y}(x, z) = \int_{-\infty}^{\infty} dk_x e^{-jk_x x} \left[h_1^U(k_x) e^{-|k_x| z_1} + h_1^D(k_x) e^{|k_x| z_1} \right]$$

The boundary conditions require that $A_{1y}(x, z) = 0$ at $z = 0$, $A_{0y}(x, z) = A_{1y}(x, z)$, $H_{0x}(x, z) - H_{1x}(x, z) = J_{sy}(x, z)$ at $z = h$, and $A_{0y}(x, z) = \Psi$

on the strip surface. Thus, we have

$$\int_{-\infty}^{\infty} dk_x e^{-jk_x x} G_h(k_x) \tilde{K}_y(k_x) = \Psi, \qquad \text{on strip} \tag{1.41}$$

where

$$G_h(k_x) = \frac{\mu_0}{|k_x|} \frac{1 - e^{-2|k_x|h}}{(1 - e^{-2|k_x|h}) + (1 + e^{-2|k_x|h})}$$

The total current, I, on the strip surface can be calculated by integrating the surface current density over strip width, and the inductance per unit length can be defined as $L = \Psi/I$. Comparing (1.40) with (1.41) while setting $\Psi = V$, $\epsilon_1 = \epsilon_0$, we have $\mu_0 I = Q_0/\epsilon_0$, and hence $LC_0 = \mu_0\epsilon_0$. By using these quasistatic approximations, propagation constant of the quasi-TEM mode can thus be derived as $k_y = \omega\sqrt{LC} = \omega\sqrt{\mu_0\epsilon_0}\sqrt{C/C_0} = \omega\sqrt{\mu_0\epsilon_0}\sqrt{\epsilon_{\text{eff}}}$. Notice that ϵ_{eff} is independent of frequency in this approximation.

11. Far-Field Approximation

In antenna and radiation problems involving layered media, the far field can be derived from the z components of fields in region (0) as

$$E_{0z}(\bar{r}) = \iint_{-\infty}^{\infty} d\bar{k}_s e^{-j\bar{k}_s\cdot\bar{r}_s} e_0(\bar{k}_s) e^{-jk_{0z}z_0}$$

$$H_{0z}(\bar{r}) = \iint_{-\infty}^{\infty} d\bar{k}_s e^{-j\bar{k}_s\cdot\bar{r}_s} h_0(\bar{k}_s) e^{-jk_{0z}z_0}$$

By applying the Cartesian-to-spherical coordinate transformation that $x = r\sin\theta\cos\phi$, $y = r\sin\theta\sin\phi$, $z_0 = r\cos\theta$, and a similar coordinate transformation in the spectral domain that $k_x = k_0\sin\alpha\cos\beta$, $k_y = k_0\sin\alpha\sin\beta$, $k_{0z} = k_0\cos\alpha$, $E_{0z}(\bar{r})$ becomes

$$E_{0z}(\bar{r}) = k_0^2 \int_{-\infty}^{\infty} d\alpha \sin\alpha\cos\alpha e^{-jk_0 r\cos\alpha\cos\theta} \int_{-\infty}^{\infty} d\beta e^{-jk_0 r\sin\alpha\sin\theta\cos(\beta-\phi)} e_0(\bar{k}_s) \tag{1.42}$$

Consider the second integral of (1.42) in the far-field region where $k_0 r$ is much greater than unity. Phase in the exponent varies dramatically as β is slightly changed, making the contribution of specific β away from ϕ almost canceled by that of nearby β. Hence, only the contribution

near $\beta = \phi$ really counts, where $\cos(\beta - \phi)$ can be approximated as $1 - (\beta - \phi)^2/2$. Thus, the second integral in (1.42) can be approximated as

$$e^{-jk_0 r \sin\alpha \sin\theta} \left[e_0(\bar{k}_s)\right]_{\beta=\phi} \int_{-\infty}^{\infty} d\beta \, e^{j(k_0 r/2)\sin\alpha \sin\theta(\beta-\phi)^2}$$

By the variable transformation that $\gamma^2 = -j(k_0 r/2)\sin\alpha\sin\theta$ with γ a complex variable and by properly detouring the integration path in the γ plane, the last integral is reduced to $\sqrt{2j\pi/(k_0 r \sin\alpha \sin\theta)}$. Hence, (1.42) is reduced to

$$E_{0z}(\bar{z}) \sim k_0^2 \sqrt{\frac{2j\pi}{k_0 r \sin\theta}} \int_{-\infty}^{\infty} d\alpha \sqrt{\sin\alpha} \cos\alpha e^{-jk_0 r \cos(\alpha-\theta)} \left[e_0(\bar{k}_s)\right]_{\beta=\phi}$$

Applying similar technique to the integration over α, we have

$$E_{0z}(\bar{r}) \sim e^{-jk_0 r} \frac{j2\pi k_0}{r} \cos\theta \left[e_0(\bar{k}_s)\right]_{\beta=\phi,\alpha=\theta}$$

where the $e^{-jk_0 r}$ and $1/r$ terms demonstrate the spherical wave nature of far fields, and the rest terms account for the angular variation of field pattern. Similarly,

$$H_{0z}(\bar{r}) \sim e^{-jk_0 r} \frac{j2\pi k_0}{r} \cos\theta \left[h_0(\bar{k}_s)\right]_{\beta=\phi,\alpha=\theta}$$

The field components tangential to the propagation direction can be derived as $E_\theta = -E_{0z}/\sin\theta$, $H_\phi = E_\theta/\eta_0$, $H_\theta = -H_{0z}/\sin\theta$ and $E_\phi = -\eta_0 H_\theta$.

12. Dyadic Green's Function

The dyadic Green's function is conceptually convenient to relate the field, $\bar{E}(\bar{r})$, and the source, $\bar{J}(\bar{r})$, as

$$\bar{E}(\bar{r}) = -j\omega\mu_o \iiint d\bar{r}' \bar{\bar{G}}(\bar{r}, \bar{r}') \cdot \bar{J}(\bar{r}')$$

where the integration is over the space occupied by the current distribution. By applying Maxwell's equations, the dyadic Green's function is proved to satisfy the wave equation as

$$\nabla \times \nabla \times \bar{\bar{G}}(\bar{r}, \bar{r}') - k^2 \bar{\bar{G}}(\bar{r}, \bar{r}') = \bar{\bar{I}}\delta(\bar{r} - \bar{r}')$$

The dyadic Green's function is related to the scalar Green's function as

$$\bar{\bar{G}}(\bar{r}, \bar{r}') = \left[\bar{\bar{I}} + \frac{1}{k^2}\nabla\nabla\right] g(\bar{r}, \bar{r}') \tag{1.43}$$

where $g(\bar{r},\bar{r}') = e^{-jk|\bar{r}-\bar{r}'|}/4\pi|\bar{r}-\bar{r}'|$ satisfies the wave equation $(\nabla^2 + k^2)$ $g(\bar{r},\bar{r}') = -\delta(\bar{r}-\bar{r}')$. Without loss of generality, consider the case with $\bar{r}' = 0$, and expand both $g(\bar{r}) = g(\bar{r},0)$ and $\delta(\bar{r})$ in the spectral domain as

$$\begin{aligned}\delta(\bar{r}) &= \frac{1}{(2\pi)^3}\iiint dk_x\, dk_y\, dk_z\, e^{-j\bar{k}\cdot\bar{r}} \\ g(\bar{r}) &= \frac{1}{(2\pi)^3}\iiint dk_x\, dk_y\, dk_z\, e^{-j\bar{k}\cdot\bar{r}}\tilde{g}(\bar{k})\end{aligned} \tag{1.44}$$

Then, the wave equation renders

$$\tilde{g}(\bar{k}) = \frac{1}{k_x^2 + k_y^2 + k_z^2 - k^2} = \frac{1}{k_z^2 - k_{0z}^2}$$

where $k_{0z} = \pm\sqrt{k^2 - k_x^2 - k_y^2}$ are the poles in the k_z plane if the integration of (1.44) is performed over k_z first by calculating

$$I_z = \frac{1}{2\pi}\int_{\infty}^{\infty} dk_z\, e^{-jk_z z}\frac{1}{k_z^2 - k_{0z}^2}$$

The integration path in the k_z plane becomes a closed contour by appending a semicircle of infinite radius in the lower k_z plane if $z > 0$ or in the upper k_z plane if $z < 0$. By applying the residue theorem, we have $I_z = -(j/2k_{0z})e^{-jk_{0z}z}$ when $z > 0$, and $I_z = -(j/2k_{0z})e^{jk_{0z}z}$ when $z < 0$. Thus, $g(\bar{r})$ is reduced to a two-dimensional integral over k_x and k_y as

$$g(\bar{r}) = \frac{-j}{(2\pi)^2}\iint d\bar{k}_s \frac{1}{2k_{0z}} e^{-jk_x x - jk_y y - jk_{0z}|z|} \tag{1.45}$$

By substituting (1.45) into (1.43) and restoring $\bar{r}'$, the dyadic Green's function becomes

$$\bar{\bar{G}}(\bar{r},\bar{r}') = -\hat{z}\hat{z}\frac{1}{k^2}\delta(\bar{r}-\bar{r}') - \frac{j}{8\pi^2}\iint d\bar{k}_s \frac{1}{k_{0z}}\bar{\bar{P}}(\bar{k}_s,\bar{r},\bar{r}')$$

The delta function, $\delta(\bar{r}-\bar{r}')$, appears when the second derivative of $g(\bar{r},\bar{r}')$ with respect to z is taken, and it accounts for the singularity when the observation point overlapps with the source point. $\bar{\bar{P}}(\bar{k}_s,\bar{r},\bar{r}') = \left[\hat{e}(k_{0z})\hat{e}(k_{0z}) + \hat{h}(k_{0z})\hat{h}(k_{0z})\right] e^{-j\bar{k}\cdot(\bar{r}-\bar{r}')}$ when $z > z'$, and $\bar{\bar{P}}(\bar{k}_s,\bar{r},\bar{r}') = \left[\hat{e}(-k_{0z})\hat{e}(-k_{0z}) + \hat{h}(-k_{0z})\hat{h}(-k_{0z})\right] e^{-j\bar{K}\cdot(\bar{r}-\bar{r}')}$ when $z < z'$. $\bar{k} = \hat{x}k_x + \hat{y}k_y + \hat{z}k_{0z}$ and $\bar{K} = \hat{x}k_x + \hat{y}k_y - \hat{z}k_{0z}$ are wave number vectors propagating in the directions of $\hat{k} = \bar{k}/k$ and $\hat{K} = \bar{K}/k$, respectively. The

polarization vectors are defined as

$$\hat{e}(k_{0z}) = \frac{\hat{k} \times \hat{z}}{|\hat{k} \times \hat{z}|} = \frac{\hat{x}k_y - \hat{y}k_x}{\sqrt{k_x^2 + k_z^2}}$$

$$\hat{h}(k_{0z}) = \bar{e}(k_{0z}) \times \hat{k} = \frac{-k_{0z}}{k\sqrt{k_x^2 + k_y^2}}(\hat{x}k_x + \hat{y}k_y) + \hat{z}\frac{\sqrt{k_x^2 + k_y^2}}{k}$$

Notice that $\hat{k}\hat{k} + \hat{e}\hat{e} + \hat{h}\hat{h} = \bar{\bar{I}}$. The dyadic Green's function can be extended to a multilayered medium by defining reflection coefficients as practiced in the previous Sections.

References

[1] J. A. Stratton, *Electromagnetic Theory,* New York: McGraw-Hill, 1941.

[2] A. Sommerfeld, *Partial Differential Equations in Physics,* New York: Academic Press, 1949.

[3] P. M. Morse and F. Feshbach, *Methods of Theoretical Physics,* New York: McGraw-Hill, 1953.

[4] R. E. Collin, *Field Theory of Guided Waves,* New York: McGraw-Hill, 1960.

[5] R. F. Harrington, *Time-Harmonic Electromagnetic Fields,* New York: McGraw-Hill, 1961

[6] R. Courant and D. Hilbert, *Methods of Mathematical Physics,* New York: Interscience, 1962.

[7] J. D. Jackson, *Classical Electrodynamics,* New York: John-Wiley, 1962.

[8] J. R. Wait, *Electromagnetic Waves in Stratified Media,* Oxford: Pergamon Press, 1962.

[9] M. Abramowitz and I. A. Stegun, *Handbook of Mathematical Functions,* New York: Dover Publications, 1965.

[10] S. Ramo, J. R. Whinnery, and T. van Duzer, *Fields and Waves in Communication Electronics,* New York: John-Wiley, 1965.

[11] R. E. Collin, *Foundations for Microwave Engineering,* New York: McGraw-Hill, 1966.

[12] R. F. Harrington, *Field Computation by Moment Method,* Malabar, FL: Krieger, 1968.

[13] C. T. Tai, *Dyadic Green's Function in Electromagnetic Theory,* New York: Intext Publishers, 1971.

[14] R. Mittra, *Computer Techniques for Electromagnetics*, ed., New York: Pergamon Press, 1973.

[15] F. B. Hildebrand, *Introduction to Numerical Analysis*, New York: McGraw-Hill, 1974.

[16] R. Mittra, *Numerical and Asymptotic Techniques in Electromagnetics*, ed., New York: Springer-Verlag, 1975.

[17] I. Stakgold, *Green's Functions and Boundary Value Problems*, New York: John-Wiley, 1979.

[18] L. Tsang, J. A. Kong, and R. T. Shin, *Theory of Microwave Remote Sensing*, New York: Wiley-Interscience, 1985.

[19] J. A. Kong, *Electromagnetic Wave Theory*, New York: John-Wiley, 1986.

[20] H. A. Haus and J. R. Melcher, *Electromagnetic Fields and Energy*, New Jersey: Prentice Hall, 1989.

[21] T. Itoh, *Numerical Techniques for Microwave and Millimeter-Wave Passive Structures*, ed., New York: Wiley-Interscience, 1989.

Chapter 2

EFFICIENT SIMULATORS AND DESIGN TECHNIQUES FOR GLOBAL MODELING OF HIGH-FREQUENCY ACTIVE MICROWAVE DEVICES

Yasser A. Hussein,[1] Muhammad D. Waliullah,[1] and Samir M. El-Ghazaly[2]

[1] *Department of Electrical Engineering*
Arizona State University, USA

[2] *Department of Electrical and Computer Engineering*
The Univeristy of Tennessee, USA

Abstract This Chapter presents two novel approaches for global modeling of high-frequency active microwave devices. The first approach considers the use of wavelets to reduce CPU time while maintaining the same degree of accuracy provided by standard global modeling simulators. The basic idea of multiresolution time domain(MRTD) technique is to adaptively refine grids at locations where the unknown variables vary rapidly, and an attractive way to implement it is to use wavelets.

The second approach presents a fully distributed equivalent circuit model for high-frequency transistors. The proposed distributed circuit model incorporates a sufficient number of segments to accurately account for wave propagation effects along the device width. For the first time, distributed circuit model having several segments is obtained in the time domain to analyze the large-signal behavior. In addition, a hybrid model that combines both physics-based and circuit-based models will be presented.

Keywords: global modeling, microwave devices, adaptive grids, wavelets, physics-based models, circuit-based models, hybrid models.

1. Introduction

Modern high-performance electronics are based on technologies such as monolithic microwave integrated circuits(MMIC's), with a large number of closely packaged passive and active devices, several levels of transmission lines and discontinuities, all operating at high speeds, high frequencies, and sometimes over very broad bandwidths. It is thus perceptible that the design of MMIC's should involve robust design tools that simulate all the circuit elements simultaneously. The possibility of achieving this type of modeling is addressed by global circuit modeling that has been demonstrated in [1]. Global modeling is a tremendous task that involves advanced numerical techniques and different algorithms. Consequently, it is computationally expensive. Therefore, there is an urgent need to develop new approaches to reduce the simulation time while maintaining the same degree of accuracy achieved by the global modeling technique. In the next Section, an approach for applying wavelets to the full-hydrodynamic model will be presented.

2. Global Modeling of Microwave Devices Using Wavelets

Multiresolution time domain(MRTD) approach has been successfully applied to finite difference time domain(FDTD) simulations of passive structures [2]. Several different approaches for solving partial differential equations(PDE's) using wavelets have been considered, and it has been noticed by several authors that nonlinear operators such as multiplication are too computationally expensive when implemented directly on a wavelet basis. One of the approaches is the interpolating wavelets technique presented in [3], in which the author deals with the nonlinearities using the sparse point representation(SPR).

Interpolating wavelets have been successfully applied to the simple drift diffusion active device model [4]. The simple drift diffusion is not suitable for modeling submicrometer devices since it leads to inaccurate estimations of device internal distributions and microwave characteristics [5]. Thus, a new approach for applying wavelets to the full-hydrodynamic model of active devices is needed.

In this Section, a new approach for applying wavelets to the full-hydrodynamic model of semiconductor devices for large-signal physical modeling is presented. The main idea is to take snap shots of the solution within the simulation and then apply biorthogonal wavelet transform to the current solution to obtain the coefficients of details. The coefficients of details are then normalized and a threshold is applied to obtain a

nonuniform grid. The grids of different variables are updated during simulation according to an updating criterion. A general updating criterion as well as threshold formula will be developed and verified.

2.1 Overview of Wavelets

Construction of biorthogonal wavelet bases relies on the notation of multiresolution analysis [6], [7]. This notation gives a formal description of the intuitive idea that every signal can be constructed by successive refinement, by adding details to an approximation iteratively. More precisely, a multiresolution analysis in $L^2(R)$ is defined to be in a sequence of nested subspaces that

$$\cdots \supset V_1 \supset V_0 \supset V_{-1} \supset \cdots$$

where

$$\bigcap_{m=-\infty}^{\infty} V_m = \{0\}, \qquad \bigcup_{m=-\infty}^{\infty} V_m \text{ is dense in } L^2(R)$$

$$x(t) \in V_m \text{ if and only if } x(2t) \in V_{m+1} \tag{2.1}$$

$$\varphi \text{ exists suh that } \{\varphi(t-n), n \in N\} \text{ form a basis set of } V_m \tag{2.2}$$

We can assign a time resolution of 2^m to each V_m, and the approximation of a signal, $x(t)$, at this resolution level is obtained by projection onto the corresponding subspace. By applying properties (2.1) and (2.2), the basis family of V_m can be derived from a basis of V_m in the following way. Starting from the scaling function, $\varphi(t)$, the basis family of V_m is obtained by taking the dilates and translates of $\varphi(t)$ as

$$\varphi_{mn}(t) = 2^{m/2}\varphi(2^m t - n)$$

Notice that the coefficients of the approximations

$$a_x[n,m] = \int_{-\infty}^{\infty} x(t)\varphi_{nm}(t)dt$$

are associated with the collection of all these bases, and share a large amount of information which is redundant. A much more economical representation consists of finding the information difference between two consecutive approximations. Then, details are added to the coarser approximation before it is passed to the finer one. This amounts to saying that for each approximation space, V_m, the details belong to a space, W_m, which is the orthogonal complement of V_m in V_{m+1} . Hence, we infer the relation

$$V_{m+1} = V_m \bigoplus W_m$$

which leads to the decomposition

$$L^2(R) = \bigoplus_{-\infty}^{\infty} W_m$$

Therefore, it is sufficient to find a proper wavelet, $\psi(t)$, so that $\{\psi(t - n), n \in Z\}$ is a basis family of W_0. Then, the set

$$\{\psi_{nm}(t) = 2^{m/2}\psi(2^m t - n), n, m \in Z\}$$

constitutes the orthonormal bases of $L^2(R)$.

Scaling Function. As the scaling function, $\varphi(t)$, is an element of V_0, it is also in V_1. Hence, there exists a set of coefficients, $h[n]$, such that

$$\varphi(t) = 2^{1/2} \sum_{-\infty}^{\infty} h[n]\varphi(2t - n)$$

$$h[n] = 2^{1/2} \int_{-\infty}^{\infty} \varphi(t)\varphi(2t - n)dt$$

$$\sum_{-\infty}^{\infty} h^2[n] = 1$$

Wavelet. As the wavelet, $\psi(t)$, is an element of V_1, it is also characterized by a discrete filter with coefficients $g[n]$ such that

$$\psi(t) = 2^{1/2} \sum_{-\infty}^{\infty} g[n]\varphi(2t - n)$$

$$g[n] = 2^{1/2} \int_{-\infty}^{\infty} \psi(t)\varphi(2t - n)dt$$

$$\sum_{-\infty}^{\infty} g^2[n] = 1$$

For orthonormal wavelet bases, filter coefficients, $h[n]$ and $g[n]$, of the scaling function and wavelet, respectively, are linked to each other by the quadrature mirror filter relation. The coefficients of the details are given as

$$d_x[n, m] = \int_{-\infty}^{\infty} x(t)\psi_{nm}(t)dt$$

While some wavelets such as Daubechies [8] are asymmetrical, it is possible to create symmetric wavelets with compact support by using two sets of wavelets, one to compose the signal and the other to construct it. Such wavelets are called orthogonal [9].

2.2 Problem Description

The transistor model used in this Chapter is a two-dimensional full-hydrodynamic large-signal physical model, which solves the Poisson equation

$$\nabla^2\phi = \frac{q}{\epsilon}(N_d - n) \tag{2.3}$$

where ϕ is the electrostatic potential, q is the charge of an electron, ϵ is the dielectric constant, N_d is the doping concentration, and n is the carrier density at any given time. It is solved in conjunction with the active device model that is based on moments of the Boltzmann's transport equation obtained by integrating over the momentum space. The integration results in a strongly coupled highly nonlinear set of partial differential equations, called the conservation equations. These equations provide a time dependent self-consistent solution for carrier density, energy and momentum.

- **current continuity**

$$\frac{\partial n}{\partial t} + \nabla \cdot (n\bar{v}) = 0 \tag{2.4}$$

- **energy conservation**

$$\frac{\partial(n\varepsilon)}{\partial t} + qn\bar{v}\cdot\bar{E} + \nabla\cdot[n\bar{v}(\varepsilon + k_B T)] = -\frac{n(\varepsilon - \varepsilon_0)}{\tau_\varepsilon(\varepsilon)} \tag{2.5}$$

- **x-momentum conservation**

$$\frac{\partial(np_x)}{\partial t} + qnE_x + \nabla\cdot(np_x\bar{v}) + \frac{\partial(nk_BT)}{\partial x} = -\frac{n(p_x - p_0)}{\tau_m(\varepsilon)} \tag{2.6}$$

In these equations, n is the electron density, $\bar{v}$ is the electron velocity, $\bar{E}$ is the electric field, ε is the electron energy, ε_0 is the equilibrium thermal energy, and p is the electron momentum. The energy and momentum relaxation times are denoted by τ_ε and τ_m , respectively. Similar expression is obtained for the y momentum. The total current density distribuation, $\bar{J}$, inside the active device at any time t is given as

$$\bar{J} = -qn\bar{v} + \epsilon\frac{\partial\bar{E}}{\partial t} \tag{2.7}$$

All the above equations can be solved in the time domain with proper discretization. Details of implementing the proposed scheme will be

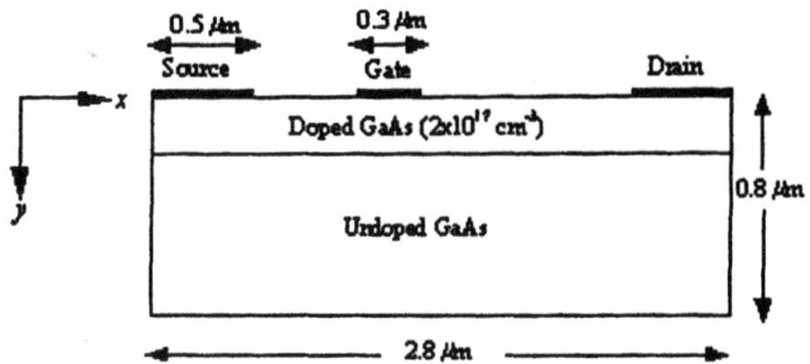

Figure 2.1. Cross section of a simulated MESFET.

given in Section 2.4. The low-field mobility is given by the empirical relation [10] as

$$\mu_0 = \frac{8000}{1 + \sqrt{N_d/10^{17}}} \ \mathrm{cm^2/Vs} \tag{2.8}$$

The above model accurately describes all the nonstationary transport effects by incorporating energy dependence into all the transport parameters such as effective mass and relaxation times. Fig.2.1 shows the cross section of a simulated structure with parameters summarized in Table 2.1.

Table 2.1. Transistor parameters used in the simulation.

Parameter	Value
Drain and source contacts	0.5 μm
Gate-source separation	0.5 μm
Gate-drain separation	1 μm
Device thickness	0.8 μm
Device length	2.8 μm
Gate length	0.3 μm
Device width	200 μm
Active layer thickness	0.2 μm
Active layer doping	2×10^{17} cm^{-3}
Schottky barrier height	0.8 V
Gate-source dc voltage	−0.5 V
Drain-source dc voltage	3 V

Next, we will demonstrate the applications of wavelets to the full-hydrodynamic simulator to accurately model submicrometer gate devices with significantly less CPU time. Ultimately, a full-hydrodynamic model should be implemented with Maxwell's equations rather than

Poisson equation to obtain self-consistent simulation of wave propagation effects.

2.3 Proposed Algorithm

Fig.2.2 shows the flow chart of the proposed algorithm. A uniform grid is defined at the beginning of simulation. Eqs.(2.3)-(2.7) are then solved in sequence as indicated in the flow chart to obtain the solution of different variables at new time instant with the following criterion

$$\frac{|\chi^{\ell}_{\max,\min} - \chi^{c}_{\max,\min}|}{|\chi^{\ell}_{\max,\min}|} \geq \gamma \tag{2.9}$$

The updating criterion simply checks if the variable, χ, has changed by γ since last iteration. The superscripts c and ℓ stand for current time and last time, respectively, when wavelet transform is performed. Max and min indicate if the maximum and minimum of the variable χ satisfy (2.9) simultaneously.

It is worth mentioning that boundary grid points are not included in the maximum or minimum checking. The value of γ used in the simulation is 0.1. If (2.9) is satisfied, wavelet transform is performed on current solution followed by thresholding to obtain an updated nonuniform grid for the variable χ. Biorthogonal wavelets are marked by BIO3.1 to indicate three vanishing moments of the mother wavelet and one vanishing moment of the scaling function. Nonuniform grids of different variables are then combined into one nonuniform grid for next iteration. The above steps are repeated until the stopping criterion is satisfied.

Note that the ranges of variables used in the simulations vary dramatically. For instance, the carrier density is on the order of 10^{17}, while the energy is on the order of 0.5. Accordingly, the threshold value should be dependent on the variable. The proposed threshold formula is

$$T = T_0 \frac{1}{n_x} \left[\sum_{i=1}^{n_x} d_i^2 \right]^{1/2} \tag{2.10}$$

where T_0 is the initial threshold value, d_i's are the coefficients of details, and n_x is the number of grid points in $\hat{x}$ or $\hat{y}$ direction. Hence, the threshold value, T, depends on the solution at any given time, no longer being fixed. The values of T_0 used in the simulation are 0.001, 0.01 and 0.05, respectively.

In this Chapter, a new technique to conceive nonuniform grids using wavelets has been developed. The main idea is to apply wavelet transform to the solution at any given time to obtain the coefficients

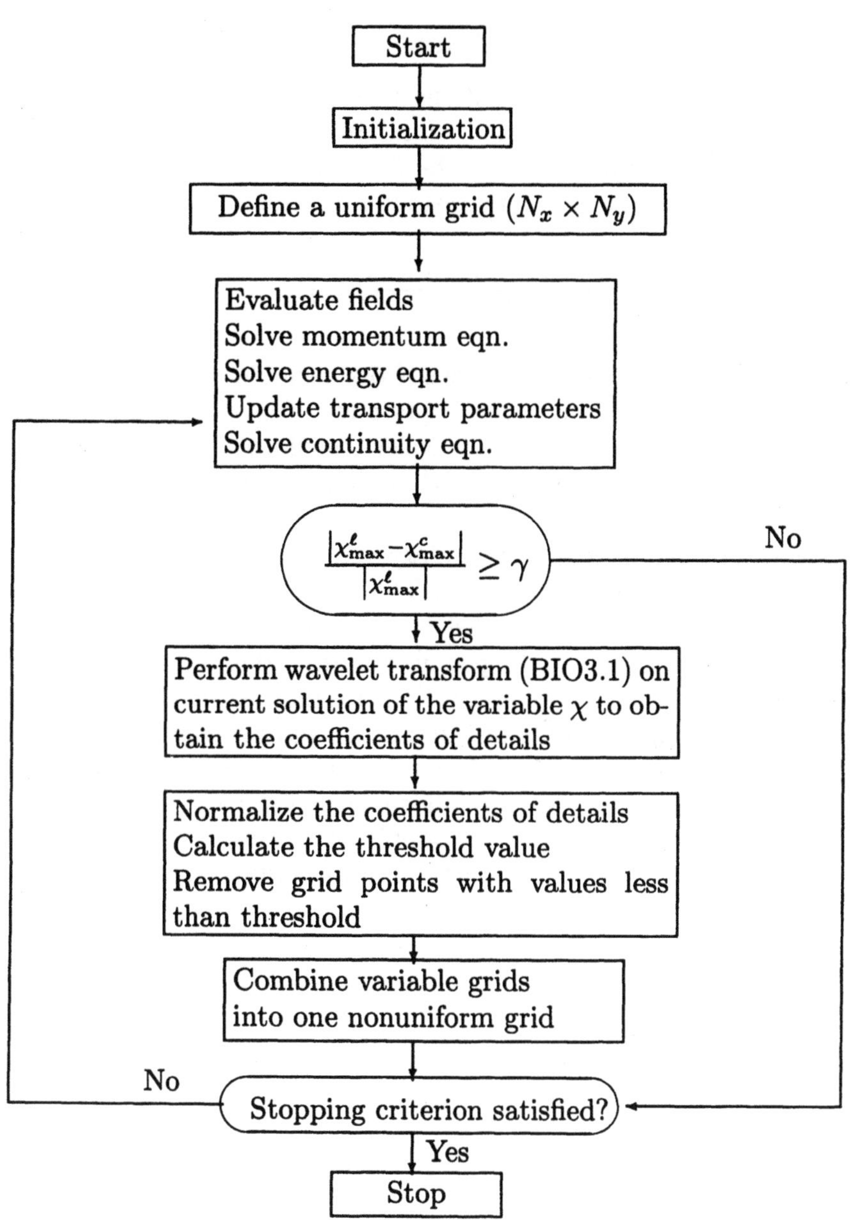

Figure 2.2. Flow chart of the proposed algorithm.

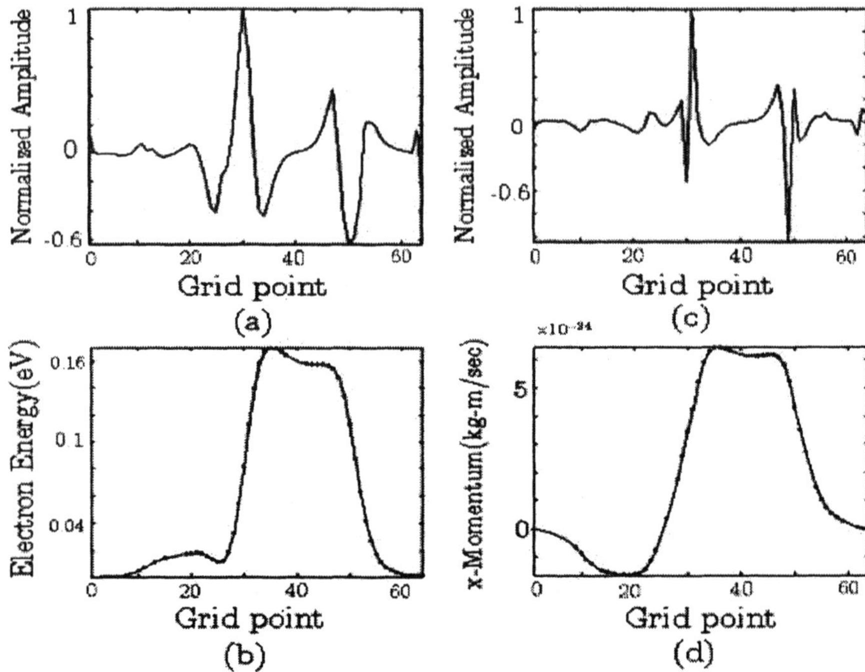

Figure 2.3. (a) Normalized details coefficients of electron energy in certain transversal cross section. (b) Grid points marked on the curve for electron energy in the same transversal cross section. (c) Normalized details coefficients of the x momentum in certain transversal cross section. (d) Grid points marked on the curve for x momentum in the same transversal cross section.

of details, which are then normalized to its maximum. Only the grid points with normalized coefficients of details larger than the threshold value are included. Figs.2.3 and 2.4 show different examples of how the nonuniform grids are obtained for electron energy and x-momentum solutions in a specific cross section. Fig.2.3 shows how the proposed algorithm obtains the nonuniform grid using transverse compression only. For instance, Fig.2.3(a) shows the normalized coefficients of details for electron energy, while Fig.2.3(b) marks the grid points remaining after thresholding the normalized coefficients of details using (2.10). It is observed that the proposed technique accurately removes grid points where the solutions change very slowly.

Fig.2.4 shows how the proposed algorithm achieves the nonuniform grid of any variable using longitudinal compression only. Comparing Figs.2.3 and 2.4, one may conclude that compression in the longitudinal cross section is much more significant than that in the transversal cross section. This complies with the fact that physical changes in the

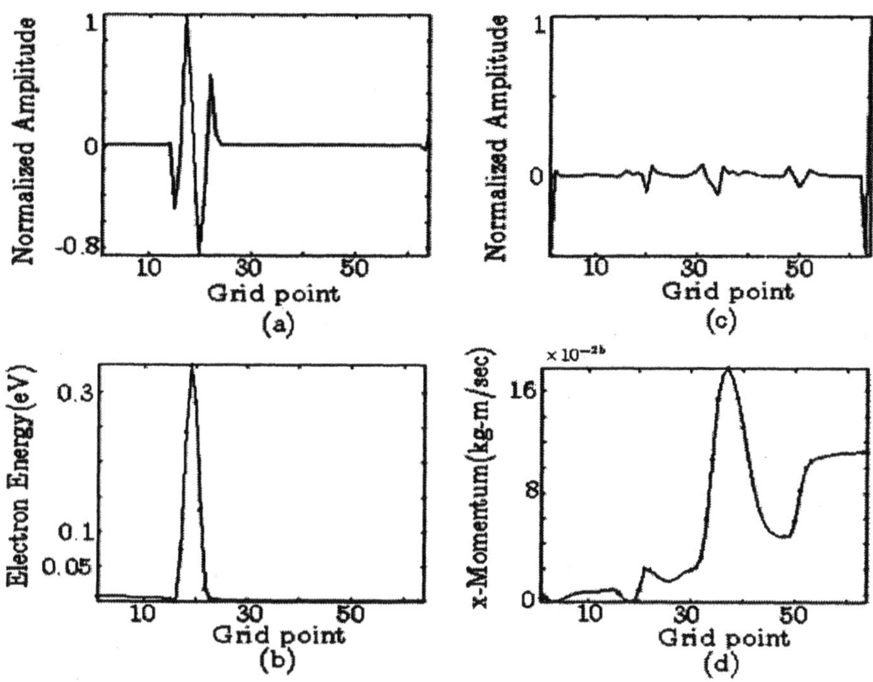

Figure 2.4. (a) Normalized details coefficients of electron energy in certain longitudinal cross section. (b) Grid points marked on the curve for electron energy in the same longitudinal cross section. (c) Normalized details coefficients of the x momentum in certain longitudinal cross section. (d) Grid points marked on the curve for x momentum in the same longitudinal cross section.

longitudinal cross sections are much slower compared to those in the transversal cross sections.

Fig.2.5 shows how the nonuniform grid for electron energy is obtained. The process is achieved by obtaining two separate grids for the transversal and longitudinal compressions, respectively. Then the two grids are combined together using logical AND to conceive the overall grid for electron energy at the given time. The same process is conducted for other variables including x momentum, y momentum, carrier density and potential. Separate grids for different variables are then combined using logical OR to obtain the overall grid for the next iteration.

The overall grid obtained needs further processing in order to define a finite difference scheme on it. The simplest way to achieve this is to have the same number of grid points for parallel cross sections, while the number of grid points in the longitudinal and transversal cross sections need not to be the same. Considering the overall grid shown in Fig.2.5(d), the percentage of unknowns remaining after adding the necessary grid points

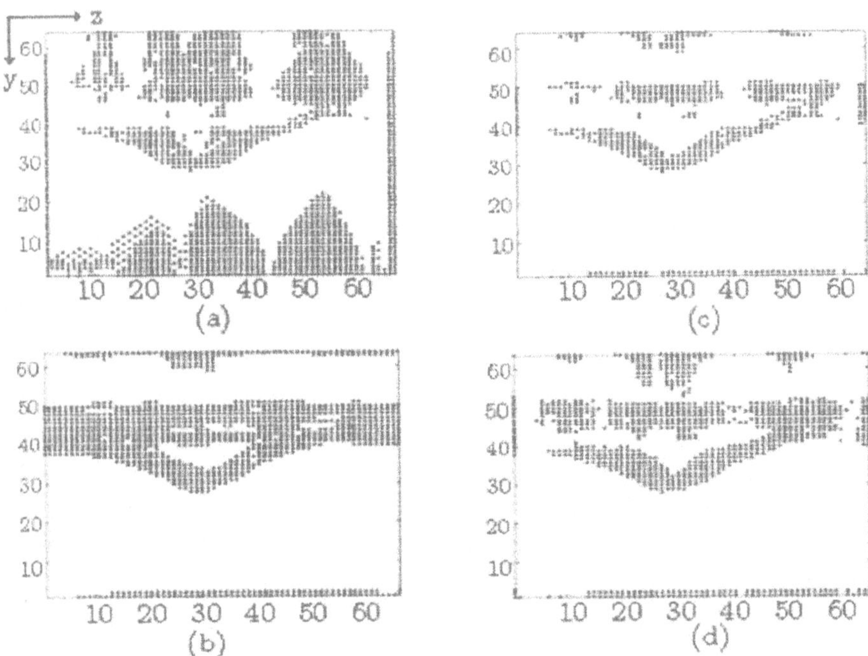

Figure 2.5. **(a) Compression of electron energy grid in transversal cross sections. (b) Compression of electron energy grid in longitudinal cross sections. (c) Final grid for electron energy. (d) Overall grid.**

is about 35%, which shows that the proposed approach has achieved a reduction in the number of unknowns despite of the overhead introduced by adding extra grid points.

Table 2.2 shows the evolution of nonuniform grids. It is observed that the number of grid points for the overall grid increases as time marches. Because at the beginning of simulation, the solution is not completely filled, and more grid points are needed to incorporate the changes in solution as time marches. Furthermore, grids of different variables should not be updated at the same rate. For instance, it is apparent that the potential needs not to be updated at the same rate as the other variables. Table 2.2 is used to demonstrate how different grids evolve. In actual simulations, the potential grid is updated a few times at the beginning of simulation, and then remains unchanged.

2.4 Results and Discussions

Dc Simulation Results. The approach presented in this Chapter is generic and can be applied to any unipolar transistor. For demonstration purpose, it is applied to an idealized MESFET structure, which is discretized by a mesh of $64\Delta x \times 64\Delta y$ and $\Delta t = 0.001$ ps. An explicit forward Euler finite difference method is adopted. In addition, upwinding is employed to have a stable finite difference scheme. The spatial step sizes are adjusted to satisfy the Debye length, while the temporal step, Δt, is chosen to satisfy the Courant-Friedrichs-Levy(CFL) condition. The zero-field mobility used in the simulation is given by (2.8), and the saturation velocity is 2×10^7 cm/s. First, dc simulations are performed according to the flow chart in Fig.2.2, and the current density is calculated using the first term in (2.7), which is the conduction current.

Fig.2.6 shows the percentage of remaining unknowns or grid points versus iteration number with different initial threshold values. Notice that as the threshold value increases, the remaining number of grid points decreases. At the end of simulation, the remaining percentage of unknowns is almost 70% for an initial threshold value of 0.001, whereas it is about 30% for an initial threshold value of 0.01. A small change of initial threshold value results in a large change in the remaining number of unknowns. Furthermore, the remaining number of unknowns changes during simulation due to adaptability of grids. In the next Section, we will study the effect of initial threshold value on the accuracy of final result as well as the trade-off between accuracy and CPU time.

Fig.2.7 shows the convergence of drain current versus CPU time for the case of uniform grid and the proposed wavelet-based adaptive grids

Table 2.2. Grid adaptability of different variables.

Variable	Unknowns remained after transversal compression (%)	Unknowns remained after longitudinal compression (%)	Total uknowns remained (%)
	Iteration # 120		
Potential	5.69	7.74	0.63
Carrier density	6.54	14.92	2.64
Energy	39.65	17.63	8.54
x momentum	43.39	16.06	9.18
y momentum	16.11	17.53	3.78
All variables	65.14	22.36	14.43
	Iteration # 250		
Potential	5.88	8.59	0.76
Carrier density	13.69	16.70	5.18
Energy	39.21	23.00	12.26
x momentum	43.65	19.09	10.64
y momentum	20.02	19.46	7.95
All variables	61.94	28.93	20.58
	Iteration # 480		
Potential	6.27	9.23	0.90
Carrier density	21.51	17.16	7.71
Energy	43.99	28.88	15.72
x momentum	38.57	23.44	12.72
y momentum	26.20	26.76	13.53
All variables	58.84	36.25	25.05
	Iteration # 590		
Potential	6.04	9.64	0.93
Carrier density	29.88	18.24	10.61
Energy	48.85	31.88	18.43
x momentum	41.91	29.08	16.21
y momentum	32.91	37.04	20.36
All variables	62.36	44.73	31.74

with different initial threshold values. It is shown that using the proposed wavelet-based time-domain approach has reduced the CPU time dramatically. For instance, there is a reduction of about 75% in CPU time over the uniform grid with the initial threshold value of 0.01, where the error of dc drain current is within 1%. In addition, increasing the initial threshold beyond certain value has a negative effect on the accuracy of final results. This is obvious with $T_0 = 0.05$, where there is no agreement between the results obtained using the uniform grid and the

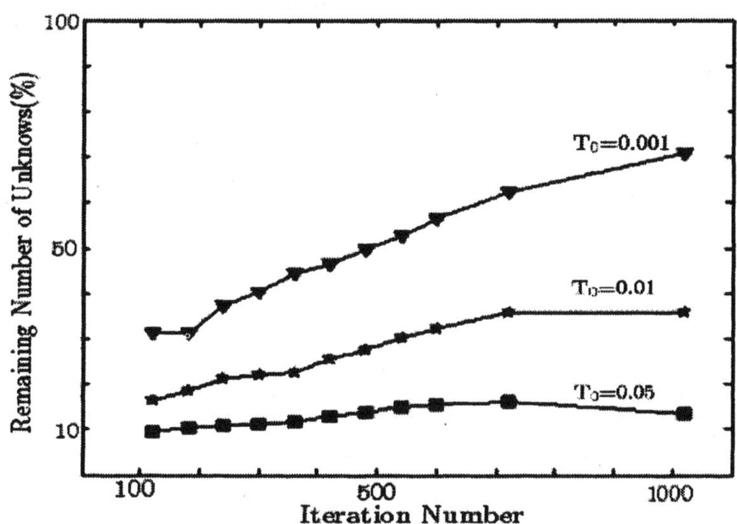

Figure 2.6. Percentage of remaining unknowns versus iteration number for different initial threshold values.

wavelet-based grids. Because using large values of T_0 implies that more grid points are removed, including some important grid points that will have a negative effect on the final result. On the other hand, using a very small threshold value implies more redundant grid points.

In summary, there should be an optimal value of T_0 such that both CPU time and error are minimized. In this Chapter, T_0 of 0.01 is suggested to have a considerable reduction in CPU time while keeping error within an acceptable level.

Fig.2.8 shows the potential distribution obtained using the proposed algorithm with T_0 equals to 0.01. It is demonstrated that the boundary conditions are satisfied at the electrodes. For instance, the potential at the gate is equal to -1.3 V, which is the applied dc voltage minus the Schottky barrier.

Ac Simulation Results. The ac excitation voltage applied to the gate electrode is given as

$$V_{gs}(t) = V_{gso} + \Delta V_{gs} \sin(\omega t) \tag{2.11}$$

where V_{gso} is the dc bias, ΔV_{gs} is the peak value of ac signal, which is 0.1 V, ω is the angular frequency of the applied signal in rad/s, and the frequency used in simulation is 60 GHz.

First, the dc solution is obtained by solving Poisson equation in conjunction with the three hydrodynamic conservation equations. Then, a new value of V_{gs} is obtained using (2.11), which is used to update the

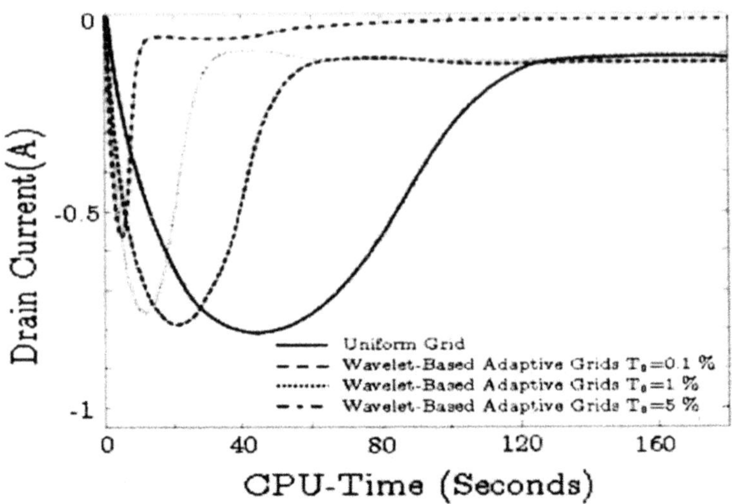

Figure 2.7. Convergence of dc drain current for uniform grid and the wavelet-based nonuniform grids with different initial threshold values.

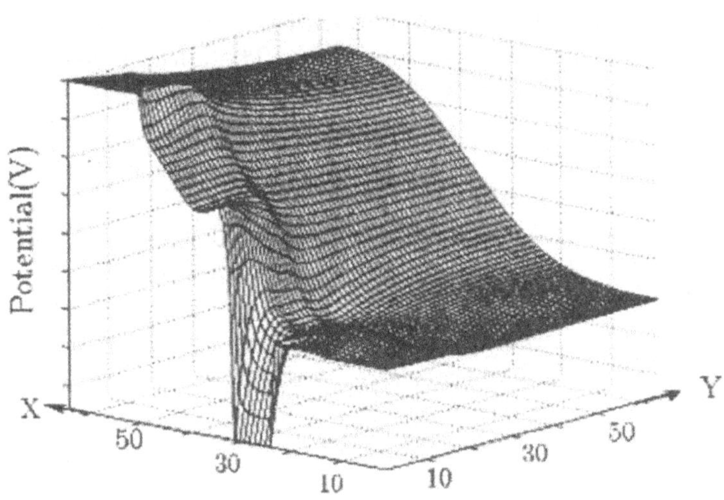

Figure 2.8. Dc potential distribution obtained by the proposed algorithm with the initial threshold value of 0.01.

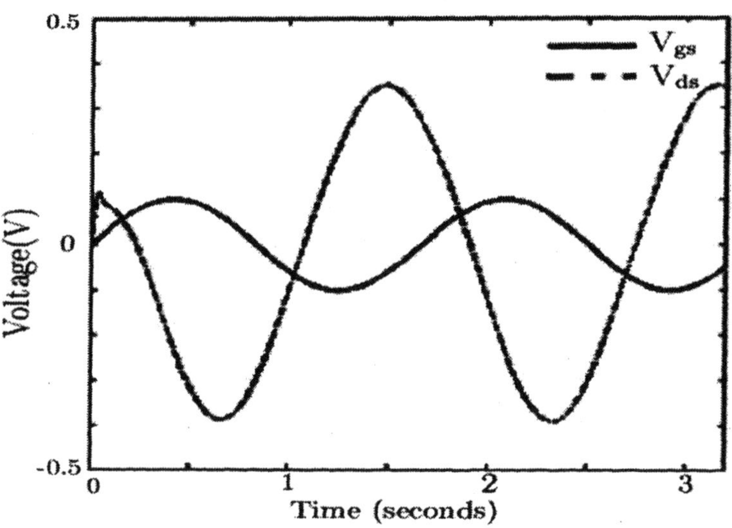

Figure 2.9. **Large-signal voltage obtained by the proposed algorithm with the initial threshold value of 0.01.**

Poisson equation to obtain a new voltage distribution, subsequently a new electric field. The electric field is then used to update the variables in the conservation equations. This process is repeated every Δt according to the proposed algorithm shown in Fig.2.2 until $t = t_{\text{max}}$. The current density is obtained using (2.7). The current density calculated on the plane located midway between the drain and gate is integrated to obtain the total current. The output voltage is calculated by multiplying the total current with the resistance associated with the dc operating point of the transistor.

Fig.2.9 shows the output voltage obtained using the proposed algorithm with $T_0 = 0.01$. The output signal is no longer a pure sinusoidal due to the existence of harmonics. A gain of about 11 dB is achieved. Furthermore, it is observed that there is an output delay of about 1 ps which is required for the transistor to respond to the input signal.

Fig.2.10 shows the output voltage with different initial threshold values. It is observed that the values of T_0 affect the accuracy of solution. For instance, using $T_0 = 0.05$ results in a completely different and inaccurate solution. Because employing a large T_0 results in removing some important grid points, which degrades the final results. Similar to dc simulations, an optimal value of T_0 is suggested. Furthermore, the mean relative errors obtained with T_0 equals to 0.001 and 0.01, respectively,

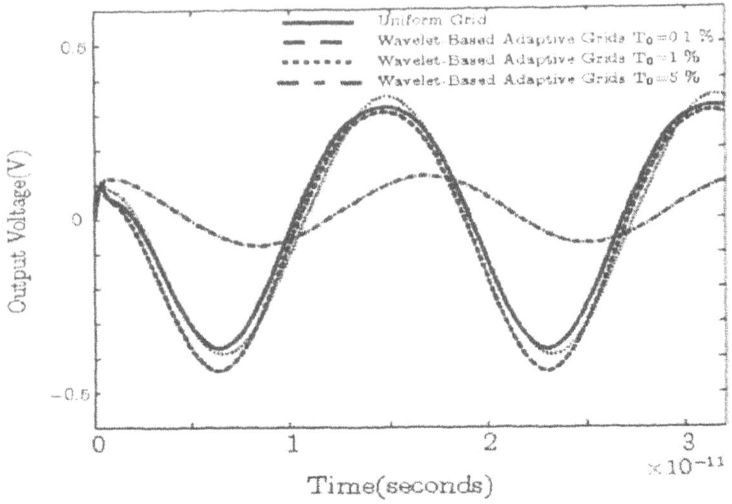

Figure 2.10. **Ac output voltage results for the uniform grid and the proposed wavelet-based nonuniform grids with different initial threshold values.**

are on the order of 3 to 4%. This suggests that setting T_0 equal to 0.01 optimizes both accuracy and CPU time.

Local truncation error for the uniform grid generally depends on the mesh size and time step used. On the other hand, local truncation error for the wavelet-based nonuniform grids depends on how the important grid points are reserved as well as the time step used. This suggests that the local truncation errors for the uniform grid and wavelet-based nonuniform grids are different. The local truncation error accumulates from iteration to iteration.

The total truncation or discretization error is thus dependent on the number of iterations used. Accordingly, one may argue that the total error due to local discretization introduced by using uniform grid may or may not be larger than that of using the proposed wavelet-based nonuniform grids, at least for the cases with $T_0 = 0.001$ and $T_0 = 0.01$. Because the number of iterations required to reach the steady state solution with the uniform grid is much larger than that with the proposed algorithm. In summary, the total error is contributed by the local truncation error and the number of iterations required to reach the final solution. This explains why it is difficult to decide which technique is more accurate from the results shown in Figs.2.7 and 2.8. Because the number of iterations required to obtain the final solution and the local discretization errors are different for each curve. The problem of identifying the most

accurate solution becomes even more difficult by the fact that we are dealing with a highly nonlinear problem.

3. Large-Signal Circuit and Physics-Based Time Domain Analysis

At high frequencies when device dimensions become comparable to wavelength, wave propagation effect need to be considered in device modeling [1]. Full-wave global modeling approach can be used to analyze wave device interaction [11]. Full-wave techniques are time consuming and need extensive computer memory. On the other hand, distributed equivalent circuit models [12] and hybrid models [13] can account for the wave propagation effects reasonably well and they are thus suitable for CAD applications. In this Section, complete equivalent circuit model and physics-based hybrid model will be developed. Time domain analysis using both techniques will also be presented.

In the proposed distributed circuit model, the simulated structure is divided into passive and active parts. The passive part consists of the source, drain and gate electrodes, forming a coplanar coupled transmission line. By applying a full-wave FDTD analysis, all the self and mutual inductances and capacitances of the transmission line can be determined. A Gaussian excitation signal is used in the full-wave analysis to extract the transmission line circuit elements in a wide frequency range. The active part is intrinsic to the device, in which a full-hydrodynamic simulation is conducted to obtain the equivalent parameters under different bias conditions.

Curve fitting technique is used to derive the Curtice large-signal parameters per unit width from the current voltage characteristics obtained by applying the full-hydrodynamic simulator to the active device [14]. In addition, scaling rule is used to develop the complete equivalent circuit model consisting of several unit segments as shown in Fig.2.11.

To simulate the large-signal behavior in a multisegment distributed circuit model, a time domain simulation technique is developed that adopts explicit iterative scheme for the nonlinear circuit elements such as transconductance, G_m, junction capacitances, C_{gs} and C_{ds}, and output conductance, G_{ds}.

In the proposed hybrid model, the active device is analyzed using a two-dimensional full-hydrodynamic simulator. The lumped elements extracted using full-wave analysis on the transmission line are then connected to the active device. To effectively model the wave propagation effects along the device width, the full device is divided into several segments. Since the proposed hybrid model uses physics-based simulator

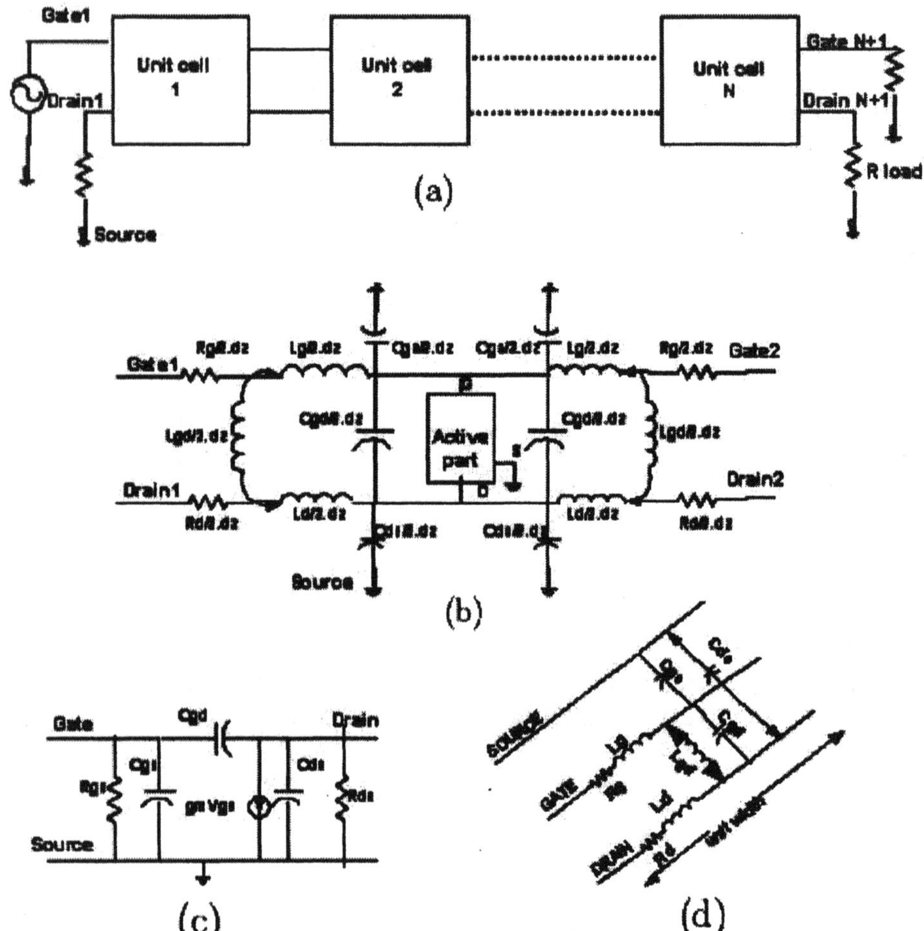

Figure 2.11. (a) Complete distributed circuit model consisting of multiple unit segments, (b) unit segment with active and passive parts, (c) equivalent circuit of the active part, (d) distributed circuit elements of the passive part.

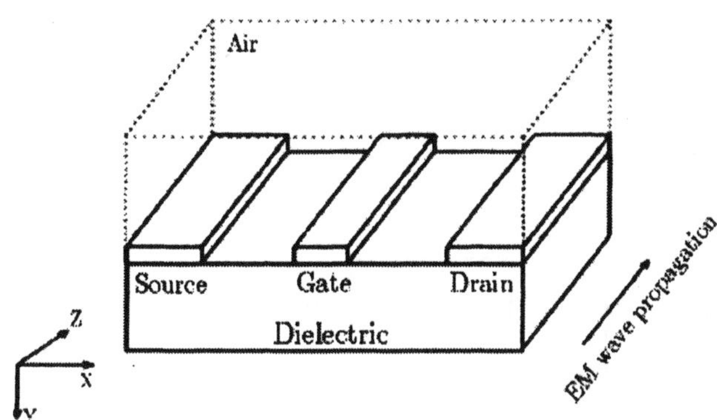

Figure 2.12. **Three-dimensional structure of MESFET used in FDTD.**

for the active part, this model is expected to be more accurate than the equivalent circuit model.

The proposed hybrid model also accounts for and gives insight into the wave propagation effects on electrons as they travel through the device. As a trade-off, more CPU time is needed for the hybrid model than for the complete circuit model which will be presented in the next Section.

3.1 Equivalent Circuit Model

Determination of Passive Circuit Parameters. To analyze wave propagation effects along the electrodes of the simulated FET, three electrode lines are viewed as coplanar transmission lines as shown in Fig.2.12, which can be represented by the distributed circuit elements as shown in Fig.2.11(d). In order to evaluate all the distributed circuit elements, both even and odd modes are solved for the passive part using FDTD approach. Line resistances are calculated considering the skin effect at high frequencies [15].

Determination of Active Circuit Parameters. Equivalent circuit of the active part is shown in Fig.2.11(c). Two-dimensional full-hydrodynamic simulation is performed for the specified FET to generate the current voltage characteristics as shown in Fig.2.13 under different bias conditions, which is then fitted to the Curtice large-signal model. In the Curtice model, the drain current, I_{ds}, is related to the gate voltage, V_{gs}, and drain voltage, V_{ds}, by

$$I_{ds} = \beta(V_{gs} - V_t)^2(1 + \lambda V_{ds})\tanh(\alpha V_{ds})$$

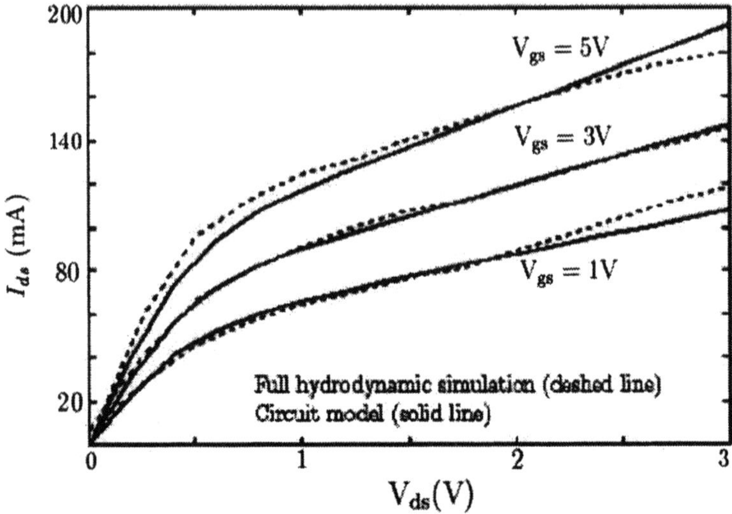

Figure 2.13. *I-V* characteristics of the MESFET obtained from simulation and curves fitting with Curtice model.

where β, λ, α and V_t are parameters in Curtice model. The transconductance, G_m, and output conductance, G_{ds}, are defined as

$$G_m = \frac{\partial I_{\mathrm{ds}}}{\partial V_{\mathrm{gs}}} = \frac{2I_{\mathrm{ds}}}{V_{\mathrm{gs}} - V_t}$$

$$G_{\mathrm{ds}} = \frac{\partial I_{\mathrm{ds}}}{\partial V_{\mathrm{ds}}} = \beta(V_{\mathrm{gs}} - V_t)^2(1 + \lambda V_{\mathrm{ds}})\frac{\alpha}{\cosh^2(\alpha V_{\mathrm{ds}})} + \beta(V_{\mathrm{gs}} - V_t)^2\lambda \tanh(\alpha V_{\mathrm{ds}})$$

The gate-source capacitance, C_{gs}, and drain-source capacitance, C_{ds}, are given by

$$C_{\mathrm{gs}} = \frac{C_{\mathrm{gso}}}{\sqrt{1 - V_{\mathrm{gs}}/V_{\mathrm{bi}}}}, \qquad C_{\mathrm{gd}} = \frac{C_{\mathrm{gdo}}}{\sqrt{1 - V_{\mathrm{gd}}/V_{\mathrm{bi}}}}$$

where V_{bi} is the built-in potential of the Schottky gate, C_{gso} and C_{gdo} are zero-bias gate-source and gate-drain capacitances, respectively. Other active circuit elements, R_{gs} and C_{ds}, are approximated as constant under all bias conditions.

Distributed Circuit Model. As shown in Fig.2.11(a), the complete distributed circuit model consists of several unit segments, each unit segment consists of an active part and a passive part. To make the unit segment symmetrical, half of the passive part is connected to

the input side and the other half to the output side. The equivalent circuit of the transmission line consists of self and mutual inductances and capacitances. The conductor loss is taken into account by the line resistances. Fig.2.11(a) also shows gate and drain terminations.

Time Domain Analysis. The complete distributed circuit model is analyzed in the time domain. In each unit segment, the input/output currents and voltages are related by four coupled discretized differential equations in the time domain. The output-input relation for the first and second stages in the time domain can be expressed in a matrix form as

$$\begin{bmatrix} I_{g2} \\ V_{g2} \\ I_{d2} \\ V_{d2} \end{bmatrix} = [A]^{-1}[B/C(1)] \begin{bmatrix} I_{g1} \\ V_{g1} \\ I_{d1} \\ V_{d1} \end{bmatrix}$$

$$\begin{bmatrix} I_{g3} \\ V_{g3} \\ I_{d3} \\ V_{d3} \end{bmatrix} = [A]^{-1}[B/C(2)] \begin{bmatrix} I_{g2} \\ V_{g2} \\ I_{d2} \\ V_{d2} \end{bmatrix} \tag{2.12}$$

By substituting the first stage expression into the second stage expression, we obtain the relation between the second stage outputs and the first stage inputs as

$$\begin{bmatrix} I_{g3} \\ V_{g3} \\ I_{d3} \\ V_{d3} \end{bmatrix} = [A]^{-1}[B/C(2)][A]^{-1}[B/C(1)] \begin{bmatrix} I_{g1} \\ V_{g1} \\ I_{d1} \\ V_{d1} \end{bmatrix} \tag{2.13}$$

Similarly, the output of segment N can be expressed in terms of the inputs of the first segment by a matrix which is the multiplication of all the individual matrices, namely

$$\begin{bmatrix} I_{g(N+1)} \\ V_{g(N+1)} \\ I_{d(N+1)} \\ V_{d(N+1)} \end{bmatrix} = [F] \begin{bmatrix} I_{g1} \\ V_{g1} \\ I_{d1} \\ V_{d1} \end{bmatrix} \tag{2.14}$$

Since the first segment is connected to a driving source and segment N is terminated with a load, there are two unknowns at the device output and two at the device input. By arranging all the four unknowns to one

side of the matrix, we have

$$\begin{bmatrix} I_{g1} \\ I_{d1} \\ I_{g(N+1)} \\ I_{d(N+1)} \end{bmatrix} = [M] \begin{bmatrix} V_{g1} \\ V_{d1} \\ V_{g(N+1)} \\ V_{d(N+1)} \end{bmatrix} \tag{2.15}$$

where the elements of matrix M are known at each time step. Hence, the input and output currents and voltages can be evaluated. Once the voltages and currents at the first stage input and the final stage output are known, all the voltages and currents at each stage can be evaluated.

Validation of Time Domain Technique. To analyze the effect of segment number chosen in the distributed model, an FET with 2 mm width is simulated. The results are shown in Fig.2.14. The gains at low frequencies are the same with different numbers of segments, as there is no wave propagation effect. At higher frequencies, however, the power gain varies with the number of segments. As the number of segments increases, the characteristics ultimately converge. For this case, the power gains with three and four segments are almost the same at high frequencies. It is conjectured that the optimal number of segments to properly model wave propagation effects depends on the width and upper operating frequencies of the device. The results presented in Fig.2.14 strongly demonstrate the necessity to incorporate wave propagation effects in the device models. The gain at high frequencies changes dramatically when the distributed effects are ignored or not properly incorporated.

3.2 Hybrid Model and Its Elements

In this Section, a hybrid technique combining physics-based and circuit models is presented. In this new model, the whole device of a given width is divided into several segments. Each unit segment consists of an active part connected to the distributed circuit elements of transmission line as shown in Fig.2.15(a). Input ac signal is applied to the gate terminal of the first segment, and the output load is connected to the drain terminal of the final segment. The full-hydrodynamic model presented in [16] is used to simulate the active part in the hybrid model. First, dc physical simulation is conducted to obtain the dc drain and gate currents of the active part. The gate and drain ac currents are found by physical simulation, given the ac gate and drain voltages at the corresponding node of each active part in the hybrid model.

These two current sources are connected to the transmission line elements to form a unit segment as shown in Fig.2.15(a). The whole device is then represented by multiple unit segments as shown in Fig.2.15(b). Applying the same techniques presented in (2.12)-(2.15), all the device

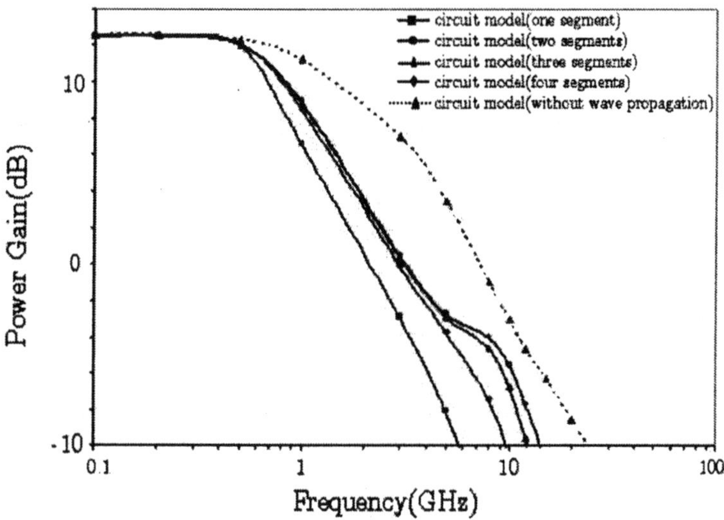

Figure 2.14. **Power gain with different numbers of segments in the circuit model.**

currents and voltages can be determined. At a specific time step, gate voltages at all segments are calculated using the old values obtained at the previous time step. The updated gate voltages and the previous drain voltages are used to determine the total gate and drain currents at each active segment by physical simulation. Then, by obtaining the ac gate and drain currents of the active current sources, the updated drain voltages and drain currents at all segments are calculated. These updated currents and voltages are used to calculate the gate voltages at all segments at the next time step. The following equations describe the technique to solve for the voltages and currents by discretizing the differential equations in the time domain.

$$\begin{bmatrix} I_{g(n+1)} \\ V_{g(n+1)} \end{bmatrix} = [F_1] \begin{bmatrix} I_{g1} \\ V_{g1} \end{bmatrix} \Rightarrow \begin{bmatrix} I_{g1} \\ I_{g(n+1)} \end{bmatrix} = [M_1] \begin{bmatrix} V_{g1} \\ V_{g(n+1)} \end{bmatrix} \tag{2.16}$$

$$\begin{bmatrix} I_{d(n+1)} \\ V_{d(n+1)} \end{bmatrix} = [F_2] \begin{bmatrix} I_{d1} \\ V_{d1} \end{bmatrix} \Rightarrow \begin{bmatrix} I_{d1} \\ I_{d(n+1)} \end{bmatrix} = [M_2] \begin{bmatrix} V_{d1} \\ V_{d(n+1)} \end{bmatrix} \tag{2.17}$$

This technique is very similar to the one used in the complete equivalent circuit approach. The main difference is that the currents and voltages at the gate and drain are solved separately by a decoupled scheme. On the gate side, current and voltage of the final stage are related to

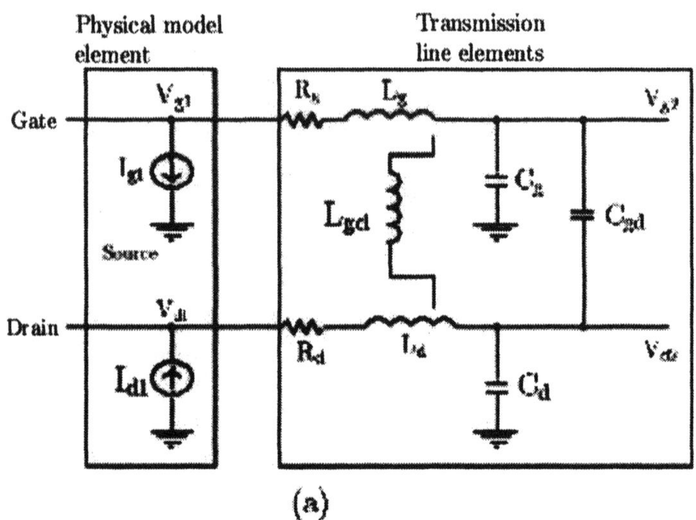

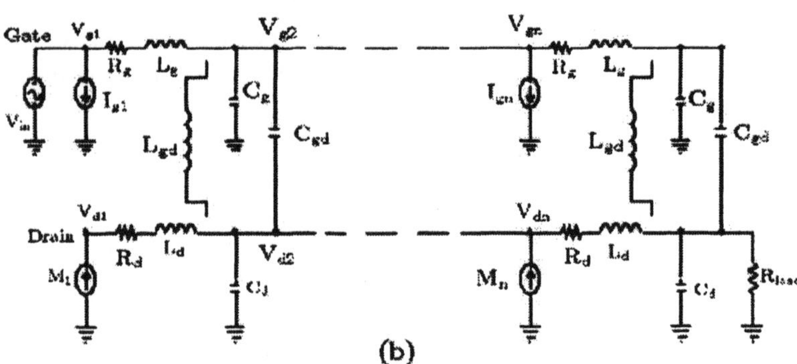

Figure 2.15. **Hybrid model: (a) unit segment, (b) multiple segments representation.**

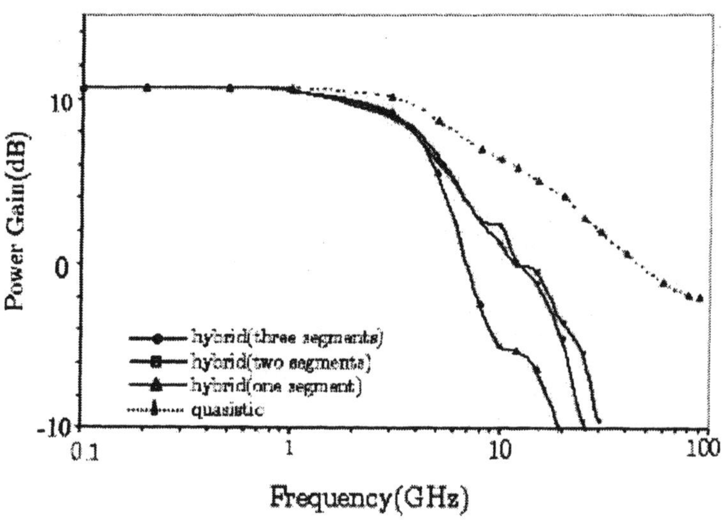

Figure 2.16. Frequency response of power gains with different numbers of segments in the hybrid model.

those of the first stage by (2.16). Knowing the terminations at two ends of the gate, the other unknowns can be calculated. Hence, the gate voltage at each node is obtained. A similar procedure is applied on the drain side. Eq.(2.17) relates the drain voltage and current of the final stage to those of the first stage. Knowing the drain end terminations, the other unknowns can be determined. Then, the drain voltages and currents at all nodes can be evaluated.

To ensure stability and convergence of the hybrid model, proper time steps for physical device simulation and circuit simulation need to be chosen [17]. In this simulation, the time steps for physical and circuit simulators are taken to be 0.001 ps and 0.1 ps, respectively. Therefore, a hundred iterations of the physical simulator are performed for each time step of the circuit simulator.

Results of Hybrid Model. The hybrid model simulation is performed at different frequencies on the same FET structure. Fig.2.16 shows the frequency response of a transistor with 2 mm width having dc bias of $V_{gs} = -0.6V$, $V_{ds} = 2V$, and terminated to a 50 Ω load.

The simulation is conducted using one, two and three segments. The results with two and three segments are almost the same at high frequencies, which indicates that the optimal number of segments for this case can be as low as two. Response of the quasistatic device model which does not include any wave propagation effects is also shown in

Fig.2.16. From the frequency response of power gain, it is evident that the wave propagation effects must be included in device modeling as the gain obtained by the hybrid model falls more obviously than the quasistatic model at high frequencies.

To study the effect of device width on the output power gain, transistors with different widths are simulated with the same dc bias and load terminations at 5 GHz. The quasistatic model, complete equivalent circuit model and hybrid model are used, keeping all parameters the same. To make a convincible comparison, a fourth simulation is conducted, in which the circuit model is used without transmission line elements, which effectively eliminates the wave propagation effects.

Fig.2.17 shows the variation of output power with the device width. For both the hybrid and equivalent circuit models, two segments are used. For a small device width below 1 mm, all models predict the same behavior. The output power increases linearly with the device width. For device width larger than 1 mm, the quasistatic and circuit models without wave propagation effects predict higher power as the device width increases. The rate of power increase with device width changes due to nonlinear effects. However, a small difference is observed between the two models due to the scaling scheme used to determine the equivalent circuit elements, which becomes inaccurate to represent the device with a larger width. On the other hand, the output power calculated using the hybrid and equivalent circuit models incorporating wave propagation effects decreases for devices wider than 1 mm.

Notice that the hybrid and equivalent circuit models exhibit significant differences for large device widths. This difference is considerably larger than that between their quasistatic counterparts. Bearing in mind that the two simulators use the same transmission line model, the difference is due to the interaction between wave and electron transport considered in the hybrid model. This observation highlights the importance of global modeling approach, and confirms the need to include the wave propagation effects with the gain-producing mechanism.

This Chapter clearly demonstrates the superiority of using the hybrid model for two reasons. First, it is very difficult to have a perfect equivalent circuit of one device under all conditions. Second, the concept of equivalent circuit is based on ignoring the impact of ac signal on the gain-producing mechanisms(electron transport). However, the circuit model has its own merits of simplicity and short computational time.

From the output power variation with width, it is evident that without including wave propagation effects in device modeling, exact performance of the device cannot be obtained, especially when the wavelength of the operating frequency becomes comparable to the device width.

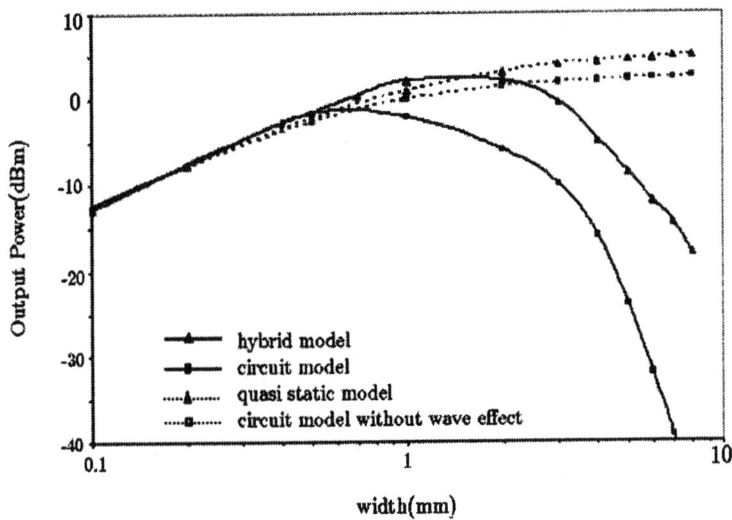

Figure 2.17. Output power versus device width with different models.

References

[1] S. El-Ghazaly and T. Itoh, "Electromagnetic interfacing of semiconductor devices and circuits," *IEEE MTT-S Int. Microwave Sym. Dig.*, pp.151-154, 1997.

[2] E. M. Tentzeris, A. Cangellaris, L. P. B. Katehi, and J. Harvey, "Multiresolution time-domain (MRTD) adaptive schemes using arbitrary resolutions of wavelets," *IEEE Trans. Microwave Theory Tech.*, vol.50, pp.501-516, Feb. 2002.

[3] M. Holmstron, "Solving hyperbolic PDE's using interpolating wavelets," Tech. Rep. No.189/1996, Uppsala University, Sweden.

[4] S. Goasguen and S. M. El-Ghazaly, "Interpolating wavelet scheme toward global modeling of microwave circuits," *IEEE MTT-S Int. Microwave Symp. Dig.*, pp.375-378, 2000.

[5] Y. K. Feng and A. Hints, "Simulation of submicrometer GaAs MESFET's using a full hydrodynamic model," *IEEE Trans. Electron. Dev.*, vol.35, pp.1419-1431, Sept. 1988.

[6] M. Fujii and W. J. R. Hoefer, "A three-dimensional Haar-wavelet-based multiresolution analysis similar to the FDTD method- derivation and application," *IEEE Trans. Microwave Theory Tech.*, vol.46, pp.2463-2475, Dec. 1998.

[7] P. Flandrin, *Time-Frequency/Time-Scale Analysis*, San Diego, CA: Academic Press, 1998.

[8] S. G. Mallat, "A theory for multiresolution signal decomposition: The wavelet representation," *IEEE Trans. Pattern Analy. Machine Intel.*, vol.11, no.7, pp.674-693, July 1989.

[9] B. Hubbard, *The World According to Wavelets: The Story of a Mathematical Technique in the Making*, Wellesley MA: A. K. Peters, Ltd., 1998.

[10] C. Hilsum, "Simple empirical relationship between mobility and carrier concentrations," *Electron. Lett.*, vol.10, no.12, pp.259-260, 1974.

[11] M. A. Alsunaidi, S. M. Imtiaz, and S. M. El-Ghazaly, "Electromagnetic wave effects on microwave transistors using a full wave high-frequency time-domain model," *IEEE Trans. Microwave Theory Tech.*, vol.44, pp.799-808, June 1996.

[12] R. LaRue, C. Yuen, and G. Zdasiuk, "Distributed GaAs FET circuit model for broadband and millimetre wave application," *IEEE MTT-S Int. Microwave Symp. Dig.*, 1984.

[13] W. Sui, D. A. Christensen, and C. H. Durney, "Extending the two-dimensional FDTD method to hybrid electromagnetic systems with active and passive elements," *IEEE Trans. Microwave Theory Tech.*, vol.40, pp.724-730, Apr. 1992.

[14] J. M. Golio, *Microwave MESFETs and HEMTs*, Boston: Artech House, 1991.

[15] S. Masuda, T. Hirose, and Y. Watanabe, "An accurate distributed small signal FET model for millimetre-wave applications," *IEEE MTT-S Int. Microwave Symp. Dig.*, pp.157-160, 1999.

[16] M. A. Alsunaidi, and S. M. El-Ghazaly, "High frequency large-signal physical modeling of microwave semiconductor devices," *IEEE MTT-S Int. Microwave Symp. Dig.*, pp.619-622, 1995.

[17] C. M. Snowden, M. J. Howes, and D. V. Morgan, "Large-signal modeling of GaAs MESFET operation," *IEEE Trans. Electron. Dev.*, vol.30, no.12, pp.1817-1824, Dec. 1983.

[18] S. M. S. Imtiaz and S. M. El-Ghazaly, "Global modeling of millimetre-wave circuits: Electromagnetic simulation of amplifiers," *IEEE Trans. Microwave Theory Tech.*, vol.45, pp.2208-2216, Dec. 1997.

[19] M. S. Hakim and S. M. El-Ghazaly, "Analysis of wave propagation effects on microwave field effect transistors," *IEEE MTT-S Int. Microwave Symp. Dig.*, pp.1303-1306, Atlanta, GA, 1993.

[20] W. Heinrich, "Limits of FET modeling by lumped elements," *Electron. Lett.*, vol.22, pp.630-632, 1986.

[21] R. O. Grondin, S. M. El-Ghazaly, and S. Goodnick, "A review of global modeling of charge transport in semiconductors and full-wave electromagnetics," *IEEE Trans. Microwave Theory Tech.*, vol.47, pp.817-829, June 1999.

[22] I. Daubechies, *Ten Lectures on Wavelets*, Philadelphia, PA: SIAM, 1992.

[23] M. Krumpholz and L. P. B. Katehi, "MRTD: New time-domain schemes based on multiresolution analysis," *IEEE Trans. Microwave Theory Tech.*, vol.44, pp.555-571, Apr. 1996.

[24] M. Werthen and I. Wolff, "A novel wavelet based time domain simulation approach," *IEEE Microwave Guided Wave Lett.*, vol.6, pp.438-440, Dec. 1996.

[25] J. Keiser, "Wavelet based approach to numerical solution of nonlinear partial differential equation," Ph.D. thesis, University of Colorado, Boulder, 1995.

[26] M. Toupikov, G. Pan, and B. K. Gilbert, "On nonlinear modeling of microwave devices using interpolating wavelets," *IEEE Trans. Microwave Theory Tech.*, vol.48, pp.500-509, Apr. 2000.

[27] K. C. Gupta, "Emerging trends in millimetre-wave CAD," *IEEE Trans. Microwave Theory Tech.*, vol.46, pp.747-755, June 1998.

[28] W. Heinrich, "Distributed equivalent-circuit model for travelling-wave FET design," *IEEE Trans. Microwave Theory Tech.*, vol.35, no.5, pp.487-491, May 1987.

[29] A. Abdipour and A. Pacaud, "Complete sliced model of microwave FET's and comparison with lumped model and experimental results," *IEEE Trans. Microwave Theory Tech.*, vol.44, no.1, pp.4-9, Jan. 1996.

[30] S. J. Nash, A. Platzker, and W. Struble, "Distributed small signal model for multi-fingered GaAs PHEMT/MESFET devices."

Chapter 3

BAND-PASS FILTER DESIGN WITH GENETIC ALGORITHMS

Jean-Fu Kiang and Hsuan-Sheng Chou
Department of Electrical Engineering and
Graduate Institute of Communication Engineering
National Taiwan University
Taipei, Taiwan, ROC

Abstract In this Chapter, genetic algorithms are applied to design multilayered end-coupled band-pass Chebyshev filters. Basic concepts of genetic algorithms and Chebyshev filters are briefly reviewed. Both local and global optimization approaches are proposed, and a quasistatic approach with transmission matrices is used to compute the fitness value of physical circuits associated with different chromosomes. Fringing capacitance is incorporated into the static model to account for the high frequency effects, full-wave solutions and measurements are also used to verify the effectiveness of this approach.

Keywords: genetic algorithm, Chebyshev, band-pass filter, multilayered structure.

1. Review of Genetic Algorithms

Genetic algorithms(GA) are heuristic optimizers inspired by the evolution and natural selection processes of living organisms to fit in their environments. Applied researchers recognize the usefulness of evolution as something worthy to be emulated and utilized [1]-[9]. Natural selection skips one big hurdle in a design task: specifying in advance all the components of a system and describing the relations among them, which are complicated and sometimes ambiguous. Genetic algorithms prove effective in solving complicated and very often, combinatorial problems.

Genetic algorithms are categorized as global optimization techniques while other methods like conjugate gradient and quasi-Newton methods are categorized as local optimization techniques. Global techniques tend to yield either global or near global extrema instead of local extrema, and often find useful solutions while local techniques fail. Global techniques are particularly useful when dealing with new problems in which the nature of solution is relatively ambiguous.

Genetic algorithms are particularly effective when only an approximate extremum is searched for in a multidimensional solution domain. Different from conventional techniques, genetic algorithms operate on a group of coexisting trial solutions(population), they normally operate on coded parameters(chromosome) rather than on the parameters themselves, and they use simple stochastic processes(crossover, mutation and selection) to explore the solution domain for an optimal solution.

Fig.3.1 shows the relevant terms related to chromosomes. Each trial solution is coded into a chromosome represented by a string of bits. A chromosome consists of several genes, each one is coded from a physical parameter. Each gene consists of several alleles with each one represented by a bit. A set of chromosomes form a population. The population size may be fixed or adaptive among different generations. If the population size is too small, the algorithm may converge too early to a suboptimal solution. If the population size is too large, the algorithm may converge too slowly.

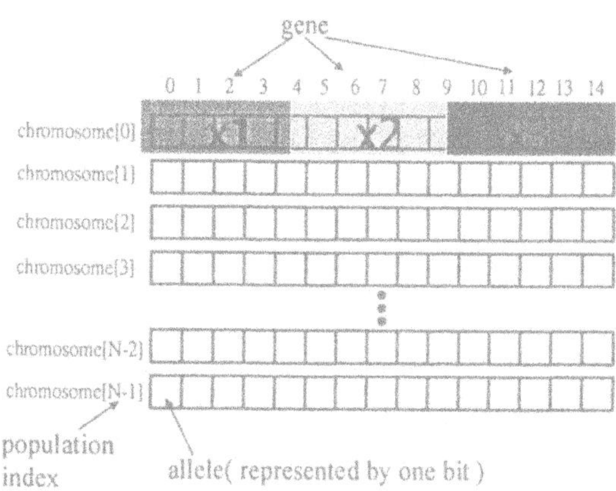

Figure 3.1. **Terms related to chromosomes, the chromosome length is 15 in this diagram, X1, X2, X3 are genes.**

A genetic algorithm is generally executed in the following steps: (1) define the chromosome to associate with trial solutions, (2) create an initial population(first generation) consisting of a predetermined number of N chromosomes, (3) assign each chromosome in the first generation a fitness value, (4) perform reproduction through crossover, mutation and selection processes to produce members of the next generation, (5) repeat the reproduction cycle until an optimal solution is reached[1].

One iteration in a genetic algorithm corresponds to one reproduction cycle to bring out the new generation from the old one. Each iteration consists of crossover, mutation and selection processes. Firstly, apply n crossover processes to create n new chromosomes where $n = \mathrm{CR} \times N$, with CR is the crossover rate. In each crossover process, extract alleles from two different parent chromosomes to create one or a pair of new child chromosomes. The mutation process may be applied to portions of the n new chromosomes where the mutated alleles(bits) are changed(flipped). Next, assign a fitness value to each of the n new chromosomes.

The selection process implements the principle of survival-of-the-fittest to ensure that the new generation contains chromosomes with higher fitness values than the old generation. The N chromosomes of the new generation are selected among the N old chromosomes plus the n new chromosomes according to their fitness values. Tournament and proportionate selections are exemplified as follows. In the tournament selection, N chromosomes are selected one at a time from the old generation plus the n new chromosomes according to their fitness values.

Fig.3.2 depicts the proportionate selection as a roulette wheel with the area of each segment proportional to the fitness value of a specific chromosome. The probability that the ith chromosome is chosen as one of the parents during a crossover process is

$$P_i = f(\text{chromosome}[i]) \left[\sum_{i=0}^{N-1} f(\text{chromosome}[i]) \right]^{-1}$$

where $f(\text{chromosome}[i])$ is the fitness value of the ith chromosome.

In most applications, a crossover process takes one pair of parents to produce one child or a pair of children. Fig.3.3 shows several different crossover processes like simple, heuristic, arithmetic, emperor-selective, best-mates-worst and adjacent-fitness-paring[2]. In the simple crossover, the crossover point is randomly selected between the first and the last alleles of the paired chromosomes as shown in Fig.3.3(a). The child chromosome duplicates the alleles from the first parent up to the crossover point, and the alleles from the second parent beyond

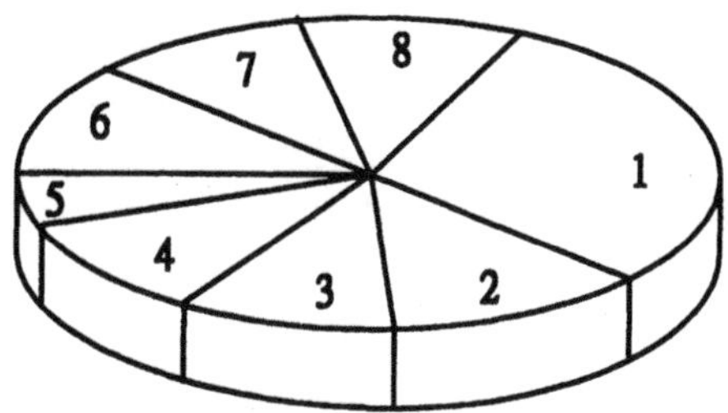

Figure 3.2. **Proportionate selection depicted as a roulette wheel with the area of each segment proportional to the fitness value of a specific chromosome.**

the crossover point. Fig.3.3(b) shows the heuristic crossover where each chromosome is translated into an integer number between 0 and $2^L - 1$. A child chromosome[k] is produced from two parents, chromosome[i] and chromosome[j], according to chromosome[k] $= r \times$ (chromosome[j] - chromosome[i]) + chromosome[j], where r is a real number between 0 and 1, and the fitness value of chromosome[j] is smaller than that of chromosome[i]. In the arithmetic crossover, a child chromosome[k] is produced as a linear combination of the parents, chromosome[i] and chromosome[j], according to chromosome[k] $= r \times$ chromosome[i] + $(1 - r) \times$ chromosome[j], where r is either a constant or a variable with $0 < r < 1$.

In the best-mates-worst crossover, the best chromosome mates with the worst chromosome, the second-best mates with the second-worst, and so forth[2]. This process has the tendency to reduce the difference in fitness values among all the new chromosomes. New species are more easily brought forth, but the convergence rate may be slow. In the adjacent-fitness-pairing crossover, the best mates with the second-best, the third-best mates with the fourth-best, and so forth. The genes are unlikely to vary drastically, and the algorithm may result in prematured convergence. In the emperor-selective crossover, the best chromosome mates with every other chromosome in the population. If a few superior chromosomes are added to the initial population as seeds, this process usually yields the best chromosome among all the processes, and converges quickly.

The crossover rate is the ratio between the number of new chromosomes and that of the original chromosomes in the old generation during one reproduction cycle. Higher crossover rate introduces new species into the population more quickly. If the crossover rate is too high, some su-

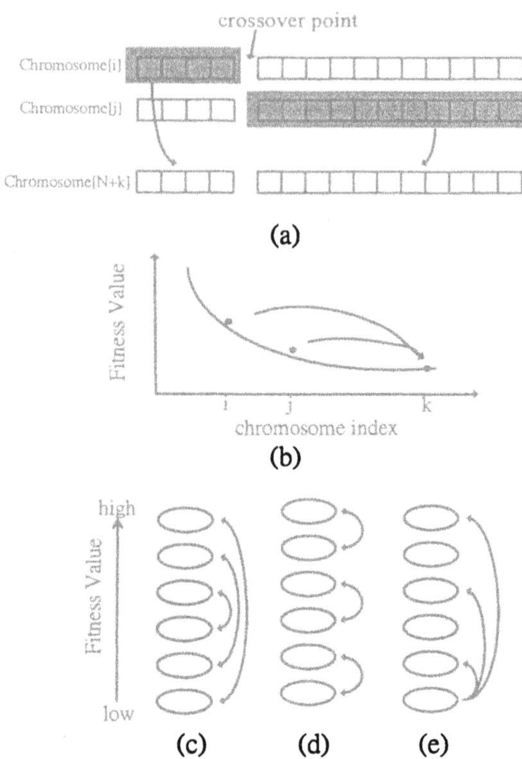

Figure 3.3. **Examples of crossover processes: (a)simple, (b)heuristic, (c)best-mates-worst, (d)adjacent-fitness-paring and (e)emperor-selective.**

perior chromosomes may be discarded too early. If the crossover rate is too low, the evolution process may stagnate.

The mutation process is a secondary process that applies to the new chromosomes to increases their versatility. Each allele in each of the n new chromosomes may be mutated(bit flipped) with a probability equal to the mutation rate(MR) which is the ratio of the number of mutated alleles to the total number of alleles in the old plus new chromosomes during a given reproduction cycle, namely, MR$\times(N + n) \times L$ mutations occur in each reproduction cycle. The mutation rate in different reproduction cycles may follow uniform or nonuniform probability distributions. In a given reproduction cycle, the location of mutated alleles may follow uniform or nonuniform probability distributions too. The mutation process helps to prevent genetic algorithms to converge to a suboptimal solution. However, high mutation rate may render slow convergence.

In general, the choice of chromosome length depends on the number and range of parameters used to describe the target problem. The population size is determined empirically, and is usually related to the chromosome length. In practice, the crossover rate is about 0.5, and the mutation rate is on the order of 1%.

2. Review of Filter Design

Fundamental microwave filter theory was based upon the analysis and design of lumped-element circuits, which proves to be effective up to 18 GHz, but is valid only over a narrow frequency band because all the lumped elements such as resonators are frequency sensitive. Lumped-element filters claim a large percentage of commercial microwave filters, and their size is usually much smaller than that of distributed filters. However, the latter are of practical interest when insertion loss and power handling capability are of concern.

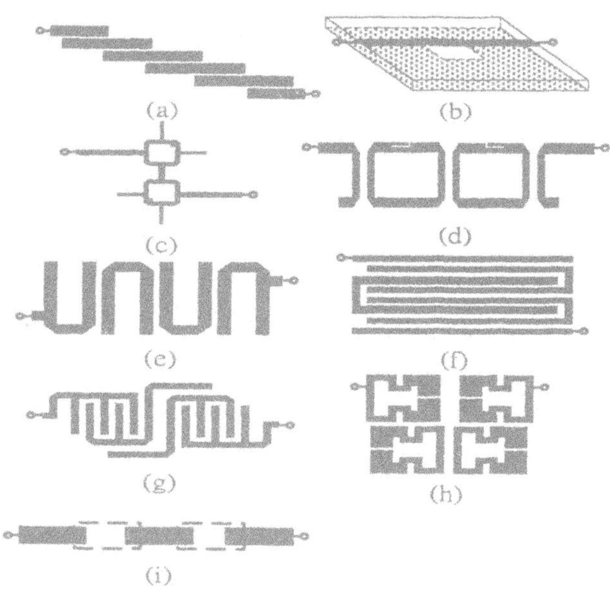

Figure 3.4. Examples of filter layout, (a) parallel-coupled lines, (b) parallel-coupled lines with perforated ground, (c) double square rings with stubs, (d) spur-lines, (e) hairpin lines, (f),(g) interdigital lines, (h) stepped impedance lines, (i) end-coupled lines.

Band-pass filters can be implemented using microstrip lines and their variations. Fig.3.4 shows examples of filter layout, including parallel-

coupled lines, ring resonators, spur-lines, hairpin lines, interdigital lines, and so on. In the early days, parallel-coupled lines were used in designing multistage band-pass filters. To implement a multipole and broadband band-pass filter with sharp band edges, strip width and spacing need to be reduced to achieve a tight coupling, and a large number of line resonators are required, which may decrease the Q factor, increase the insertion loss, and incur extra performance degradation due to fabrication tolerance. In [10], parallel-coupled line band-pass filters consisting of half-wavelength resonators are presented. There are several advantages in this design. The overall length of the filter is approximately half that of the end-coupled type, the spacings are wide enough to bear fabrication tolerance, the insertion loss is symmetrical with respect to the center frequency, and the third harmonic is away from the center frequency. Formulas have been derived for the design of Chebyshev, maximally flat and other physically realizable band-pass filters.

In [11], a modified version is proposed in which an aperture is carved in the ground plane to enhance the capacitive coupling between two parallel-coupled lines. The coupling section is characterized by an equivalent J inverter and two line segments. The extracted parameters indicate that the coupling is frequency dependent and increases rapidly as the aperture is widened. Sharper band edges can usually be acquired by concatenating multiple resonators.

The advantages of ring resonators include size compactness, production economy, high Q factor and low radiation loss. Two degenerate modes in a ring are coupled together to achieve band-pass characteristics. At the first resonant frequency, the circumferential length of a ring is about one wavelength, and the coupling length between two adjacent rings is about a quarter wavelength. In [12], wideband band-pass filters composed of square rings, tuning stubs and perturbation stubs are proposed. The tuning stubs are series resonators used to sharpen band edges of the filter. The perturbation stub excites both degenerate modes on the ring resonator to increase the bandwidth.

A spur-line is made by etching away portion of a microstrip line to form an open stub. The open stub acts as a notch filter to suppress higher order harmonics by absorbing the frequency component with wavelength roughly equal to four times of the stub length[13]. These spur-line filters are usually attached to the point where maximum amplitude of the undesired harmonic appears. In [13], two spur-line filters are proposed to enhance the performance of a coupled ring resonator.

Hairpin line filters are constructed by folding each half-wavelength resonator line of a parallel-coupled line filter into an U shape[14]. How-

ever, folding the resonator lines reduces the coupling between adjacent resonators due to reduction of coupling length as compared to the original parallel-coupled lines. The two arms of a given hairpin resonator may function as a pair of quarter-wavelength coupled lines themselves if they are spaced too tightly and the bottom of the U-shaped hairpin is grounded.

Interdigital line filters are more compact than conventional parallel-coupled line filters [15]. Each interdigital line is shorter than a quarter-wavelength at the center frequency. The frequency response in the pass-band is fairly flat and the band edges are sharp due to multiple poles in the pass-band and an attenuation pole at the edge of the upper stop-band. One of their disadvantages is the use of grounding vias.

Another three-stage interdigital line band-pass filter is presented in [16]. It is observed that the band edge of a parallel-coupled line filter is sharper at the lower frequency end of the pass-band than at the upper frequency end. An interdigital line filter usually has sharper band edges than a parallel-coupled line filter of the same order. While a lumped-element band-pass filter has a unique pass-band, distributed filters always have spurious pass-bands due to higher order modes incurred in each resonator. Because the finger length of an interdigital line filter is about half that of a line resonator in a comparable parallel-coupled line filter, spurious bands of the former are farther away from the center frequency than the latter. Notice that the interdigital line filter is a slow wave structure due to the relatively large capacitance between fingers. Pass-band and stop-band characteristics are also expected from its periodic layout.

In [17], filters of stepped impedance microstrip resonators are presented where proximity coupling of electric or magnetic fields can be controlled by properly orienting the resonators. In [18], an end-coupled band-pass filter implemented in a two-layered structure is proposed. No other discontinuities appear in this layout except the overlapping gaps serving to couple adjacent half-wavelength resonators. The gaps between resonators can be modeled by a π-network of capacitors. Tighter coupling between adjacent resonators renders a wider pass-band, and the overlapping gaps provide a stronger coupling than the end coupling in the single-layered structure.

The performances of band-pass filters just reviewed are summarized in Table 3.1 for comparison. Full-wave approaches are usually applied to design band-pass filters at high frequencies. When the order of filters is increased due to performance requirements, the physical layout becomes more complicated, and more geometrical parameters need to be optimized. Using full-wave approaches in such occasions prove to

be time consuming. A systematic procedure using a more efficient and reasonably accurate approach will be desirable in such a design task. Genetic algorithms are chosen as optimization tools for their versatility. A quasistatic approach with high frequency behaviors accounted for by fringing capacitance is used to assess the filter performance in each iteration, which is much faster than the full-wave approaches.

Table 3.1. Summary of band-pass filter performance, [11]: parallel-coupled, PG two-pole, [11]: parallel-coupled, PG four-pole, [12]: single square ring with stubs, [13]: spur-line, [14]: hairpin, [15]: interdigital line, [16]: interdigital line, [17]: stepped impedance, [18]: end-coupled; BW: bandwidth, IL: insertion loss, RL: return loss, f_0: center frequency, PG: perforated ground.

Ref.	3 dB BW (%)	10 dB BW (%)	Max. IL (dB)	Min. RL (dB)	f_0 (GHz)
[11]	73	150	1.2	15	6
[11]	73	93	0.6	16	6
[12]	53.5	56.5	0.7	12	6
[12]	50.5	53.3	1	12	6
[13]	6	15	2.5	15	2
[14]	3.5	5.3	3	15	0.9335
[15]	10	19	2	10	1.1
[16]	4	8.3	0.5	8	2.44
[17]	2.5	4.1	2.8	12	1.95
[18]	33.65	37.5	1.3	10	4

Compactness is a critical factor in designing wireless communication products, which is usually associated with less power consumption and longer battery life. Package technology is usually applied to miniaturize modules composed of several integrated circuit chips. Interconnections between different chips are embedded in a multilayered substrate using silicon-based thin film technology. Since silicon materials have high losses at high frequencies, low-loss ceramic materials become attractive alternatives as packaging materials. Due to the available high-permittivity ceramic materials, passive components and interconnections can be significantly miniaturized.

Multilayered packages provide circuit designers more flexibility to design compact modules with better performance[19], [20]. Two-dimensional passive components like resistors, capacitors, inductors, couplers, filters and transmission lines can be converted to three-dimensional configurations embedded in LTCC to reduce module size[21], [22]. Chips are then mounted atop to form a multichip module [23], [24].

3. Chebyshev Approach for Filter Design

Filter specification can be expressed in terms of insertion loss over a specific frequency band. Design approach based on Chebyshev polynomials offers a sharp edge between pass-band and stop-band as shown in Fig.3.5(a) in which the insertion loss characteristics can be expressed as [25]

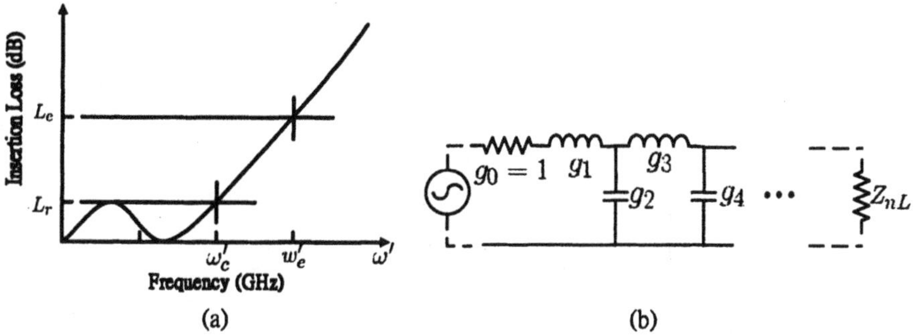

Figure 3.5. Normalized Chebyshev low-pass filter,(a) frequency response, (b) equivalent circuit.

$$L(\omega') = \begin{cases} 10\log_{10}\left\{1+\epsilon\cos^2\left[N\cos^{-1}\left(\omega'/\omega_c'\right)\right]\right\}, & \omega' \le \omega_c' \\ 10\log_{10}\left\{1+\epsilon\cosh^2\left[N\cosh^{-1}\left(\omega'/\omega_c'\right)\right]\right\}, & \omega' \ge \omega_c' \end{cases}$$

where $\epsilon = 10^{L_r/10} - 1$, N is the degree of Chebyshev polynomial, L_r is the tolerance of insertion loss in the pass-band, ω' is the frequency variable of the normalized low-pass filter, and $\omega_c' = 1$ is the normalized cutoff frequency. The specifications shown in Fig.3.5(a) can be implemented by the equivalent circuit shown in Fig.3.5(b) in which the normalized impedances are derived as [25]

$$g_0 = 1, \qquad g_1 = 2a_1/\sinh\left(\frac{\beta}{2N}\right)$$

$$g_i = \frac{4a_{i-1}a_i}{b_{i-1}g_{i-1}}, \qquad 2 \le i \le N$$

$$g_{N+1} = \begin{cases} 1, & N \text{ is odd} \\ \coth(\beta/4), & N \text{ is even} \end{cases}$$

where

$$a_i = \sin\left(\frac{2i-1}{2N}\pi\right), \qquad 1 \le i \le N$$

$$b_i = \sinh^2\left(\frac{\beta}{2N}\right) + \sin^2\left(\frac{i\pi}{N}\right), \qquad 1 \le i \le N$$

$$\beta = \log_e\left(\coth\frac{Lr}{17.372}\right)$$

From the low-pass filter prototype, band-pass filter specifications can be derived as shown in Fig.3.6(a). If ω_1 and ω_2 denote the corner frequencies of pass-band, the band-pass frequency response can be mapped by using the transformation

$$\frac{\omega'}{\omega'_c} = \frac{1}{\xi}\left(\frac{\omega}{\omega_o} - \frac{\omega_o}{\omega}\right)$$

where $\xi = (\omega_2 - \omega_1)/\omega_o$ and $\omega_o = \sqrt{\omega_1\omega_2}$ are the fractional bandwidth and center frequency, respectively, of the band-pass filter; $\omega'$and ω are the frequency variables of normalized low-pass and band-pass filters, respectively. Each shunt capacitor in the low-pass prototype shown in Fig.3.5(b) is converted to a shunt LC circuit shown in Fig.3.6(b), having element values given by

$$L_i = \frac{g_i Z_o}{\xi\omega_o}, \qquad C_i = \frac{\xi}{\omega_o g_i Z_o} \tag{3.1}$$

where Z_0 is the feed line impedance of the realized band-pass filter. Each series inductor of the low-pass prototype is converted to a series LC circuit having element values given by

$$L_i = \frac{\xi Z_o}{\omega_o g_i}, \qquad C_i = \frac{g_i}{\xi\omega_o Z_o} \tag{3.2}$$

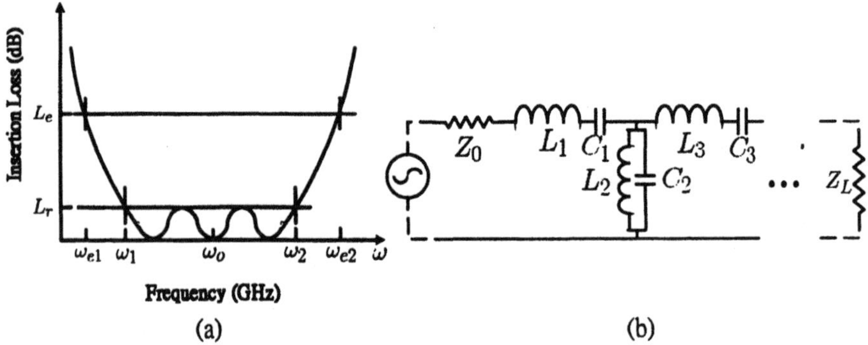

Figure 3.6. Chebyshev band-pass filter transformed from normalized low-pass prototype, (a) frequency response, (b) equivalent circuit.

Fig.3.7(a) shows a section of lossless transmission line having a characteristic impedance Z_i, phase constant β_i, and phase velocity v_i. At the center frequency $\omega = \omega_o$, the section length is $\ell_i = \lambda_i/2$. Let $\omega = \omega_o + \delta\omega$ with $\delta\omega \ll \omega_0$, we have

$$\beta_i \ell_i = \frac{\omega \ell_i}{v_i} = \frac{(\omega_o + \delta\omega)\ell_i}{v_i} = \pi\left(1 + \frac{\delta\omega}{\omega_o}\right)$$

The input impedance of this open-circuited transmission line section is

$$Z_{\text{in}}^{TL} = \frac{Z_i}{j\tan\beta_i\ell_i} = \frac{Z_i\omega_o}{j\delta\omega\pi}$$

Similarly, input impedance of the shunt LC circuit shown in Fig.3.7(b) is

$$Z_{\text{in}}^{LC} = \left(j\omega C_i + \frac{1}{j\omega L_i}\right)^{-1} = \left(\frac{\omega^2 L_i C_i - \omega_o^2 L_i C_i}{j\omega L_i}\right)^{-1} \cong \frac{1}{j2C_i\delta\omega}$$

Based on the same form of input impedance, the transmission line section in Fig.3.7(a) can be used to implement the shunt LC circuit in Fig.3.7(b).

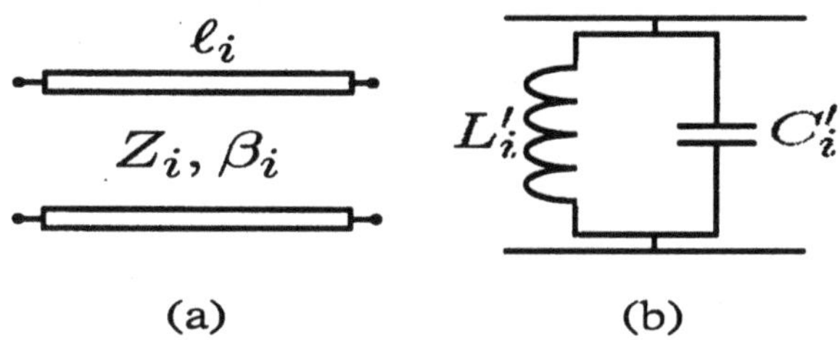

Figure 3.7. A transmission line section close to $\lambda/2$ in (a) used to implement a shunt LC circuit in (b).

To demonstrate the implementation of band-pass filter using coupled transmission line sections, consider the example shown in Fig.3.8(a) with $N = 3$. The admittance of the series combination of L_1 and C_1 is

$$Y_{01}^{LC} = \frac{j\omega C_1}{1 - \omega^2 L_1 C_1}$$

The series circuit of L_1 and C_1(L_3 and C_3) is converted to the shunt circuit of L_{01} and C_{01}(L_{03} and C_{03}) in parallel with two admittance inverters(J) to form a JSJ(inverter-shunt-inverter) subcircuit. The admittance inverter, J, is defined to convert the admittance of the JSJ

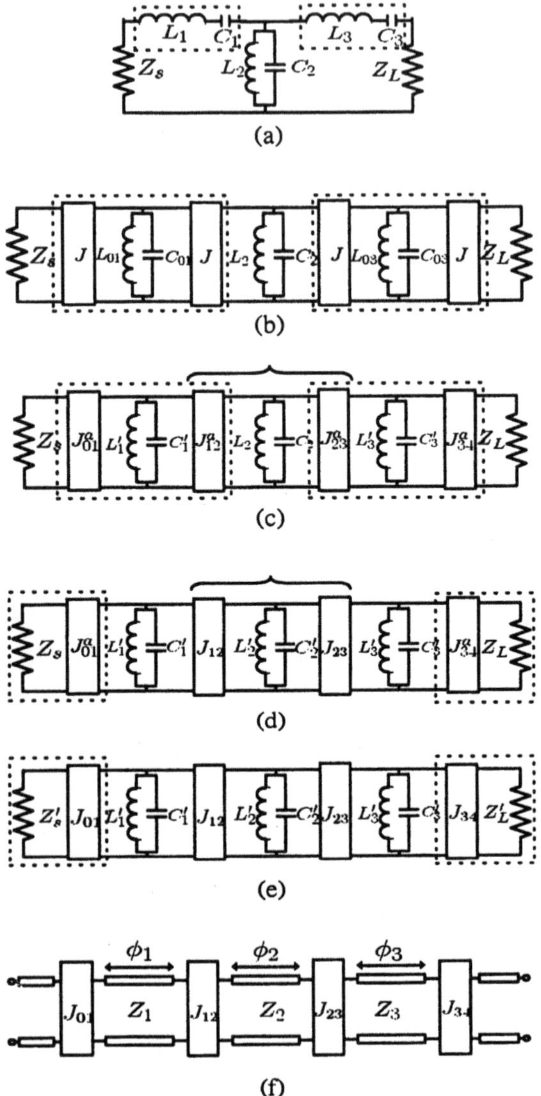

Figure 3.8. **Transform a band-pass filter to a realizable version by using admittance inverter. (a) A prototype band-pass filter, (b) series LC circuits replaced by JSJ subcircuits, (c) convert L_{01}, C_{01}, L_{03} and C_{03} to practical values of L_1', C_1', L_3' and C_3', (d) convert L_2 and C_2 to practical values of L_2' and C_2', (e) convert Z_s and Z_L to practical values of Z_s' and Z_L', (f) each shunt LC circuit is implemented by a transmission line section of half wavelength at the center frequency, namely $\phi_1 = \phi_2 = \phi_3 = \pi$ at the center frequency.**

subcircuit into

$$Y_{01}^{JSJ} = \frac{J^2}{(j\omega L_{01})^{-1} + j\omega C_{01}} = \frac{J^2 j\omega L_{01}}{1 - \omega^2 L_{01} C_{01}}$$

In the subsequent discussions, J is set to unity. Imposing that $Y_{01}^{LC} = Y_{01}^{JSJ}$, we have $L_{01} = C_1$ and $C_{01} = L_1$. Similarly, we have $L_{03} = C_3$ and $C_{03} = L_3$.

Next, convert L_{01}, C_{01}, L_{03} and C_{03} in Fig.3.8(b) to practical values of L_1', C_1', L_3' and C_3' in Fig.3.8(c), with the constraint that the admittance of each JSJ subcircuit remains the same, namely

$$\frac{(J_{01}^a)^2}{2jC_1'\delta\omega} = \frac{J^2}{2jC_{01}\delta\omega}, \qquad \frac{(J_{23}^a)^2}{2jC_3'\delta\omega} = \frac{J^2}{2jC_{03}\delta\omega}$$

With $J = 1$, we have

$$J_{01}^a = J_{12}^a = \sqrt{\frac{C_1'}{L_1}}, \qquad J_{23}^a = J_{34}^a = \sqrt{\frac{C_3'}{L_3}} \tag{3.3}$$

Next, convert L_2 and C_2 in Fig.3.8(c) into practical values of L_2' and C_2' in Fig.3.8(d) by requiring that the admittance of shunt J_{12}^a-L_2-C_2-J_{23}^a be equal to that of shunt J_{12}-L_2'-C_2'-J_{23}. Thus, we have

$$\frac{(J_{12})^2}{2jC_2'\delta\omega} = \frac{(J_{12}^a)^2}{2jC_2\delta\omega}, \qquad \frac{(J_{23})^2}{2jC_2'\delta\omega} = \frac{(J_{23}^a)^2}{2jC_2\delta\omega} \tag{3.4}$$

By substituting (3.3) into (3.4), we have

$$J_{12} = \sqrt{\frac{C_1'C_2'}{L_1C_2}}, \qquad J_{23} = \sqrt{\frac{C_3'C_2'}{L_3C_2}}$$

Finally, convert Z_s and Z_L in Fig.3.8(d) to practical values of Z_s' and Z_L' in Fig.3.8(e) by imposing that the admittance of shunt Z_s-J_{01}^a(J_{34}^a-Z_L) be equal to that of shunt Z_s'-J_{01}(J_{34}-Z_L'), namely

$$\frac{(J_{01})^2}{(Z_s)^{-1}} = \frac{(J_{01}^a)^2}{(Z_s')^{-1}}, \qquad \frac{(J_{34})^2}{(Z_L)^{-1}} = \frac{(J_{34}^a)^2}{(Z_L')^{-1}}$$

Thus, we have

$$J_{01} = \sqrt{\frac{Z_sC_1'}{Z_s'L_1}}, \qquad J_{34} = \sqrt{\frac{Z_LC_3'}{Z_L'L_3}}$$

Fig.3.8(e) shows a realizable band-pass filter consisting of shunt LC circuits and admittance inverters, where practical values of L_i' and C_i' can be chosen, and the inverters are related to these L_i' and C_i' by

$$J_{01} = \sqrt{\frac{Z_sC_1'}{Z_s'L_1}}$$

$$J_{i(i+1)} = \begin{cases} \sqrt{\dfrac{C_i'C_{i+1}'}{L_iC_{i+1}}}, & 1 \le i < N,\ i \text{ is odd} \\[2ex] \sqrt{\dfrac{C_i'C_{i+1}'}{C_iL_{i+1}}}, & 1 \le i < N,\ i \text{ is even} \end{cases}$$

$$J_{N(N+1)} = \begin{cases} \sqrt{\dfrac{Z_LC_N'}{Z_L'L_N}}, & N \text{ is odd} \\[2ex] \sqrt{\dfrac{C_N'}{Z_LZ_L'C_N}}, & N \text{ is even} \end{cases} \tag{3.5}$$

Next, each shunt LC circuit shown in Fig.3.8(e) is implemented by a transmission line section as discussed in Fig.3.7. The final schematic shown in Fig.3.8(f) consists of lumped elements and transmission line sections.

The characteristic impedance of the transmission line sections connected to the source and load can be conveniently chosen as a typical transmission line impedance Z_o, namely $Z_s' = Z_L' = Z_o$. Hence, C_i' in each line section is related to the characteristic impedance Z_i as

$$C_i' = \frac{\pi}{2Z_i\omega_0} \tag{3.6}$$

Substituting (3.1), (3.2) and (3.6) into (3.5), the admittance inverters are reduced to

$$J_{01} = \sqrt{\frac{\pi\xi}{2g_1Z_oZ_1}}$$

$$J_{i(i+1)} = \frac{\pi\xi}{2}\sqrt{\frac{1}{Z_{i+1}Z_ig_{i+1}g_i}}, \qquad 1 \le i < N$$

$$J_{N(N+1)} = \sqrt{\frac{\pi\xi}{2g_NZ_oZ_N}}$$

Fig.3.9(a) shows a broadside end-coupled filter to implement the schematic shown in Fig.3.8(f). Each overlapping gap between two transmission line sections can be modeled as a π-network of capacitors as shown in Fig.3.10(a) where Z_i is the characteristic impedance of the ith transmission line section with electrical length of $\theta_i = \beta_i\ell_i$.

To estimate the capacitance of these π-network elements, capacitance per unit length of the uniform lines as shown in Fig.3.9(b) is calculated by using an integral equation approach, assuming that strips S_i and S_{i+1} in Fig.3.9(b) are of infinite length among the $\hat{y}$ direction. Define

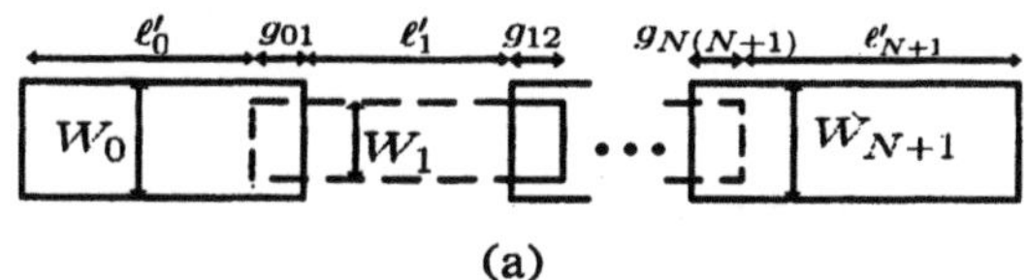

(a)

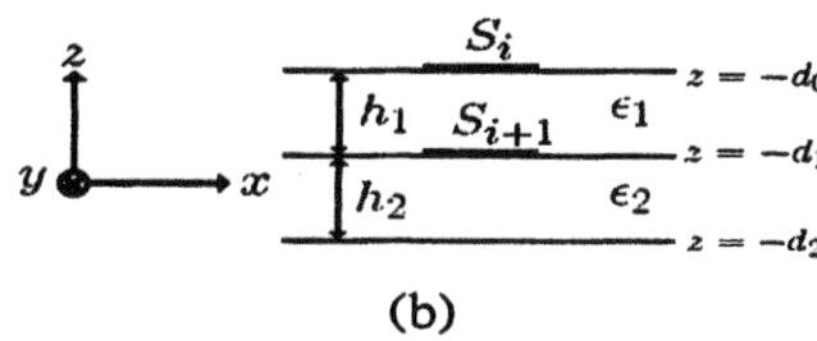

(b)

Figure 3.9. Layout of a multilayered end-coupled band-pass filter,(a) top view,(b) cross sectional view.

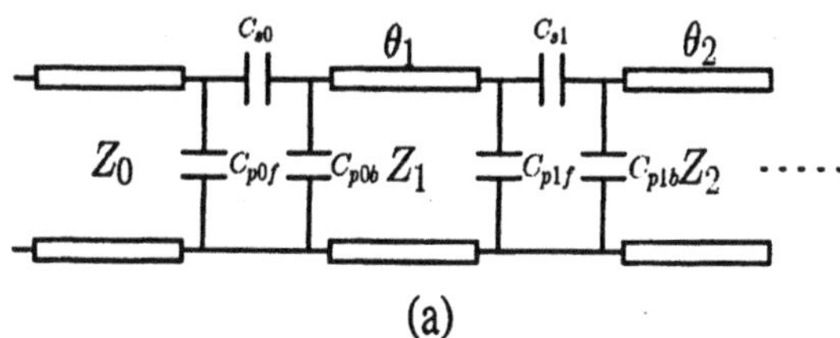

(a)

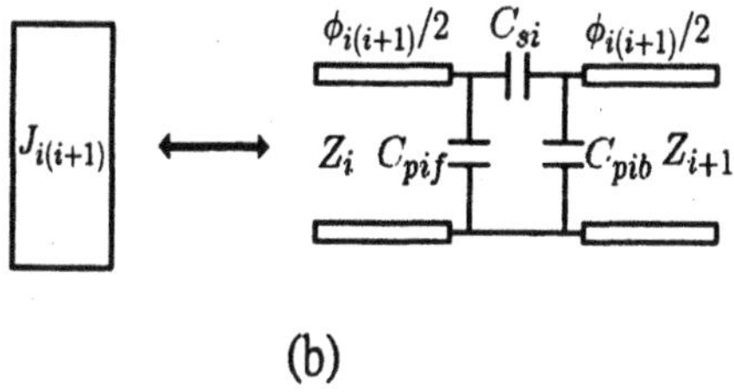

(b)

Figure 3.10. (a) Schematic derived from the end-coupled band-pass filter shown in Fig.3.9, (b) equivalent schematic of an admittance inverter.

the coupling capacitance per unit length between S_i and S_{i+1} as $\tilde{C}_{si}$, the capacitance per unit length between $S_i(S_{i+1})$ and ground as $\tilde{C}_{pif}(\tilde{C}_{pib})$. Next, define $C_{si} = \tilde{C}_{si}g_{i(i+1)}$, $C_{pif} = \tilde{C}_{pif}g_{i(i+1)}$ and $C_{pib} = \tilde{C}_{pib}g_{i(i+1)}$ where $g_{i(i+1)}$ is the length of overlapping gap between sections i and $i+1$.

In order to use the schematic in Fig.3.10(a) to implement that in Fig.3.8(f), the admittance inverter, $J_{i(i+1)}$, must be implementable by a π-network of capacitors with attached transmission line sections as shown in Fig.3.10(b). By equating the transmission matrices of both schematics, we have $K_i^{1a} = K_i^{1b}$ and $K_i^{2a} = K_i^{2b}$ with[18]

$$K_i^{1a} = \frac{Z_i}{Z_{i+1}}, \qquad K_i^{1b} = \frac{B_{pib} + B_{si}}{B_{pif} + B_{si}}, \qquad K_i^{2a} = \frac{1}{J_i Z_i Z_{i+1}} - J_i$$

$$K_i^{2b} = B_{pif} + B_{pib} + \frac{B_{pif}B_{pib} + (Z_i Z_{i+1})^{-1}}{B_{si}}$$

where $B_{pif} = \omega_o C_{pif}, B_{pib} = \omega_o C_{pib}, B_{si} = \omega_o C_{si}$ with ω_o the center frequency. The electrical length, $\phi_{i(i+1)}$, associated with the ith overlapping gap can be expressed as

$$\phi_{i(i+1)} = \tan^{-1}\left[\frac{2(B_{si} + B_{pib})}{Z_i(B_{pif}B_{pib} + B_{pib}B_{si} + B_{si}B_{pif}) - (Z_{i+1})^{-1}}\right] \tag{3.7}$$

Next, replace each admittance inverter in Fig.3.8(f) by its equivalent schematic depicted in Fig.3.10(b) to obtain the target schematic shown in Fig.3.10(a) with $\theta_i = \phi_{(i-1)i}/2 + \phi_i + \phi_{i(i+1)}/2$.

The physical length of section i is determined as $\ell_i = (\theta_i c)/(\omega_o \sqrt{\epsilon_i^{\text{eff}}})$, where c is the speed of light in free space and ϵ_i^{eff} is the effective permitivity of a uniform microstrip line with the same cross section as line section i.

An open-ended microstrip line can be represented by an equivalent circuit with an excess capacitance which can be further translated into an incremental line section with length $\Delta\ell_i$. In other words, it is treated as if an equivalent open end is located at a distance $\Delta\ell_i$ extended from the physical open end, and the electric fields underneath the equivalent strip exhibit no fringing effect.

4. Optimization Approaches

At the ith overlapping gap, the width W_{i+1} and gap length $g_{i(i+1)}$ need to be optimized to determine the proper susceptances(B_{pif}, B_{pib} and B_{si}), characteristic impedance, Z_{i+1}, and admittance inverter, $J_{i(i+1)}$. The first and last transmission line sections are chosen to have a typical impedance of 50 Ω. To have an impedance of 50 Ω in a two-layered FR4 substrate(2.9 mm thick, ϵ_r=4.3), the width of transmission line is 5.8 mm. The variables to be optimized are $g_{i(i+1)}$ and W_{i+1} with $0 \leq i < N$ and $g_{N(N+1)}$. Fig.3.11 shows a typical genetic algorithm which can be

applied consecutively on each overlapping gap to optimize the gap length and width of the next line section using the following fitness function

$$\text{fitness function} = \left|K_i^{1a} - K_i^{1b}\right| + w\left|K_i^{2a} - K_i^{2b}\right|$$

where w is a weighting factor.

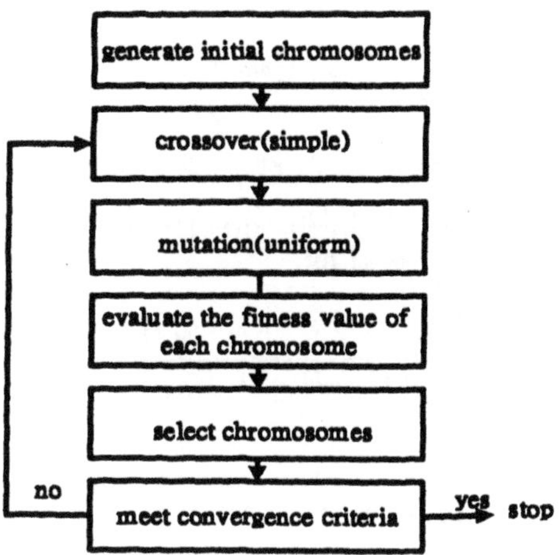

Figure 3.11. A typical genetic algorithm.

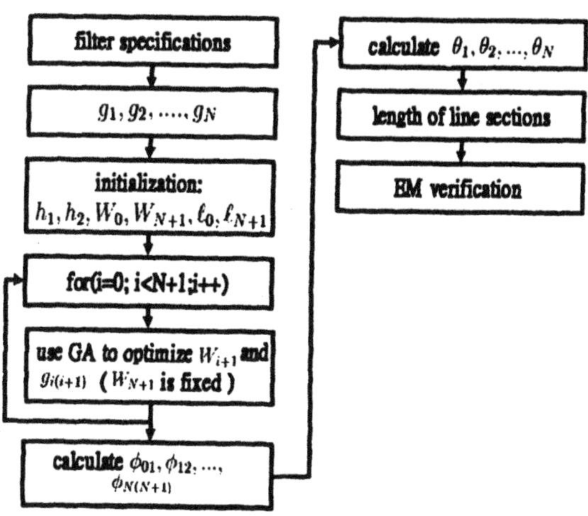

Figure 3.12. Design procedure of a multilayered end-coupled band-pass filter.

The overall design procedure is outlined in Fig.3.12. Simple crossover, uniform random mutation and tournament selection processes are used

to reproduce new generations. Once the overlapping gap lengths and line section widths are determined, electrical lengths of the overlapping gaps defined in (3.7) and the incremental length due to fringing fields are used to correct the physical length of all line sections. Full-wave approach is then applied to verify the frequency response of the designed filter as compared with specifications.

Global optimization approach is an alternative to design the same type of band-pass filters using genetic algorithm. The lengths and widths of all the transmission line sections and the overlapping gap lengths as marked in Fig.3.13(a) are optimized concurrently. The fitness function is defined as the difference of scattering parameters between the equivalent circuit and specifications.

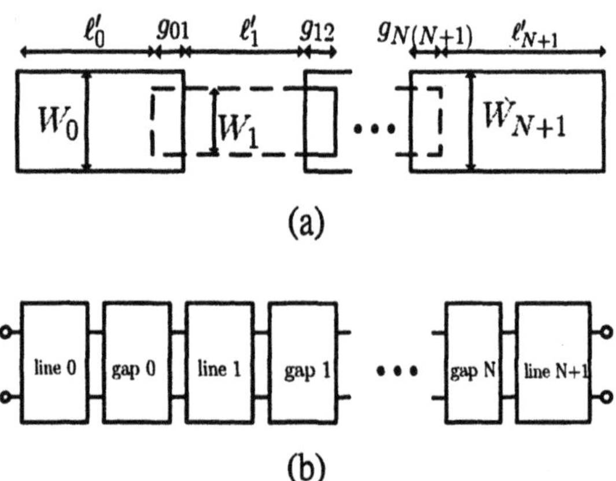

Figure 3.13. **(a) A broadside end-coupled band-pass filter and (b) its block diagram.**

Band-pass filters are composed of linear passive components which can be modeled by lumped circuits over certain frequency range. The order of a filter determines its topology. The physical layout can be represented as a cascade of block diagrams as shown in Fig.3.13(b) where $N+2$ line sections and $N+1$ overlapping gaps are used to implement an Nth order Chebyshev filter. Each block diagram can be further translated into an equivalent circuit.

The total transmission matrix associated with the block diagram shown in Fig.3.13(b) is then obtained by multiplying the transmission matrices of all the cascaded block diagrams as

$$\begin{bmatrix} A & B \\ C & D \end{bmatrix}_{\text{total}} = \begin{bmatrix} A & B \\ C & D \end{bmatrix}_{\text{gap } 0} \begin{bmatrix} A & B \\ C & D \end{bmatrix}_{\text{line } 1} \cdots \begin{bmatrix} A & B \\ C & D \end{bmatrix}_{\text{gap N}}$$

The scattering parameters are then derived from the total transmission matrix.

In this approach, the parameters to be optimized are line widths(W_1, W_2,..., W_N), line lengths($\ell_1, \ell_2, ..., \ell_N$) and overlapping gap lengths(g_{01}, g_{12},...,$g_{N(N+1)}$). All these parameters are included in the definition of chromosome. Fig.3.14 shows the procedure to optimize the overlapping gap lengths, line widths and line lengths of filter using the following fitness function

$$\text{fitness function} = w_{21}^{SB} \sum_{\text{lower SB}} \|S_{21}^{GA} - S_{21}^{LSB}\| + w_{21}^{PB} \sum_{\text{PB}} \|S_{21}^{GA} - S_{21}^{PB}\| + w_{11}^{PB} \sum_{\text{PB}} \|S_{11}^{GA} - S_{11}^{PB}\| + w_{21}^{SB} \sum_{\text{upper SB}} \|S_{21}^{GA} - S_{21}^{USB}\|$$

where all the scattering parameters are in dB.

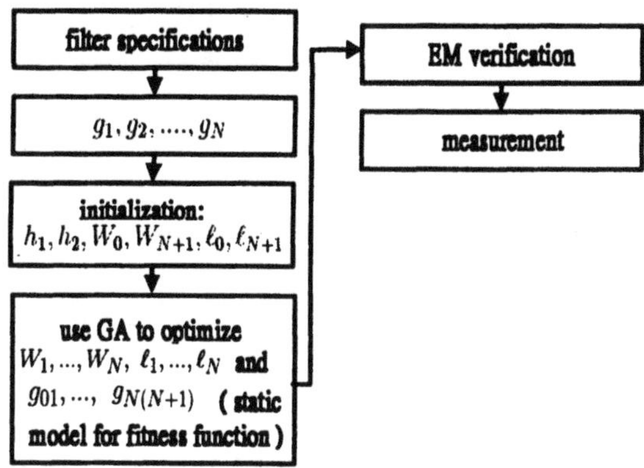

Figure 3.14. Design procedure of a multilayered end-coupled band-pass filters using global optimization approach.

Referring to Fig.3.15, w_{21}^{PB} is the weighting factor on S_{21} in the pass-band, w_{11}^{PB} is the weighting factor on S_{11} in the pass-band, and w_{21}^{SB} is the weighting factor on S_{21} in the stop-bands, $S_{21}^{LSB}, S_{21}^{PB}, S_{21}^{USB}$ are the insertion loss thresholds in the lower stop-band, pass-band and upper stop-band, respectively, S_{11}^{PB} is the return loss threshold in the pass-band. The norm $\|S_{21}^{GA} - S_{21}^{LSB}\|$ is equal to $|S_{21}^{GA} - S_{21}^{LSB}|$ if $S_{21}^{GA} < S_{21}^{LSB}$, and is equal to zero if $S_{21}^{GA} \geq S_{21}^{LSB}$. The other three norms are defined in a similar manner. The return loss threshold in the pass-band is related to the ripple level of insertion loss specification. Simple crossover, uniform random mutation and tournament selection processes are used to reproduce new generations.

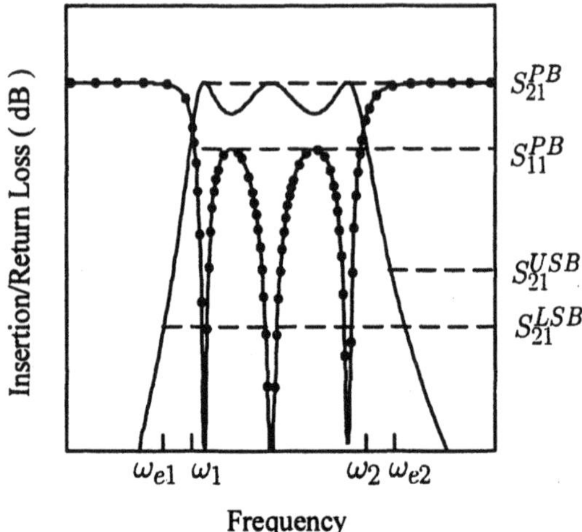

Figure 3.15. Filter specifications and thresholds used in defining the fitness function, S_{21}^{PB}: insertion loss threshold in pass-band, S_{11}^{PB}: return loss threshold in pass-band, S_{21}^{USB}: insertion loss threshold in upper stop-band, S_{21}^{LSB}: insertion loss threshold in lower stop-band, pass-band is $\omega_1 < \omega < \omega_2$, upper stop-band is $\omega > \omega_{e2}$, lower stop-band is $\omega < \omega_{e1}$.

Once the overlapping gap lengths, line section lengths and widths are determined, incremental lengths due to fringing fields are used to correct the physical length of all the line sections. Full-wave approach and measurement are then applied to verify the frequency response of the designed filter as compared with specifications.

5. Results and Discussions

Both local and global optimization approaches are used to design a broadside end-coupled band-pass filter. A similar layout in [18] is chosen as a benchmark for comparison, in which the center frequency is 4 GHz, the 3 dB bandwidth is about 35% and the insertion loss at 3 GHz is at least 20 dB. A third order Chebyshev filter is designed using an FR4 substrate. The width of microstrip feed line is 5.8 mm to have a characteristic impedance of 50 Ω.

Local optimization approach is first used to design a band-pass filter. Proper ranges of overlapping gap length and line width for each line section are selected to define a chromosome. It is important that the range of variables be wide enough to cover the optimal value, yet limited to save computation time. Next, the ranges of overlapping gap lengths and line widths are discretized into an integer to fit in the genetic algorithm. Under these circumstances, the proper range of overlapping gap length

is between 0.1 mm and 12.8 mm, and the proper range of line width is between 4 mm and 16.7 mm. Each variable is expressed by a 7-bit string, corresponding to 128 possible values.

The initial population consists of 100 chromosomes with each allele determined in a random manner. A reproduction cycle begins with mating the top 50 chromosomes to create another 50 child chromosomes by using a simple crossover process. Next, a uniform random mutation is applied to the child chromosomes, which is executed by randomly flipping 0 to 1% of the bits. The mutation process facilitates the genetic algorithm to converge to an optimal solution. However, high mutation rate may render a slow convergence rate. Next, the top 100 chromosomes are chosen using tournament selection process, and are retained as the new generation. The reproduction cycle of crossover, mutation and selection processes is repeated until a convergent solution is claimed. If the convergence criterion is set too loose, the algorithm may converge too early to a meaningless solution.

Final layout of the filter is shown in Fig.3.16(a). Figs.3.16(b) and 3.16(c) show the filter performance simulated by using a static model and the IE3D package. The static model is based on a cascaded transmission matrix approach as described in the last Section. The IE3D is based on an integral equation approach and method of moments. The simulation results are summarized in Table 3.2.

Table 3.2. Summary of band-pass filter performance verified using IE3D, BW: bandwidth, IL: insertion loss, RL: return loss.

Approach	Filter order	3 dB BW (%)	10 dB BW (%)	Max. IL (dB)	Min. RL (dB)
local	3	37.5	55	2.8	8.3
local	5	35	50	1.8	8
global	3	32	41	5	11.5
global	5	28	33	6	8.5
[18]	5	33.65	37.5	1.3	10

The insertion loss using IE3D drops around the lower corner of pass-band because the equivalent circuit model of band-pass filter is not accurate enough to cover all the wave mechanisms that may occur in the physical layout. The sharp null around 5.8 GHz in the insertion loss curve using IE3D suggests that an attenuation pole appears near the edge of upper stop-band. This pole may be contributed by the proximity effect of two open ends at the same level, which is not accounted for in our equivalent circuit model. In Fig.3.16(c), the return loss in the pass-band simulated using IE3D is higher than specification because

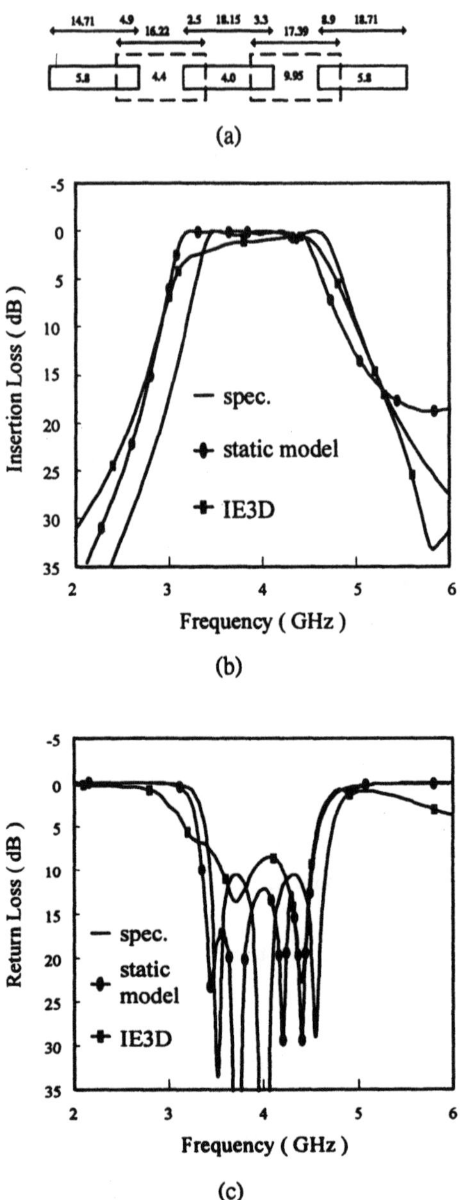

Figure 3.16. A third order Chebyshev band-pass filter designed with local optimization approach, ripple is 0.25 dB, bandwidth is 35%,(a) physical layout(in mm), (b) insertion loss, (c) return loss.

full-wave approach covers reflection mechanism at the open ends of strips while the static model does not. The computation time to design a fil-

ter is less than one second using a personal computer with Pentium III CPU, 600MHz clock rate. The required iteration number is between 2 to 12 at each gap.

The minimum return loss in the pass-band can be derived from the pass-band ripple if the substrate is lossless. An insertion loss of 0.1 dB is equivalent to a return loss of 16 dB in the pass-band. A higher-order Chebyshev filter usually has a smaller ripple and sharper band edges than a lower-order one. A fifth-order Chebyshev filter is also designed with 0.1 dB ripple in the pass-band. More line sections and overlapping gaps may incur more severe reflections at the open ends. Hence, insertion loss increases and return loss decreases in the pass-band.

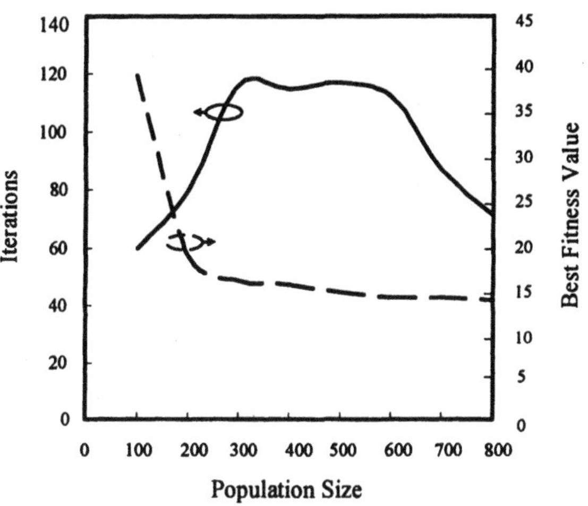

Figure 3.17. Effect of population size on the convergece rate and best fitness value.

Next, the global optimization approach is used to design a band-pass filter. Effects of population size on the iteration number and best fitness value are shown in Fig.3.17. The algorithm converges in relatively less iterations when the population size is smaller than 300. When the population size is greater than 600, the iteration is accelerated again. This implies that when the population size is large enough, an optimal solution has a better chance to show up earlier during an evolution. But a larger population size takes longer computation time. If the population size is smaller than 200, the best fitness value indicates that the solution is likely to be premature. If the population size is larger than 200, the best fitness value converges to about the same level, indicating that the optimal solution becomes insensitive to population size any more. Based on these observations, the population size of 400 is used in our global

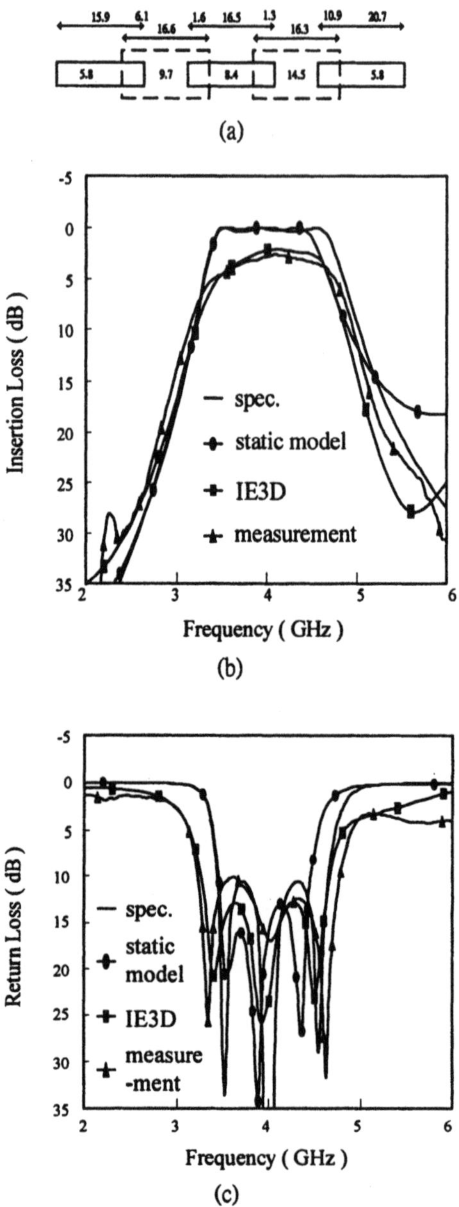

Figure 3.18. **A third order Chebyshev band-pass filter designed using global optimization approach, ripple is 0.25 dB, bandwidth is 35%, (a) physical layout(in mm), (b) insertion loss, (c) return loss.**

optimization approach. The crossover rate is chosen to be 50%, and the mutation rate is uniformly chosen between 0 and 5%.

The filter is targeted to have a bandwidth of 35%, centered at 4 GHz. The specifications of a third-order Chebyshev filter with a pass-band ripple of 0.25 dB are chosen. Final layout of the filter is shown in Fig.3.18(a). Figs.3.18(b) and (c) show the filter performance using static model, IE3D and measurement. The results are also summarized in Table 3.2. The computation time taken to design this filter varies between five and ten minutes, and the iteration number required is between 70 and 125.

As shown in Fig.3.18(b), measured insertion loss in the pass-band is lower than that predicted by the static model, but is close to that simulated using IE3D. The deviation can be explained as follows. The loss tangent of typical FR4 substrates is about 0.018, which is not considered in our static model. The width of feed line is determined based on a lossless model. Thus, losses in the substrates also incurs impedance mismatch and power reflection. Also, the connector does not match perfectly with the feed line.

In the upper stop-band, the insertion loss using the static model is higher than that using IE3D and the measurement because the threshold in the upper stop-band is lower as described in the last Section. In Fig.3.18(c), the measured return loss is similar to that using IE3D. The return loss using IE3D simulation and measurement in both stop-bands are higher than that using the static model because open-end microstrip may radiate like patch antenna.

The global optimization approach is also used to design a band-pass filter having a bandwidth of 35%, centered at 4 GHz. The specifications of a fifith-order Chebyshev filter with pass-band ripple of 0.1 dB are chosen.

References

[1] J. M. Johnson and R. S. Yahya, "Genetic algorithm optimization for aerospace electromagnetic design and analysis," *Conf. Aerospace Appl.*, vol.1, pp.87-102, Taiwan, 1996.

[2] Y. Lu, X. Cai, and Z. Gao, "Optimal design of special corner reflector antennas by the real-coded genetic algorithm," *Asia-Pacific Microwave Conf.*, pp.1457-1460, Singapore, 2000.

[3] F. J. Ares-Pena, J. A. Rodriguez-Gonzalez, E. Villanusva-Lopez, and S. R. Rengarajan, "Genetic algorithms in the design and optimization of antenna array patterns," *IEEE Trans. Antennas Propagat.*, vol.47, no.3, pp.506-510, Mar. 1999.

[4] J. M. Johnson and Y. Rahmat-Samii, "Genetic algorithms and method of moments (GA/MOM) for the design of integrated anten-

nas," *IEEE Trans. Antennas Propagat.*, vol.47, no.10, pp.1606-1614, Oct. 1999.

[5] M. H. ÖKtem and B. Saka, "Design of multilayered cylindrical shields using a genetic algorithm," *IEEE Trans. Electromagn. Compat.*, vol.43, no.2, pp.170-176, May 2001.

[6] E. E. Alshuler, "Design of a vehicular antenna for GPS/Iridium using a genetic algorithm," *IEEE Trans. Antennas Propagat.*, vol.48, no.6, pp.968-972, June 2000.

[7] S. Chakravarty and R. Mittra, "Application of the micro-genetic algorithm to the design of spatial filters with frequency-selective surfaces embedded in dielectric media," *IEEE Trans. Electromagn. Compat.*, vol.44, no.5, pp.338-346, May 2002.

[8] R. Thamvichai, T. Bose, and R. L. Haupt, "Design of 2-D multiplierless IIR filters using the genetic algorithm," *IEEE Trans. Circuits Syst. I*, vol.49, no.6, pp.878-882, June 2002.

[9] E. Michielssen, J.-M. Sajer, S. Ranjithan, and R. Mittra, "Design of lightweight, broad-band microwave absorbers using genetic algorithms," *IEEE Trans. Microwave Theory Tech.*, vol.46, no.10, pp. 1024 -1031, Oct. 1998.

[10] S. B. Cohn, "Parallel-coupled transmission-line-resonator filters," *IRE Trans. Microwave Theory Tech.*, pp.223-231, 1958.

[11] L. Zhu, H. Bu, and K. Wu, "Broadband and compact multi-pole microstrip bandpass filters using ground plane aperture technique," *IEEE Trans. Antennas Propagat.*, vol.49, pp.71-77, Feb. 2002.

[12] L. H. Hsieh and K. Chang, "Compact, low insertion loss, sharp rejection wideband bandpass filters using dual-mode ring resonators with tuning stubs," *Electron. Lett.*, vol.37, no.22, pp.1345-1347, Oct. 2001.

[13] A. Griol, J. Marti, and L. Sempere, "Microstrip multistage coupled ring bandpass filters using spur-line filters for harmonic suppression," *Electron. Lett.*, vol.37, no.9, pp.572-573, Apr. 2000.

[14] D. Lee, D. Cargill, and P. Pramanick, "Computer aided design of shielded microstrip line hairpin-line filters," *Elect. Computer Eng.*, vol.2, pp.1148-1151, Canada, 1993.

[15] J. S. Hong and M. J. Lancaster, "Development of new microstrip pseudo-interdigital bandpass filters," *IEEE Microwave Guided Wave Lett.*, vol.5, pp.261-263, Aug. 1995.

[16] K. C. Huang, D. Hyland, A. Jenkins, D. Edwards, and D. D. Hughes, "A miniaturized interdigital microstrip bandpass filter,"

IEEE Trans. Appl. Superconduct., vol.9, no.2, pp.389-392, June 1999.

[17] H. Nam, H. Lee, and Y. Lim, "A design and fabrication of bandpass filter using miniaturized microstrip square SIR," *Proc. IEEE Reg. 10 Int. Conf. Electr. Electron. Technol.*, pp.395-398, 2001.

[18] C. Cho and K.C. Gupta, "Design of end-coupled bandpass filters in multilayer microstrip configurations," *IEEE MTT-S Int. Microwave Symp. Dig.*, pp.711-714, Baltimore, 1999.

[19] C. Cho and K. C. Gupta, "Design methodology for multiplayer coupled line filters," *IEEE MTT-S Int. Microwave Symp. Dig.*, pp.785-788, June 1997.

[20] W. Schwab and W. Menzel, "Compact bandpass filters with improved stop-band characteristics using planar multilayer structures," *IEEE MTT-S Int. Microwave Symp. Dig.*, pp.1207-1209, June 1992.

[21] Y. J. Huang, F.-C. Fang, K.-H. Hsiau, and S. L. Fu, "A packaging system for built-in microcircuits in multilayer substrates," *IEEE Elect. Manuf. Technol. Symp.*, pp.460-463, Dec. 1995.

[22] S. A. Raby and A. C. Cangellaris, "Interconnect properties and multilayer bandpass filter design in LTCC substrates," *IEEE Conf. Wireless Commun.*, pp.187-192, Aug. 1997.

[23] A. Sutono, A. Pham, J. Laskar, and W. R. Smith, "Development of three dimensional ceramic-based MCM inductors for hybrid RF / microwave applications," *IEEE Radio Freq. Integ. Circuits Symp.*, pp. 175-178, 1999.

[24] A. H. Khalid, T. Gokdemir, S. Economides, A. A. Rezazadeh, and I. D. Robertson, "Multilayer techniques for MMICs," *IEE Colloq. Adv. Develop. Microelect. Eng.*, pp.6/1-6/5, Nov. 1996.

[25] G. Matthaei, L. Young, and E. M. T. Jones, *Microwave Filters, Impedance-Matching Networks, and Coupling Structures*, Dedham, MA: Artech House, 1980.

Chapter 4

DESIGN OF MULTISTAGE COMBLINE BAND-PASS FILTERS IN LAYERED STRUCTURES

Ikuo Awai

Department of Electrical and Electronic Engineering
Yamaguchi University
Tokiwadai Ube, Japan

Abstract A new design method is proposed for multistage combline band-pass filter made of strip line resonators. Grayzel transformation is used to separate the coupling of adjacent resonators and to decompose the filter into pairs of coupled resonators. After the design is finished, similar resonators of the same length but a little different width are equally spaced and accompanied by auxiliary capacitors, which has the simplest structure to fabricate using low-temperature cofired ceramics(LTCC).

Keywords: band-pass filter, multistage, combline, strip line resonator, Grayzel's identity.

1. Introduction

Filters in a portable equipment for mobile communication are required to be as small as possible. A multistage band-pass filter(BPF) is chosen to have sharper out-of-band characteristics if more efficient use of frequency resource is required. A combline BPF in a laminated ceramic structure is one of the most widely used filters because of its reasonable dimension and pass-band attenuation. Though a two-stage BPF is easily designed based on the even and odd mode analysis, it is difficult to design a multistage combline BPF.

Matthaei reported a design method of multistage combline BPFs many years ago [1]. It is a pioneer work that elucidates the design theory of combline BPF made of ideal transmission line coupled resonators and realized in practical distributed resonators, the rectangular bar res-

onators. He used the Getsinger's equations and graphs to determine the filter dimensions [2]. Although it is only applicable to a narrow band BPF, its basic concept has been incorporated into the design theories proposed later.

In 1964, Crystal pointed out that a circular cylindrical rod has better features in manufacturing, with the same loss property as Matthaei's rectangular bar resonators [3]. A similar design of interdigital BPF is demonstrated. Then, it was improved by Vadopalas with dimensions determined using graphical approach [4]. Later in 1977, Milligan proposed an algorithm with graphs for numerical determination of dimensions of microstrip resonators that became more popular to microwave engineers [5].

Since the input and output coupling lines of Matthaei's structure take space, a design of tapping I/O structure was proposed by Dishal in 1965 [6]. It was extended to a BPF with wider bandwidth by Crystal ten years later [7]. A purely analytical procedure was proposed by Caspi and Adelman using tapping structures in 1988 [8].

Recent demands for combline BPFs in a laminated LTCC structure are miniaturization, moderate loss and good out-of-band attenuation. Thus, based on Matthaei's theory, we choose a capacitively coupled I/O port which is good for miniaturization and easy to fabricate. An LTCC module has multiple ground planes to support layered strip line structures. Therefore, thin rectangular strip segments are used in our BPF design, which has been analyzed by Getsinger [2]. An analytical procedure for obtaining the filter dimensions is given, which is more appropriate for computerized design. A theoretical supplement is added to Matthaei's theory, which extends the equivalent circuit theory of two coupled resonators to a chain of coupled resonators. The Grayzel's identity is used to prove its validity [9].

The most important feature of this Chapter is a new method for controlling the attenuation poles intrinsic to combline BPF. It is useful for realizing better out-of-band characteristics.

2. Transformation of Coupled Resonators

In the equivalent circuit of the coupled transmission lines shown in Fig.4.1, C_i denotes shunt capacitance per unit length to the ground, and $C_{i,i+1}$ is the mutual capacitance between adjacent lines. In [9], Grayzel proved that the configurations in Fig.4.1 are exactly equivalent to those in Fig.4.2 when

$$C_i = C_i' + C_i'', \qquad 2 \le i \le n-1$$

This implies that each line can be divided into two lines, taking into account the right and left-neighbor coupling to each line separately.

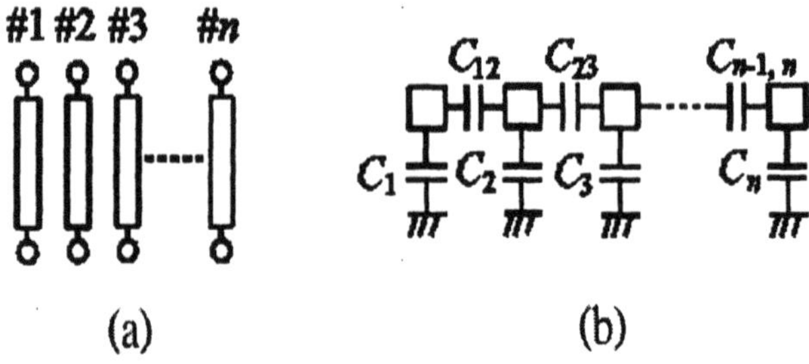

Figure 4.1. Coupled transmission lines and their lumped element model.

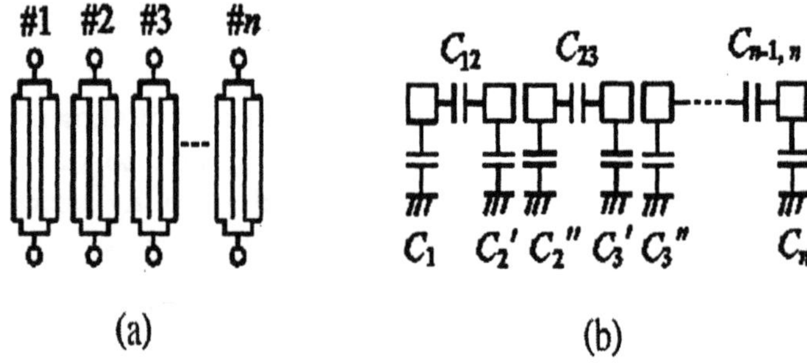

Figure 4.2. Equivalent coupled transmission lines and lumped element models to those in Fig.4.1.

Fig.4.3 shows a pair of coupled lines. The input admittances of the circuit in Fig.4.3(a) are given as [1]

$$Y_{\alpha e} = -jY_{0e}^{\alpha} \cot\theta, \qquad Y_{\alpha o} = -jY_{0o}^{\alpha} \cot\theta, \qquad \alpha = a, b$$

for the even and odd mode, respectively, where $\theta = (\omega/v)\ell$, $Y_{0e}^{a} = vC_a$, $Y_{0o}^{a} = v\,(C_a + 2C_c)$, $Y_{0e}^{b} = vC_b$, $Y_{0o}^{b} = v\,(C_b + 2C_c)$, and $v = 1/\sqrt{\epsilon\mu_0}$. Hence, the circuits in Figs.4.3(a) and 4.3(b) are easily transformed into that in Fig.4.3(c), considering both even and odd mode excitations.

Next, consider a set of coupled $\lambda/4$ resonators shown in Fig.4.4(a). Each resonator is divided according to the Grayzel's identity as shown in Fig.4.4(b). Now, each coupled line resonator in Fig.4.4(b) is replaced

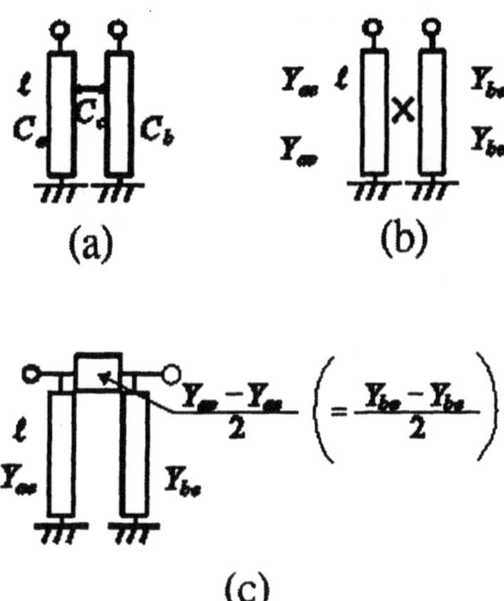

Figure 4.3. Transformation of a distributed coupling resonator.

by the circuit in Fig.4.3(c) to obtain a ladder circuit shown in Fig.4.4(c). Two neighboring shunt resonators are combined to one resonator as shown in Fig.4.4(d). Thus, the distributed coupling between adjacent resonators shown in Fig.4.4(a) is accounted for by simple lumped elements. Notice that the coupling admittance, $(Y'_{i+1,o} - Y'_{i+1,e})/2$, and the input admittance, Y_{ie}, of each resonator are

$$\frac{Y'_{i+1,o} - Y'_{i+1,e}}{2} = -jvC_{i,i+1}\cot\theta$$
$$Y_{ie} = Y'_{ie} + Y''_{ie} = -jvC_i\cot\theta$$

respectively. These results were obtained by Matthaei using a different approach [1].

3. Multistage Distributed Coupling Resonators in Strip Lines

The structure shown in Fig.4.4(a) is the essential part of a multistage combline BPF. Hence, it is important to find the critical dimensions of the coupled resonators when C_{is}'s and $C_{i,i+1s}$'s are given. The design procedure is simplified when all the transmission lines constituting the resonators have the same shunt and mutual capacitances per unit length. It is assumed that the dimensions and spacings of all the trans-

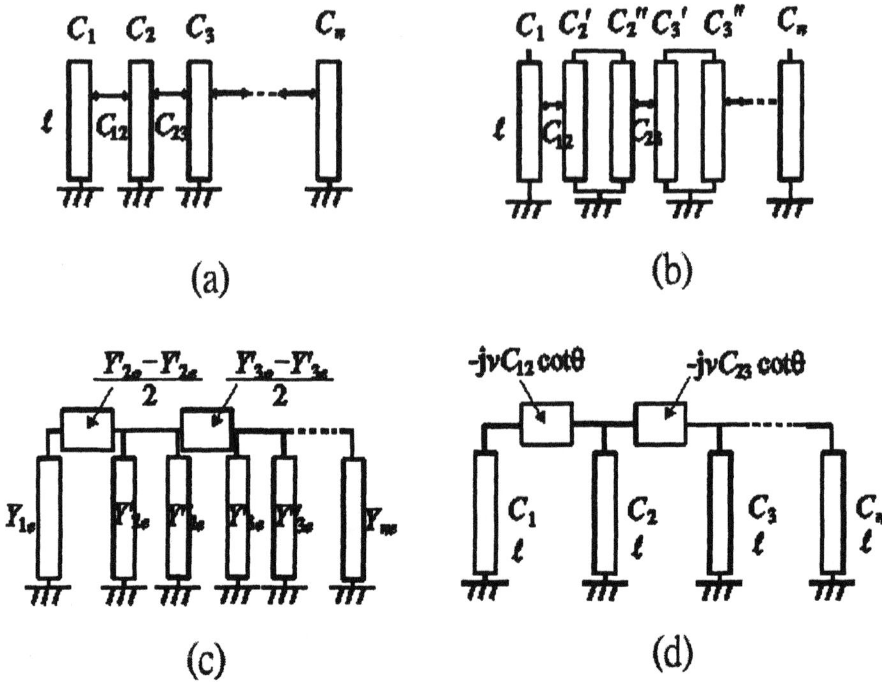

Figure 4.4. Transformation of distributed coupling resonators to lumped element coupling resonators.

mission lines are the same. Fig.4.5(a) shows the schematic of coupled $\lambda/4$ resonators and Fig.4.5(b) shows the cross sectional view of strip line implementation. Due to their alignment, it can be assumed that coupling takes place only between adjacent resonators.

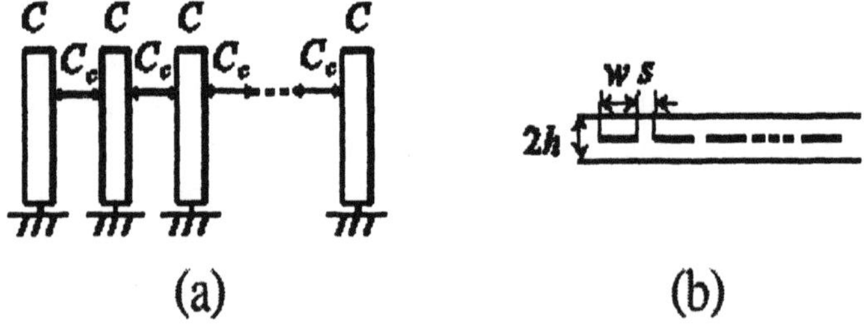

Figure 4.5. Coupled $\lambda/4$ resonators implemented in strip lines.

When the same voltage is applied to each resonator in Fig.4.5(b), the shunt capacitance per unit length between each resonator and the ground plane is C. As shown in Fig.4.6(a), electric field distributions around the outer resonators are different from those around the inner ones in spite of the same dimensions, which results in different capacitances of C and C' with the inner and outer resonators, respectively. Combining the rightmost and leftmost resonators, one obtains a two-stage coupled stripline resonators as shown in Fig.4.6(b). Therefore, the relation between width of the outermost strip line resonators in Fig.4.6(a) and its shunt capacitance per unit length is the same as that for the even mode of the two-stage resonator shown in Fig.4.6(b).

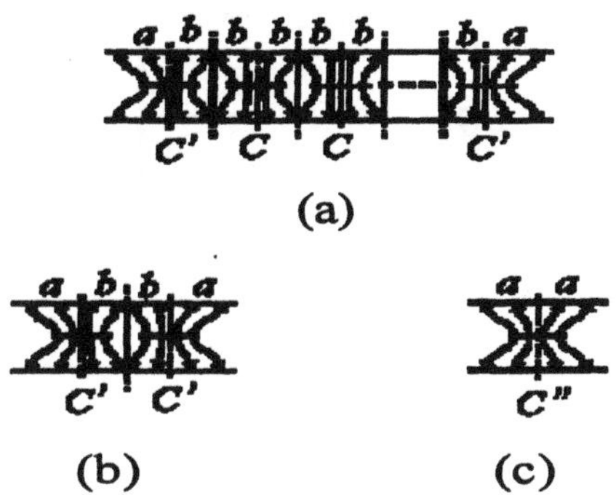

Figure 4.6. Electric field distributions of the even mode.

For two coupled strip lines, we have [10]

$$C' = \frac{Y_{0e}}{v} = 4\epsilon\frac{K(k_e)}{K(k_e')}, \text{ with } k_e = \tanh\left(\frac{\pi}{4}\frac{w}{h}\right)\tanh\left(\frac{\pi}{4}\frac{w+s}{h}\right) \quad (4.1)$$

where $k_e' = \sqrt{1-k_e^2}$. The width of the outermost resonator, w, is determined by solving this equation for a given C'. The electric field distribution in a single resonator is depicted in Fig.4.6(c), as the combination of the right and leftmost parts in Figs.4.6(a) or 4.6(b). Comparing the electric fields in Figs.4.6(a), 4.6(b) and 4.6(c), we obtain $C = 2C' - C''$. Since C'' is the shunt capacitance per unit length of a single strip line, it can be calculated as

$$C'' = 4\epsilon\frac{K(k)}{K(k')}, \text{ with } k = \tanh\left(\frac{\pi}{4}\frac{w}{h}\right) \quad (4.2)$$

where $k' = \sqrt{1-k^2}$. Thus, the shunt capacitance of the inner resonator is given by

$$C = 4\epsilon \left[2\frac{K(k_e)}{K(k'_e)} - \frac{K(k)}{K(k')} \right] \tag{4.3}$$

The mutual capacitance between two resonators, C_c, is calculated assuming that the voltages on two adjacent resonators are V_0 and $-V_0$, respectively. The C_c in Fig.4.5(a) is calculated referring to Fig.4.7(b), based on the similarity of electric field distributions between two resonators shown in Figs.4.7(a) and 4.7(b). Thus, capacitance of the odd mode is given by $C_{oo} = C_{oe} + 2C_c$, where C_{oe} and C_{oo} are the even and odd mode capacitances per unit length, respectively, as shown in Figs.4.6(b) and 4.7(b). Note that $C_{oe} = C''$. Thus, C_c is determined as

$$C_c = \frac{1}{2}(C_{oo} - C_{oe}) = 2\epsilon \frac{K(k_o)}{K(k'_o)} - \frac{C'}{2}$$
$$\text{with } k_o = \tanh\left(\frac{\pi}{4}\frac{w}{h}\right)\coth\left(\frac{\pi}{4}\frac{w+s}{h}\right) \tag{4.4}$$

where $k'_o = \sqrt{1-k_o^2}$.

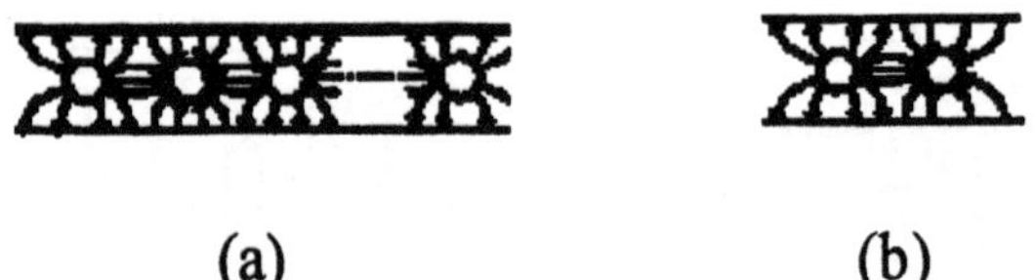

Figure 4.7. Electric field distributions of the odd mode.

4. Design of Combline Band-Pass Filter

4.1 Capacitor Coupled Parallel *LC* Resonator BPF

The procedure to design a combline band-pass filter is listed in steps as follows.

Step 1. Based on the insertion loss method, one begins with defining the filter type(Wagner, Chebyshev, and so on), number of elements, center frequency and bandwidth.

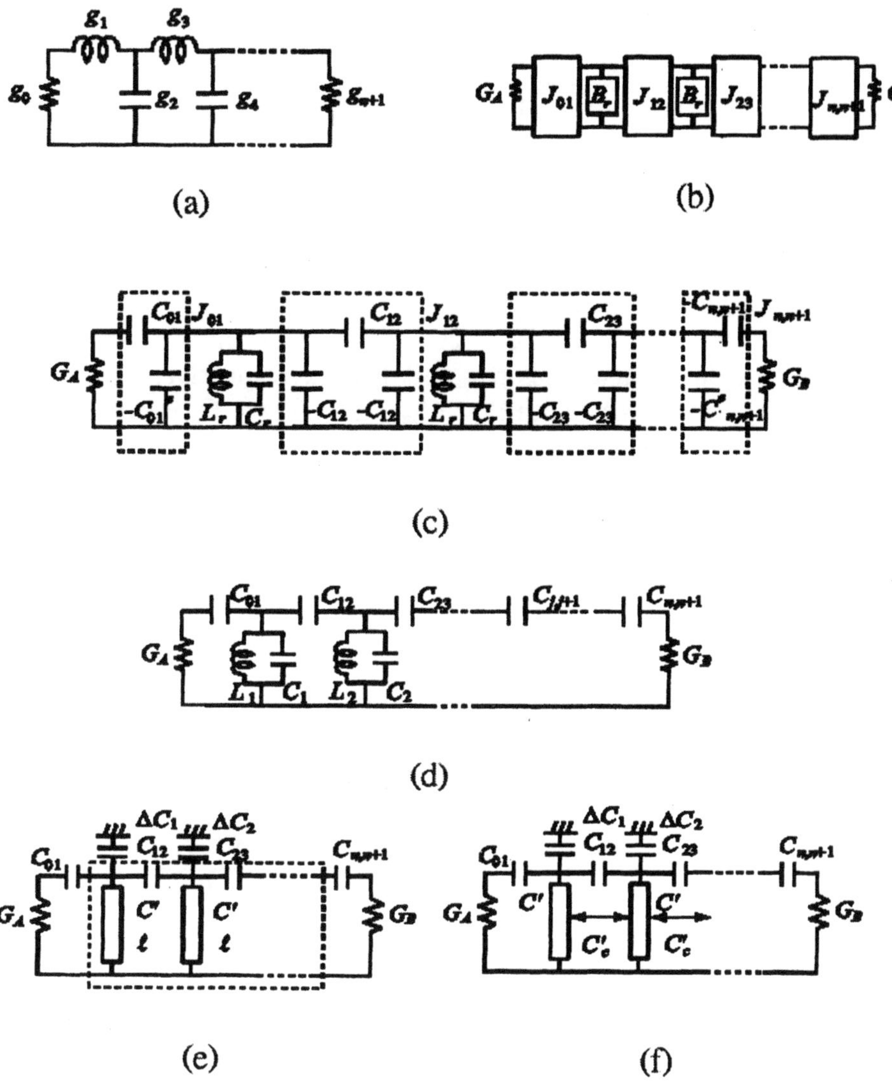

Figure 4.8. Design of capacitor coupled combline BPF with shunt capacitors.

Step 2. Circuit elements of the prototype low-pass filter is determined as shown in Fig.4.8(a) using the information in step 1.

Step 3. Choose the structure of coupled combline resonators in strip line configuration, and choose material and cross section of the resonator. All the resonators are assumed to have the same dimensions and spacings for simplicity. The resonator length will be determined later.

Step 4. Take the constituent single resonator in step 3 as the first working model. J inverters are inserted between resonators which can be expressed as $B_r(\omega)$ shown in Fig.4.8(b) with

$$B_r(\omega) = -vC\cot\left(\frac{\omega}{v}\ell\right) \tag{4.5}$$

Step 5. The value of J inverters are calculated as

$$J_{01} = \sqrt{\frac{G_A b_1 B_f}{g_0 g_1 \omega_1'}}, \qquad J_{n,n+1} = \sqrt{\frac{G_B b_n B_f}{\omega_1' g_n g_{n+1}}}$$

$$J_{j,j+1} = \frac{B_f}{\omega_1'}\sqrt{\frac{b_j b_{j+1}}{g_j g_{j+1}}}, \qquad 1 \le j \le n-1$$

where B_f is the fractional bandwidth, ω_1' is the cutoff frequency of the prototype LPF, b_j is the susceptance slope of each resonator, defined as

$$b_j = \frac{\omega_0}{2}\left.\frac{dB_r(\omega)}{d\omega}\right|_{\omega=\omega_0}$$

which has the same value irrespective of the index j, and ω_0 is the center frequency of the designed BPF.

Step 6. Convert the circuit in Fig.4.8(b) into a lumped element circuit in Fig.4.8(c) where the distributed resonators, B_r's, are implemented by lumped element LC circuits, and the J inverters are realized by the capacitor networks encircled by broken lines in Fig.4.8(c). Then, the element values are expressed as

$$C_r = \frac{b_j}{\omega_0}, \qquad L_r = \frac{1}{C_r\omega_0^2}$$

$$C_{01} = \frac{J_{01}}{\omega_0\sqrt{1-(J_{01}/G_A)^2}}, \qquad C_{n,n+1} = \frac{J_{n,n+1}}{\omega_0\sqrt{1-(J_{n,n+1}/G_B)^2}}$$

$$C_{j,j+1} = \frac{J_{j,j+1}}{\omega_0}, \qquad 1 \le j \le n-1$$

$$C_{01}^e = \frac{C_{01}}{1+(\omega_0 C_{01}/G_A)^2}, \qquad C_{n,n+1}^e = \frac{C_{n,n+1}}{1+(\omega_0 C_{n,n+1}/G_B)^2}$$

Step 7. The circuit in Fig.4.8(c) is further transformed into the circuit in Fig.4.8(d), where C_j is the sum of parallel capacitances as

$$C_1 = C_r - C_{01}^e - C_{12}, \qquad C_n = C_r - C_{n-1,n} - C_{n,n+1}^e$$
$$C_j = C_r - C_{j-1,j} - C_{j,j+1}, \qquad 2 \le j \le n-1 \tag{4.6}$$

4.2 Capacitor Loaded Combline BPF in Strip Lines

Step 8. In order to apply the approach described in Section 2, extract a common component, C_0, from all resonators in Fig.4.8(d) and denote the residual part as ΔC_j, namely

$$C_j = C_0 + \Delta C_j \tag{4.7}$$

Step 9. Now, we have identical parallel LC resonators accompanied by small parallel capacitance ΔC_j's. The LC parallel resonators are transformed to $\lambda/4$ resonators as shown in Fig.4.8(e). The capacitance per unit length, C, in (4.5) should be slightly reduced to C'. The input admittance of each resonator with a residual capacitance becomes

$$Y_j = -jvC' \cot\left(\frac{\omega}{v}\ell\right) + j\omega\Delta C_j$$

In order to make Y_j and its slope equal to those of the LC circuit in Fig.4.8(d) at the center frequency, ω_0, the following equations are imposed

$$\begin{aligned} -vC' \cot\left(\frac{\omega_0}{v}\ell\right) + \omega_0\Delta C_j &= \omega_0 C_0 - \frac{1}{\omega_0 L_r} + \omega_0 \Delta C_j \\ vC'\frac{\ell}{v}\sec^2\left(\frac{\omega_0}{v}\ell\right) + \Delta C_j &= C_0 + \frac{1}{\omega_0^2 L_r} + \Delta C_j \end{aligned} \tag{4.8}$$

Define $x_r = (\omega_0/v)\ell$, one obtains the following equation for x_r after eliminating C'

$$\frac{\sin 2x_r}{2x_r} = \frac{\omega_r^2 - \omega_0^2}{\omega_r^2 + \omega_0^2} \tag{4.9}$$

The resonator length, ℓ, can be derived from x_r as

$$\ell = \frac{v}{\omega_0} x_r \tag{4.10}$$

Substituting (4.10) into (4.8), the shunt capacitance per unit length of the $\lambda/4$ resonator is determined as

$$C' = \frac{\omega_0^2 + \omega_r^2}{vL_r\omega_0\omega_r^2}\frac{\sin^2 x_r}{x_r} \tag{4.11}$$

Notice that both ℓ and C'' are the same for all resonators.

Step 10. Comparing the circuit encircled by broken line in Fig.4.8(e) with that in Fig.4.4(d), one realizes that the coupled resonator in Fig.4.4(a)

can be used to implement the present design. Notice that C_j's and $C_{j,j+1}'$s are the same irrespective of j. Finally, the target BPF configuration is shown in Fig.4.8(f). Shunt capacitance of the outermost resonators are given in (4.1) in terms of resonator dimensions. With the resonator spacings determined in step 3, one obtains the resonator width, w, using (4.1).

Substitution of w into (4.4) gives the capacitance between resonators, C_c. Considering that the mutual coupling between adjacent resonators is reduced to a series admittance, $-jv\cot\theta$, the capacitance between resonators, $C_{j,j+1}$, in Fig.4.8(e) can be modified to be

$$C_{j,j+1}' = C_{j,j+1} + \frac{vC_c}{\omega_0}\cot\left(\frac{\omega_0}{v}\ell\right), \qquad 1 \leq j \leq n-1$$

Step 11. In order to have the same shunt capacitance, the inner resonators should be a little wider than the outermost ones. The adjustment can be carried out by solving (4.3) in terms of w, setting C equal to C' derived in step 9.

5. Design Example

The design procedure is demonstrated with an example in this section.

Step 1. A three-stage Wagner BPF with the center frequency of 2 GHz and a bandwidth of 100 MHz is specified.

Step 2. The g factors are $g_0 = 1$, $g_1 = 1$, $g_2 = 2$, $g_3 = 1$ and $g_4 = 1$ for the filter type specified in step 1.

Step 3. Coupled strip lines in a triplate structure is assumed with $w = 1.047$ mm, $h = 0.5$ mm, $s = 0.2$ mm and $\epsilon_r = 10$.

Step 4. The parameters in (4.5) are calculated as $v = v_0/\sqrt{\epsilon_r} = 9.487 \times 10^7$ m/s, $C = 527.1$ pF/m using (4.3). The resonator length is approximately 11.86 mm, assuming that it resonates at $f_0 = 2$ GHz.

Step 5. $b_1 = b_2 = b_3 = \pi f_0 C\ell = 0.03928$, and

$$J_{01} = J_{34} = \sqrt{\frac{0.02 \times 0.03928 \times 0.05}{1 \times 1 \times 1}} = 6.267 \times 10^{-3}$$

$$J_{12} = J_{23} = 0.05 \times \sqrt{\frac{(0.03928)^2}{1 \times 2}} = 1.389 \times 10^{-3}$$

Step 6.

$$C_r = 0.03928/(2\pi \times 2 \times 10^9) = 3.126 \times 10^{-12}$$

$$L_r = \left[3.126 \times 10^{-12} \times (2\pi \times 2 \times 10^9)^2\right]^{-1} = 2.026 \times 10^{-9}$$

$$C_{01} = C_{34} = \frac{6.267 \times 10^{-3}}{2\pi \times 2 \times 10^9 \sqrt{1 - (6.276 \times 10^{-3}/0.02)^2}} = 0.5252 \times 10^{-1}$$

$$C_{12} = C_{23} = 1.389 \times 10^{-3}/(2\pi \times 2 \times 10^9) = 0.1105 \times 10^{-12}$$

$$C_{01}^e = C_{34}^e = \frac{0.5252 \times 10^{-12}}{1 + (2\pi \times 2 \times 10^9 \times 0.5252 \times 10^{-12}/0.02)^2}$$
$$= 0.4736 \times 10^{-12}$$

Step 7. The C_js are calculated using (4.6).

$$C_1 = C_3 = (3.126 - 0.4736 - 0.1105) \times 10^{-12} = 2.542 \times 10^{-12}$$
$$C_2 = (3.126 - 2 \times 0.1105) \times 10^{-12} = 2.905 \times 10^{-12}$$

Step 8. In order to have the same parallel LC resonators, one may choose any value for C_0 in (4.7) as long as it is less than each C_j. Here, C_0 is chosen to be the same as C_1. Thus, we have $C_0 = 2.542 \times 10^{-12}$, $\Delta C_1 = \Delta C_3 = 0$, $\Delta C_2 = 0.3631 \times 10^{-12}$.

Step 9.

$$\omega_r = \frac{1}{\sqrt{2.026 \times 10^{-9} \times 2.542 \times 10^{-12}}} = 1.393 \times 10^{10}$$

Hence, (4.9) is reduced to $\sin(2x)/(2x) = 0.1030$, which gives $x_r = 1.422$. By applying (4.10) and (4.11), we have

$$\ell = \frac{9.487 \times 10^7}{2\pi \times 2 \times 10^9} \times 1.422 = 1.0735 \times 10^{-2}$$

$$C' = \frac{(1.5791 + 1.9417) \times 10^{20}}{9.487 \times 10^7 \times 2.026 \times 10^{-9} \times 2\pi \times 2 \times 10^9 \times (1.394 \times 10^{10})^2}$$
$$\times \frac{0.9780}{1.422} = 515.9 \times 10^{-12}$$

Step 10. Use (4.1) to give

$$\frac{K(k_e)}{K(k_e')} = \frac{515.9 \times 10^{-12}}{4 \times 10 \times 8.854 \times 10^{-12}} = 1.457$$

Thus, we have $k_e = 0.9213$. With $s = 0.2$ mm and $h = 0.5$ mm, one obtains $w = 1.153 \times 10^{-3}$, which is the strip width of the outermost resonators. Next, the argument in (4.4) is calculated as

$$k_o = \tanh\left(\frac{\pi \times 1.153}{4 \times 0.5}\right) \coth\left(\frac{\pi \times 1.353}{4 \times 0.5}\right) = 0.9753$$

and the mutual capacitance, C_c, is obtained as

$$C_c = 2 \times 10 \times 8.854 \times 10^{-12} \times 1.836 - \frac{1}{2} \times 515.9 \times 10^{-12} = 67.1 \times 10^{-12}$$

Then, the mutual capacitance between resonators becomes

$$C'_{12} = C'_{23} = 0.1105 \times 10^{-12} + \frac{9.487 \times 10^7 \times 67.1 \times 10^{-12}}{2\pi \times 2 \times 10^9}$$
$$\cot\left(\frac{2\pi \times 2 \times 10^9 \times 1.074 \times 10^{-2}}{9.487 \times 10^7}\right) = 0.1861 \times 10^{-12}$$

Step 11. Width of the inner resonators is determined by solving (4.3). Substitution of numerical values gives a transcendental equation which is solved to have $w = 1.289 \times 10^{-3}$.

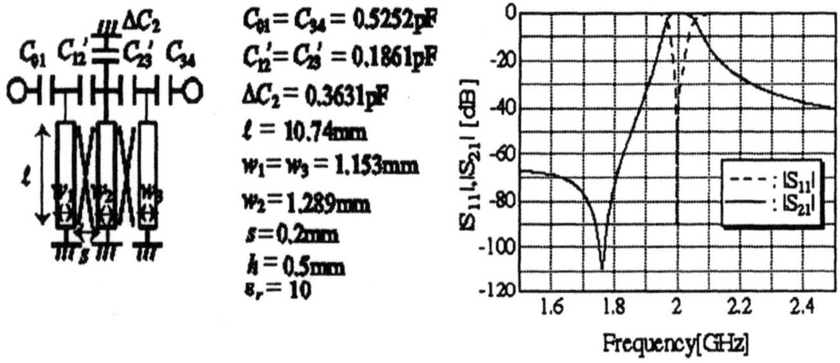

Figure 4.9. Designed three-stage Wagner combline BPF in strip line structure, $f_0 = 2$ GHz, $B_f = 0.05$.

The simulated frequency response in Fig.4.9 shows that the bandwidth is a little narrower than the specified value of 100 MHz, which is due to an unexpected attenuation pole at 1.76 GHz. The attenuation pole appears at the frequency where the effective bridging susceptance between resonators becomes zero, namely,

$$\omega C'_{j,j+1} + \frac{Y_o - Y_e}{2j} = \omega C'_{j,j+1} - vC_c \cot\left(\frac{\omega}{v}\ell\right) = 0 \tag{4.12}$$

which is the same condition as in the two-sage combline BPF [11]. Control of attenuation poles will be briefly reviewed in the next Section based on (4.12).

6. Control of Attenuation Poles

6.1 Attenuation Poles below Pass-Band

Eq.(4.12) for the attenuation poles can be rewritten as

$$\frac{C_{j,j+1}}{vC_c} + \frac{1}{\omega_0}\cot\left(\frac{\omega_0}{v}\ell\right) = \frac{1}{\omega_a}\cot\left(\frac{\omega_a}{v}\ell\right) \tag{4.13}$$

where ω_a denotes the attenuation pole frequency to be determined. The relation among ω_a, ω_0 and ω_r depicted in Fig.4.10 can be used to facilitate the solution.

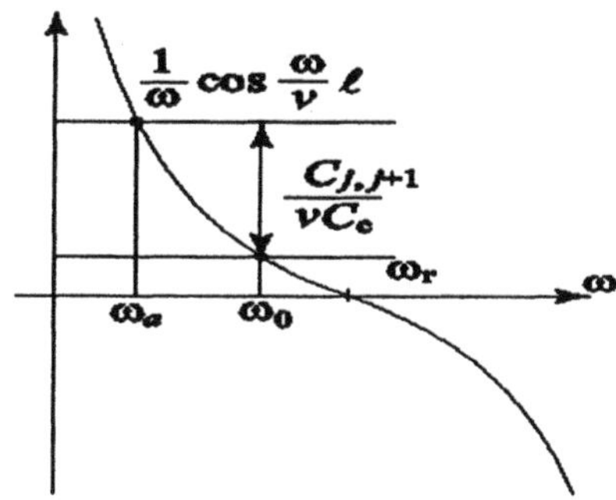

Figure 4.10. Relation among attenuation pole frequency, ω_a, ω_0 and ω_r.

The difference $\omega_0 - \omega_a$ becomes smaller if the mutual capacitance per unit length, C_c, becomes larger. In other words, the attenuation pole moves closer to the pass-band with narrower spacing between resonators, without affecting the center frequency and bandwidth of the BPF. Fig.4.11 illustrates an example in which the attenuation pole frequency is closer to the pass-band compared with that in Fig.4.9 by reducing the resonator spacing from 0.2 mm to 0.1 mm. As a result, the bandwidth is decreased to 80 MHz from the specified value of 100 MHz. The lower band edge has been eroded by the attenuation pole.

6.2 Attenuation Poles above Pass-Band

It is elucidated that attenuation poles can be placed above the pass-band by choosing a proper combination of loading and connecting capacitors to each resonator for a two-stage combline BPF [11]. An inductive coupling between resonators is found useful to achieve this goal. In the case of multistage BPF, (4.13) gives a clue to the solution. Inductive coupling can be realized by using the inverter shown in Fig.4.12(a) in-

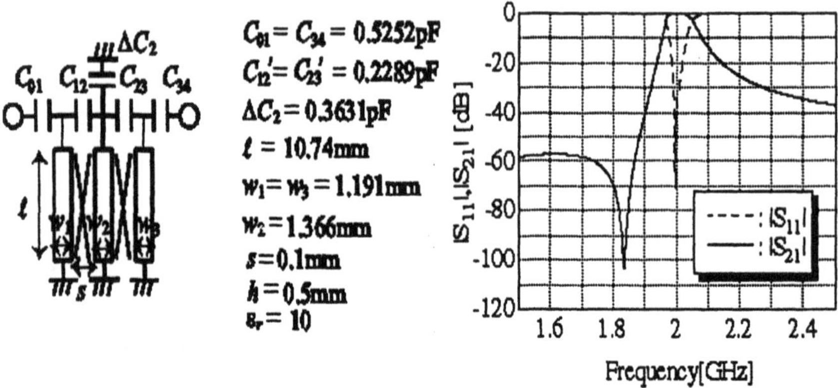

Figure 4.11. Design example with an attenuation pole closer to the pass-band.

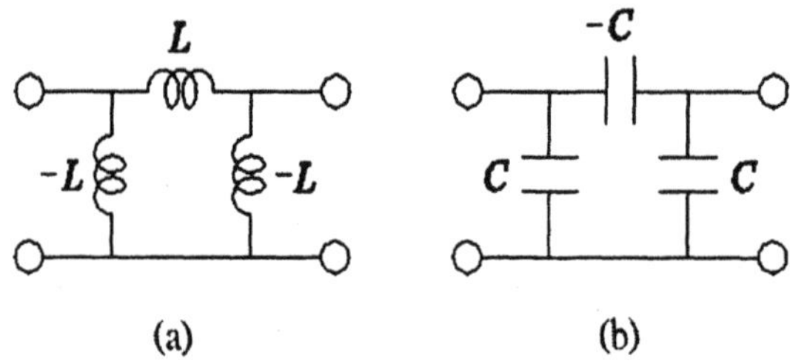

Figure 4.12. Inverters to be used in BPF with an attenuation pole above the pass-band.

stead of that used in Fig.4.8(c). However, an inductor is more difficult to implement than a capacitor in LTCC configuration. Hence, one can adopt the inverter in Fig.4.12(b), which plays a similar role as that in Fig.4.12(a).

Fig.4.8(c) is converted successively into Fig.4.13(a) with the element values of $L_1 = L_2 = \cdots = L_r$, $C_1 = C_r + C_{12} - C_{01}^e$, $C_j = C_r + C_{j-1,j} + C_{j,j+1}$ and $C_n = C_r + C_{n-1,n} - C_{n,n+1}^e$. Next, extract C_0 from each C_j to leave a residual, ΔC_j, as shown in Fig.4.13(c). Each parallel LC resonator in Fig.4.13(c) is transformed to a λ /4 resonator as shown in Fig.4.13(d) with the same procedure as in Fig.4.8. The final transformation to the distributed coupling resonator in Fig.4.13(e) concludes the

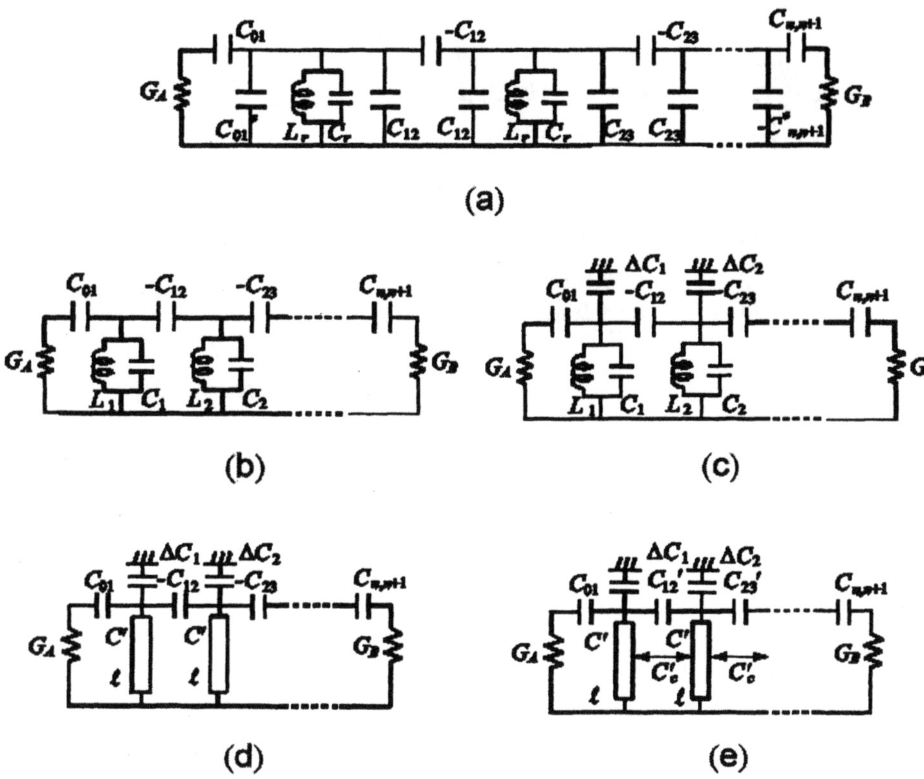

Figure 4.13. **Design of combline BPF with attenuation poles above the pass-band.**

design. The coupling capacitances are

$$C'_{j,j+1} = C_{j,j+1} + \frac{vC_c}{\omega_0} \cot\left(\frac{\omega_0}{v}\ell\right) \tag{4.14}$$

Observing the lumped elements with negative value in Fig.4.13(d), one should choose ΔC_j large enough to realize positive $C'_{j,j+1}$, positive and implementable coupling capacitance. Substituting (4.14) into (4.12), the attenuation pole is determined by solving

$$-\frac{C_{j,j+1}}{vC_c} + \frac{1}{\omega_0} \cot\left(\frac{\omega_0}{v}\ell\right) = \frac{1}{\omega_a} \cot\left(\frac{\omega_a}{v}\ell\right)$$

Compared with (4.13), one can easily verify that ω_a in this case is larger than ω_0.

Fig.4.14 shows an example which has double poles above the pass-band. The inductive coupling between resonators gives double poles

above the pass-band as in a two-stage combline BPF [11]. Adjustment of attenuation pole frequency is possible following the same procedure as before. By decreasing resonator spacing, one can move the pole frequency closer to the pass-band.

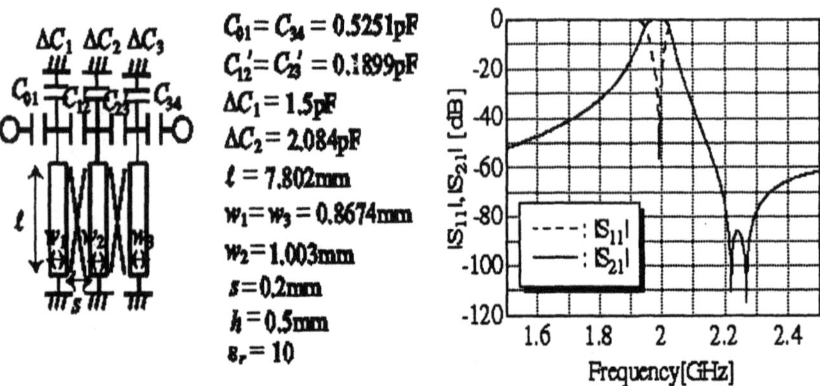

Figure 4.14. Design example with an attenuation pole above the pass-band.

7. Conclusions

We have demonstrated a design procedure for multistage combline BPFs in strip lines. Although a similar design approach was already proposed many years ago, it was not popularly used. This procedure is well fitted to coaxial or laminated ceramic BPFs widely used for mobile communication systems, the attenuation poles can also be controlled.

References

[1] G. Matthaei, "Comb-line band-pass filter of narrow or moderate bandwidth," *Microwave J.*, vol.6, no.8, pp.82-96, 1963.

[2] W. J. Getsinger, "Coupled rectangular bars between parallel plates," *IRE Trans. Microwave Theory Tech.*, vol.10, no.1, pp.65-72, 1962.

[3] E. G. Cristal, "Coupled circular cylindrical rods between parallel ground planes," *IEEE Trans. Microwave Theory Tech.*, vol.12, no.7, pp.428-439, 1964.

[4] P. Vadopalas, "Coupled rods between ground planes," *IEEE Trans. Microwave Theory Tech.*, vol.13, no.3, pp.254-255, 1965.

[5] T. A. Milligan, "Dimensions of microstrip coupled lines and interdigital structures," *IEEE Trans. Microwave Theory Tech.*, vol.25, no.5, pp.405-410, 1977.

[6] M. Dishal, "A simple design procedure for small percentage bandwidth round-rod interdigital filters," *IEEE Trans. Microwave Theory Tech.*, vol.13, no.9, pp.696-698, 1965.

[7] E. G. Cristal, "Tapped-line coupled transmission lines with applications to interdigital and combline filters," *IEEE Trans. Microwave Theory Tech.*, vol.23, no.12, pp.1007-1012, 1975.

[8] S. Caspi and J. Adelman, "Design of combline and interdigital filters with tapped-line input," *IEEE Trans. Microwave Theory Tech.*, vol.36, no.4, pp.759-763, 1988.

[9] A. I. Grayzel, "A useful identity for analysis of a class of coupled transmission-line structures," *IEEE Trans. Microwave Theory Tech.*, vol.22, no.10, pp.904-906, 1974.

[10] R. E. Collin, *Foundations for Microwave Engineering*, McGraw-Hill, p.173, 1992.

[11] I. Awai, "Recent advance in microwave filter circuit design technique: Control of attenuation poles in two-stage combline BPF," *Asia-Pacific Microwave Conf.*, pp.485-491, Taipei, Taiwan, Dec. 2001.

Chapter 5

MILLIMETER WAVE FILTERS FOR LOW-COST MASS FABRICATION

Ruediger Vahldieck and Erdem Ofli
Swiss Federal Institute of Technology, IFH
ETH Zentrum, Gloriastrasse 35, CH - 8092 Zurich

Abstract Traditional market for millimeter wave filters has been almost exclusively in the military arena for many years. Major requirement for this market was performance, followed by the potential for medium-volume production and last but not least, the price. With the advent of new broadband communication systems for private consumer market, satellite communication has moved up into the Ka-band, and broadband terrestrial local area networks are being established at frequencies as high as 60 GHz as the last-mile solutions to provide a wireless interface for the vast bandwidth available on the fiber optic backbone. The emerging consumer market is drawing heavily on technology that has initially been developed for military applications. Today, however, the key to success is the price associated with the potential for medium to large-volume production, trading off performance along the way. This Chapter will give an overview of millimeter wave filters and diplexers suitable for mass fabrication. As most millimeter wave components are based on waveguide technology, the emphasis will be on structures easily realized in waveguide or implemented in waveguide housings such as E-plane filters, quasi-planar filters and suspended substrate filters. The use of plastic injection molding technology for large-volume production will also be discussed with respect to tolerances and the potential to optimize the price-performance ratio.

Keywords: low cost, mass fabrication, tolerance analysis, E-plane, quasi-planar, waveguide, cavity filter, milling, molding.

1. Introduction

Consumer-focused wireless broadband communication will take place mainly in the higher frequency range from about 17 to 65 GHz. In

particular, the millimeter wave range from 30 GHz and up will play a major role in satellite communication and local area networks, with the 60 GHz range being the highest frequency allocated so far for short-haul broadband links. Among the spectrum available to the consumer market today, this frequency range offers the widest bandwidth to accommodate the high-capacity wireless interfaces for terrestrial fiber optic distribution network.

For the broadband communication systems to reach consumers soon, low-cost customer premises equipment(CPE) must become available. One of the major cost drivers for CPE's is the millimeter wave(MMW) front-end and therein to a large extent, the antenna and filters/diplexers. In addition, integration techniques play a pivotal role in low-cost manufacturing, and must be considered in the early stages of front-end design. Manufacturing tolerances are critical as dimensions of the millimeter wave circuits reach the limits of what is possible with low-cost fabrication techniques. This is in particular a problem in filter design where the acceptable tolerances of filter dimensions can become so small that the resulting frequency shift or deterioration of insertion and return loss characteristics cannot be tolerated. In these cases, fine tuning of filters becomes necessary which is prohibitive for low-cost manufacturing. Since filters and diplexers are crucial for the overall system performance, and because manufacturing requires a high degree of accuracy to avoid fine tuning, they are in many cases the most expensive single component of a front-end.

To lower their cost in a large-volume production, only certain filter structures are suitable to accommodate the tolerance margins dictated by low-cost manufacturing. Furthermore, in almost all cases, the filter bandwidth must be sufficiently wider than the required signal bandwidth to accommodate the frequency shift introduced by manufacturing tolerances. Needless to say, the more insensitive a filter structure becomes with respect to manufacturing tolerances, the closer the signal bandwidth and filter bandwidth can coincide and thus allow for more efficient spectrum use.

Low insertion loss, high return loss, high slope selectivity, often also harmonic suppression and making all these possible at low cost are concurrent requirements on the filters in millimeter wave mass market. Among the wide variety of possible filter structures potentially satisfying some or all of the above criteria, only few are really suitable for low-cost mass fabrication. For most broadband radio applications in the upper microwave and millimeter wave bands, the waveguide type of cavity filters are the preferred choice. These filters are compatible

with waveguide technology, but can also be configured as drop-in filters if appropriate transmission line/waveguide transitions are incorporated.

Besides the direct-coupled cavities in single resonance, dual-mode cavities are also possible under certain circumstances. Traditionally, waveguide cavity filters are fabricated by milling techniques which, in a mass-fabrication process, can at best deliver $\pm 15\mu m$ of accuracy. Better overall accuracy is possible by using metal or plastic injection molding techniques(and subsequent metal plating on plastic surface), since the molding tool can be optimized with respect to the strain characteristics of the plastic material used and specific geometrical details of the filter, which are particularly sensitive to tolerances.

An alternative to pure waveguide structures are filters based on quasi-planar technology, in which the actual filter performance is mainly determined by an insert in the middle of a rectangular waveguide. The insert is placed either in parallel to the electric field of the fundamental TE_{10} mode or perpendicular to it. The inserts can be made by a specially structured metallic sheet of sufficient thickness to ensure mechanical stability, or a thin dielectric substrate of low permittivity, which supports a very thin metallic layer. The advantage of the latter is that a thin metal layer can be etched with better accuracy than a thick metal sheet. In both cases, a split-block waveguide housing must be milled to accommodate the insert. Since the waveguide dimensions affect the filter performance(mainly the center frequency), the milling tolerances still cause a problem, although not as serious as in a pure waveguide filter.

One way to reduce the effect of waveguide housing on the filter characteristics is to use either ridge waveguide structures or to increase the thickness of dielectric slab and use a material of high permittivity to support the metallic pattern. In both cases, the electric field is drawn into the center region of the waveguide, away from the sidewalls, thus reducing the effects of milling tolerances. The drawback of ridge waveguide is that the length of ridge and the gap between them are simultaneously subject to etching tolerances. For a thick, high-permittivity substrate, tolerances of the dielectric constant will affect the filter performance as well. Another advantage of using a dielectric substrate to support the filter structure is that efficient integration of RF/millimeter wave front-ends and filter banks become possible by putting both circuits on the same substrate.

The waveguide filters are known for their high unloaded Q factor(Q_u) which ranges from 5,000 to 10,000 or higher. This allows the design of narrowband filters with low insertion loss and high slope selectivity. However, this feature normally comes at the expense of increased

sensitivity to manufacturing tolerances. To alleviate this problem to certain extent, E-plane structures can be used. In these filters, the actual filter performance is mainly determined by a ladder-shaped insert, which is fabricated using etching or electrodeposition techniques. The latter technique provides a much higher accuracy than is achievable with the former. Furthermore, since waveguide-based filters are highly sensitive to the nonlinear behavior of guided wavelength with respect to frequency, it is always advisable to choose the waveguide dimensions such that the pass-band of filter is located far away from the waveguide cutoff frequency.

In contrast to waveguide filters, suspended substrate filters operate on the basis of TEM mode. The waveguide housing is typically designed to operate below its cutoff frequency and is used only as an electromagnetic shield to the substrate carrier. The advantage is that the housing tolerances have less important or even negligible effects on filter performance. The filter response is determined by the metallic pattern on the supporting substrate. The unloaded Q factor, however, is typically limited to below 1,000 to 1,500, which may not be high enough for many applications.

In the following Sections, various filter structures including waveguide cavity type, E-plane, quasi-planar and suspended substrate filters will be discussed as to performance and suitability for mass fabrication.

2. Cavity Filters

2.1 Iris-Coupled Filters

Among the wide variety of cavity filters, only few are suitable for mass fabrication. Of particular interest are iris-coupled cavity filters where the iris extends over the full height of waveguide. Fig.5.1 shows two examples of this filter type: direct-coupled and cross-coupled iris filters [1]. These structures are simple but effective since the number of geometrical parameters that determine the filter performance is reduced to just three: the iris width, cavity width and separation between irises. From the design point of view, opening of iris over the full waveguide height allows not only fast and accurate simulation/optimization of the filter transfer function based on TE_{mo} mode expansion [2], but is also more appropriate for fabrication by milling techniques.

To keep the assembly as simple as possible, typical layout of an iris-coupled filter consists of the filter body and a top plate to seal the structure. As shown in Fig.5.2, a tolerance analysis on this type of filter shows that the milling tolerances of $\pm 30\mu$m at 33GHz, which is about 0.5% of the waveguide width at Ka-band, are acceptable as long as the

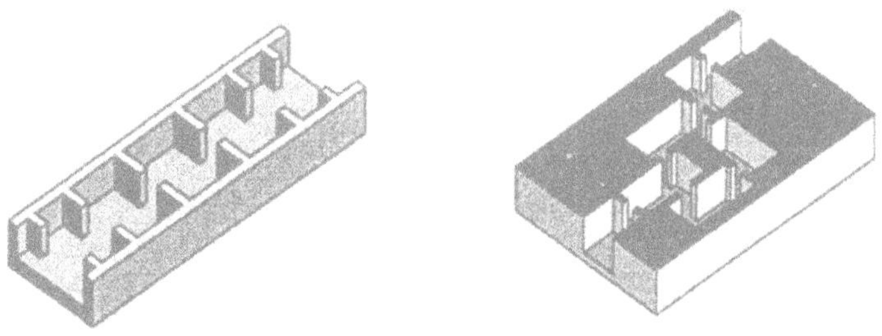

Figure 5.1. (a) A five-pole direct-coupled iris filter, (b) a five-pole cross-coupled iris filter.

relative bandwidth of pass-band does not exceed 5%. For narrow band applications, the milling tolerances must be reduced to below 0.1% or about $\pm 5\mu m$ at 33 GHz in order to avoid fine tuning on the production floor, which is prohibitive for low-cost manufacturing.

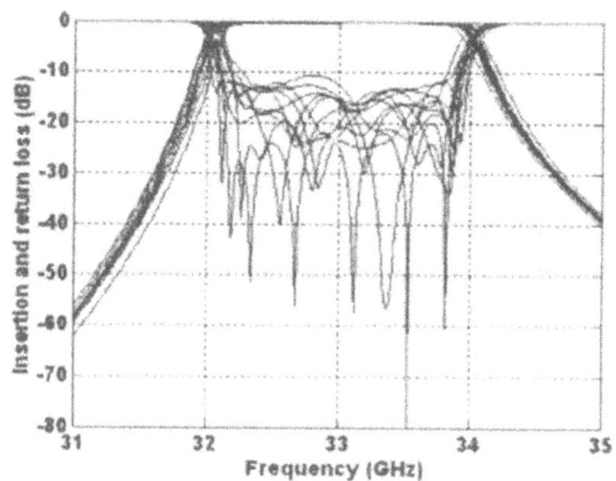

Figure 5.2. Tolerance analysis($\pm 30\mu m$) on an iris filter.

Problem also occurs at corners of a cavity which, for ease of manufacturing, can never be a right angle due to the finite radius of milling tool. As shown in Fig.5.3, the effect of rounded corners can be significant and must be accounted for in the design.

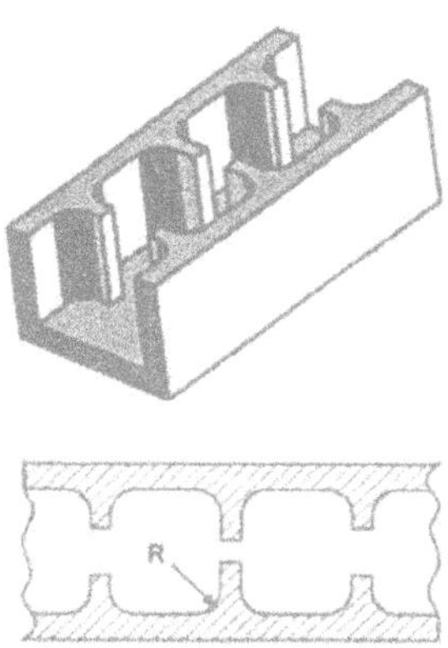

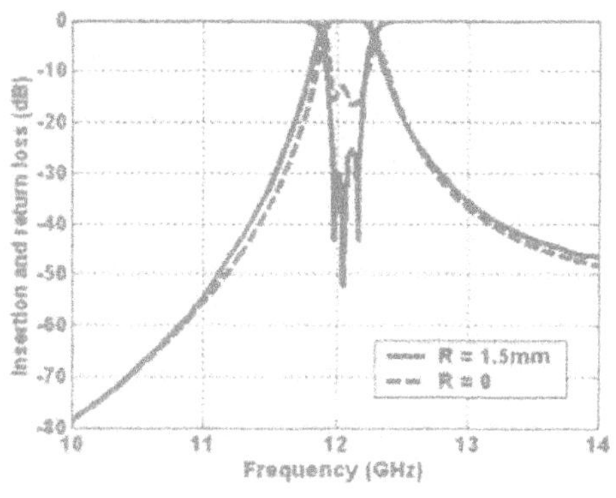

Figure 5.3. (a) Iris filter with rounded corners and (b) effect of finite radius on a three-cavity H-plane filter.

When it comes to quantities of more than 10,000 pieces, a low-cost alternative to milling is to use either metal powder or plastic injection molding techniques. The latter must be combined with metal plating or gold plating after molding, to passivate the surface against oxidization. Besides the price, injection molding techniques have several advantages over milling techniques. Once the molding tool is developed, the fabrication process is much faster and does not require fixing of each metal block which is necessary with milling. Fine tuning elements can be implemented within the molding tool to compensate for dimensional changes due to material shrinkage during cooling as well as mechanical tolerances of the molding tool itself. In this way, the effective tolerances of molding techniques can be comparable to those of high-quality milling techniques. An example of successful realization of an iris-coupled diplexer using plastic injection molding technique is shown in Fig.5.4 [3].

By replacing the inductive irises with short-circuited T-junctions as inverter circuits, stop-band poles in the frequency response are created and can be placed at either side of the filter pass-band [4]. For a given out-of-band attenuation, especially close to the filter pass-band, this measure requires less number of physical resonators, and therefore leads to a lower insertion loss compared to standard filter designs. The diplexer in Fig.5.4 shows an insertion loss of less than 1 dB and a rejection of better than 60 dB in the pass-bands as shown in Fig.5.5.

Figure 5.4. Picture of a diplexer using plastic injection molding technique.

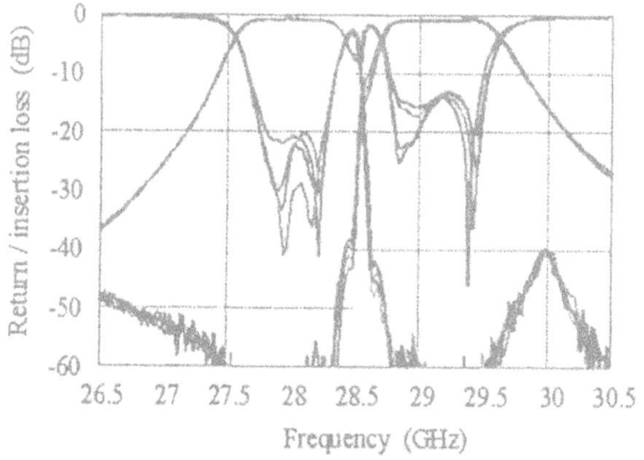

Figure 5.5. Frequency response of four diplexer samples using plastic injection molding technique.

2.2 Plastic Injection Molding of Filters

For low-cost, high-volume fabrication, plastic injection molding combined with electroplating is an attractive approach to fabricate waveguide circuits. As shown in the example of Fig.5.4, a fully metallized printed circuit board is soldered to the top of filters after electroplating to enhance their mechanical stability. To cope with the challenges of this production method, plausible modifications must be considered during circuit design. To avoid tensions within material as much as possible, the wall thickness must be kept constant throughout the design. For example, the iris and walls between waveguide sections should be assigned the same thickness. In addition, to enhance the stability, the basic filter

structure should be duplicated symmetrically below the bottom plate of the filter.

Other problems associated with tolerances may stem from the production process itself. Shrinkage of the injection molded plastic parts during cooling can be precalculated to some extent, although it is not accurate enough for the fabrication of filters and some initial trial-and-error runs are unavoidable. Fig.5.6 shows that in designing the diplexer filters in Fig.5.4, cylindrical posts are included in the center of cavities as tuning elements. These posts function like screws with adjustable depth through the holes. Samples of filters are fabricated with the post height being zero, then external screws are used to trim the filter response. Resulting heights of the posts are then transferred to the design. Once the filters produced with this modified form possess acceptable performance, mass production can be launched.

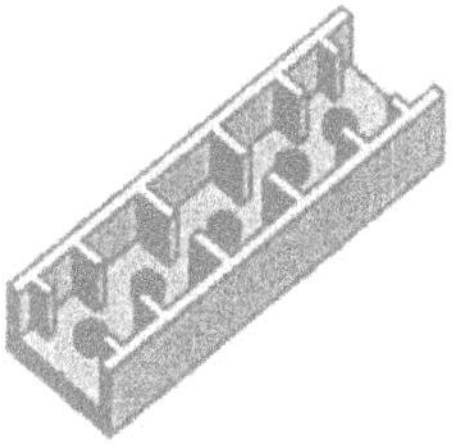

Figure 5.6. **Three-dimensional schematic of an iris-coupled filter with tuning posts included in the injection molding fabrication process.**

Proven material series for metallized plastic waveguides and cavities include polyetherimide(PEI) like Ultem(GE Plastics), liquid crystal polymers(LCP) like Vectra(Ticona) and polyphenylenesulfone(PPS). In order to be reliably metallized, these materials are compounded with certain amount of mineral filler particles that form microscopic, open, recessed caverns after aggressive surface etching. Solder joints, use of conductive epoxy and mechanical mounting methods like screwing are possible, but specific thermoplast compatible methods like press fitting, clamping and nonconductive epoxy joints are preferred [3]. The diplexer shown in Fig.5.4 is fabricated with LCP which is known for its anisotropic feature. However, no excessive temperature dependence is observed as shown in Fig.5.7.

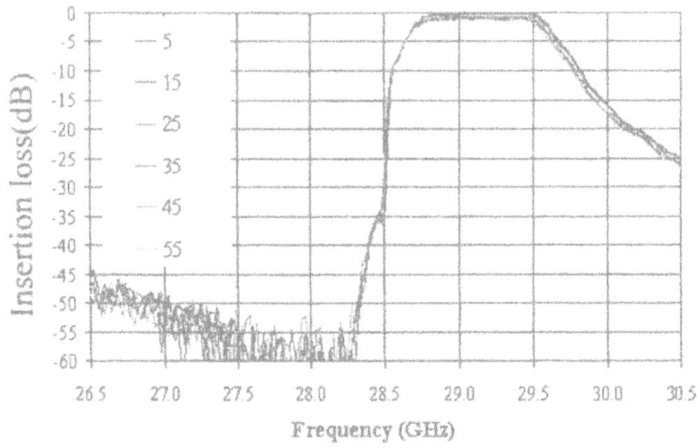

Figure 5.7. Temperature characteristics of the diplexer in Fig.5.4.

2.3 Cross-Coupled and Dual-Mode Cavity Filters

From the assembly point of view, cross-coupled iris filters are also well suitable for mass fabrication either by milling or by molding techniques if they are built as shown in Fig.5.8 [5], [6]. The same fine tuning techniques as used for the plastic diplexer shown in Fig.5.4 can be used to prepare the structure for mass fabrication. The sensitivity of cross coupling to manufacturing tolerances is on the same order as that for the other filter dimensions.

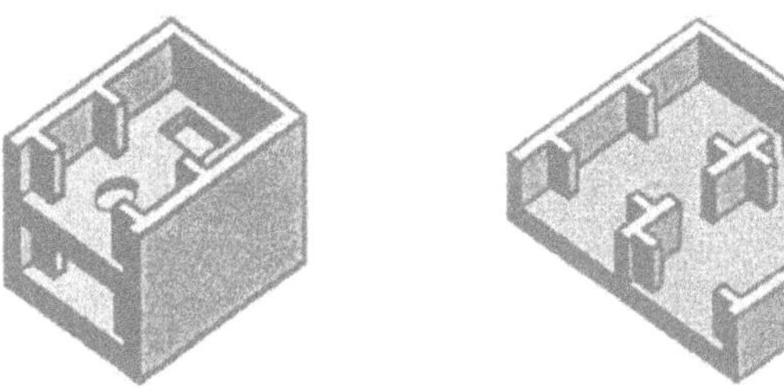

Figure 5.8. Cross-coupled iris filters.

Dual-mode filters are made possible by enlarging the cavity of one resonator such that two modes can be supported. By arranging the excitation asymmetrically, coupling takes place between the TE_{102} and

TE_{201} modes. As shown in Fig.5.9, a six-pole filter with three transmission zeros has been produced at 39 GHz [7]. The input and output cavities will each generate two transmission poles and two transmission zeros where the latter are placed at the same frequency. As shown in Fig.5.10, tolerance analysis on a three-pole filter with one transmission zero verifies that this configuration is an excellent candidate for low-cost millimeter wave filters.

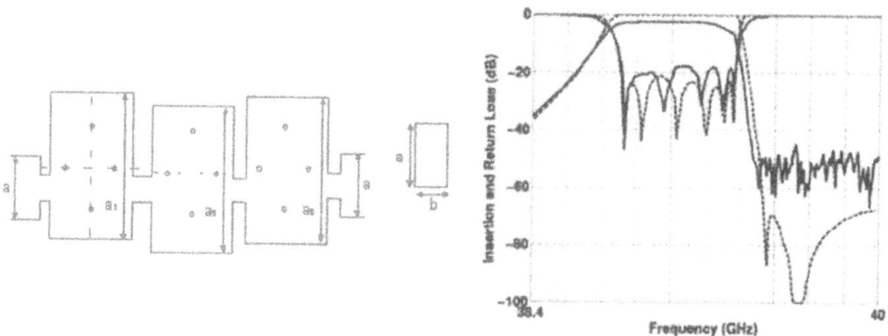

Figure 5.9. Iris-coupled dual-mode filter suitable for mass fabrication.

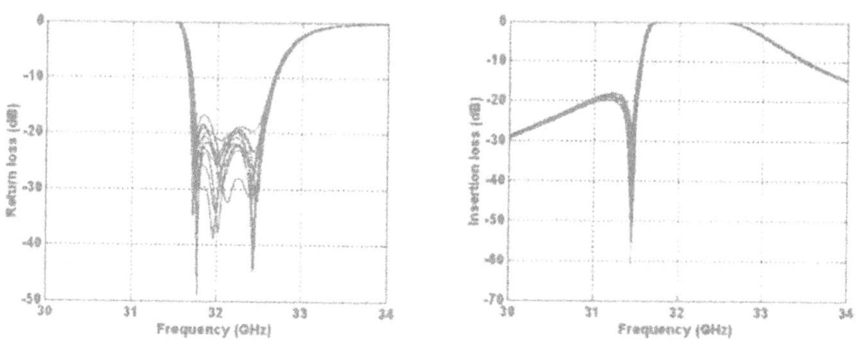

Figure 5.10. Tolerance analysis ($\pm 20\mu m$) on the filter in Fig.5.9.

3. *E*-Plane Metal Insert Filters

As shown in Fig.5.11, *E*-plane filters have been suggested in the past for mass fabrication. Performance of these filters is essentially determined by the metallization pattern of insert. The insert thickness ranges

typically from 50 to 100 μm and can be fabricated by photolithographic or electroforming techniques, where the latter has high accuracy and repeatability.

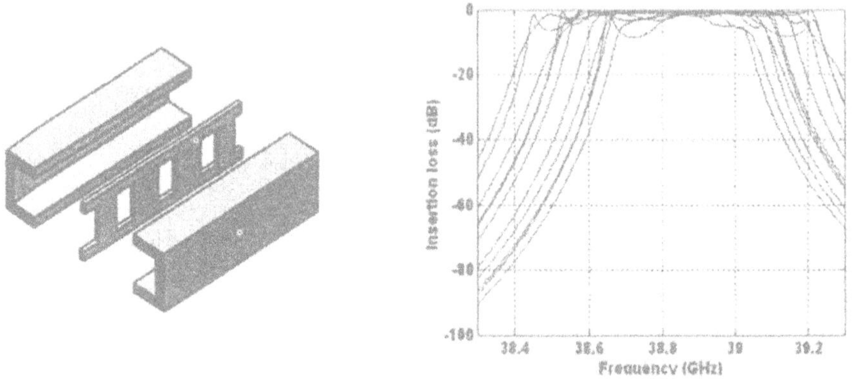

Figure 5.11. **(a) Single metal insert E-plane filter and (b) sensitivity analysis with $\pm 30\mu$m tolerances of the waveguide width and insert dimensions.**

However, several problems exist in this type of filter. For high-slope selectivity, in particular for diplexer application with sharp cutoff between channels to avoid crosstalk, direct-coupled E-plane filters must consist of many resonators to meet the specifications. This in turn increases the insertion loss, makes the filters quite long and increases the sensitivity with respect to manufacturing tolerances of the waveguide housing as well as the metal insert. An example of a 38 GHz E-plane diplexer with quite demanding requirements is shown in Fig.5.12. Both eight-resonator filters are fabricated on the same metal insert made from a 50 μm thick silver-plated nickel sheet using electrodeposition process to achieve low tolerances of less than 3 μm. Reduced tolerances of the inserts make it possible to meet the specifications of this filter, which include a return loss of higher than 15 dB and an insertion loss of less than 2 dB with a signal bandwidth of 300 MHz within a 500 MHz window. Fabricating the insert with normal etching techniques would have created tolerances of $\pm$ 20 μm, which can not meet the electrical requirements.

Other E-plane filter structures are possible to reduce the number of resonators. One solution is to add inductively coupled stop-band sections in front and behind the filter as shown in Fig.5.13(a) [8]. This increases the slope selectivity significantly at the expense of more complicated waveguide housing design. Another solution is to use folded E-plane

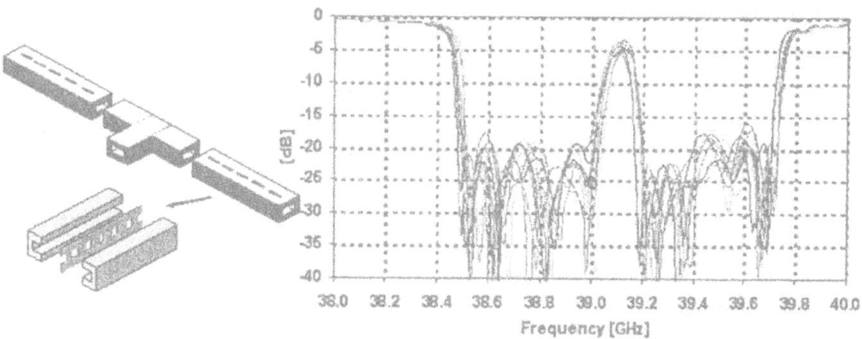

Figure 5.12. **E-plane metal insert diplexer in the 38 GHz range and frequency response of ten samples with less than $\pm10\mu$m tolerances in waveguide width and less than 3 μm in insert dimensions.**

metal insert filters as shown in Fig.5.13(b) which introduces cross couplings between resonators. This solution leads to a very compact filter. Transfer characteristics of these filters show a much steeper slope selectivity than the direct-coupled E-plane filters. Thus, filter specifications can be met with a lower number of resonators and hence lower insertion loss.

For low-cost mass fabrication, the separation wall between waveguides is made from the same thin metal sheet as the inserts so that the cross-coupling between resonators can also be fabricated with high precision using electrodeposition techniques [9]. Positioning and opening of the cross couplings are calculated using mode matching technique, following a similar procedure as outlined in the design of E-plane filters [10]. Fig.5.14 sketches typical building blocks of a cross-coupled filter section. The cross coupling between resonators consist of an opening in the waveguide wall extending over the full height of waveguide. Thus, only the TE_{m0} modes need to be considered. The bifurcations left and right to region 5 are described by generalized three-port S-matrices connecting the generalized S-matrices of the E-plane filters in different waveguide channels. The overall filter design is then optimized to meet the specifications.

4. Quasi-Planar Filters

Another E-plane filter structure having improved slope selectivity is shown in Fig.5.15 for the case of a fourth-order filter with one transmission zero. Instead of folding the structure as in Fig.5.13(b), the two filter halves are offset in lateral direction by one waveguide width. The

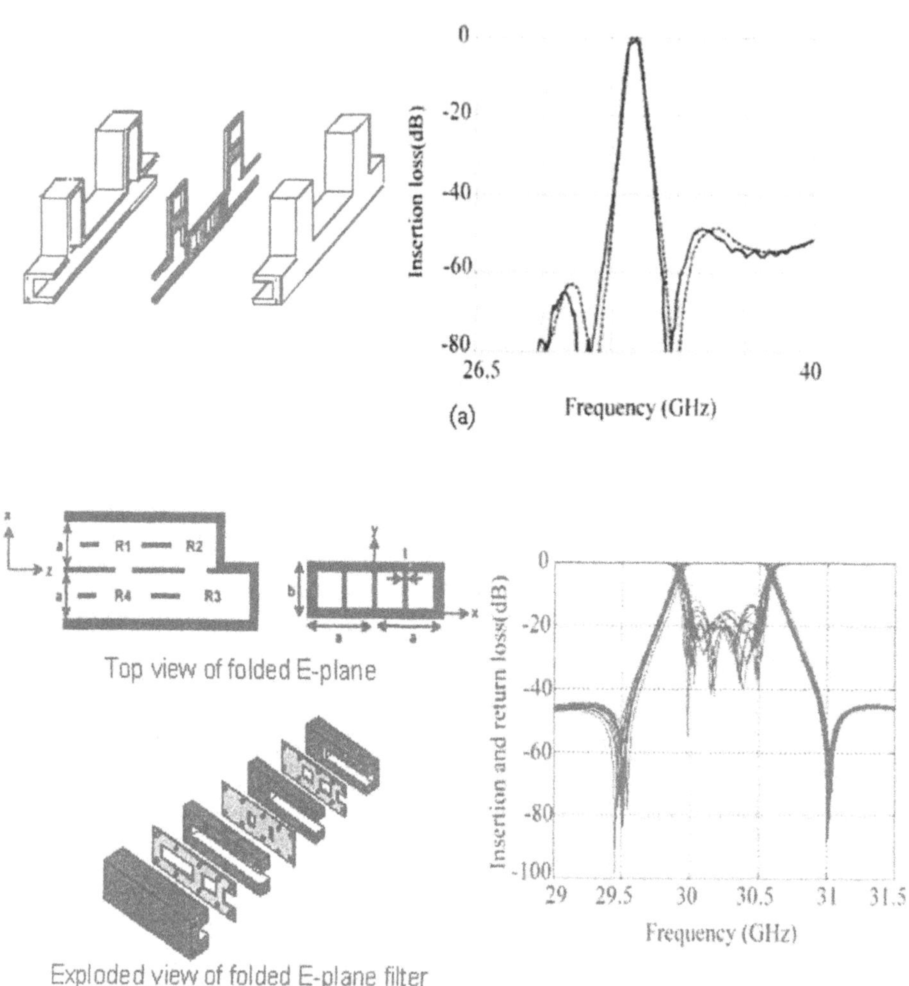

Figure 5.13. **Two different solutions to improve the skirt selectivity of *E*-plane filters. (a) Utilizing inductively coupled stop-band sections or (b) utilizing cross-coupled resonators(tolerance of $\pm$ 10 μm).**

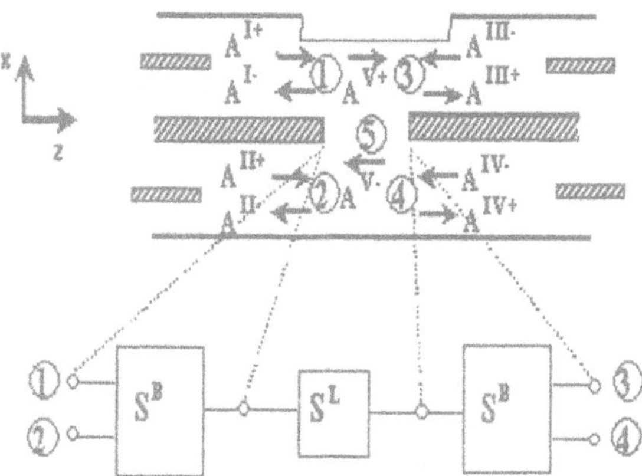

Figure 5.14. Typical building blocks: waveguide bifurcations 1-2 and 3-4 for cross-coupling, and homogeneous waveguide section 5.

structure shown in Fig.5.15 has four direct-coupled cavities whereby the second resonator is coupled to the third through a lateral wall opening which again equals the waveguide height. The third cavity is longer than all the other cavities and acts as a strongly dispersive stub to introduce one transmission zero. Tolerance analysis on the filter indicates that the tolerance of waveguide width and insert dimensions must stay well below $\pm$ 10 μm to achieve repeatable results over a large number of samples.

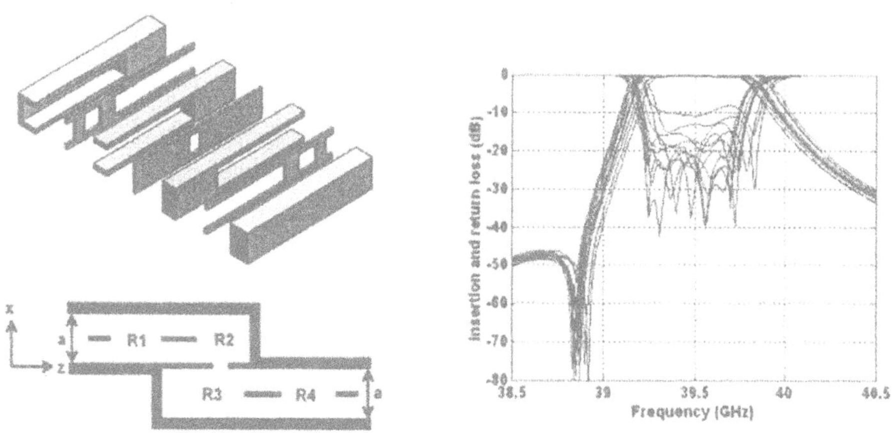

Figure 5.15. Tolerance analysis ($\pm 10\mu$m) on a direct-coupled E-plane metal insert filter structure with a strongly dispersive stub section and four resonators.

The term quasi-planar implies that all the metal structures are printed on a low-loss substrate and embedded in a split-block waveguide housing. Since a thick metal sheet is difficult to etch with high accuracy, a thin metal sheet is used instead. The function of substrate is then reduced to that of supporting structure only. There are different classes of quasi-planar filters. In some cases, the planar structure is positioned parallel to the electric field of the TE_{10} waveguide mode, and in some cases perpendicular to it. In the former, the planar structure affects the fundamental mode of the waveguide in such a way that the required filtering effect is achieved. In the latter, the waveguide housing provides a shielding ground only. The filter function is solely determined by the printed circuit suspended in the waveguide, whose dimensions must be chosen such that the operating frequency of filter is always below the cutoff frequency of the fundamental waveguide mode. These filters are based on TEM mode operation. In contrast, the E-plane structures are based on TE_{10} mode operation and show strongly dispersive behavior.

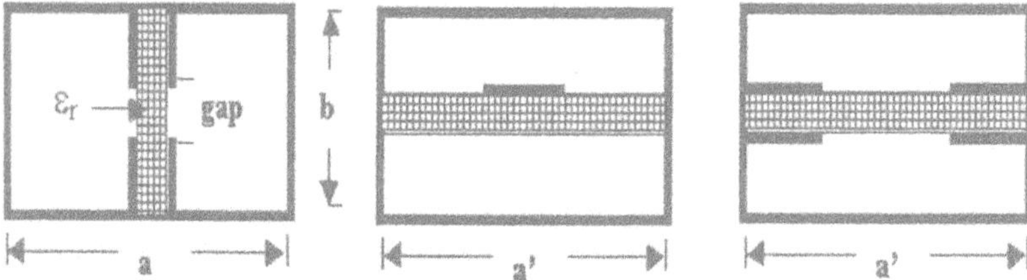

Figure 5.16. **Cross sections of quasi-planar filters: (a) bilateral, (b) uniplanar, (c) antipodal.**

As shown in Fig.5.16, fin line filters made from bilateral, uniplanar or antipodal metallization are most frequently used. Due to their high unloaded Q factor($\simeq$1,500) and the potential for integration with active devices, they are frequently used as band-pass filters in integrated front-ends. Fig.5.17 shows an example of a bilateral, large-gap fin line filter fabricated on the same substrate with the other circuit elements of a receiver front-end.

An overview of quasi-planar filters can be found in [11]. A third class of quasi-planar filters is neither based on TEM mode nor on TE_{10} mode operation, but utilizes $HE_{01\delta}$ mode. An interesting design of highly compact filters which is suitable for mass fabrication is given in [12]. Here, the TE_{010} mode is excited in a circular hollow patch which is formed on a high-permittivity substrate with bilateral metallization as shown in Fig.5.18. The hollow patch is fabricated by etching techniques.

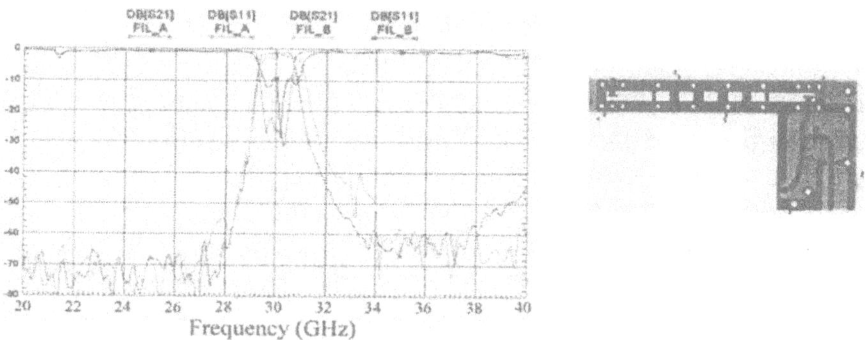

Figure 5.17. **Example of a large-gap fin line filter at 30 GHz integrated into a receiver front-end and measurements on two different samples.**

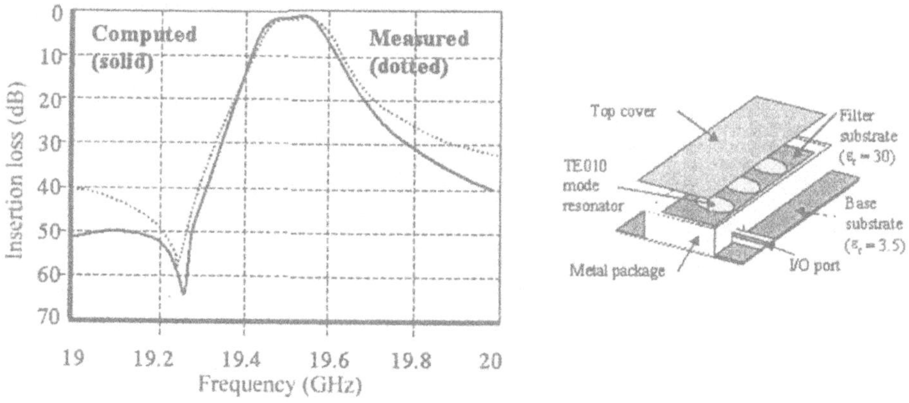

Figure 5.18. **Example of TE_{010} mode filter [2].**

The electromagnetic field is highly confined within the substrate area not covered by metallization, thus very small filter components can be made. Also, the unloaded Q factor has been reported in the range of about 1,500, which is sufficient for a wide range of applications. Furthermore, this structure is highly insensitive to housing tolerances and thus are suitable for mass fabrication. For drop-in applications, the input/output coupling is realized through microstrip lines.

Figure 5.19. Different configurations of suspended strip line, (a) strips on top or bottom of substrate, (b) strips on both sides of substrate, and (c) microstrip line.

5. Suspended Substrate Filters

Suspended substrate filters as shown in Fig.5.16(b) are based on TEM mode operation. The structure is enclosed by a waveguide with reduced dimensions so that only the quasi-TEM mode can be guided. In this case, the accuracy with which the waveguide housing can be manufactured is of minor or no effects on the filter performance. Repeatability of filter response depends solely on the accuracy of etching technique which defines the metal pattern on substrate. Thus, these structures are the prime candidates for mass production with medium Q_u factor in the range of 1,000-1,500.

Suspended substrate strip lines(SSS) have been widely used in microwave and millimeter wave filters. Three different configurations are shown in Fig.5.19 [13]. In addition, suspended strip line filters are insensitive to temperature variations(depending on the substrate used), have wider bandwidth and the possibility to incorporate broadband, low-loss transitions directly on the filter substrate to connect to surrounding planar circuits or rectangular waveguides. Since both sides of the substrate can be used for circuit layout, a very wide range of coupling coefficients is available, for example, by adjusting the distance and overlapping area of lines on different sides of the substrate.

Suspended substrate band-pass filters using parallel or end-coupled lines have been extensively investigated in the literatures [14], [15]. For example, the response of a seven-element parallel-coupled filter for Ka-band is shown in Fig.5.20. In this structure, a broadband transition between rectangular waveguide and suspended strip line is utilized. Very low insertion loss, wide pass-bands and wide rejection bandwidth have been obtained. End-coupled suspended substrate filter with a pass-band of 32-35 GHz [16] and a three-resonator end-coupled suspended substrate

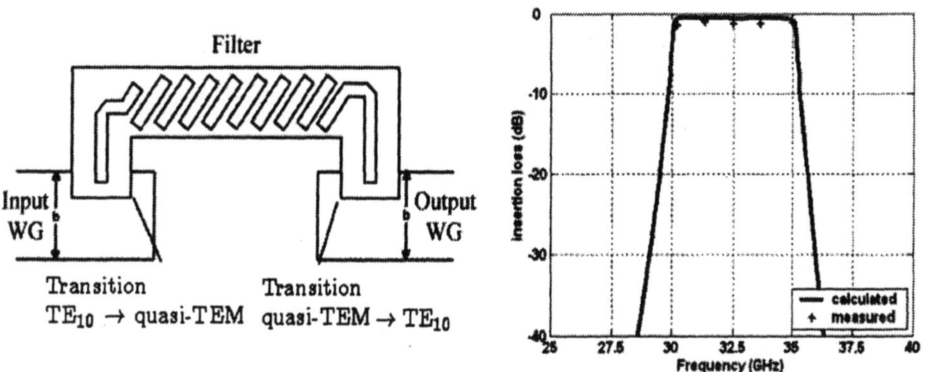

Figure 5.20. Measured and computed response of a Ka-band coupled-line suspended substrate band-pass filter.

filter operating at 45 GHz [15] have been reported as well. Using metallization patterns on both side of the substrate, very compact, parallel or end-coupled suspended substrate filter structures are plausible [17], [18].

The performance of a Ku-band double-sided and parallel-coupled filter structure is shown in Fig.5.21. This configuration provides significant size reduction and wide bandwidths [18]. End-coupled suspended substrate filters using both sides of substrate at 10 GHz have been reported in [17]. Measured response of these filters shows 0.4-0.5 dB of insertion loss in the pass-band as illustrated in Fig.5.22. To suppress the second pass-band which typically occurs at twice the center frequency, the resonators can be modified to have an attenuation of about 35 dB at twice the center frequency [17].

Another suspended substrate filter potentially suitable for mass fabrication was recently proposed in [19] with the layout shown in Fig.5.23(a). The coupling structure which is used as an inverter element provides one or two return-loss zeros. The resulting band-pass filter is compact and features a broad pass-band of about 30 %.

The measured and simulated responses of a five-resonator filter is shown in Fig.5.23(b). Spurious response at the first harmonic of the pass-band can be removed by extending the filter to have six resonators and choosing a shunt inductance as the central inverter. Combining low-pass and high-pass suspended substrate filters, broadband band-pass filters with wide spurious-free stop-band can be realized [13]. The low-pass filter is designed utilizing the extremely different impedance levels of suspended strip line and microstrip line. A broadband high-pass

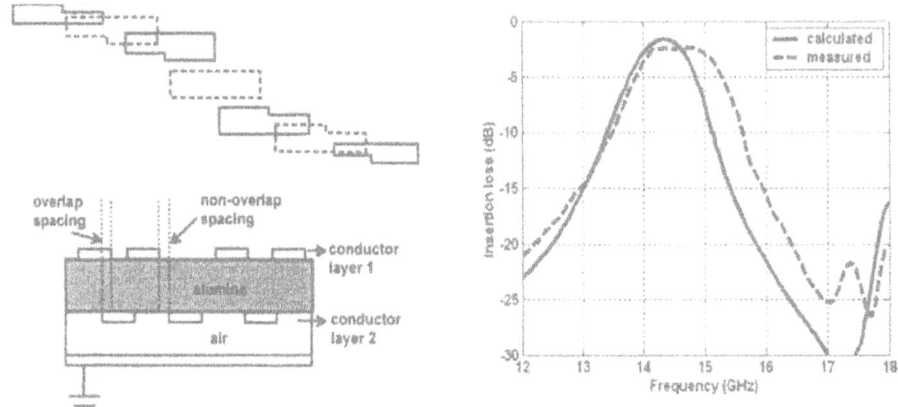

Figure 5.21. (a) Layout and cross section, and (b) performance of a double-sided suspended substrate filter.

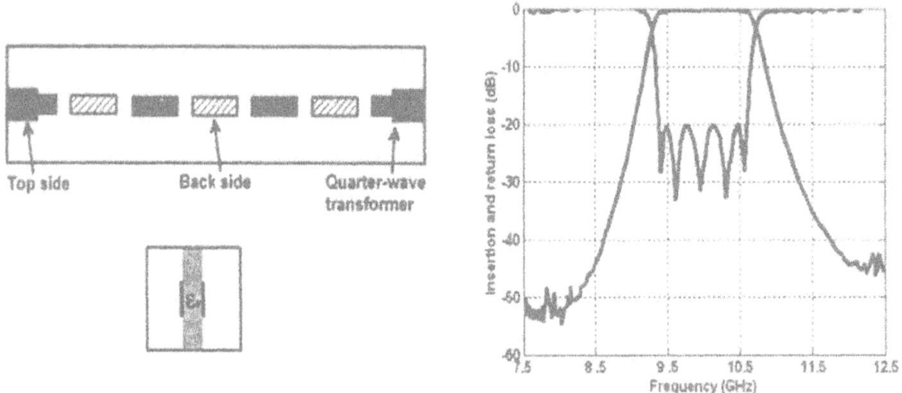

Figure 5.22. (a) Basic layout of a suspended strip line configuration and (b) measured filter characteristics.

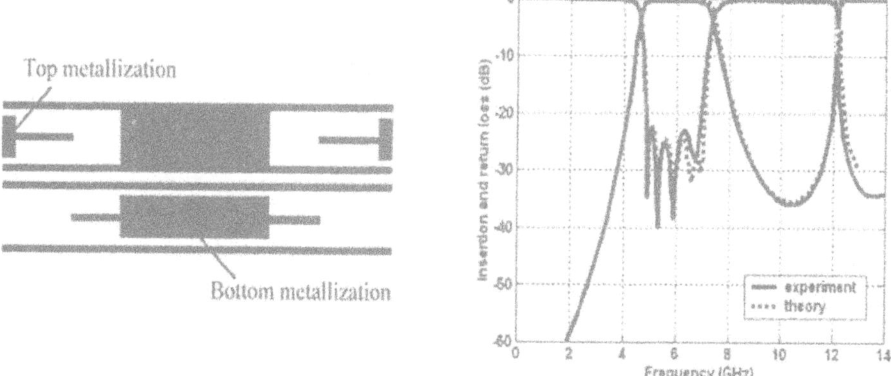

Figure 5.23. (a) Layout of a suspended substrate filter, and (b) measured and simulated filter characteristics.

filter is designed using the low impedance of odd-mode of the coupled suspended substrate line and a high impedance suspended strip line shunt-stub. The return loss in the pass-band of the filter(3.5-7.5 GHz, 80 % bandwidth) is higher than 13 dB, whereas the insertion loss is about 0.15-0.3 dB.

6. Conclusions

Low-cost and mass producible millimeter wave filters and diplexers are key components to lower the cost of future broadband communication systems. Various classes of plausible low-cost filters have been described in terms of their Q_u factor as well as ease of manufacturing. Particular fabrication aspects and sensitivity issue have been discussed. Since the millimeter wave technology is largely based on waveguide technology, the emphasis is on waveguide cavity filters, quasi-planar filters and suspended substrate filters. Although the suspended substrate filters are probably the best choice in terms of low manufacturing tolerances and cost, their moderate Q_u factor may not be suitable for many applications. In comparison, some of the waveguide filters discussed also show low sensitivity to housing tolerances and much higher Q_u factors. In conjunction with plastic injection molding waveguide housings, these filters may be attractive alternatives to the suspended substrate technology, in particular for high-volume production.

References

[1] U. Rosenberg, "New planar waveguide cavity elliptic function filters," *Eur. Microwave Conf.*, pp.524-527, Sept. 1995.

[2] J. Bornemann, S. Amari, and R. Vahldieck, "A combined mode-matching and coupled-integral-equations technique for the design of narrow-band H-plane waveguide diplexers," *IEEE AP-S Int. Symp. Dig.*, pp.950-953, 1999.

[3] W. Menzel, J. Kassner, and U. Goebel "Innovative packaging and fabrication concept for a 28GHz communication front-end," *Trans. IEICE*, vol.E82-C, no.11, pp.2021-2028, 1999.

[4] W. Menzel, F. Alessandri, A. Plattner, and J. Bornemann, "Planar integrated waveguide diplexer for low-loss millimeter wave applications," *Eur. Microwave Conf.*, pp.676-680, Sept. 1997.

[5] J. Kocbach and K. Folgero, "Design procedure for waveguide filters with cross-couplings," *IEEE MTT-S Int. Microwave Symp. Dig.*, pp.1449-1452, 2002.

[6] S. Amari and J. Bornemann, "CIET-analysis and design of folded asymmetric H-plane waveguide filters with source-load coupling," *Eur. Microwave Conf.*, pp.270-273, Oct. 2000.

[7] M. Guglielmi, O. Roquebrun, P. Jarry, E. Kerherve, M. Capurso, and M. Piloni, "Low-cost dual-mode asymmetric filters in rectangular waveguide," *IEEE MTT-S Int. Microwave Symp. Dig.*, pp.1787-1790, 2001.

[8] J. Bornemann, "Selectivity-improved E-plane filter for millimeter-wave applications," *Electron. Lett.*, vol.27, pp.1891-1893, Oct. 1991.

[9] E. Ofli, R. Vahldieck, and S. Amari, "Analysis and design of mass-producible cross-coupled, folded E-plane filters," *IEEE MTT-S Int. Microwave Symp. Dig.*, pp.1775-1778, 2001.

[10] R. Vahldieck, J. Bornemann, F. Arndt, and D. Grauerholz, "Optimized waveguide E-plane metal insert filters for millimeter wave applications," *IEEE Trans. Microwave Theory Tech.*, vol.31, pp.65-69, Jan. 1983.

[11] R. Vahldieck, "Quasi-planar filters for millimeter wave applications," *IEEE Trans. Microwave Theory Tech.*, vol.37, pp.324-334, Feb. 1989.

[12] Y. Ishikawa, T. Hiratsuka, S. Yamashita, and K.Iio, "Planar type dielectric resonator filter at millimeter wave frequency," *Trans. IEICE*, vol.E79-C, no.5, pp.679-684, 1996.

[13] W. Menzel, "Broadband filter circuits using an extended suspended substrate transmission line configuration," *Eur. Microwave Conf.*, pp.459-463, Aug. 1992.

[14] L. Q. Bui, Y. C. Shih, and T. N. Ton, "MM-wave harmonic reject filter," *Microwave J.*, vol.30, pp.119-122, July 1987.

[15] R. M. Dougherty, "MM-wave filter design with suspended stripline," *Microwave J.*, vol.29, pp.75-84, July 1986.

[16] C. Nguyen and K. Chang, "Design and performance of millimeter-wave end-coupled bandpass filters," *Int. J. Infrared Millim. Waves*, vol.6, pp.497-509, 1985.

[17] W. Schwab, F. Boegelsack, and W. Menzel, "Multilayer suspended stripline and coplanar filters," *IEEE Trans. Microwave Theory Tech.*, vol.42, pp.1403-1407, July 1994.

[18] S. Uysal and L. Lee, "Ku-band double-sided suspended substrate microstrip coupled-line bandpass filter," *Electron. Lett.*, vol.35, pp.1088-1090, June 1999.

[19] W. Menzel, L. Zhu, K. Wu, and F. Boegelsack, "Compact broad-band planar filters," *Eur. Microwave Conf.*, pp.41-44, Sept. 2001.

Chapter 6

HIGH-TEMPERATURE SUPERCONDUCTING PLANAR FILTERS FOR WIRELESS COMMUNICATION

Janina Mazierska and Mohan Jacob
Electrical and Computer Engineering
James Cook University
Townsville, Australia

Abstract Exquisite properties of HTS filters employed in the base station receivers of cellular systems lead to reduced adjacent channel interference, increased coverage and better spectrum utilization, and hence increase the revenue for the communication service providers. There are approximately 1,500 HTS front-end mobile radio receivers working commercially in the USA and Asia, and they are predicted to make a significant impact on the third and fourth generations of wireless communication. This Chapter presents a brief history of superconducting materials and microwave properties of high-temperature superconducting(HTS) films, reviews designs and properties of HTS resonators and filters, shows cryogenic HTS base station receivers available on the market, and summarizes possible trends in this field.

Keywords: HTS, superconducting planar filter, base station, receiver, out-of-band rejection, cryogenic.

1. Introduction

To provide high-quality communication services and efficient use of the allocated bandwidth, mobile radio base stations need highly selective, very sensitive and small-size filters in their receivers. Such high-performance filters can ensure little adjacent channel interference, better spectrum utilization, increase the coverage area without increasing antenna height, eliminate areas with poor reception(as illustrated in

Fig.6.1) or enable reduction in power requirements of telephone handsets.

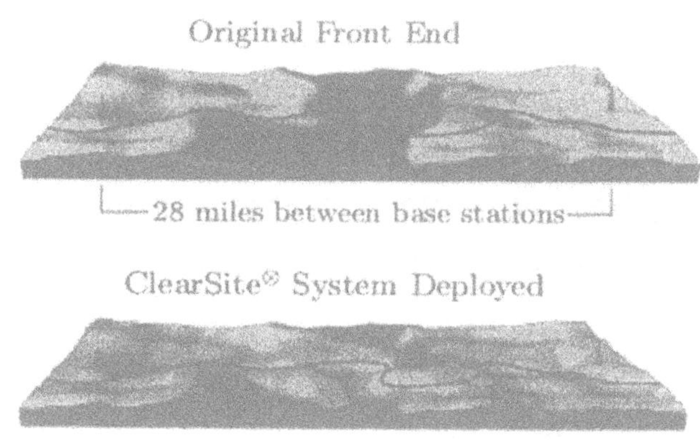

Figure 6.1. Comparison of reception without and with a superconducting filter [1], dark spots indicate the areas with poor reception.

High selectivity and very low insertion loss demand filters with extremely high Q factors. Air cavity filters exhibit a very high Q factor, but their size and weight prevent them from being mounted on top of towers. Small size of filters favors planar circuit technologies, but metallic microstrip filters have insertion loss too high and Q factors too low to be practical. Hence, conventional filter technologies cannot fulfil conflicting requirements for providing high-quality communication services. However, if superconducting thin films are used in planar resonators deposited on low-loss substrates, filters of small size, high Q factors and high skirt slopes of 30 dB/100 kHz or higher can be designed. A comparison between superconducting and conventional filters for cellular base stations based on field data of Conductus is shown in Fig.6.2 [2].

The high Q factors of superconducting filters are due to the very small surface resistance(R_s) of superconducting materials at high frequencies up to 200 GHz [3]. The R_s of $YBa_2Cu_3O_{7-\delta}$ films(one of the two major superconducting materials used for microwave applications) is approximately 1,000 times smaller than that of copper at 890 MHz and at the temperature of 77°K, and 330 times smaller at 1.85 GHz. When a low-noise amplifier is incorporated with a superconducting filter inside a cryocooler, the noise figure of cellular receiver can be reduced by approximately 2 dB [4].

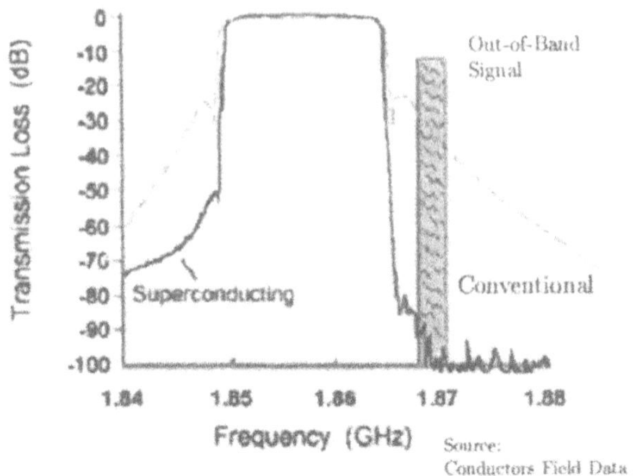

Figure 6.2. Measured characteristics of conventional and HTS cellular filters [2].

Exclusive properties of superconducting materials have generated considerable commercial interest for applications in front-end receivers for terrestrial and satellite mobile telecommunications soon after the discovery of high-temperature superconductors(HTS), based on copper oxides, in 1986. The first successful field trials in commercial receivers for cellular base stations using superconducting filters occurred in 1995.

By the middle of 2002, there were over 1,500 cryogenic front-end receivers(CRFE) with HTS filters for CDMA, GSM, TDMA, 2.5G and 3G networks working commercially in base stations in USA and Asia [5], manufactured by three companies in USA: STI, Conductus and ISCO Int.

In Europe, a consortium, SCENET, was formed in 1999, with the aim to develop HTS filters and antennas for terrestrial and satellite communications. The SCENET consists of several universities, research centers and industrial partners, including two small HTS companies: Cryoelectra and CryoSystems [6]. Several big companies like Denso, Toshiba, Fujitsu, LG, NEC and DuPont are developing their own HTS filter designs with an intention to enter the market when it is big enough.

The development of suitable HTS filters for wireless communication went through several phases. The initial designs were based on conventional planar filter, simply with the YBCO films replacing metallic films. This approach results in poorer performance than expected. Hence, new design paradigms have been created to take full advantage of the properties of superconducting materials. In designing HTS filters, it is important to minimize its physical size, and hence the size of cryocooler, while

providing large out-of-band rejection(OBR). This is being achieved by novel resonator designs, and in some cases by elliptic or quasi-elliptic filter response approximations. Availability of inexpensive and small-size cooling systems with verified reliability was another major issue during the development stage of HTS receivers, which was successfully solved except the cost.

The performance of currently available commercial HTS receivers for cellular base stations is excellent. Their deployment results in increased traffic(up to 100%), better coverage with fewer cells(up to 100% increase in coverage area), better audio quality, filling of coverage gaps, smaller number of dropout calls(up to 40%) and increased minutes-of-use(MOU). The increased revenues resulting from the HTS receivers over lifetime are assessed to be $280,000 in the rural areas(for 60% coverage increase), $350,000 in the suburban areas(for 25% coverage increase) and $350,000 in the urban areas [7].

The total number of deployed HTS receivers is still very small compared to that of base stations. The superiority of HTS technology over conventional technology will become more obvious when the third and fourth generations of communication systems are implemented, especially with the WCDMA systems. The requirements of high band efficiency, high quality of voice and high data rate transmission may cause utilization of the HTS filters absolutely necessary for some systems. This may result in a dramatic increase of penetration rate of the HTS receivers. However, in order to satisfy future market demand for HTS receivers, issues of high yield in mass production still need to be resolved.

2. Properties of Superconducting Materials

Superconducting materials exhibit very unique properties as compared to conventional conductors, namely zero resistance to dc current flow, perfect diamagnetism(resulting in levitation effect) and lack of dispersion [7]. These exquisite properties have fascinated scientists since the discovery of superconducting phenomenon in mercury below 4.2°K by the Dutch scientist, K. Onnes, in 1911 [8], [9]. It was soon found that surprisingly many elements, compounds and alloys, including metals(except the noble metals), intermetallic compounds and semiconductors(for example, aluminium, cadmium, indium, lead, niobium, tantalum, tin, titanium, Nb_3Al, AuBe, $AuSb_2$, TiCo, $ZrAl_2$) become superconducting below certain temperature referred to as the critical temperature, T_c [10]. Until 1986, critical temperatures of all the known superconduct-

ing materials were below 30°K, and the problems of cooling with liquid helium(to achieve sufficiently low temperature) have prevented the wide applications of superconductors.

It was only after copper oxide-based superconductors with relatively high critical temperatures(and hence referred to as high-temperature superconductors) were discovered, many possible applications of superconducting materials are becoming reality as only liquid nitrogen is needed for their operation. Superconducting phenomenon in the copper oxide compounds was discovered in $LaSrCuO_{4-x}$ by B. Muller and K. Bednorz from IBM, Zurich in 1986 [11]. As the T_c of this material is 30°K, higher than the previous record of 23.2°K for Nb_3Ge, this discovery created a furor in the scientific world and resulted in not only the Nobel Prize for those two scientists in 1987, but also a frantic search for new materials with even higher T_c.

In early 1987, superconductivity was reported in $YBa_2Cu_3O_{7-\delta}$ with the critical temperature of 90°K [12]. Since then, several new superconductors have been discovered with higher T_c(TlCaBaCuO family with T_c up to 125°K [13] and $HgBa_2Ca_2Cu_3O_8$ with T_c of 135°K [14]). In 2001, Nagamatsu et al. announced the discovery of superconductivity in MgB_2 at 39°K, the first non-copper oxide-based superconductor with high critical temperature [15]. A brief history of discovering superconducting materials is demonstrated in Fig.6.3.

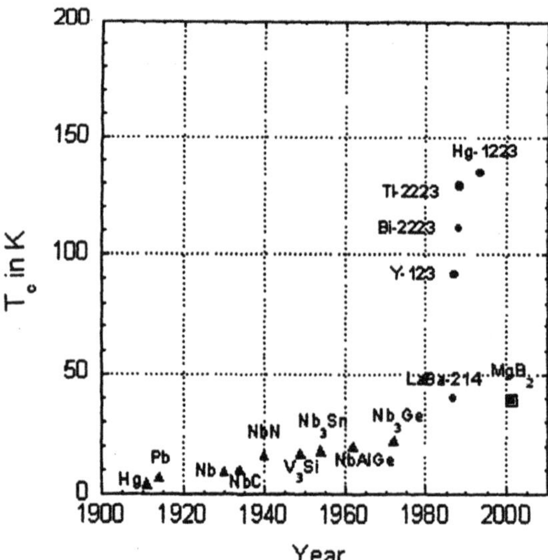

Figure 6.3. Evolution of T_c of various superconducting materials.

The superconducting properties can only be exhibited if the temperature, external magnetic field, current density in the material and the operating frequency are below certain critical values of T_c, H_c, J_c and f_c, respectively, of the given superconductor. Values of the critical parameters are not constant, but depend on the other parameters as illustrated in Fig.6.4. The critical frequency for HTS materials is very high, usually in the THz range. A typical characteristics of dc resistance versus temperature of $YBa_2Cu_3O_7$ films, with no external magnetic field applied, is shown in Fig.6.5 [16].

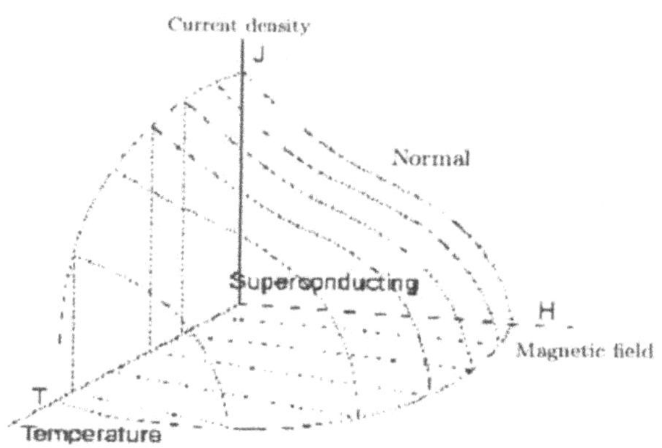

Figure 6.4. Interdependence of critical values of T_c, H_c and J_c, assuming $f_c = 0$.

The losses of HTS materials to ac current are not zero as in the case of dc current. However, they are much smaller than the losses of conventional conductors for a wide range of frequencies. The microwave impedance of superconducting materials is defined in the same way as for conventional conductors, namely

$$Z_s = \left.\frac{E_x}{H_y}\right|_{z=0} = \sqrt{\frac{j\mu\omega}{\sigma}} = R_s + jX_s$$

The propagation constant, γ, can be expressed in the same way as for conductors, namely $\gamma^2 = j\omega\mu\sigma$. The basic difference between superconductors and conductors is that the conductivity, σ, of superconductors is a complex number, $\sigma = \sigma_1 - j\sigma_2$, where σ_1 and σ_2 are the conductivities of normal and superconducting electron pairs, respectively. Real part of the complex conductivity involves normal electrons only, while the imaginary part involves both normal and superconducting types of electrons. Both σ_1 and σ_2 are strongly temperature dependent, and $\sigma_2 \gg \sigma_1$ for $T < T_c$. In contrast, the conductivity of conductors is exclusively real.

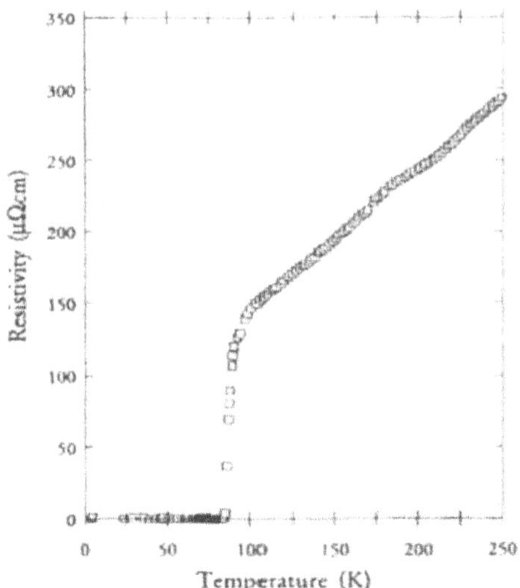

Figure 6.5. Measured R_{dc} of $YBa_2Cu_3O_7$ films versus temperature [16].

Various theories were proposed to describe and predict the electromagnetic and thermal properties of conventional superconductors. Well known examples are London theory, BCS model, Ginzburg-Landau theory and the two-fluid model [9]. None of these theories can fully explain the behavior of high-temperature superconductors. However, they are very useful in understanding the principle of current conduction and major properties of superconducting materials. The two-fluid phenomenological model is based on the concept of two types of charge carriers that coexist in a superconducting material, namely normal electrons and superconducting electrons. The latter flow through the material without any resistance while the former encounter resistance due to collisions with other electrons or interactions with the crystal lattice. The total current density, $\bar{J}$, in the HTS material predicted by the two-fluid model is the sum of superconducting and normal currents as $\bar{J} = \bar{J}_n + \bar{J}_s = (\sigma_1 - j\sigma_2)\bar{E}$, where σ_1 and σ_2 are given by

$$\sigma_1 = \frac{n_n e^2 \tau}{m(1+\omega^2\tau^2)}, \qquad \sigma_2 = \frac{n_s e^2}{m\omega} + \frac{n_n e^2 (\omega\tau)^2}{m\omega(1+\omega^2\tau^2)}$$

where e is the fundamental unit of charge, n_n and n_s are the densities of normal and superconducting electrons, respectively, m is the mass of an electron, ω is the angular frequency, and τ is the momentum relaxation time.

The normal current, $\bar{J}_n$, obeys Maxwell's equations and Ohm's law [9], while the superconducting current, $\bar{J}_s$, needs to obey additional London equations

$$\nabla \times \bar{J}_s = \frac{1}{\mu_0 \lambda_L^2} \bar{B}, \qquad \bar{J}_s = \frac{1}{\mu_0 \lambda_L^2} \bar{E}$$

The London equations are based on the phenomenon that superconductors do not allow exterior magnetic flux to penetrate into their interior. Superconductors were considered perfect conductors until the perfect diamagnetism(Meissner) effect was discovered in 1933 [17]. While the magnetic flux density, $\bar{B}$, in a normal conductor is equal to $\mu_r \mu_0 \bar{H}_{\text{ext}}$ when exposed to an external magnetic field, $\bar{H}_{\text{ext}}$, the magnetic flux density inside superconducting materials is zero. In practice, magnetic fields can penetrate a small distance into a superconductor. The penetration distance is very small because the field decays exponentially with distance.

The distance at which the field falls to e^{-1} of the surface field is referred to as the London penetration depth, λ, which is much smaller than the skin depth of normal conductors. The penetration depth is related to the imaginary part of conductivity as $\lambda^2 = (\omega\mu\sigma_2)^{-1}$. It is strongly temperature dependent, but it does not vary with frequency. The lack of frequency dependence in λ results in the propagation constant being a constant for all frequencies as

$$\gamma^2 = j\omega\mu(\sigma_1 - j\sigma_2) = \frac{j}{\delta_1^2} + \frac{1}{\lambda^2} \cong \frac{1}{\lambda^2}$$

As a result, all the frequency components in a signal can travel through the superconductors at the same speed. In practice, superconducting transmission lines exhibit certain dispersion due to the presence of supporting structures. However, the level of dispersion is considered negligible for frequencies below 150 GHz [18].

The penetration depth also determines the magnitude of surface reactance, X_s, of superconductors as $X_s(T) = \omega\mu_o\lambda(T)$, which is much larger than the surface resistance, R_s, unlike conventional conductors. Surface resistance of a superconductor can be expressed as [19]

$$R_s = \frac{\sigma_1}{2\sigma_2}\sqrt{\frac{\omega\mu}{\sigma_2}} = \frac{1}{2}\omega^2\mu_o^2\lambda^3\sigma_1 \tag{6.1}$$

which is temperature dependent. Typical temperature variation of the surface resistance of $YBa_2Cu_3O_7$ thin film is shown in Fig.6.6. For temperature above T_c, superconducting materials behave like normal

conductors with $\sigma_2 = 0$, $\sigma_1 = \sigma_{1(n)}$, $R_s = X_s = \sqrt{\omega\mu_o/(2\sigma_{1(n)})}$, $\gamma^2 = j\omega\mu\sigma_{1(n)} = j/\delta^2_{1(n)}$, where $\delta_{1(n)}$ is the skin depth.

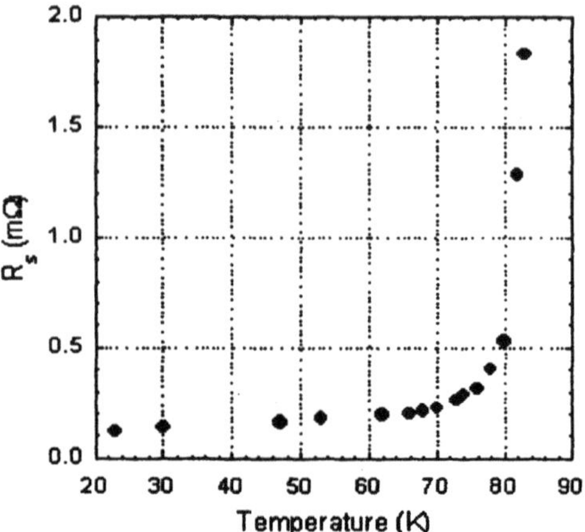

Figure 6.6. Surface resistance of $YBa_2Cu_3O_{7-\delta}$ thin film on $LaAlO_3$ substrate at the frequency of 10 GHz [19].

The surface resistance of superconducting materials increases with frequency squared as indicated in (6.1). Fig.6.7 shows the frequency dependence of R_s of YBCO films, as measured by various research centers, in comparison with copper and superconducting niobium. Although the losses of HTS materials increase with f^2, they are smaller than those of copper for frequencies up to 200 GHz [2].

HTS thin films exhibit increase of surface resistance at high microwave power. This nonlinear behavior of R_s is attributed to one or more mechanisms affecting the supercurrent flow, such as grain boundaries in the films, local or global heating, flux creep, flux flow, phase slippage, vortex-antivortex depairing, vortex creation and its motion, and nonequilibrium pair breaking [10]. Such nonlinear effect restricts the applications of HTS thin-film circuits to lower RF power.

Critical temperatures of bulk HTS reported by different manufacturers are practically the same. The T_c is 92°K for $YBa_2Cu_3O_{7-\delta}$, 125°K for $Tl_2Ba_2Ca_2Cu_2O_{10}$, 105°K for $PbBi_2Sr_2Ca_2Cu_3O_x$, and 135°K for $HgBa_2Ca_2Cu_3O_y$ [10]. However, the T_c of HTS thin films can vary significantly, depending on the method of deposition, optimization process, substrate used and deposition temperature. For microwave applications, the generally used $YBa_2Cu_3O_{7-\delta}$ and $Tl_2Ba_2Ca_2Cu_2O_8$ thin films have their T_c varies from 85°K to 92°K and 100°K to 106°K, respectively [10]. The surface resistance of $YBa_2Cu_3O_{7-\delta}$ and $Tl_2Ba_2Ca_2Cu_2O_8$ thin films

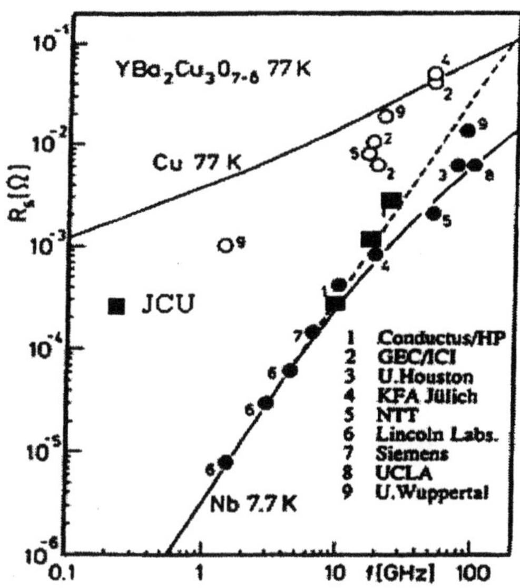

Figure 6.7. Surface resistance as a function of frequency of $YBa_2Cu_3O_7$ thin films and copper at 77°K measured by various groups(modified version of [3].)

at 77°K and 10 GHz varies from 300 to 500 mΩ and from 220 to 500 mΩ, respectively. The $Bi_2Sr_2CaCu_2O_8$ bulk material has found applications in cables and power, and the Hg-based superconductor has found no practical applications so far.

Substrates used for the deposition of HTS films include $LaAlO_3(\epsilon_r = 23.4)$, Ce-buffered sapphire$(\epsilon_r = 9.3)$ and $MgO(\epsilon_r = 9.8)$. The $LaAlO_3$ is a preferred substrate for Chebyshev or dual-mode designs over cheaper sapphire and MgO substrates due to smaller size of resonators. For pseudo-elliptic and elliptic filters using $Tl_2Ba_2Ca_2Cu_2O_{10}$, MgO substrates are often used as they allow for deposition of thicker films and do not have twinning or anisotropy problems as $LaAlO_3$ or Al_2O_3 substrates. Table 6.1 shows some commonly used substrates for HTS thin film fabrication and their dielectric loss at the temperature of 77°K [20].

3. HTS Filters for Wireless Communication

Development of microwave HTS filters started almost immediately after $YBa_2Cu_3O_7$ material became available in film form, and its surface resistance is found to be much lower than that of bulk YBCO [21]-[24]. Initially, superconducting planar filters were based on coupled lines and

Table 6.1. Commonly used substrates for HTS thin film deposition and their loss tangent.

Substrate	Frequency (GHz)	Loss tangent
Sapphire	9	1.5×10^{-6}
$LaAlO_3$	10	7.6×10^{-6}
MgO	8	2×10^{-6}
$SrLaAlO_4$	12	2×10^{-6}
Alumina	7.7	2×10^{-5}

conventional hairpin designs [25]. Fig.6.8 presents one of the early HTS filter designs by STI and the frequency responses of filter fabricated using superconducting and gold thin films, respectively, at the temperature of 80°K [26].

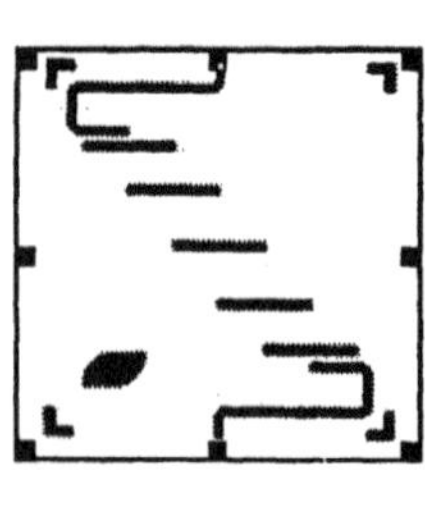

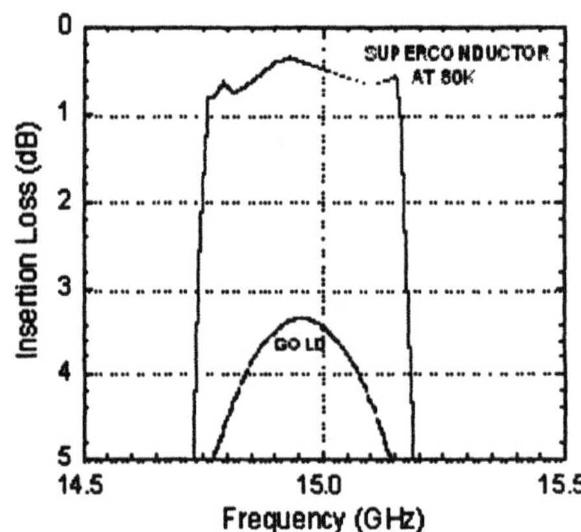

Figure 6.8. Comparison of a five-pole filter fabricated using superconducting material and gold, respectively, at 80°K [26].

However, early HTS filters exhibited higher insertion loss and lower Q factors than expected. In order to take advantage of the extremely low loss of superconducting materials, previously neglected phenomena such as coupling losses, radiation effects and dispersion have to be taken into account during the design phase. Most popular CAD tools do not have the capabilities to account for these previously considered second order effects. Hence, accurate simulation and optimization of HTS filters becomes a difficult task. Another important issue for HTS filters, as mentioned in the Introduction, is to minimize the physical size of filter

while providing large out-of-band rejection(OBR). Efforts to reduce the size of HTS filters result in novel shapes of meander line, spiral and miniature hairpin resonators as shown in Fig.6.9 [27]-[39]. To improve power handling capabilities of the HTS resonators, patch resonators have been proposed, often in dual-mode configurations to reduce the number of resonators [40]-[44].

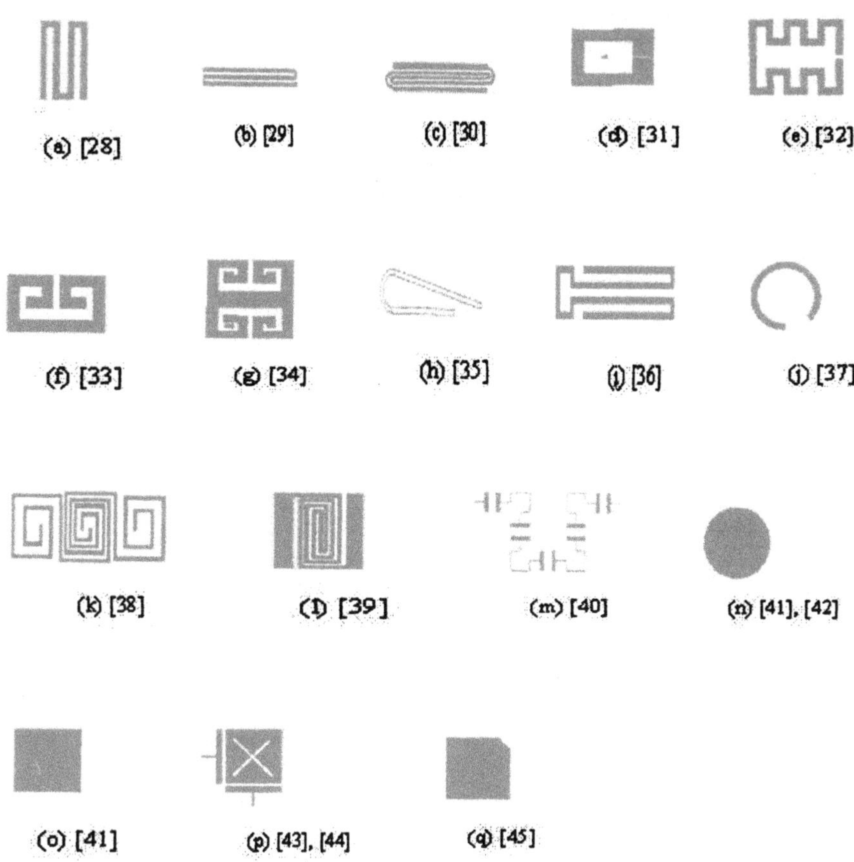

Figure 6.9. Layouts of resonators used for superconducting filters.

To reduce the required number of resonators while retaining sharp skirts, pseudo-elliptic and elliptic designs have been employed in some HTS filter designs instead of the Chebyshev response approximation [30]. This approach allows reduction of poles needed to obtain the required OBR with the disadvantage of poorer delay characteristics. Coupling between nonadjacent resonators is achieved either through appropriate

arrangement of resonators and capacitive coupling or through transmission lines [45].

Characteristics of several HTS filters developed in the past few years using some of the resonators shown in Fig.6.9 are illustrated below in chronological order. A nine-pole quasi-elliptic filter design based on the modified hairpin resonator of Fig.6.9(g) is shown in Fig.6.10 [33]. Two filters are fabricated based on this design using YBCO films on an LAO substrate(39 mm × 22.5 mm), with the bandwidth of 10 MHz and 15 MHz, respectively, and the measured insertion loss of 0.25 dB.

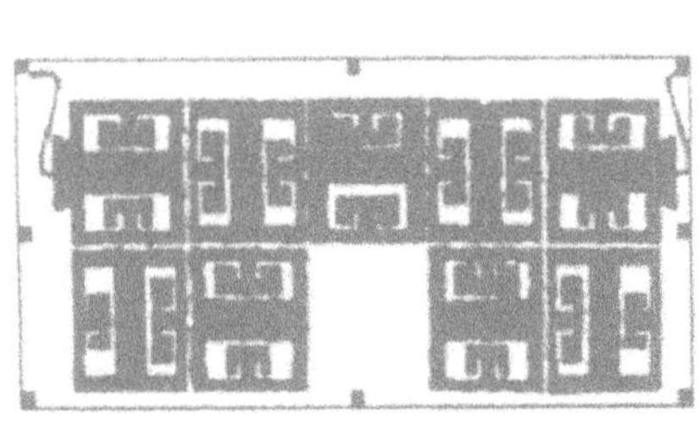

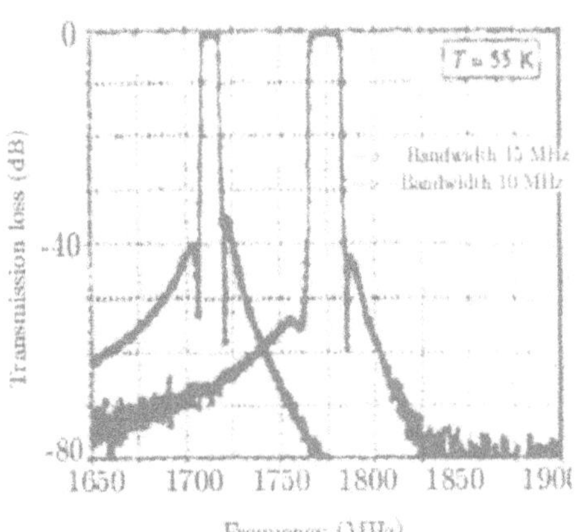

Figure 6.10. **Layout of a nine-pole filter and its frequency response [33].**

Fig.6.11 shows an eight-pole quasi-elliptic filter fabricated on an LAO substrate(39 mm × 12 mm), which is based on the hairpin resonator of Fig.6.9(a), its bandwidth is 15 MHz at the center frequency of 1.8 GHz, the insertion loss is 0.3 dB, and the measured out-of-band rejection is approximately 60 dB/5 MHz [27].

Fig.6.12 shows the layout and measured characteristics of an eight-pole modified hairpin[Fig.6.9(d)] YBCO filter fabricated on an LAO substrate(39 mm × 23.5 mm) [46]. The filter is designed to have a bandwidth of 15 MHz at the center frequency of 1.777 GHz. The OBR is 30 dB at 17.5 MHz from the band edge.

Fig.6.13 shows the quadruplet geometry and transmission line coupling used to obtain the cross coupling between resonators in a nine-pole Chebyshev filter together with its insertion loss and return loss characteristics [39]. This filter is fabricated using YBCO films on an LAO substrate(46 mm × 18 mm) and has a bandwidth of 11 MHz. The insertion loss is 0.8 dB and the out-of-band rejection is 34 dB/MHz [39]. Another ten-pole HTS filter design, based on the modified hairpin

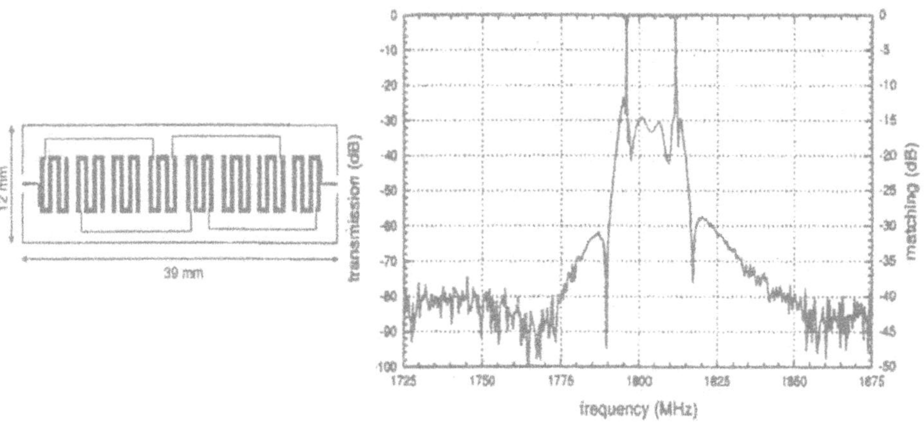

Figure 6.11. Layout of an eight-pole filter and its return loss and insertion loss characteristics [27].

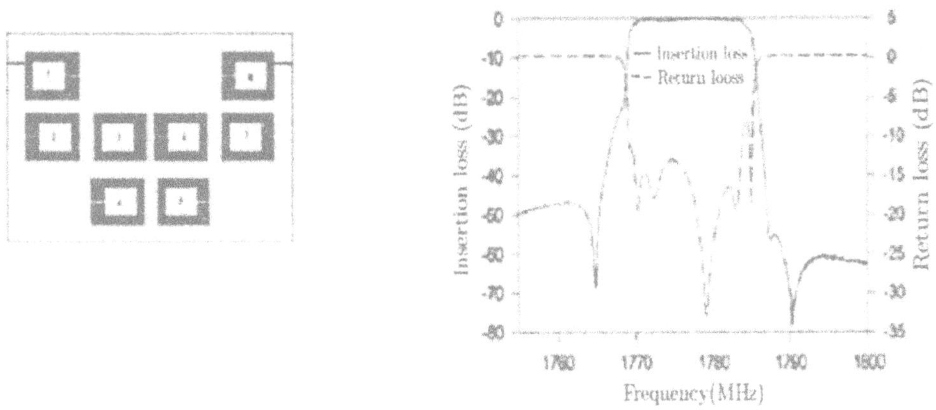

Figure 6.12. Layout and characteristics of an eight-pole HTS filter [46].

resonator[Fig.6.9(b)] is shown in Fig.6.14 together with the characteristics of a 12-pole filter of the same type [28]. This filter is designed with the center frequency of 1.775 GHz and bandwidth of 10 MHz, and is fabricated using YBCO films on an LAO substrate(35mm × 20 mm). The insertion loss is 0.5 dB, and the OBR is 75 dB/2 MHz.

Fig.6.15 shows a 32-pole Chebyshev filter based on J-shaped resonator [Fig.6.9(h)] fabricated on a 3" MgO wafer and its frequency response [34]. This filter is designed with the center frequency of 1.93 GHz and 20 MHz of bandwidth, exhibits an insertion loss of 0.3 dB at the center

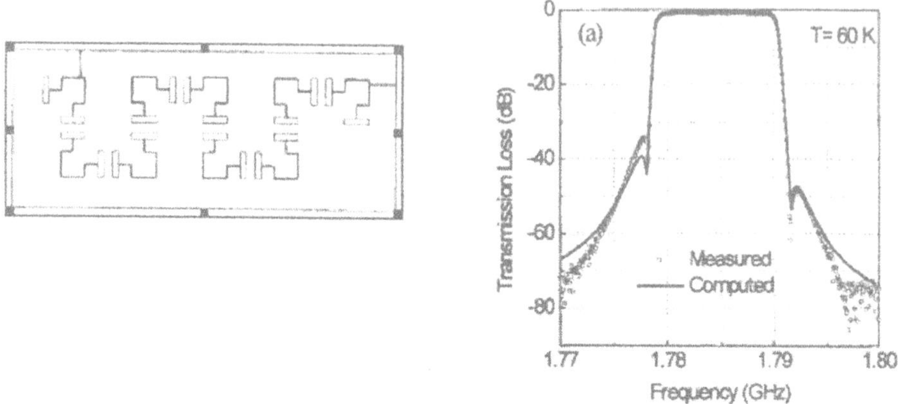

Figure 6.13. A nine-pole Chebyshev filter and its simulated and measured characteristics [39].

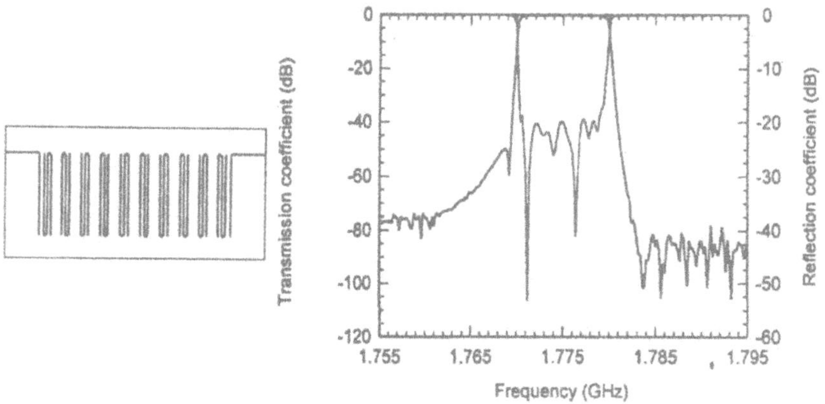

Figure 6.14. Layout of a ten-pole HTS filter and the characteristics of a 12-pole filter of the same type [28].

frequency and 1.1 dB at the band edges, and over 20 dB of return loss. The out-of-band rejection is approximately 30 dB/400 kHz.

Fig.6.16 shows the characteristics of a 13-pole filter with 5 MHz of bandwidth at the center frequency of 1.95 GHz and with very sharp skirts, fabricated by Cryoelectra [45]. The OBR of this filter is 80 dB/MHz.

HTS filters based on patch resonators allow for much higher power due to the lack of current peaks at edges. Designs up to the eighth order have

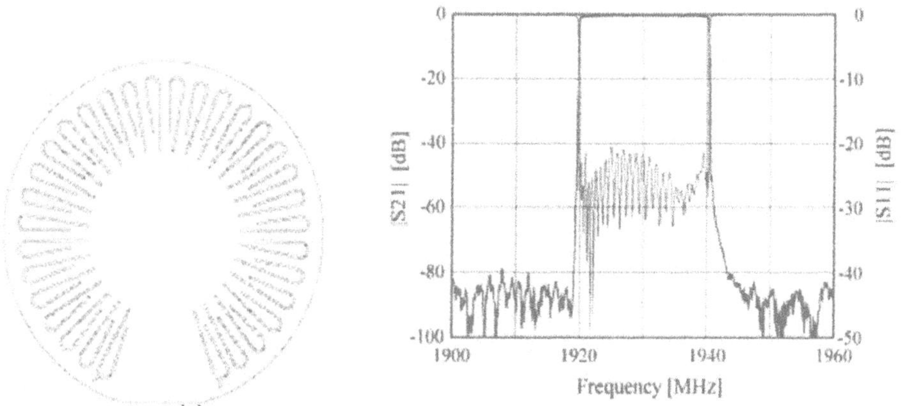

Figure 6.15. **Layout and frequency response of a 32-pole Chebyshev filter [34].**

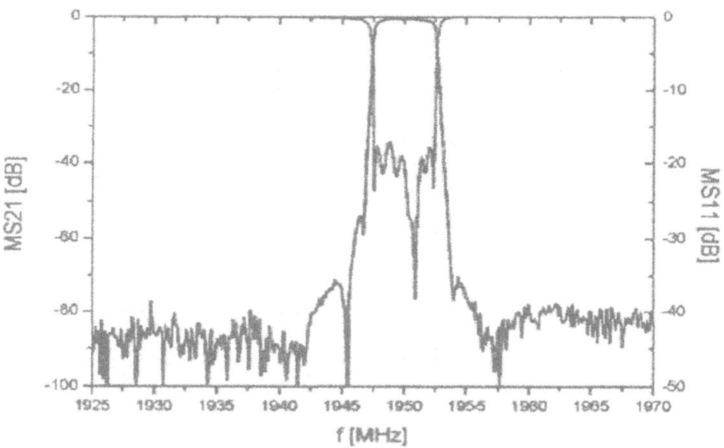

Figure 6.16. **Return loss and insertion loss characteristics of a 13-pole HTS filter fabricated by Cryoelectra [45].**

been published so far to the authors' knowledge. Fig.6.17 shows a four-pole Chebyshev disc filter[Fig.6.9(n)], fabricated on an LAO substrate with 14 mm of diameter and its frequency response [41]. The measured bandwidth is 24 MHz and the out-of-band rejection is 30 dB/40 MHz. A dual-mode four-pole filter based on the configuration of Fig.6.9(p) is realized using 22.3 mm $\times$ 10 mm YBCO thin film on an LAO substrate, and its measured frequency response is shown in Fig.6.18 [48]. The

bandwidth is 1% at the center frequency of 3.32 GHz, the insertion loss is 0.9 dB, and the OBR is 50 dB/20 MHz.

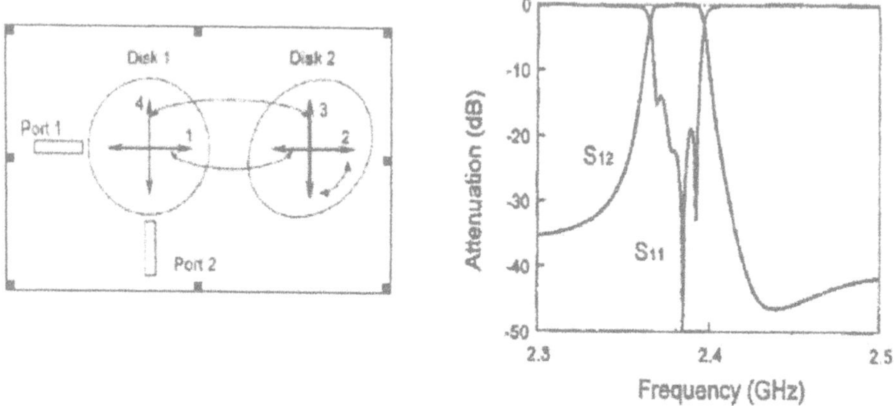

Figure 6.17. Layout and filter response of a four-pole Chebyshev disc filter [41].

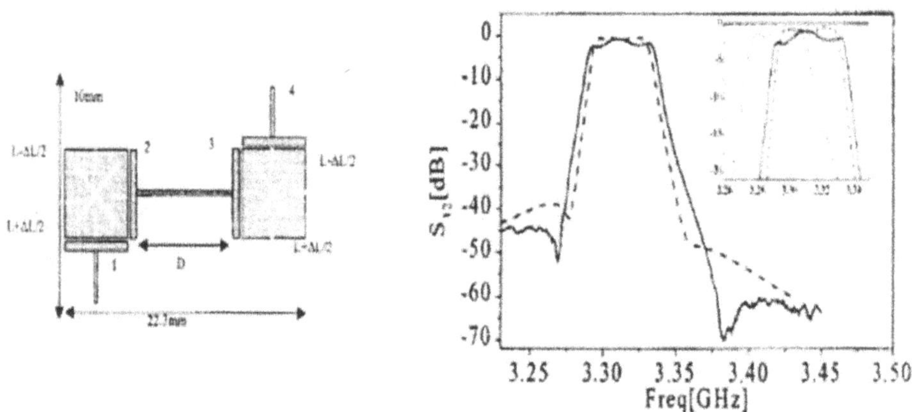

Figure 6.18. Layout and characteristics of a four-pole dual-mode filter [48].

A very recent design of 22-pole HTS filter with ten transmission zeros based on a new concept of clip resonator is shown in Fig.6.19 [35]. This ultra-selective filter is designed for applications in 3G wireless communication at the center frequency of 1.95 GHz and 20 MHz of bandwidth. The clip resonators are 9.6 mm long and 2.15 mm wide with 0.3 mm of line width, and the filter fits on a 2" YBCO on MgO wafer. Coupling

diagram of this filter is illustrated in Fig.6.19 with the solid and dashed lines denoting major and quadruplet cross coupling, respectively.

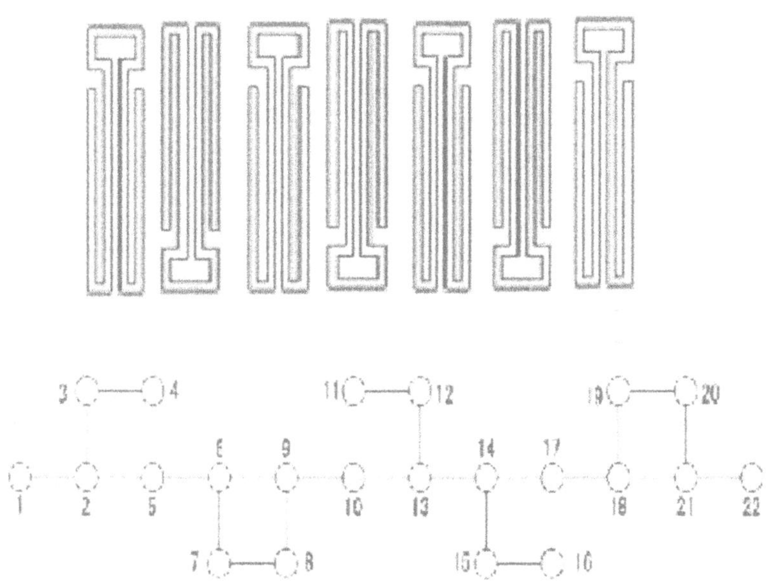

Figure 6.19. **Layout and coupling diagram of a 22-pole HTS filter [35].**

The measured insertion and return losses shown in Fig.6.20(a) exhibit excellent selectivity with skirt slopes larger than 30 dB/100 kHz, 90 dB of OBR at 350 kHz from the band edges, and insertion loss of 0.2 dB at the center frequency. The performance is superior to that of a 50-pole Chebyshev filter in terms of insertion loss and takes much less volume than the 50-pole Chebyshev filter. Measured group delay of this filter is shown in Fig.6.20(b). Delay at the center frequency is 210 ns, and is 2,890 ns and 3,350 ns, respectively, at the band edges.

This short review of the state-of-the-art HTS filters and the summary presented in Table 6.2 are by no means complete and as informative as we would have expected. Initial designs and full specifications of superconducting filters are made public. Recently, HTS filters are considered to have market potential. While some groups still share results of their researches, designs of commercial products are not usually made available.

Table 6.2. Specifications of some HTS filters. P: pole number, E: elliptic, QE: quasi-elliptic, C: Chebyshev, BW: bandwidth, IL: insertion loss, OBR: out-of-band rejection.

Ref.	P	Type	Substrate	Size (mm)	f_0 (GHz)	BW (MHz)	IL (dB)	OBR (dB/MHz)
[4]	9	E	Sapphire		1.93	20	0.1	15
[27]	8	QE	LAO	39 × 12	1.8	15	0.3	60 / 5
[28]	12	QE	LAO	35 × 20	1.775	10	0.5	75 / 2
[33]	4	QE	LAO	20 × 20	1.972	5	0.4	-
[33]	9	QE	LAO	39 × 22.5	1.715	10	0.25	-
[33]	9	QE	LAO	39 × 22.5	1.777	15	0.25	-
[35]	22	C	LAO	dia. 2"	1.95	20	0.2	30 / 0.1
[34]	16	C	MgO	half of 3"	1.93	20	0.14	30 / 1.3
[34]	32	C	MgO	dia. 3"	1.93	20	0.32	30 / 0.5
[38]	9	C	LAO	46 × 18	1.784	11	0.8	34 / 1
[43]	2	C	LAO	10 × 10	3.7	1/10%		30 / 45
[43]	7	C	LAO	dia. 50"	1.76	75		30 / 5
[41]	4	C	LAO	dia. 14	2.38	24	0.21	30 / 40
[45]	13	QE			1.95	5		80 / 1
[45]	17	QE		22 × 70	1.93	20		30 / 1
[46]	8	QE	LAO	39 × 23.5	1.778	15	-	30 / 17.5
[46]	8	QE	MgO(0.3)	39 × 23.5	1.778	15	15.5	36.5 / 2
[47]	10	QE	MgO	34 × 18	842	15	0.4	60 / 2
[48]	4	C	LAO	22.3 × 10	3.32	1%	0.9	50 / 20
[49]	6	C	Sapphire	43 × 10	1.857	20	0.2	60 / 15
[50]	11	C	LAO	dia. 50	1.778	11.5	0.6	80 / 20
[51]	8	QE	LAO	dia. 50	2.260	68	0.7	55 / 10
[52]	11	C	MgO	10 × 10	1.92	15	0.1	60 / 20

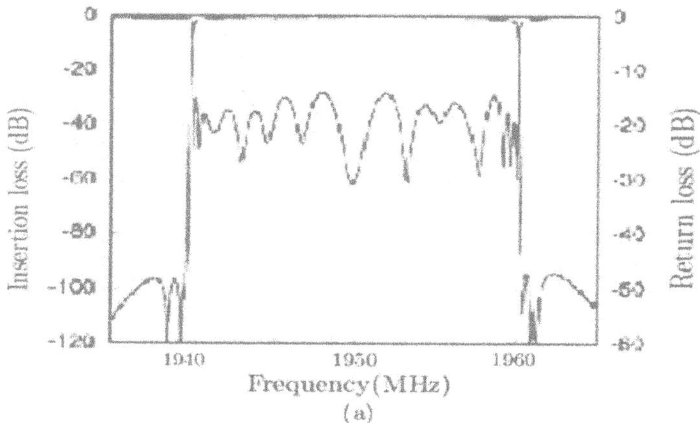

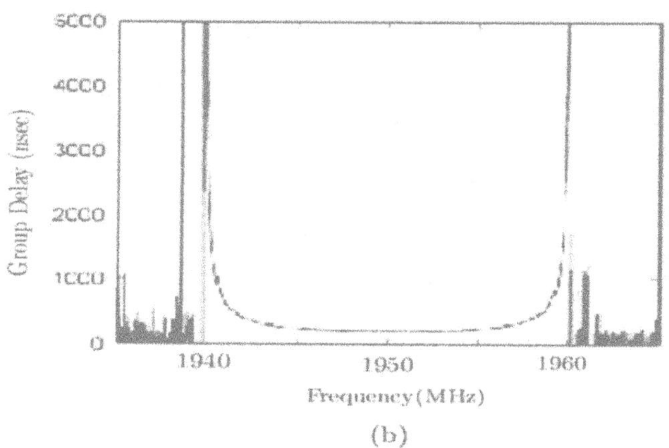

Figure 6.20. **Measured (a) insertion and return losses and (b) group delay of the 22-pole HTS filter shown in Fig.6.19 [35].**

4. Commercial HTS Base Station Receivers

As mentioned in the Introduction, at the middle of 2002, there were three American companies(STI, Conductus and ISCO) supplying commercial front-end HTS receivers to wireless communication service providers for omni or three-sectored CDMA, GSM, TDMA, 2.5G and 3G systems.

Superconductor Technologies Inc. (STI), headquartered in Santa Barbara, CA, was founded in 1987 and launched their first commercial HTS receiver, SuperFilter^{TM}, in 1997 [53]. This HTS filter is based on

the resonator shown in Fig.6.9(c), fabricated using $Tl_2Ba_2Ca_2Cu_3O_{10}$ films on MgO substrate(34 mm × 18 mm). The frequency characteristics is shown in Fig.6.21 [47]. The return loss is below −30 dB and the slope of skirts is 60 dB/2 MHz.

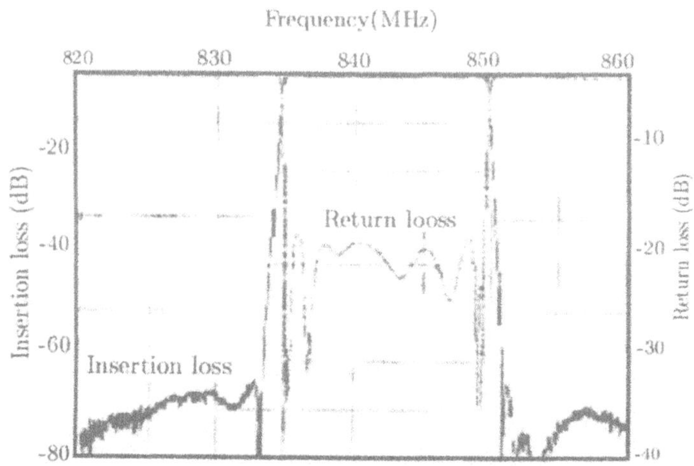

Figure 6.21. Frequency characteristics of the SuperFilter® by STI [47].

Later, SuperFilterII® and SuperFilter® PCS as illustrated in Fig.6.22 were introduced. The latter system has either 15 MHz or 5 MHz bandwidth, with noise figure of 0.75 and 0.9 dB, respectively, as shown in Fig.6.23, and uses a 50 W Stirling cryocooler to maintain HTS filters(based on $Tl_2Ba_2Ca_2Cu_3O_{10}$ films on MgO substrates) and LNAs at the temperature of 77°K.

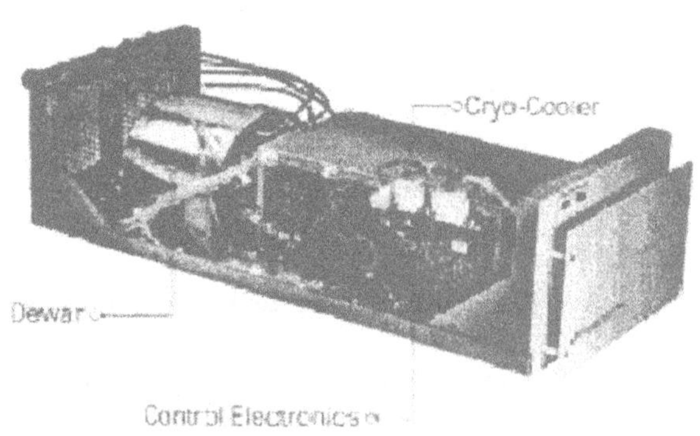

Figure 6.22. SuperFilter® PCS and SuperLink™ Rx 850.

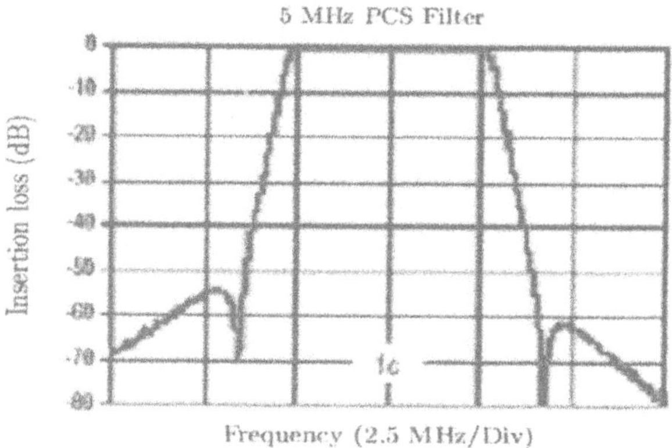

Figure 6.23. **Insertion loss of HTS filter by STI with 5 MHz of bandwidth [53].**

Conductus Inc. was founded in 1987 and is based in Sunyvale, CA. The company's HTS front-end receivers include ClearSite® 2100 and 2300 systems as shown in Fig.6.24 [2], [55], employing filters based on $YBa_2Cu_3O_7$ superconducting thin films on $LaAlO_3$ substrates, cooled low-noise amplifiers and Gifford-McMahon cryocoolers. Their mean time between failures is longer than fifteen years.

Figure 6.24. **Conductus' ClearSite® 2300 Metro [55].**

The Conductus performed field trials with KDDI and Hitachi in Japan that demonstrated the elimination of interference caused by personal handiphone system(PHS) as shown in Fig.6.25 [1]. In this trial, a WCDMA cell site was observed to essentially collapse in the presence of frequent

but moderate-level interferences. The ClearSite® fully restored the capacity, coverage and data rate of the cell site, and improved the bit error rate by three orders of magnitude. Recently, the Conductus received an order for two HTS prototype filter systems to be delivered to a 4G customer in Japan.

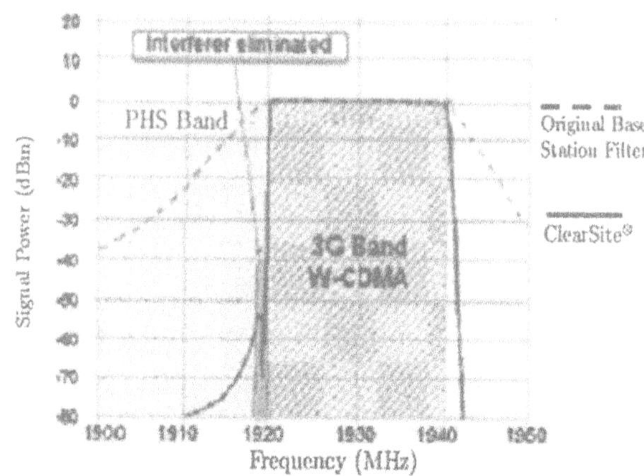

Figure 6.25. Filtering of PHS interference by ClearSite® [1].

Illinois Superconductor Corporation, (ISCO) was founded in 1989 and entered the market with HTS thick-film filters in 1996 [56]. Use of Stirling cycle cryocoolers and significant size reduction of filters have enabled product size reduction as shown in Fig.6.26 [57]. The ISCO Int. markets HTS front-end receivers for the 800 MHz and 1,900 MHz bands based on thick-film superconducting technology. An ISCO Int. system for the 1,900 MHz band used for 3G services in Japan is shown in Fig.6.27 [57]. This indoor unit contains two filter paths, each with 32 resonators and a low-noise amplifier.

5. Conclusions

High-temperature superconducting materials exhibit exceptionally low surface resistance, allowing for manufacturing of small-size microstrip planar filters with very sharp skirts, Q factors up to 90,000 and insertion loss of 0.2 dB. HTS planar filters based on novel hairpin, spiral and patch resonators, enclosed in air evacuated cryocoolers together with low-noise amplifiers make superior cryogenic front-end receivers for wireless communication as compared to conventional hardware. The HTS receivers mounted on top of base station towers provide excellent voice quality for CDMA, GSM and TDMA systems by blocking interference

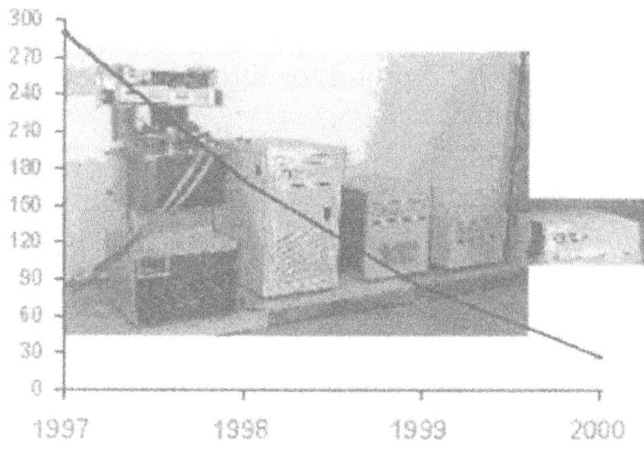

Figure 6.26. Volume of ISCO OMNI receivers [57].

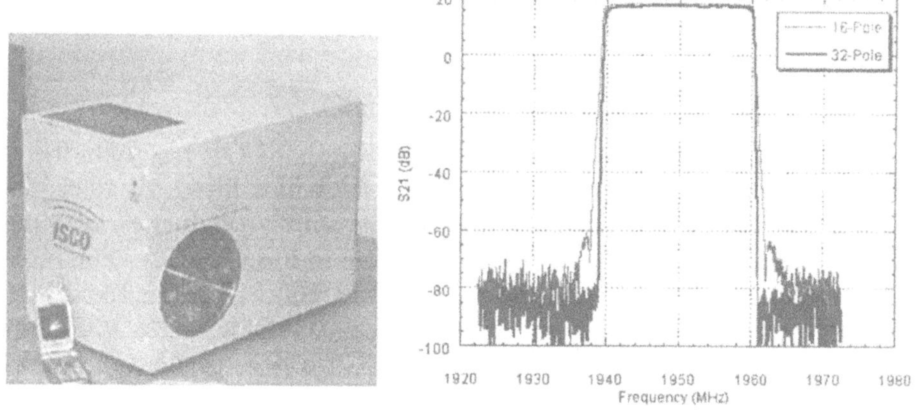

Figure 6.27. The OMNI unit from ISCO and the 16 and 32-pole filter characteristics [57].

and increasing sensitivity. Deployment of HTS receivers by several top-ten US service providers results in significant increase in minutes-of-use and reduction of dropped call rates. More than 1,500 CRFE systems have been deployed worldwide, logging in excess of twenty million hours of cumulative operation. The HTS receivers for wireless base stations are now a mature technology.

For wideband wireless communication, the benefits of superconducting cryogenic receivers are even more pronounced. Field trials of 2.5G and 3G systems have demonstrated that the HTS receivers are able to

restore the capacity and coverage in cell sites that were previously degraded by interference without a need for guardbands. The property of eliminating interference is very useful, especially for the WCDMA systems whose sensitivity to interference is greater than that of conventional CDMA systems. The HTS receivers also support higher data rates and improve the bit error rates for wireless data transmission.

Although the cost of CRFE systems is higher than that of conventional receivers, it is smaller than one tenth of a new base station, which may otherwise be needed to counteract all negative effects of interference for 3G and 4G systems. Alternative solutions to enhance the performance of base stations using tower-mounted amplifiers and smart antennas do not solve the problem of interference and are more expensive than the superconducting receiver.

Even though the market penetration of HTS CRFE receivers is currently low, this may change rapidly when this new technology receives wide acceptance by 3G and 4G service providers. Manufacturers of HTS receivers need to be ready for such a turn of tide.

Acknowledgements

This work has been supported by the ARC Large Grant No.00000799, James Cook University.

References

[1] R. Simon, "Superconductor systems for wireless networks," www.conductus.com, 2002.

[2] Conductus Website, 1997.

[3] H. Weinstock and M. Nisenoff, *Microwave Superconductivity*, NATO Science Series, Kluwer Academic Publishers, 1999.

[4] M. Klauda et al., "Superconductors and cryogenics for future communication systems," *IEEE Trans. Microwave Theory Tech.*, vol.48, no.7, pp.1227- 1239, 2000.

[5] S. J. Berkowitz, "Current state of HTS for wireless applications and future needs," *7th Symp. High-Temp. Superconduct. High-Freq. Fields*, Woods Hole, MA, June 9-12, 2002.

[6] SCENET Website http://orchidea.maspec.bo.cnr.it/homepage.html, 2002.

[7] B. A. Willemsen, "HTS filter subsystems for wireless telecommunications," *IEEE Trans. Appl. Superconduct.*, vol.11, no.11, pp.80, 2001.

[8] H. K. Onnes, "The resistance of pure mercury at helium temperatures," Leiden Comm., 120b, 1911.

[9] A. C. Rose-Innes and E. H. Rhoderick, *Introduction to Superconductivity*, Pergamon Publishers, 1994.

[10] Z.-Y. Shen, *High Temperature Superconducting Microwave Circuits*, Artech House, 1994.

[11] J. G. Bednorz and K. A. Muller, "Possible high Tc superconductivity in the Ba-La-Cu-O system," *J. Phys. B.*, vol.64, pp.189-193, 1986.

[12] M. K. Wu, J. R. Ashburn, C. J. Torng, P. H. Hor, R. L. Meng, L. Gao, Z. J. Huang, Y. Q. Wang, and C. W. Chu, "Superconductivity at 93K in a new mixed phase YBCO compound system at ambient pressure," *Phys. Rev. Lett.*, vol.58, pp.908-910, 1987.

[13] C. C. Torardi, M. A. Subramanian, J. C. Calabrese, J. Gopalakrishnan, K. J. Morissey, T. R. Askew, R. B. Flippen, U. Chowdhry, and A. W. Sleight, "Crystal structure of TlBaCaCu at 125K superconductor," *Science*, vol.242, pp.249-252, 1988.

[14] A. Schilling, M. Cantoni, G. D. Guo, and H. R. Ott, "Superconducitivity above 130 K in the HgBaCaCuO system," *Nature*, vol.363, pp.56-58, 1993.

[15] J. Nagamatsu, N. Nakagawa, T. Muranaka, Y. Zenitani, and J. Akimitsu, "Superconductivity at 39 K in magnesium diboride," *Nature*, vol.410, pp.63-64, 2001.

[16] Measurements performed at Stanford University, 1991.

[17] based on [9]; W. Meissner and R. Ochsenfeld, Naturwissenschaften, vol.21, pp.787-788, 1933.

[18] Z. Kresin and S. A. Wolf, *Fundamentals of Superconductivity*, Plenum Press, 1990.

[19] M. V. Jacob and J. Mazierska, "Microwave characterisation of YBCO films on Sapphire and LAO substrates," *J. Low Temp. Phys.*, in press, and also JCU microwave surface resistance measurement data.

[20] M. J. Lancaster, *Passive Microwave Device Applications of High Temperature Superconductors*, Cambridge University Press, 1997.

[21] M. Sagawa, K. Takahashi, and M. Makimoto, "Miniaturised hairpin resonator filters and their applications to receiver front end MIC's," *IEEE Trans. Microwave Theory Tech.*, vol.37, no.12, pp.1991-1997, 1989.

[22] G. L. Matthaei and G. L. Hey-Shipton, "Novel staggered resonator array superconducting 2.3 GHz bandpass filter," *IEEE MTT-S Int. Microwave Symp. Dig.*, pp.1269-1272, 1993.

[23] G. C. Liang, D. Zhang, C. F. Shih, M. E. Johanson, R. S. Withers, D. E. Oates, and A. C. Anderson, "High power HTS microstrip filters for cellular base station application," *IEEE Trans. Appl. Superconduct.*, vol.5, no.2, pp.2652-2655, 1995.

[24] D. Zhang, G. C. Liang, C. F. Shih, Z. H. Lu, and M. E. Johansson, "A 19-pole cellular bandpass filter using 75mm diameter HTS thin films," *IEEE Microwave Guided Wave Lett.*, vol.5, no.11, pp.405-407, 1995.

[25] J. Mazierska and M. Jacob, "Progress in high temperature superconducting panar filters for wireless communication," *Asia-Pacific Microwave Conf.*, Nov.30 - Dec.3, pp.1004-1007, Taipei, 2001.

[26] STI brochure, 1997.

[27] M. Reppel and J. C. Mage, "Superconducting microstrip bandpass filter on LAO with high out of band rejection," *IEEE Microwave Guided Wave Lett.*, vol.10, no.5, pp.180-182, 2000.

[28] I. B. Vendik, A. N. Deleniv, V. O. Sherman, A. A. Svishchev, V. V. Kondratiev, D. V. Kholodniak, A. V. Lapshin, P. N. Yudin, B. C. Min, Y. H. Choi, and B. Oh, "Narrownband YBCO filter with quasi elliptic characteristic," *IEEE Trans. Appl. Superconduct.*, vol.11, no.1, pp.477-480, 2001.

[29] E. R. Soares, J. D. Fuller, P. J. Marozick, and R. L. Alvarez, "Applications of high temperature superconducting filters and cryoelectronics for satellite communication," *IEEE Trans. Microwave Theory Tech.*, vol.48, no.7, pp.1190-1197, 2000.

[30] J. S. Hong and M. J. Lancaster, "Design of highly selective microstrip bandpass filters with a single pair of attenuation poles at finite frequencies," *IEEE Trans. Microwave Theory Tech.*, vol.48, no.7, pp.1098-1107, 2000.

[31] J. S. Hong et al., "Thin film high T_c passive microwave components for advanced communication systems," *IEEE Trans. Appl. Superconduct.*, vol.9, pp.3893-3896, 1999.

[32] M. Barra, A. Cassinese, M. Cirillo , M. V. Jacob, R. Russo, and R. Vaglio, "Miniaturized hairpin superconducting filters for telecommunication application," *J. Microwave Opt. Technol. Lett.*, Dec. 2002.

[33] M. Repel and H. Chaloupka, "Novel approach for narrowband superconducting filters," *IEEE MTT-S Int. Microwave Symp. Dig.*, TH2D-6, 1999.

[34] G. Tsuzuki, M. Suzuki, and N. Sakakibara, "Superconducting filter for IMT-2000 band," *IEEE Trans. Microwave Theory Tech.*, vol.48, no.12, pp.2519-2525, 2000.

[35] G. Tsuzuki, S. Ye, and S. Berkowitz, "Ultra selective 22-pole, 10-transmission zero superconducting bandpass filter supasses 50-pole Chebyshev filter," *IEEE Trans. Microwave Theory Tech.*, Dec. 2002.

[36] G. Tsuzuki, M. Suzuki, N. Sakakibara, and Y. Ueno, "Narrow band 2 GHz superconducting filter," *Asia-Pacific Microwave Conf.*, pp.1509-1512, Yokohama, Japan, 1998.

[37] C. K. Ong et al., "HTS bandpass spiral filter," *IEEE Microwave Guided Wave Lett.*, vol.9, no.10, pp.407-409, 1999.

[38] H. Kayano, H. Fuke, F. Aiga, Y. Terashima, R. Kato, and Y. Suzuki, "2 GHz superconducting bandpass filter with parallel resonators structure," *Asia-Pacific Microwave Conf.*, pp.592-595, Sydney, Australia, 2000.

[39] B. C. Min, Y. H. Choi, S. K. Kim, and B. Oh, "Cross coupled band pass filter using HTS microstrip resonators," *IEEE Trans. Appl. Superconduct.*, vol.11, no.1, pp.485-488, 2001.

[40] H. Chaloupka, "Concepts and systems using disc resonators," *5th Symp. High-Temp. Superconduct. High-Freq. Fields*, Stockholm, Sweden, June 22-25, 1998.

[41] K. Setsune and A. Enokihara, "Elliptic disc filters of HTS films for power handling capacity over 100 W," *IEEE Trans. Microwave Theory Tech.*, vol.48, no.7, pp.1256-1264, 2000.

[42] L. Zhu, P. M. Wecowski, and K. Wu, "New planar dual mode filter using cross slotted patch resonator for simultaneous size and loss reduction," *IEEE Trans. Microwave Theory Tech.*, vol.47, no.5, pp.650-654, 1999.

[43] A. Andreone, A. Cassinese, M. Iavarone, P. Orgiani, F. Palomba, G. Pica, M. Salluzzo, and R. Valio, "Development of L-band and C-band superconducting planar filters for wireless systems," *Asia-Pacific Microwave Conf.*, Sydney, Australia, pp.581-586, 2000.

[44] Z. Y. Shen, "Planar high temperature superconducting filters with backside coupling," US Patent 5750473, May 1998.

[45] T. Kasser, private communication, 2002.

[46] J. S. Hong, M. J. Lancaster, D. Jedamzik, R. B. Greed, and J. C. Mage, "On the performance of HTS microstrip quasi-elliptic function filters for mobile communications applications," *IEEE Trans. Microwave Theory Tech.*, vol.48, no.7, pp.1240-1246, 2000.

[47] G. H. Shipton, "Superconducting filters for base stations in the USA," *IEEE MTT-S Int. Microwave Symp.*, Workshop Notes on high-performance and emerging filter technologies for wireless, Phoenix, Arizona, 2001.

[48] M. Barra, A. Cassinese, M. Cirillo, G. Panariello, R. Russo, and R. Vaglio, "Superconducting dual mode dual stage cross slotted filters," *Microwave Opt. Technol. Lett.*, vol.33, no.6, pp.389-392, 2002.

[49] R. S. Kwok, D. Zhang, Q. Huang, T. S. Kaplan, J. Lu, and G. C. Liang, "Superconducting quasi-lumped element filter on R-plane sapphire," *IEEE Trans. Microwave Theory Tech.*, vol.47, no.5, pp.586-591, 1999.

[50] H. T. Kim, B. C. Min, Y. H. Choi, S. H. Moon, S. M. Lee, B. Oh, J. T. Lee, I. Park, and C. C. Shin, "A compact narrowband HTS microstrip filter for PCS applications," *IEEE Trans. Appl. Superconduct.*, vol.9, no.2, pp.3909-3912, 1999.

[51] B. Lippmeier and M. Sinclair, "HTS microwave filter for ANTF," CSIRO, Australia, private communication, 2001.

[52] H. Kanaya, T. Shinto, K. Yoshida, T. Uchiyama, and Z. Wang, "Miniaturised HTS coplanar waveguide bandpass filters with highly packed meaderlines," *IEEE Trans. Appl. Superconduct.*, vol.11, pp.481-484, 2001.

[53] STI Website: www.suptech.com, 2002.

[54] STI, private communication, 2002.

[55] Conductus, private communication, 2002.

[56] ISCO Int. Website: www.iscointl.com, 2002.

[57] ISCO private communication, 2002.

[58] F. Huang, "Ultra-compact superconducting narrow-band filters using single and twin spiral resonators," *IEEE Trans. Microwave Theory Tech.*, vol.51, pp.487-491, 2003.

Chapter 7

MICROMACHINED DEVICES FOR WIRELESS COMMUNICATION

Yo-Sheng Lin[1] and Shey-Shi Lu[2]

[1] *Department of Electrical Engineering*
National Chi-Nan University
Puli, Taiwan, ROC

[2] *Department of Electrical Engineering*
National Taiwan University
Taipei, Taiwan, ROC

Abstract Recently, RF micromachining technology has received more attention because it can improve the quality factor of on-chip devices, enhance RF power distribution circuits, innovate RF circuit design, and reduce system cost and size significantly. An overview of research and development of micromachined devices for wireless communication subsystems is presented in this Chapter. Specific devices described include micromachined inductors with high Q factor, tunable micromachined capacitors, low-loss micromechanical switches, microscale vibrating mechanical resonators, thin-film bulk acoustic resonators(FBARs), micromachined low-loss microwave and millimeter wave filters, and miniaturized antennas for millimeter wave applications.

Keywords: micromachining, miniature, membrane, switch, inductor, capacitor, resonator, filter, antenna.

1. Introduction

Exponential market growth in portable wireless communication continuously demands the development of vantage systems with small size, high operating frequency, high degree of integration, low cost, low power consumption, low signal distortion and low noise level. Modern process for silicon radio-frequency integrated circuit(RFIC), including comple-

mentary metal-oxide-semiconductor(CMOS), BiCMOS, and so on, endow transistors with high f_T and f_{max}. This progress enables silicon RFIC to compete with GaAs RFIC of which the operating frequency is 5 GHz or even higher [1]. In addition, from the perspective of cost and potential of integration with baseband circuits, silicon RFIC process is better than GaAs RFIC process. Therefore, the former process seems to be the most promising candidate to meet the design requirements.

Performance-limiting components in an RFIC include voltage-controlled oscillator(VCO), low-noise amplifier(LNA), frequency synthesizer, and so on. Using silicon RFIC technology to design and integrate these components requires the use of RF inductors with high Q factor and other reactive components to meet the design requirements. However, due to higher loss of the silicon substrate than that of the GaAs substrate, it is difficult to realize RF inductors with high Q factor and other reactive components. Many researchers have reported various layouts, fabrication techniques and materials to improve the performance of RF passive devices and IC's on silicon substrates. For instance, the state-of-the-art monolithic CMOS LNA's with 10 mW dissipation on a thin substrate($\simeq$ 20 μm) has shown a noise figure of 2.17 dB at the 5 GHz band. It is achieved mainly by the improved Q factor of on-chip inductors due to lower substrate loss and coupling noise [2].

In this Chapter, MEMS technologies for many important microwave and millimeter wave devices, including high Q inductors, tunable capacitors, low-loss switches, vibrating mechanical resonators, thin-film bulk acoustic resonators(FBAR), low-loss filters, antennas, and so on, are briefly reviewed. Background information, major issues, current trend and future challenges of each device will be discussed.

2. High-Q Micromachined Inductors

Many efforts have been spent to apply traditional Si-based RFIC technologies to implement spiral inductors, of which the Q factor is not satisfactory up to now [1]. Although spiral inductors have been improved by adopting patterned ground shields [3] or directly using bonding wires as inductors [4], the Q factor achieved are insufficient for many applications. In addition to conductor related loss that is common to all kinds of substrates, implementation of spiral inductors by using Si-based RFIC technologies suffer from two other substrate related dissipative mechanisms: capacitive current and induced Eddy currents in lossy substrate, which are difficult to mitigate [5]. Hence, for inductance ranging from about 2 to 12 nH, achievable Q factors usually fall in the range of 5 to

15, owing to the combination of thin conductors and dissipative silicon substrates.

The loss induced by capacitive current is related to the parasitic capacitance, C_{ox}, between the planar spiral inductor and the top surface of substrate as shown in Fig.7.1(a). The thin insulating oxide between them is typically several μm thick. At high frequencies, the current flows through this capacitance into the lossy substrate can be modeled by a parallel RC network. Such dissipation adds a real component to the inductive reactance, hence degrading its Q factor. This component can be significant at frequencies where the skin depth exceeds the substrate thickness. When frequencies or substrate doping levels increase so that the skin depth is comparable to or smaller than the substrate thickness, Eddy currents in the substrate become more critical. This magnetically induced loss can be depicted by the transformer analog with the secondary terminated in a resistive load as shown in Fig.7.1(b).

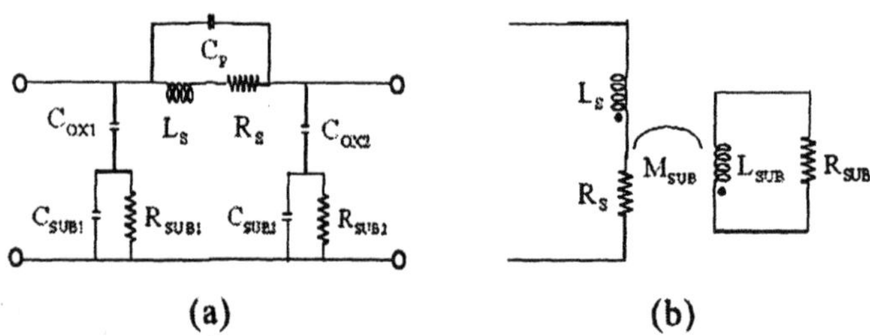

Figure 7.1. **(a) Simple model for a planar spiral inductor without considering Eddy current loss, (b) simple model of Eddy current loss without other parasitics.**

To implement a state-of-the-art inductor with high Q factor on a silicon substrate, one needs to minimize the conductor loss and the power loss associated with the induced Eddy current in the substrate. Hence, there are intensive ongoing researches on novel micromachining technologies to improve the performance of on-chip inductor. In the subsequent discussions, major approaches and their drawbacks will be discussed.

The first approach enhances the coupling among inductive coils by using magnetic cores. The inductor in Fig.7.2(a) has an NiFe core under a planar metal spiral to increase the magnetic flux linkage. For a ten-turn inductor with $W_{\text{wind}} = 50$ μm and $h_{\text{wind}} = 10$ μm in an area of 2× 10 mm^2, inductance of 2.7mH can be attained with a Q factor of 6.6 at 4 MHz [6]. The configuration combining metal coils with magnetic cores is found to have the highest Q factor relative to three other configurations(air-core coil, a NiFe core confined to the center of

NiFe windings). However, this type of inductor is still inadequate for applications in communication systems because the magnetic coil only responds in certain frequency range. Its permeability drops to an impractical value beyond that range. Methods to solve this problem are under investigation.

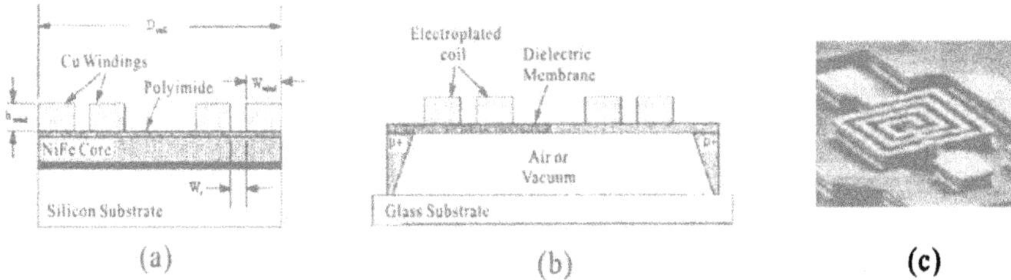

Figure 7.2. **Integrated inductors utilizing micromachining processes, (a) spiral inductor over an NiFe core to increase its inductance, (b) spiral inductor with the backside silicon substrate removed and mounted on a glass substrate [7], (c) spiral inductor with the silicon wafer around it etched away through selective etching to reduce Eddy current loss [10].**

The second approach removes the lossy substrates by fabricating inductors on suspended membranes via well known backside etching techniques. Fig.7.2(b) shows the schematic of a miniaturized spiral inductor fabricated on a membrane or a suspended substrate, which achieves an overall inductance of 115 nH, with a Q factor of 22 at 275 MHz. Microwave/millimeter wave inductors fabricated by using similar techniques have attained inductances on the order of 1.2 nH. The associated self-resonant frequencies are 70 GHz with substrate removal, and 22 GHz with substrate intact, and the predicted Q factors are 60 to 80 at 40 GHz [7]. These inductors and the like have achieved some of the highest Q factors among on-chip inductors at their respective operating frequencies. In spite of these promising results, this type of inductor is still unsuitable for certain applications in communication systems. Even if the substrate is removed, parasitic self-resonance problems still remain to limit their frequency range. Self-resonance is generally not a problem for inductance of less than 10 nH, but can be significant when the inductance is greater than 20 nH.

The third approach removes the silicon wafer around a spiral inductor through selective etching, as shown in Fig.7.2(c). For an 1.5 nH inductor, high resonant frequency of 38 GHz and maximum Q factor of about 30 can be achieved [8], [9]. In addition, high linearity is also available. Compared to standard printed circuit technology, these micromachined

spiral inductors can be used to design X-band dividers, K-band hybrids and Ka-band couplers. The results turn out to be almost two orders of magnitude reduction in size, as well as improved electrical performance in terms of insertion loss and bandwidth. X and K-band power dividers with wide bandwidth(25% bandwidth at 15 dB return loss) and low loss(0.6 dB) have been demonstrated [8], [9]. It is also shown that by using spiral inductors with high Q factor, the loss induced by lumped Wilkinson power divider is even lower than some distributed designs [10].

3. Tunable Micromachined Capacitors

Although standard *pn*-junction diodes can be built by using a source/ drain diffusion and n(or p) well, high resistivity of the well generally degrades the Q factor to an unacceptable low level. Another approach is to adopt the gate capacitance of a standard enhancement mode MOSFET, making use of the change in capacitance due to variation of gate bias. A minor modification to standard MOS varactor results in a large improvement in Q factor. As shown in Fig.7.3, by replacing the $p+$ source/drain diffusion of a PMOS transistor with the $n+$ diffusion of an NMOS transistor, one can construct a MOS capacitor that may operate in the accumulation mode. Basically, fabrication of this structure does not require process change, only a design rule adjustment is necessary. The capacitance of such a device exhibits a large change near the flat-band voltage. Typically, the achievable capacitance tuning range is close to 2:1. Because of the lowered channel resistance, operation in the accumulation mode can be advantageous. As a result, high Q-frequency products in excess of 100 GHz are achievable with current processes, and continue to improve as technology evolves [11].

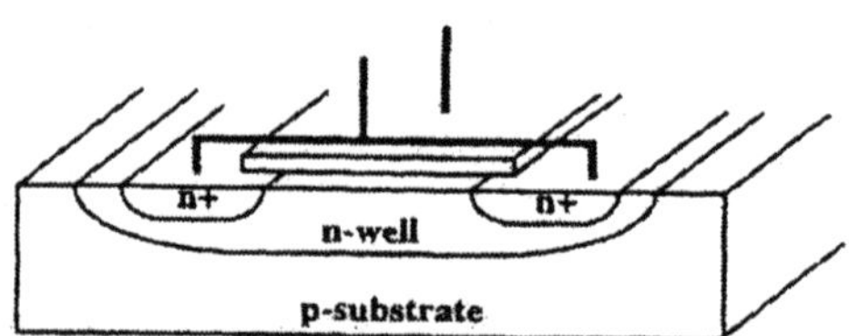

Figure 7.3. A MOS capacitor operated in the accumulation mode [1].

Good performance of on-chip varactors has been demonstrated. However, compared with their off-chip counterparts, the on-chip diodes still exhibit excessive series resistance, which leads to insufficient Q factor for some applications such as the VCO used in synthesizers. In addition,

tuning range of such capacitors beyond available supply voltages is often limited, and trimming is required to set a reference capacitance value.

Voltage-tunable micromachined high-Q capacitor is presently under close examination, because it has the potential to miniaturize the VCO used in synthesizers that generates local oscillator signal. Currently, such VCO's are implemented by combining off-chip inductors(with Q factor of at least 30) with off-chip voltage-tunable *pn* diode capacitors(frequently hyperabruptly doped and with Q factor of at least 40). These relatively low Q factors are justified for VCO applications. Because the oscillator in synthesizers is locked up to a much more stable reference crystal oscillator, which cleans up the phase noise spectrum of VCO at low f_m through a phase locked loop(PLL). Therefore, in a given transceiver system, the phase noise specifications for VCO are the easiest to satisfy. A typical number is −100 dBc/Hz at 10 kHz offset from the carrier for European GSM system.

However, attempts at miniaturization based on conventional IC technologies are not satisfactory. In particular, the CMOS ring or relaxation oscillators attain only −60 dBc/Hz at 10 kHz offset [12], far from adequate for the GSM system. Furthermore, slightly modified silicon CMOS and bipolar processes that include on-chip spiral inductors [13] or bonding wire inductors yield VCO's with unsuitable phase noise performance of about −85 dBc/Hz. A large portion of this problem arises from the implementation of tunable capacitor. Specifically, the on-chip diodes used in place of the off-chip varactor diodes exhibit excessive series resistance, much higher than their off-chip counterparts. As a result, their Q factors are severely reduced.

Recent demonstrations of voltage-tunable capacitors, composed of micromachined movable plates, show substantial improvements over the on-chip diode-based capacitors. Therefore, it has the potential to miniaturize the VCO used in synthesizers. Fig.7.4 shows the schematic of a voltage-tunable micromachined capacitor, which consists of an aluminum top plate suspended by soft flexures above a bottom plate [14]. The separation, and thus the capacitance, between plates is electrostatically controlled by the underlying electrode. The prototype capacitor in Fig.7.4 exhibits a tuning range of about 16% over a 5.5V range of applied bias, with capacitance values on the order of 2.2 pF(achieved by four of these devices wired in parallel, each with $L_p = 200$ μm and $d = 1.5$ μm) and a Q factor of 62 at 1 GHz, all on a par with the off-chip varactor diodes [14].

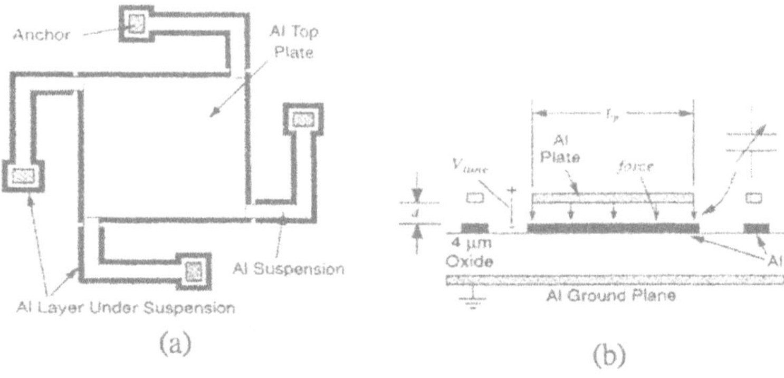

Figure 7.4. **(a) Top view and (b) cross sectional view of a voltage-tunable micromechanical capacitor [7].**

4. Low-Loss Micromechanical Switches

Many researches have focused on the implementation of micromechanical switches which can be applied to antenna or filter bank selection in multiband communication systems, and in the phased array antennas operating beyond Ka-band where antennas are small. Micromechanical switches are often characterized by the metrics describing switching speed and off/on impedance. Up to now, electrostatic actuation is the main operation mechanism of switches for communication applications.

Micromechanical switches are particularly attractive in low-power RF/ microwave/millimeter wave communication systems because of their low distortion. In addition, micromechanical switches are also very suitable for active antenna applications, such as phased arrays, power combining, and so on, due to their low insertion loss(which releave the requirement on power amplifiers) and very little dc power consumption. Current research emphasis on micromechanical switches includes lower actuation voltage and faster response time.

There are a number of ways to categorize micromachined switches. In this Section, the switches are divided into three groups: fixed-fixed-beam switches, cantilever-beam switches and compliant-beam switches. Each of them can be further described by switch type(series versus shunt), type of contact(metal-to-metal versus capacitive) and actuation voltage required for switch operation.

4.1 Fixed-Fixed-Beam Switches

All the fixed-fixed-beam architectures that have been reported nowadays are of capacitive type, which may operate in series or shunt con-

figuration. The first few designs were demonstrated by Goldsmith et al. [15], [16]. The switch consists of a lower electrode fabricated on the surface of an integrated circuit and a thin aluminum membrane suspended over the electrode, as shown in Fig.7.5. This membrane switch is of the shunt type and operates normally on a coplanar waveguide(CPW). The membrane is directly connected to the ground of CPW, while a thin dielectric covers the lower electrode to avoid direct metal-to-metal contact. The metal membrane moves between the up and down positions driven by an externally applied dc voltage between 30 and 55 V. The air gap between the two conductors determines the on-to-off capacitance ratio and specifies the switch isolation in the frequency band of interest.

The fixed-fixed-beam switches normally outperform those implemented with PIN diodes or GaAs field effect transistors(FET's) in the on-state insertion loss and off-state isolation. Unlike their solid-state counterparts which consume finite amount of current when activated, the fixed-fixed-beam switches consume no power when activated. However, the latter act much slower than the PIN or FET switches(4-20 ms versus 1-40 ns), and require relatively higher actuation voltage(20-60 V versus 3-5 V).

Micromechanical switches are also prone to stiction problems in the metal-to-metal type and dielectric charging problems in the capacitive type with a nitride film between two electrodes. Research is also being conducted to evaluate their lifetime which is on the order of billions of cycles today. However, the observation that micromechanical switches are extremely linear device makes them favorable for communication systems. A 2 GHz switch in a two-tone experiment results in negligible intermodulaton and an extrapolated IP3 of 36 dBm(4W).

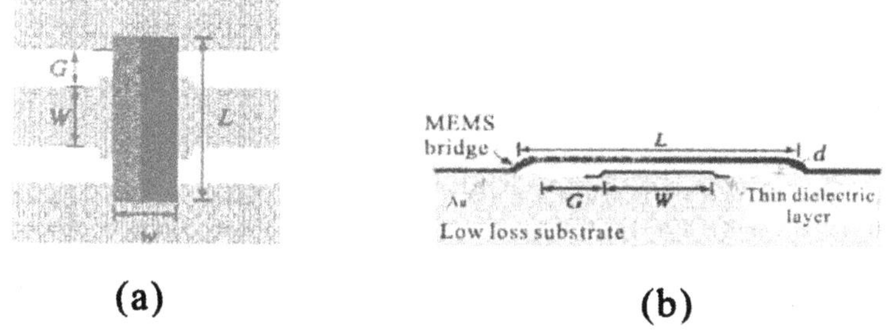

Figure 7.5. (a) Top view of a typical shunt capacitive membrane switch over a CPW. This type of switch has W of the same order as L. (b) Cross sectional schematic of a typical air-bridge switch over a CPW. This type of switch has W much less than L [5].

4.2 Cantilever-Beam Switches

Many results on cantilever-beam switches have been reported [17], [18]. Schematic of such a switch is shown in Fig.7.6. Primary structural difference between this architecture and the fixed-fixed beam switch lies in the use of dielectric layers to support the beam and to provide isolation between dc and RF electrodes. This allows for a metal-to-metal contact in series or shunt switch architecture. The cantilever-beam switches require lower activation voltage than the fixed-fixed-beam switches and exhibit higher isolation.

Measurements of scattering parameters show a very low insertion loss over a broad frequency range from a few MHz to at least 10 GHz(X-band). The series switch has a very low insertion loss of 0.2 dB up to 40 GHz, the isolation is −40 dB at 500 MHz and degrades to −15 dB at 40 GHz. Detailed description of design, fabrication and experimental verification for this switch can be found in [17], [18]. To extend the switch performance into higher frequencies while keeping the isolation below −30 dB, it is plausible to combine with resonant switches.

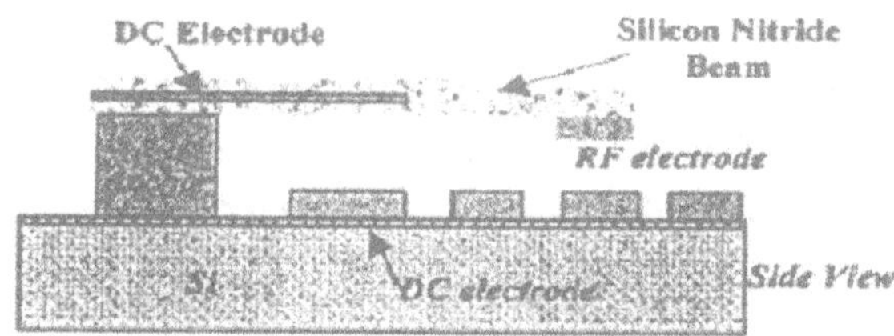

Figure 7.6. **Cross sectional schematic of a typical cantilever-beam switch [5].**

4.3 Compliant-Beam Switches

Major disadvantage of the switches presented above is the need to operate under relatively high dc voltages which vary from 30 to 120 V. For MEMS switches to become useful for wireless communication, actuation voltages less than 6 V are required. Therefore, research on the design of electrostatic shunt microwave switches with low actuation voltage is ongoing. Ultra low-loss and low-actuation-voltage RF MEMS switches have been successfully fabricated [19]. These switches are designed to be used with finite ground coplanar waveguide(FGCPW), but the design methodology and fabrication technique are generic and can be extended for the development of any low-actuation-voltage switch. In order to

reduce the pull-in voltage of the structures, the following three design approaches may be pursued: design a structure with low spring constant, increase the area of actuation, or reduce the gap between switch and the bottom electrode.

5. Microscale Vibrating Micromechanical Resonators

From a designer's perspective, micromechanical resonators are extremely flexible. Design and implementation of complicated resonator systems become much easier due to the following features: (1) the Q factor is large and the usable frequency range is wide, (2) instinctive voltage-controlled frequency tenability and switchability [20], (3) adaptability to trimming, (4) flexibility in the choice of geometries and materials, which leads to a wide variety of possible designs, (5) flexibility in the type of transduction such as electrostatic, piezoelectric and magnetostrictive which have been applied in the past.

Crucial features are usually defined in a single masking step, which is part of a planar process and is hence highly compatible with conventional IC processes. Therefore, construction of resonators is relatively straightforward.

For low-frequency applications, flexural-mode micromechanical resonators with various kinds of structural materials have been implemented by using planar IC-compatible micromachining processes. Excellent results from low frequency(LF) to very high frequency(VHF) have been demonstrated. Q factors exceeding 80,000 in vacuum have been reported for LF flexural-mode resonators constructed in surface micromachined polysilicon [21]. In addition, Q factors on the order of 20,000 have been achieved at 70 MHz(VHF) in single-crystal silicon materials [22].

Micromechanical resonator devices have their drawbacks, including requirement of vacuum to achieve high Q factor, higher temperature coefficient(-10 ppm/°C) than quartz, and uncertainties in dynamic range and power handling capacity. Research on micromechanical resonators and their high frequency applications does not start until recently. Operation in the GHz range is predictable. However, its frequency limit is still unknown [20].

6. Thin-Film Bulk Acoustic Resonators

In the past, Q factor of thousands is required to obtain more accurate oscillators and channel-select filters. Hence, one must consider vibrating

mechanical resonators like quartz crystal resonators and surface acoustic wave(SAW) resonators. Since such Q factors are attainable by using the aforementioned macroscopic vibrating mechanical tanks, there have been many researches on how to develop novel thin film technologies so that miniaturized versions of vibrating mechanical resonators can be produced to operate on the principles similar to their macroscopic counterparts. This type of piezoelectric devices is called thin-film bulk acoustic resonator(FBAR), which is the only commercialized micromachining product today.

Fig.7.7 shows a schematic of FBAR which is composed of deposited piezoelectric films(for example, aluminum nitride) between conductors in a pattern similar to quartz crystals, and the whole structure is suspended on a thin membrane with low stress(for example, silicon nitride), acoustically isolated from the substrate. Such resonators have been demonstrated with measured Q factors over 1,000 and resonant frequencies ranging from 1.5 to 7.5 GHz, with sizes of less than 400 × 400 μm^2 [23]. The device shown in Fig.7.7 is promising, but the structural integrity of supporting membrane could be a problem. For example, strain in the film can lead to breakage [24]. Hence, continuous improvements in process and trimming technologies are necessary.

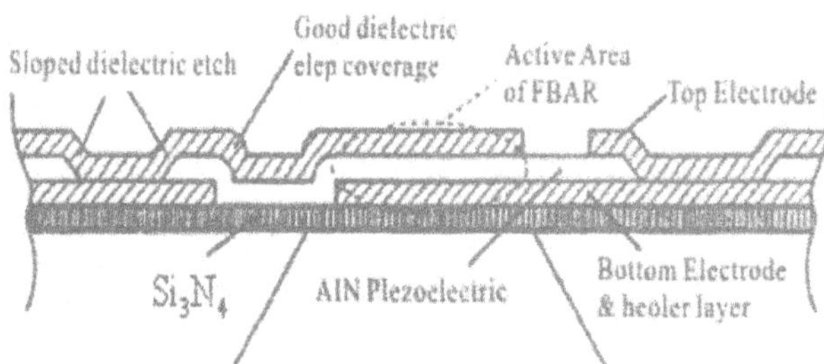

Figure 7.7. **Cross section of a membrane-supported FBAR resonator [23].**

One possible solution to the above problem is to avoid the fragile membrane support and to build the thin film resonator directly on top of a solid substrate, as shown in Fig.7.8. In this scheme, energy loss to the substrate can be reduced by acoustically separating the substrate from the piezoelectric resonator. Proper number and thickness of thin-film layers can be selected to transform the relative impedance between resonator and substrate, so that isolation between these two media can be achieved [24]. Although the implementation of such a solidly mounted resonator(SMR) requires careful deposition of layers, it is expected to

improve the resilience of thin-film bulk acoustic resonators to a large extent.

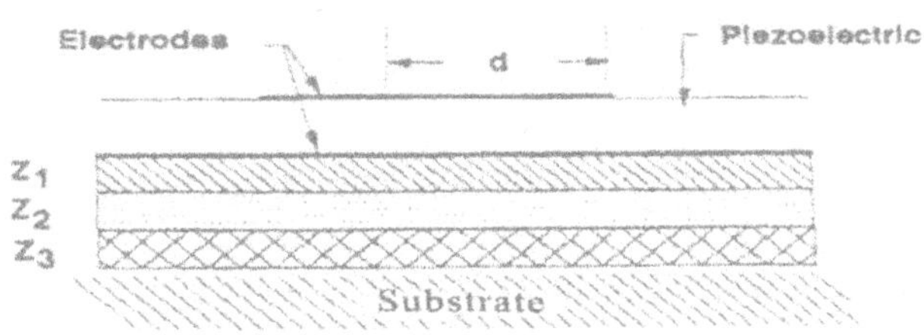

Figure 7.8. Cross section of a solidly mounted resonator using quarter-wavelength reflecting layers [24].

Although thin-film bulk acoustic resonators exhibit many advantages, there are still several major drawbacks that restrain their use at the present time. First, the resonators developed in the GHz range are too thick and cumbersome when extended to lower frequencies. Second, a trimming technique is necessary, especially if groups of these resonators are used to implement narrow-band filters with small shape factors. However, rapid and effective approach for trimming and tuning is not available up to now.

7. Micromachined Microwave and Millimeter Wave Filters

Fig.7.9(a) shows a two-resonator micromechanical band-pass filter with frequencies up to 14.5 MHz, bandwidth on the order of 0.2% and insertion loss of less than 1 dB [25]. In addition, Fig.7.9(b) shows a three-resonator filter with frequencies near 455 kHz, bandwidth of 0.09%, very low insertion loss, and more than 64 dB of stop-band rejection [26]. This filter features balanced comb transduction for feedthrough suppression, low velocity coupling, frequency tuning electrodes, and skillfully demonstrates the complexity and flexibility of MEMS technology.

For transceivers operating in the K-band and higher frequencies, preselect or image-reject filters with bandwidths of 1-8% are implemented by using microwave and millimeter wave components. Because such front-end filters are placed just behind the antenna and before any amplifier, the low loss feature is important because the loss contributes directly to noise figure of the receiver and reduces the effective radiated power from the transmitter. Meanwhile, in order to minimize loss, high-Q resonators are required. Again, this can only be achieved by using external

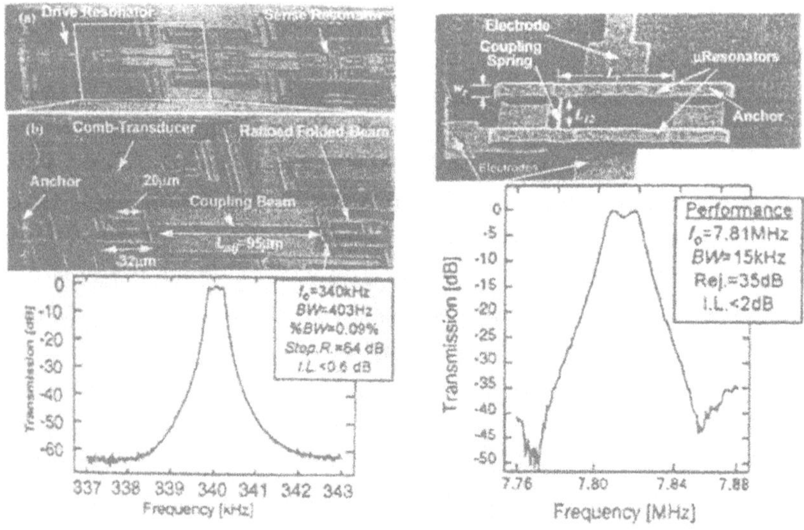

Figure 7.9. **(a) SEM of a high-frequency surface micromachined two-resonator filter and measured frequency response. (b) SEM of a medium-frequency surface micromachined three-resonator filter and measured frequency response [7].**

elements such as dielectric resonators or waveguide components, which increase the size and cost of the receiver.

To solve this problem, filter implementation using monolithic transmission-line resonators has been studied. Dielectric constants of Si and GaAs substrates are 11.9 and 13.1, respectively. Major problems of transmission line resonators on high-permittivity substrates are the radiation loss into dielectric and the increased Ohmic loss due to small dimensions. Such losses reduce the Q factor of planar resonators to about 40-60 at 30-60 GHz, not acceptable for low-loss filters. An interim solution is to integrate the planar filters on Teflon or quartz substrates with dielectric constant of 2.2-4, which increases the Q factor to 200-300 at 30-60 GHz. The drawback of this approach is being incompatible with Si/GaAs integrated circuits. Hence, transitions between millimeter wave active circuits and external filters are required.

It has been demonstrated that by integrating the resonator on a thin dielectric membrane about 1-1.5 μm thick, and enclosing the transistor in a microshielded cavity about 100-500 μm high, the Q factor of a planar resonator on Si/GaAs substrates can be improved considerably. Standard micromachining processes are used to fabricate such a

resonator, which include the following four steps: deposition of a silicon nitride membrane, back etching of the wafer, etching and filling of via holes with gold, and attachment of several wafers to form a miniature stack.

The transmission line resonators are effectively suspended in free space and are limited only by Ohmic loss. Furthermore, the resonators are wider than their counterparts on Si/GaAs substrate in order to obtain the same filter impedance, while considerably reducing their Ohmic loss. The measured Q factors of quarter-wavelength resonators with $W = 700$ μm and $h = 200$ μm at 60 GHz are around 600, which is ten times higher than comparable resonators on Si/GaAs substrates and only three to four times lower than that of waveguide resonators.

Micromachined transmission lines fabricated on thin dielectric membranes have been used extensively in filter designs from 14 to 250 GHz. Standard elliptic and Chebyshev filters can be designed, and typical transmission lines such as coplanar waveguide, strip line, suspended microstrip line can be used for implementation. Excellent performances have been demonstrated. Much attention has been paid to the transition between Si/GaAs substrate and membrane transmission lines, and a transition with lower than -20 dB of return loss is achievable up to 60 GHz. For a four-pole elliptic filter at 60 GHz, an insertion loss of only 1.5 dB and a rejection level of lower than -40 dB have been achieved, as shown in Fig.7.10 [26]. This state-of-the-art filter is only 0.5-0.7 dB more lossy than its waveguide counterpart, but its size reduction is one thousand folds.

Another application of micromachining is the fabrication of three-dimensional cavities to synthesize miniaturized waveguide components at 10-60 GHz. A resonant waveguide cavity with reduced height and dimensions of $\lambda/2$ by $\lambda/2$ has been etched in a silicon wafer and fed either by a slot transition or by bonding wires. Since the fields are not confined to the planar resonators, the structures have a Q factor of around 500 at 10 GHz, and up to 1,100 at 30 GHz [27]. These miniaturized waveguide cavities are under investigation for use in low-phase noise oscillators and satellite filters.

Another advantage of micromachined lines is their micropackaging feature. As shown in Fig.7.10, filter or any component inside the microcavity is completely shielded, and does not couple any RF energy to the outside. Typical measured isolation between adjacent transmission lines is -50 dB at 30 GHz and -45 dB between closely spaced filter banks at 10 GHz [28]. Such micropackaging techniques are expected to facilitate the integration of millimeter wave transmitters and receivers on the same chip.

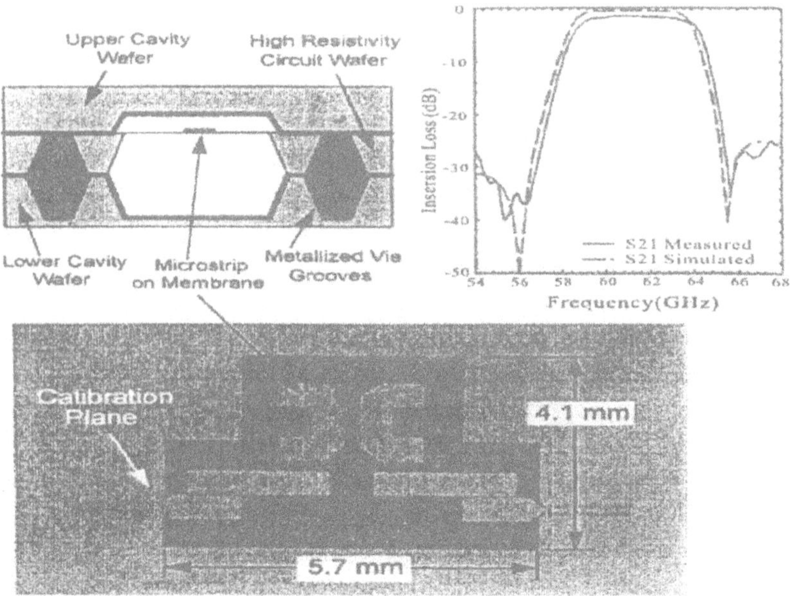

Figure 7.10. Four-pole micromachined elliptic filter with an 8% bandwidth at 60 GHz, the resonator width is 500 μm, its Q factor is 500, and insertion loss in the pass-band is 1.5 dB [7].

8. Miniature Antennas for Millimeter Wave Applications

At millimeter wave and THz frequencies, the skin depth is smaller than the thickness of underlying substrate. Therefore, antennas on Si/GaAs or even quartz substrates are ineffective radiators at these frequencies. Hence, micromachining has been developed to implement antennas at millimeter wave and THz frequencies to reduce the loss and size constraints. Micromachining has been used to suspend dipole or slot radiators on thin dielectric membranes without shielding cavities. Based on the same reason, suspended planar antennas are expected to radiate effectively in free space. Anisotropic etching has also been used to integrate silicon pyramidal cavities which act as miniature integrated horns around dipole antennas, directing the radiated energy efficiently in the forward direction [29].

At microwave frequencies, membrane technology with $\epsilon_r \simeq 1$ results in physically large antennas and is thus impractical. Hence, micromachining other than membrane technology can be applied to synthesize

a required dielectric constant for the substrate underneath the antennas. In particular, by etching a portion of the substrate or by properly choosing the density of via holes in the substrate, dielectric constant between one(complete removal) and ϵ_r(full) can be synthesized, as shown in Fig.7.11(a). Using such techniques, the efficiency of microstrip antennas at 12-13 GHz has been increased from 55%(ϵ_r=10) to 85%(ϵ_{syn} =2.3) [30]. Similar techniques can also be applied at 77 GHz for automotive radar applications. Dielectric constant around the planar antenna can be further modified to enhance its radiation efficiency and bandwidth.

Micromachining techniques have also been used to synthesize photonic band gap(PBG) materials on high-resistivity silicon substrates at 30 GHz by etching a periodic pattern of hexagonal holes with $d = 1{,}050$ μm and period of 1,700 μm in the silicon substrate, as shown in Fig.7.11(b) [31]. The hole array suppresses wave modes in the substrate and therefore increase the radiation efficiency of planar antennas, from 30% without holes to 85% with holes. Photonic band gap materials may play an important role in the integration of high-efficiency antennas on Si and GaAs substrates.

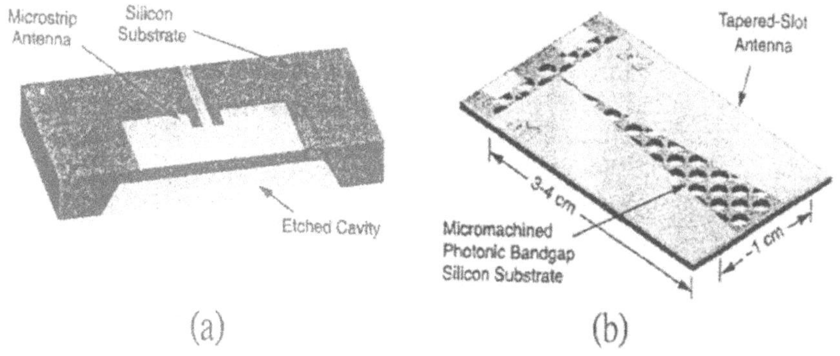

Figure 7.11. **(a) Microstrip antenna fabricated on a synthesized dielectric by partially removing the silicon substrate. (b) High-efficiency tapered slot antenna operating at 30 GHz, fabricated on a micromachined photonic band gap substrate [31].**

References

[1] T. H. Lee and S. S. Wang, "CMOS RF integrated circuits at 5 GHz and beyond," *Proc. IEEE*, vol.88, no.10, pp.1560-1571, Oct. 2000.

[2] H. W. Chiu and S. S. Lu, "A 2.17 dB NF, 5 GHz band monolithic CMOS LNA with 10mW DC power consumption," *IEEE VLSI Circuits Symp.*, June 2002.

[3] C. P. Yue and S. S. Wong, "On chip spiral inductors with patterned ground shields for Si-based RF IC's," *IEEE J. Solid-State Circuits,* vol.33, pp.743-752, May 1998.

[4] J. Craninckx and M. S. J. Steyaert, "A 1.8 GHz CMOS low-phase-noise voltage controlled oscillator with prescaler," *IEEE J. Solid-State Circuits,* vol.30, pp.1474-1482, Dec. 1995.

[5] S. S. Mohan, "The design, modeling and optimization of on-chip inductor and transformer circuits," Ph. D. thesis, Stanford Univ., Stanford, CA, 1999.

[6] J. A. von Arx and K. Najafi, "On-chip coils with integrated cores for remote inductive powering of integrated microsystems," *Int. Conf. Solid-State Sensors, Actuator,* pp.999-1002, 1997.

[7] C. T. C. Nguyen, L. P. B. Katehi, and G. M. Rebeiz, "Micromachined devices for wireless communications," *Proc. IEEE,* vol.86, no.8, pp.1756-1768, Aug. 1998.

[8] L. H. Lu, P. Bhattacharya, G. Ponchak, and L. P. B. Katehi, "X-band and K-band lumped Wilkinson power dividers with a micromachined technology," *IEEE MTT-S Int. Microwave Symp. Dig.*, pp.287-290, 2000.

[9] J. S. Rieh, L. H. Lu, L. P. B. Katehi, P. Bhattacharya, E. T. Croke, G. E. Ponchak, and S. A. Alterovitz, "X- and Ku- band amplifiers based on Si/SiGe HBT's and micromachined lumped components," *IEEE Trans. Microwave Theory Tech.,* vol.42, pp.685-694, May 1998.

[10] D. Kother, B. Hopf, T. Sporkman, and I. Wolf, "MMIC Wilkinson couplers for frequencies up to 110 GHz," *IEEE MTT-S Int. Microwave Symp. Dig.*, pp.663-666, Dec. 1995.

[11] H. Samavati et al., "Fractal capacitors," *IEEE J. Solid-State Circuits,* vol.33, pp.256-257, Dec. 1998.

[12] T. C. Weigandt, B. Kim, and P. R. Gray, "Analysis of timing jitter in CMOS ring oscillators," *Proc. ISCAS,* pp.27-30, June 1994.

[13] N. M. Nguyen, "A 1.8-GHz monolithic LC voltage-controlled oscillator," *IEEE J. Solid-State Circuits,* vol.27, pp.444-450, Mar. 1992.

[14] D. J. Young and B. E. Boser, "A micromachined variable capacitor for monolithic low-noise VCO's," *Int. Conf. Solid-State Sensor, Actuator,* pp.86-89, Hilton Head Island, SC, 1996.

[15] C. Goldsmith, J. Randall, S. Eshelman, and T. H. Lin, "Characteristics of micromachined switches at microwave frequencies," *IEEE MTT-S Int. Microwave Symp. Dig.*, pp.1141-1144, 1996.

[16] C. Goldsmith, A. Malczewski, Z. J. Yao, S. Chen, J. Ehmke, and D. H. Hinzel, "RF MEMS variable capacitors for tunable filters," *RF Microwave Computer-Aided Eng.*, vol.9, no.4, pp.362-374, July 1999.

[17] Y. Cai and L. P. B. Katehi, "Wide band series switch fabricated using metal as sacrificial layer," *Eur. Microwave Conf.*, pp.32-34, Paris, France, 2000.

[18] D. Hyman et al., "Surface-micromachined RF MEMS switches on GaAs substrates," *RF Microwave Computer-Aided Eng.*, vol.9, no.4, pp.348-361, July 1999.

[19] S. P. Pacheco, L. P. B. Katehi, and C. T.-C. Nguyen, "Design of low actuation voltage RF MEMS switch," *IEEE MTT-S Int. Microwave Symp. Dig.*, pp.165-168, 2000.

[20] C. T.-C. Nguyen, "Frequency selective MEMS for miniaturized communication devices," *IEEE Aerospace Conf.*, pp.445-460, 1998.

[21] C. T.-C. Nguyen and R. T. Howe, "Quality factor control for micromechanical resonators," *IEEE Int. Electron. Dev. Meeting*, pp.505-508, 1992.

[22] A. N. Cleland and M. L. Roukes, "Fabrication of high frequency nanometer scale mechanical resonators from bulk Si crystals," *Appl. Phys. Lett.*, vol.69, no.18, pp.2653-2655, Oct. 28, 1996.

[23] S. V. Krishnaswamy, J. Rosenbaum, S. Horwitz, C. Yale, and R. A. Moore, "Compact FBAR filters offer low-loss performance," *Microwave RF*, pp.127-136, Sept. 1991.

[24] K. M. Lakin, G. R. Kline, and K. T. McCarron, "Development of miniature filters for wireless applications," *IEEE Trans. Microwave Theory Tech.*, vol.43, pp.2933-2939, Dec. 1995.

[25] F. D. Bannon, III and C. T. C. Nguyen, "High frequency microelectromechanical IF filters," *IEEE Int. Electron. Dev. Meeting*, pp.773-776, 1996.

[26] K. Wang and C. T. C. Nguyen, "High-order micromechanical electronic filters," *IEEE Int. MEMS Workshop*, pp.25-30, Nagoya, 1997.

[27] J. Papapolymerou, J. C. Cheng, J. East, and L. Katehi, "A micromachined high-Q X-band resonator," *IEEE Microwave Guided Wave Lett.*, vol.7, pp.168-170, June 1997.

[28] A. R. Brown, P. Blondy, K. Hong, and G. M. Rebeiz, "Low-loss millimeter-wave filters and high-Q micromachined cavity resonators," *U.S. Army Res. Office Int. Rep.*, Dec. 18, 1997.

[29] W. Y. Ali-Ahmad and G. M. Rebeiz, "An 86-106 GHz quasi-integrated low-noise receiver," *IEEE Trans. Microwave Theory Tech.*, vol.41, pp.558-564, Apr. 1993.

[30] J. Papapolymerou, R. F. Drayton, and L. Katehi, "Micromachined patch antennas," *IEEE Trans. Antennas Propagat.*, vol.46, pp.275-283, Feb. 1998.

[31] T. J. Ellis and G. M. Rebeiz, "MM-wave tapered slot antennas on micromachined photonic bandgap dielectrics," *IEEE MTT-S Int. Microwave Symp. Dig.*, pp.1157-1160, June 1996.

Chapter 8

INTRODUCTION TO LOW-TEMPERATURE COFIRED CERAMICS TECHNOLOGY

I-Nan Lin
Materials Science Center
National Tsing-Hua University
Hsin-Chu, Taiwan, ROC

Abstract In this Chapter, low-temperature cofired ceramics(LTCC) technology is briefly reviewed for its processes, advantages, design rules, potential applications and future developments.

Keywords: package, miniaturization, multilayer, green tape, ceramics, screen printing, shrinkage.

1. Advantages of LTCC Technology

Driving factors like pocket size, light weight, low power, and above all, low cost, have triggered a revolution in the development of wireless communication systems which demand better performance than earlier systems in a multipath environment. Integration of state-of-the-art integrated circuits embedding both active and passive components is required.

Fast progress of active devices has created pressure on the integration of passive components. In most modern electronic systems, passive components outnumber active devices by several folds. In wireless communication devices such as cellular phones, wireless LANs and satellite GPS, the ratio of passive to active components can be greater than a hundred. This high ratio restricts further reduction of size, weight and cost. A new approach to integrate passive components to keep pace with the evolution of IC technology is urgently needed.

1.1 Miniaturization of Wireless Communication Devices

Volume reduction requirements on personal electronic devices for communication and surveillance demand new technology. Packaging technologies available for microelectronics in the early 1980s were thick-film and thin-film circuits hermetically sealed in a package made of ceramics or metal with glass-to-metal feedthrough. Required functions of interconnection, packaging and component integration complicated the assembly of hybrid microcircuits and increased its volume and weight. None of the existing technologies at that time were able to fulfill all these three functions.

High-temperature cofired ceramics(HTCC) technology provides a durable hermetic package, but the refractory metals are too lossy and the thick-film dielectrics can not easily provide the required thickness or thickness tolerance to obtain the designed impedance of transmission line. Low-temperature cofired ceramics(LTCC) is expected to provide a new integrated packaging technology from a combination of thick film and cofired dielectrics. One major advantage of LTCC is the use of low-loss conductors for microwave circuit applications.

1.2 Advantages of Passive Components Integration

LTCC has a distinct advantage in making multilayered structures where each layer is processed separately, and resolution can be maintained throughout the process. In contrast, resolution of thick film gets worse due to the accumulation of thickness irregularities with more layers. LTCC offers the capability of finer pitch for interconnections. In addition to the high density of lines, RF grounding and shielding may also be deployed throughout the structure by using vias.

More than 40 layers can be stacked in a single firing process with LTCC, which is impractical with most other technologies, such as thick film, thin film or MCM-D. The cost of processing layers sequentially becomes prohibitive. With LTCC, the processing costs are only fractionally increased. LTCC has another advantage of being able to replace defective layers, which is also useful for making changes in the circuitry.

The LTCC provides the capability to construct various geometries of interconnections by layer cutouts, which enables the implementation of RF strip line transitions as well as other high-frequency circuitries at lower cost and without potential transition losses. In addition, some LTCC tapes have improved loss properties that make them more desirable than traditional technologies for RF applications. The

LTCC technology offers more design flexibility over conventional thick-film, thin-film and HTCC technologies in integrating microstrips, strip lines, coplanar waveguides and dc lines into one multilayered structure. Impedance control and shielding of RF and microwave signals through low-loss medium are critical in designing RF and microwave packages, submodules and modules. The three-dimensional layout of LTCC offers capabilities to accomplish these goals. Dc and power feed lines are contained within the same structure and are isolated from each other by ground planes or arrays of grounding vias. Fig.8.1 shows a schematic of integrated RF module in LTCC.

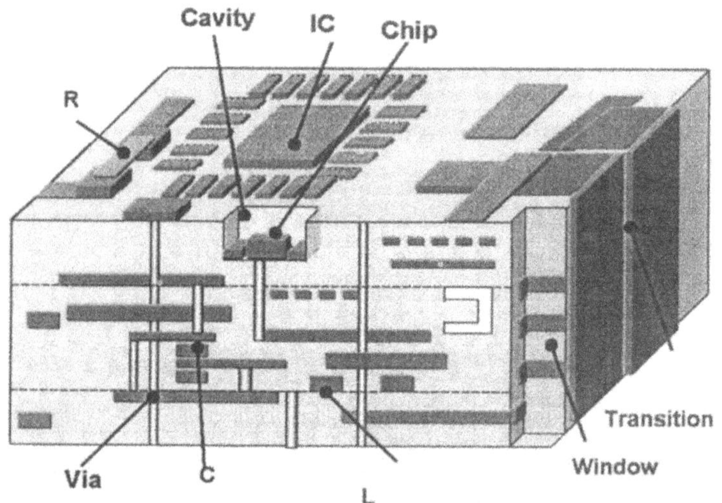

Figure 8.1. **An integrated RF module in LTCC package.**

Packaging technology is another area where the LTCC claims a distinct advantage. IC chips may be placed in a cavity to lower the package profile. With available hermetic brazing processes, large packages can be built using only LTCC and a lid, or using a brazed seal ring on all the LTCC packages. Such unique capability for LTCC to form a hermetic package using noble metals eliminates the need for fabricating separate packages and substrates. Another advantage of this packaging technology is the reduced number of interconnections and processing steps. The interconnection capabilities for high-frequency microwave and RF transmission lines and transitions has made it possible to design and build integrated package of RF circuits.

An obvious advantage of passive component integration is the number reduction of contacts or transitions, the associated losses are thus

reduced significantly. Another advantage is the increased reliability by merely reducing the number of transitions, as faults occur primarily at the transitions or interfaces between materials. An additional advantage of integrated components in LTCC is cost reduction. A large number of assembly steps are eliminated with a few additional steps required for component integration. Also, footprint constraints are removed by embedding components in multiple layers within the package. In some cases, components can be distributed vertically by using vias.

In summary, LTCC technology has a clear advantage over other multichip module(MCM) technologies in integrating passive components. A much wider range of component feature becomes available compared to other MCM technologies, and most, if not all, passive structures can be integrated in a single processing step. Thus, the cycle time of technology evolution can be dramatically shortened.

2. LTCC Technology

LTCC is conceived as a technology that possesses the advantages of both thick-film and HTCC technologies and has excellent properties for packaging and MCM applications, rendering simpler processes and high layer count. The green tapes are easily fabricated. They are soft, pliable and easily dissolved and abraded. Once the material is fired and fully sintered, it becomes highly rigid. Small structures can be carved and machined using diamond tools, and excimer laser can be utilized to ablade alumina with little tolerances.

In summary, there are several reasons for using the LTCC technologies. By optimizing steps of the manufacturing process, mass production methods can be applied. The fabrication techniques are relatively simple and inexpensive for automation. Design and manufacture of three-dimensional circuits is feasible, and the number of signal layers is almost unlimited. There is a good match to semiconductor TCEs. Tapes of different compositions can be manufactured with desired layer properties. The tapes can be used at frequencies over 30 GHz. The substrate has very good hermeticity. It has high resistance against hostile environment, and its thermo-physical properties can be modified.

2.1 Manufacturing Processes

LTCC tapes are glass ceramics composite materials. Alumina(Al_2O_3) or any other ceramics can be used as ceramic filler. A glass frit binder is usually used to lower the processing temperature as well as to render the material compatible with thick-film technology. Doctor blade machine is used to obtain controlled tape thickness. Casting is accomplished by

spreading the desired slurry formulation to form a paste on a moving carrier substrate(usually a film of cellulose acetate, Teflon, Mylar or cellophane). By a soft bake procedure, one obtains a thin and flexible tape that can be cut and stamped to the desired configuration prior to firing. The doctor blade procedure requires careful control to avoid warpage, thickness tolerance violation and other defects. A variety of other additives are included in the formulation for both aqueous and nonaqueous doctor blade systems.

One of the important features of green tape technology is the possibility of fabricating three-dimensional structures using multiple layers of green tapes. Each layer is fabricated in green tape with whatever features needed to achieve the overall function of the three-dimensional structure. Each layer may embed vias, cavities, channels, capacitors, resistors and interconnections. Individual layers are then stacked in proper order and placed in a registry to yield the desired structure. Location holes for registry and vias are usually abraded or cut. The next step is moving the stacked blanks to the press where heat and pressure are applied to complete the lamination process. Then, the structure is ready for sintering in an air furnace.

A complete processing sequence of LTCC packages is depicted in Fig. 8.2. The number of layers can reach 80 or even higher. During stacking, successive layers should be rotated by 90° to compensate for the texture(preferential orientation) induced by fabrication process. Nominal thickness tolerance after casting is 1.5 μm with commercially available materials. In green tapes, the smallest punched cavity or via hole is on the order of 25 μm in diameter.

LTCC shrinks upon sintering. Nominal DuPont alumina LTCC formulation shrinks 12% in the tape plane and 15% perpendicular to the tape plane. If shrinkage is uniform and predictable, one can compensate for the shrinkage during the design phase. The fracture strength of an alumina formulation is 320 MPa, which renders the fired structure extremely rigid. Porosity of LTCC is very small due to the use of vitreous material such as alumina binder.

There are several metallization schemes that are shrinkage matched to LTCC tapes. Metals such as Au, Ag(air fired) or Cu(reducing or neutral atmosphere) are used for interconnections, electrodes and via filling. Resistors and dielectric formulations are available with complete shrinkage compatibility to LTCC materials. After sintering, the LTCC tapes become very stiff and partially resistant to glass leaching although it is possible to deposit thick films(cermets or polymeric) and perform further sintering at a lower temperature. Finally, these materials can be machined using laser or diamond tools to build internal structures.

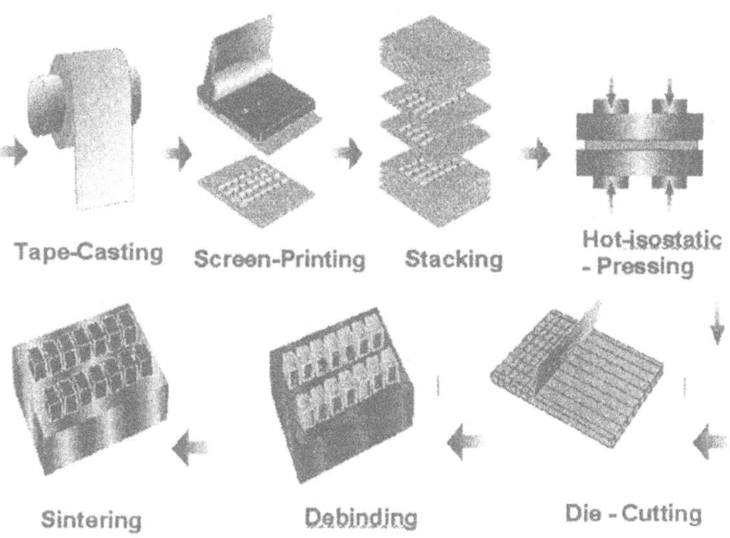

Figure 8.2. Processing sequence of LTCC packages.

LTCC can be viewed as a multilayered circuit fabricated by laminating green tapes with printed conductor lines or patches on their surfaces and firing them all together in one step. This process is similar to that of HTCC. Major difference between these two technologies is the firing temperature: in excess of 1,000°C for HTCC and below 1,000°C for LTCC, mostly 850 to 875°C, which makes it possible to use silver for conductor lines. Major advantage of LTCC is the possibility to use low-resistivity conductors like silver, gold, copper and alloys with palladium and platinum instead of tungsten and molybdenum, which is of critical importance for microwave applications. Some merits of LTCC are listed in Table 8.1.

The LTCC process includes several steps as shown in Fig. 8.3 and are described as follows.

Design. Design of multilayered circuits for RF applications requires simulation, modeling and layout tools.

Preparation. Green tapes are usually shipped on a roll, and have to be unrolled on a clean, stainless steel table. The sheet is cut with a razor, laser or a punch into parts.

Forming and filling vias. Vias may be punched or drilled with a low-power laser. Vias can be filled with a conventional thick-film screen printer or an extrusion via filler. Both methods require a mask which is made of stainless steel or Mylar foil 150-200 μm thick.

Table 8.1. Merits of LTCC for microwave applications.

- High density of interconnections with three-dimensional design
- Integration of antennas, resistors, inductors and capacitors
- Packaging capabilities
- Cost effective manufacturing by parallel process steps and reduction of SMT components
- Robust against mechanical stress and high temperature
- TCE close to semiconductor material, suitable for integration
- Potential in radio-frequency applications(< 30 GHz)

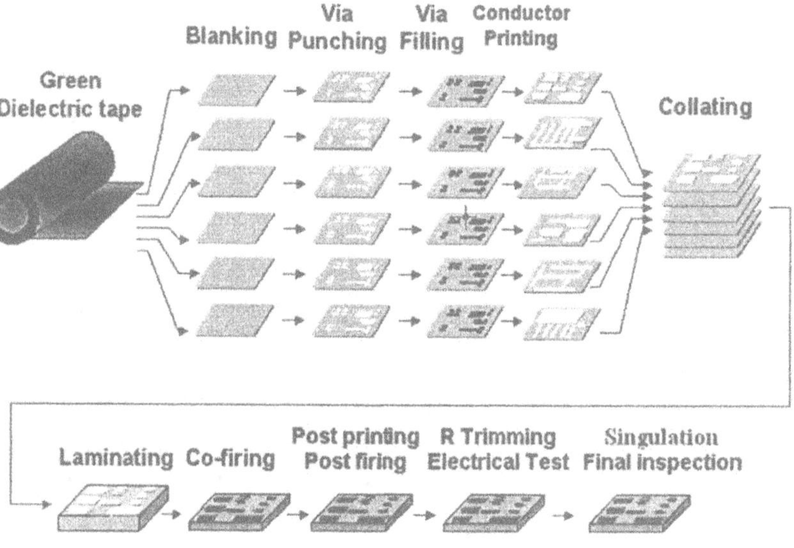

Figure 8.3. Typical LTCC process.

Screen printing. Cofireable conductors are printed on the green tape using a conventional thick-film screen printer with standard(250 - 325 mesh) emulsion type of screens. Screen printing is easier and of higher resolution than standard thick-film on alumina due to the flatness and solvent absorption of tape.

Lamination. Each layer is placed in turns over tooling pins, one on top of the other. Some processors use heat pliers to fix the sheets. There are two ways of laminating tapes. The first is uniaxial lamination

in which the tapes are pressed between heated plates at 70°C and 200 bar for 10 minutes, and a 180° rotation is required after 5 minutes. The second way is to use an isostatic press. The stacked tapes are vacuum packaged in a foil and pressed in hot water.

Cofiring. The laminates are fired in one step on a smooth and flat setter tile. The firing profile requires a programmable box kiln. A typical profile shows a slow temperature rising rate of about 2-5°C per minute up to about 450°C with a dwell time of about one to two hours in which organic burnout of binder takes place. Then, the temperature is raised up to 850 to 875°C with a dwell time of about 10 to 15 minutes. The whole firing cycle lasts between three and eight hours, depending on the materials.

Postprocessing. There are three different ways to cut fired parts into smaller pieces or other shapes. The first way is to use a post-fire dicing saw. It holds tight dimension tolerances to have high quality edges, and works very well for rectangular shapes. The second way is to use an ultrasonic cutter which is very slow and expensive, the final part shows low tolerances and may have irregular shapes. The third way is to use a laser to cut the fired tape. The tolerances are tight, but the quality of edges is very poor.

2.2 Design Rules

Synthesis of LTCC tapes is an extremely complicated process. Not only the interaction between glass and ceramic constituents, but also the compatibility between ceramic tape and conductive paste need to be stringently controlled. Major companies have developed LTCC tapes with good characteristics for various applications, some of them are listed in Table 8.2. Due to the shrinkage of LTCC tapes during firing, it is necessary to observe various design rules exemplified in Table 8.3, where the features listed in the table are indicated in Fig.8.4. A few design guidelines published by manufacturers are also listed in Table 8.4.

3. LTCC Applications

LTCC technology provides a monolithic, three-dimensional, cost-effective solution for microwave circuits and packages. This new technology is developed for firing temperature typically lower than 1,000°C, below the melting points of many good conductors such as gold, silver and copper. The use of low-loss metallization is a significant improvement over the HTCC technology.

LTCC is based on glass ceramics composite that provides a versatile approach to integrate high-performance electronic packaging. This

Table 8.2. Suppliers of LTCC tapes.

Company	Materials
DuPont	951: ϵ_r = 7.5 (standard tapes: CT, AT, A2, AX) 934: ϵ_r = 7.8 (low-loss tape)
ESL(Electro-Science Laboratories)	41110-25C: ϵ_r = 4 - 5 (transfer tape: zero shrink) 41010-25C: ϵ_r = 7.2 - 8.2 (transfer tape: zero shrink) 41020-25C: ϵ_r = 8 - 10 (transfer tape: zero shrink) 41110-70C: ϵ_r = 4.3 - 4.7 41020-70C: ϵ_r = 7 - 8
Ferro	A6M: ϵ_r = 5.9 (microwave tape) A6S: ϵ_r = 5.9 (low-cost microwave tape)
Heraeus	CT2000: ϵ_r = 9.1 CT700: ϵ_r = 7.5 - 7.9 CT800: ϵ_r = 7.5 - 7.9 (zero-shrink tape)
Kyocera	GL550: ϵ_r = 5.6 - 5.7 GL660: ϵ_r = 9.4 - 9.5
Nikko	Ag2: ϵ_r = 7.8 Ag3: ϵ_r = 7.1
NG(Northrop Grumman)	Low K: ϵ_r = 3.9
Samsung	TCL-6A: ϵ_r = 6.3 TCL-7A: ϵ_r = 6.8

Table 8.3. Example of design rules offered by some LTCC foundries.

Features	Standard value	Possible value
Edge to feature(μm)	500	250
Line width(μm)	200	100
Line width(μm) (fine line process)	50	50
Line spacing(μm)	200	100
Line spacing(μm) (fine line process)	50	50

Table 8.3 (continued)

Line pitch(μm)	400	200
Line to cavity wall spacing(μm)	400	
Via diameter(μm)	250	100
Via coverpad/catchpad	2 × via diam.	via diam. + 25 μm
Via spacing(μm)	200	175
Via pitch(μm)	400	300
Via stagger	1.5 × via diam.	
Via to cavity wall spacing(μm)	400	
Via center to part edge(μm)	550	300
Via geometry	round + rectangular	round + rectangular
Via type	blind, buried, stacked	blind, buried, stacked
Thermal via coverage(%)	15	30
Min. resistor dimension(mm)	1 × 1	0.5 × 0.5
Min. resistor overlap(μm)	300	200
Min. width after trim(μm)	300	200
Tolerance after trim (surface/buried)	±2 %/±25 %	±1 %/
Ground/power plane coverage	50%	75%
Opening for feedthrough (length in μm)	750	350
Max. number of layers	depends on thickness (10-50)	depends on thickness (10-50)
Min. number of layers	depends on thickness (2-4)	depends on thickness (2-4)
Circuit shape	rectangular, special shapes possible	rectangular, special shapes possible
Size tolerance	±0.5 %	±0.25 %
Cavity shape	rectangular, circular, special shapes possible	rectangular, circular, special shapes possible
Min. floor thickness(μm)	450	200
Max. cavity depth(μm)	1,500	2,500
Min. cavity size(μm)	1,000	500
Max. cavity size(μm)	depends on layout	depends on layout
Cavity wall to part edge spacing (μm)	$\geq$ 1,000	

Table 8.3 (continued)

Bond pad/conductor discoverage(μm)	750	
Cavity wall to conductor spacing(μm)	400	
Cavity wall to via spacing(μm)	2.5 × via size	
Cavity wall to pad spacing(μm)	250	
Spacing between two cavities (μm)	1,270	
Window shape	rectangular, circular, special shapes possible	rectangular, circular, special shapes possible
Window depth(μm)	< 2,000	> 2,000
Min. window size (length in μm)	1,000	500
Max. window size (length in μm)	depends on layout	depends on layout
Pitch of BGA/LGA I/Os(mm)	2.54	0.635
Wire bonding pitch(μm)	500	150
Solder pad pitch(μm)	400	250
Flip-chip bonding pitch(μm)	400	250

Table 8.4. Design guidelines provided by major manufacturers.

- DuPont: LTCC technology design and layout guidelines: green tape system (EKP, 12.8.1998)
- CTS Microelectronics: Low-temperature cofired ceramics design and layout guidelines for the fabrication of networks, packages and multichip modules
- National Semiconductor: Design rules for physical layout of low-temperature cofired ceramics modules(Ver.8.1)
- Scrantom Engineering Inc.: Low-temperature cofired ceramics design guidelines(Rev.C)
- Siegert Electronic GmbH: Design rules for LTCC(Rev.C)
- Sorep-Erulec: General design guide for LTCC substrates(8-Jan-99)

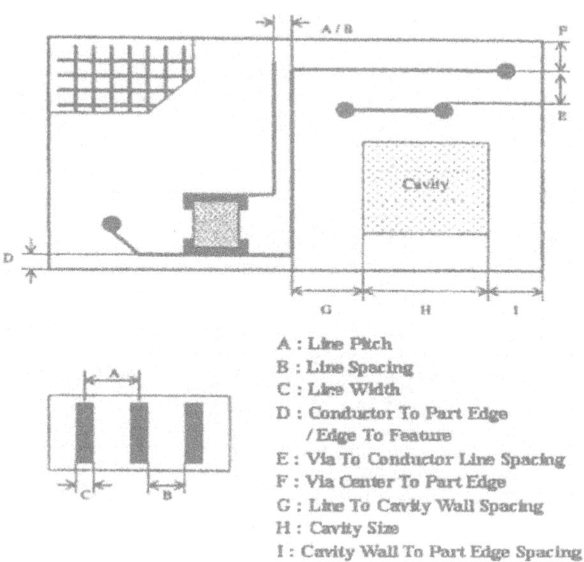

Figure 8.4. Definitions of features listed in Table 8.3.

technology offers significant benefits in terms of design flexibility, density and reliability. Multilayered capability of the LTCC technology allows multiple circuitries to be designed in a self-contained, hermetic package. Microstrips, strip lines, coplanar waveguides and dc lines can be incorporated into a single module. The capability to integrate digital, analog, RF, microwave and passive components reduces the assembly complexity and improves the overall system reliability by reducing the number of parts and interconnections. In addition, reduced weight of the LTCC packages and low-loss characteristics of the dielectrics and conductors make LTCC an ideal technology for high-performance commercial and military electronic systems.

Unlike thick-film process where successive lamination and firing steps cause bowing and line degradation, the single-step lamination and firing of LTCC produces circuit patterns with fine and high-quality layout. In addition, elimination of costly firing steps greatly increases the number of conductor layers. Current LTCC technology allows high-density circuitry(as fine as 50 μm in lines and spacings) to be interconnected with conductive vias(as fine as 50 μm in diameter).

3.1 Miniaturized LTCC Filters

LTCC technology reserves all the design advantages of strip lines. Built-in circuits can be directly wire bonded to other ceramics-based hybrids. The use of LTCC technology also helps to eliminate manufacturing defects that may be present in conventional strip line assemblies, which will enhance the performance of circuits while maintaining the advantages in size, weight, cost and mechanical simplicity.

Miniaturized LTCC filters can be used in portable telephone and other devices. Figure 8.5 shows the layout of a two-pole combline band-pass filter which is made up of five layers of low-temperature cofired ceramics. BiZnNbO composite has relative permittivity of 58, and is cofired with silver paste at 950°C. The band-pass filter is made of strip line resonators, and the temperature coefficient of resonant frequency is +20 ppm/°C. The planar line layer contains parallel-coupled strip line resonators a_1 and a_2 on one side. The coupling capacitor layer contains four electrodes, of which b_1 and b_2 are the input/output coupling capacitors, c is a shunt capacitor connected to ground, and d is an inter-resonator coupling capacitor which introduces an attenuation pole near the pass-band to improve its frequency response. Dimension of the filter is 4.5 mm × 3.2 mm × 2.0 mm. The measured insertion losses are less than 1.8 dB at 950 MHz and less than 1.2 dB at 1.9 GHz. The return losses are smaller than −20 dB in both pass-bands. The attenuation poles are located at 100 MHz below 950 MHz and 300 MHz above 1.9 GHz.

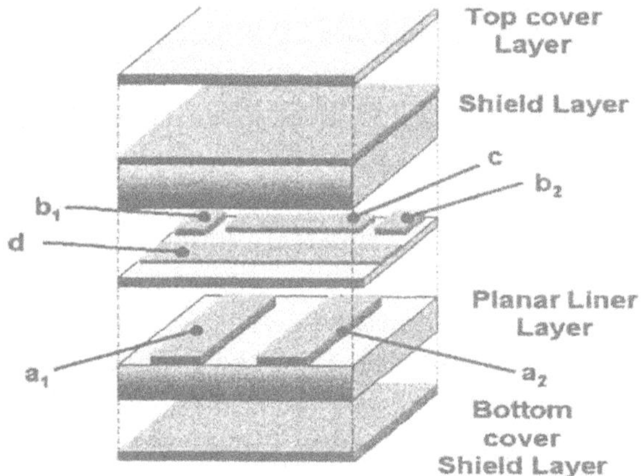

Figure 8.5. Layout of a two-pole combline band-pass filter.

3.2 Miniaturized RF Modules

The LTCC possesses well controlled, low-loss dielectric properties. The cofired process has a lower cost than conventional thick-film multilayered process and allows innovative implementation of circuits like baluns and filters. The ability to integrate inductors, resistors and capacitors facilitates miniaturization of circuitries, and the use of cavities and integrated heat sinks is beneficial to the chips possibly attached. The LTCC process can be used to implement parts in low-cost SMT packages like BGA to reduce their footprint size. Incorporation with IC technology further opens a wide range of functionalities and modularization.

Layer thickness of the green tape can be adjusted between 25 μm and 250 μm, and the layer count can reach as high as 80. Relative dielectric constant can be varied between 5 and 300, and loss tangent can be varied between 0.001 and 0.005. Silver has been used extensively as conductor materials. The sintered LTCC structure is hermetic, hence conductors embedded within it is not prone to migration. Surface materials used are a combination of platinum, palladium, silver and gold, depending on the mounted components and interconnections.

Microstrip, strip line and coplanar strip line elements can be easily fabricated at the cost of additional conductor materials only. Couplers, baluns and filters can also be realized within the LTCC substrate which can support, on the external surfaces, passive components such as printed thick-film resistors and capacitors. Parallel-plate ceramics capacitors, inductors and transformers can be embedded within layers. Cavities and tunnels can be made hermetically by sealing seam or soldering a lid over the top. In addition, it is possible to solder a stainless steel of Kovar ring frame to the substrate surface. A lid can be welded to the ring by means of laser welding after wire bonding. Such integration of passive components provides a three-dimensional configuration, leaving the external surfaces free for attaching active components. Fig.8.6 shows the structure of an LTCC substrate integrated with various kinds of components.

Use of LTCC in RF modules provides many advantages. In portable applications, for example, there are a large number of passive components within RF circuits, which can be embedded within LTCC to reduce size, weight and cost. Couplers, filters and impedance matching networks can also be incorporated. Active components can be reliably connected by wire bonding or flip-chip with adequate thermal consideration.

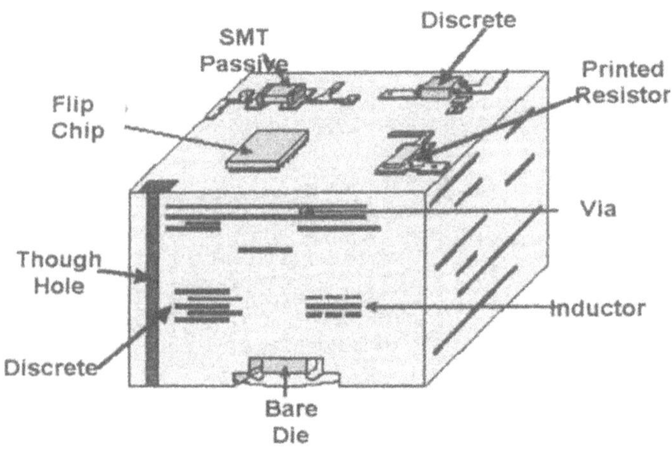

Figure 8.6. An LTCC substrate integrated with various kinds of components.

Ceramics ICs of voltage-controlled oscillator(VCO), power amplifiers (PA) and frequency synthesizer have been developed, embedding resonator and several passive elements of loop filter. They can be used in the RF front-end of transceivers for wireless phones, direct satellite broadcast systems, wireless LANs and pagers. This embedded technology reduces the size to one quarter and the cost to one half that of a discrete version. Fig.8.7 shows how the LTCC process is applied to reduce the size of an RF device or module.

Other modules to be integrated include filters, EMI shields, controlled impedance networks and transmission lines. Key improvements are the impedance matching between circuits or modules, which can improve the performance and power efficiency of the power amplifiers. Concurrent design of active RFICs with passive LTCC circuitries will simplify IC design while improving the power efficiency, which implies longer battery life or reduced battery size.

Future development of complete LTCC-based RF front-ends allows for the implementation of wireless functions without dedicated IC design. Early versions of these circuits are currently being developed.

4. Future Developments

LTCC uses the same processes as thick film to layout circuits. Thus, both technologies have similar tolerances to embedded components. The achievable precision imposes a limit on the size and density of components and interconnections. Coplanar waveguides and microstrip lines

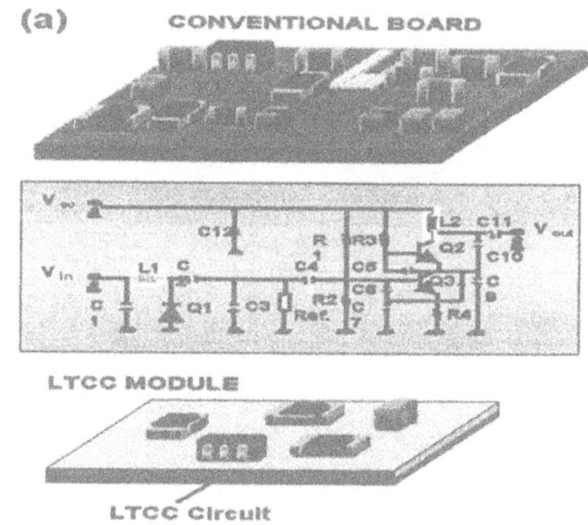

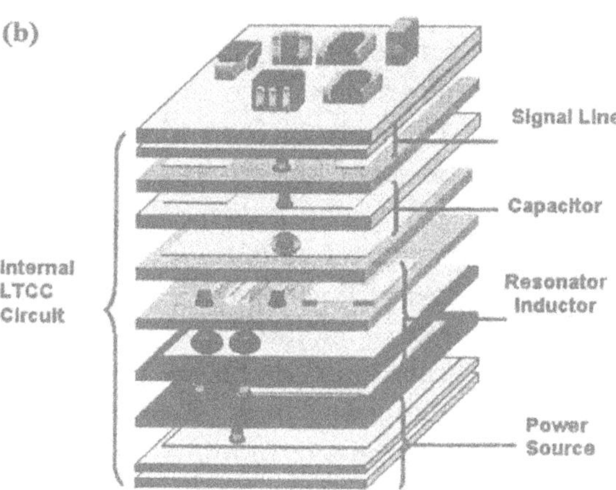

Figure 8.7. **(a) LTCC module versus conventional circuit board, (b) multilayered package consisting of LTCC modules and interconnections.**

are two-dimensional structures which are difficult to implement in LTCC since the separation between ground plane and transmission line is comparable to the tolerances achievable with screen printing. In other words, major limitation on LTCC technology is not imposed by materials, but by the precision and tolerance associated with manufacturing processes.

Capability of processing 2-mil lines and spacings improves the tolerance of LTCC transmission line structures, reduces the size, the amount of dielectrics required and the cost. An additional benefit is that narrow

lines with high precision allow for the design and manufacture of higher impedance lines such as 100 Ω lines. Finally, self-resonant frequency of the embedded components is affected by interactions with adjacent components. An example is the parasitic capacitance of embedded inductors. The larger the size of an inductor, the greater is the parasitic capacitance between the inductor and nearby ground planes. This imposes an upper limit on practical inductors. Finer lines will allow for the design and manufacture of spiral inductors with high inductance and self-resonant frequency. Due to frequency dependence of the Q factor, practical frequency range is well below the self-resonant frequency.

Use of photoformable LTCC materials and processes is one of the approaches for increasing the precision and tolerance associated with a manufacturing process. photoformable LTCC materials are glass ceramics composite with conventional composition except that in the organics, a negative photoresist is included besides plasticizers and anifloculants. A photosensitive tape is about 150 μm thick. The tape is exposed using contact photolithography. In order to develop the tape, a sodium carbonate spray shower is used. After developing, the tape is removed from glass and finally laminated and fired. One can control the partial etching depth by modifying process parameters like shower pressure and developing time. Feature sizes smaller than 30 μm may be achievable, but the resolution limit in turbid media seems to be imposed by the diffusive nature of light scattering within filler alumina grains.

Key issues in the future LTCC technology include zero shrinking, high-permittivity dielectric pastes, buried(active) components, TB-BGA, optical components, liquid and gas interfaces/conductors, brick box system and high-definition patterning process. These techniques will become standard processes for manufacturing circuits using LTCC.

Zero shrinking. Due to(uneven) shrinking of the LTCC tapes during firing, the printing widths and spacings are limited. With the use of nonshrinking tape, it will be possible to process finer structures with high precision like ball grid arrays(BGAs).

High-permittivity dielectric pastes. To realize high-density circuitries, ceramics with relative dielectric constant of 200 is known and is being developed for standard processes.

Buried(active) components. With this technology, it will be possible to integrate active elements(chips) which cannot be cofired with the LTCC substrate. An LTCC carrier/substrate with a cavity is processed, chip is then mounted into the cavity, and an LTCC lid is soldered on top to close the cavity.

TB-BGA. The top-bottom ball grid array allows different circuit modules to be mounted on top of each other by soldering, thus forming a highly functional device.

Optical components. It is possible to manufacture LTCC substrates/boxes with optical interfaces.

Liquid and gas interfaces/conductors. It is possible to integrate sensors and pipes for cooling liquid or gas into the LTCC substrate.

Brick box system. Separate LTCC modules(electrical, optical, fluidical) are connected for example, by soldering, to become a larger module. Active chips are then mounted on surfaces of the whole connected module.

5. Conclusions

LTCC technology has a versatile capability for integrating interconnections, hermetic packaging and passive components into a single monolithic structure. The benefits are reduced size, improved reliability, improved performance and lower cost. New materials and processes under development will further improve the range and performance of modules and devices.

References

[1] www.acxc.com.tw, ACX Corp.

[2] www.atceramics.com, ATC.

[3] www.cismst.de, Institute fur Mikrosensorik GmbH.

[4] http://classifications.wipo.int, World Intellectual Property Organization.

[5] www.cmac.ca, C-MAC.

[6] www.ctscorp.com, CTS.

[7] www.depanet.de, Deutsches Patent und Markenamt.

[8] www.dupont.comwww.dupont.com/mcm, DuPont.

[9] www.electroncircuitsystems.com, Electronic Circuit Systems.

[10] www.electroscience.com, Electro-Science Laboratories, Inc.

[11] www.epcos.de, EPCOS.

[12] www.ferro.com, Ferro.

[13] www.green-tape.com, Green Tape Application Group.

[14] www.4hcd.comwww.4cmd.com, Heraeus.

[15] www.ikts.fhg.de/v1.ceramics.html, Fraunhofer Institut Keramik Technologien und Sinterwerkstoffe.

[16] www.imaps.org, International Microelectronis and Packaging Society.

[17] www.imst.de, www.ltcc.de, IMST GmbH.

[18] www.irti.loyola.edu, International Technology Research Institute, Maryland.

[19] www.kyocera.co.ip, Kyocera.

[20] www.laminaceramics.com, Lamina Ceramics.

[21] www.mse-microelectronics.de, Micro Systems Engineering GmbH Co.

[22] www.murata.com, Murata.

[23] www.national.com, National Semiconductor.

[24] www.neg.co.jp/neg-ken/kenzaie/index.hwm, Nippon Electric Glass.

[25] www.nikkos.co.jp, Nikko.

[26] www.patents.ibm.com, Patent listing by IBM.

[27] www.rsc.rockwell.com, Rockwell Science Center.

[28] www.sem.co.krwww.sem.samsung.com, Samsung ThinkCera.

[29] www.sarnoff.com, Sarnoff Corp.

[30] www.scrantom.com, Scrantom Engineering Inc.

[31] http://sensor.northgrum.com, Northrop Grumman Corp.

[32] www.sorep.com www.thalesgroup.com, Sorep-Erulec, Thales Microelectronics.

[33] www.via-electronic.de, VIA Electronic.

[34] www.visprocorp.com, VisPro.

Chapter 9

ANALYSIS OF SPIRAL INDUCTORS EMBEDDED IN LAYERED MEDIA

Jean-Fu Kiang and Chi-Yu Peng
Department of Electrical Engineering and
Graduate Institute of Communication Engineering
National Taiwan University
Taipei, Taiwan, ROC

Abstract In this Chapter, basic concepts and generic models of sprial inductors implemented in microwave circuits are briefly reviewed. Challenging issues regarding analysis, design and improvement of integrated inductors are discussed. Simulation results are also presented to illustrate various designs and explanations.

Keywords: spiral inductor, layered medium, patterned ground, via, Eddy current.

1. Review of Inductor Models

Compactness is a critical factor for reducing power consumption in wireless communication products [1]. Multilayered structures offer more flexibility to design compact products with better performance [2]. Passive components like resistors, capacitors, inductors, filters and couplers usually occupy large area in integrated circuits(IC). Their size can be reduced by using multilayered topology [3]. If a planar inductor is fabricated in three-dimensional configuration, its area and parasitic capacitance can be significantly reduced [4].

Multichip module(MCM) technology can be used to reduce the cost and size of transceivers. However, extending its application to millimeter wave regime is difficult because silicon has high Ohmic loss at high frequencies [5]. Hence, low-loss and high-permittivity ceramic materials are given more consideration as substrates. In recent years, much progress

has been made in developing low-temperature cofired ceramics(LTCC), which has the advantages of low loss, low cost and easy fabrication. It also becomes viable packaging technology for multichip modules [6].

Inductors are important passive components in many circuits like linear filters, baluns, narrow-band impedance matching, low-noise feedback, power splitting and combining, VCO tank, low-noise amplifiers(LNA) and power amplifiers(PA). Most practical inductors are of spiral form emdedded in multilayered substrate. Usually, via or air bridge is needed to tap the inner port of an inductor to another component. Characteristics of spiral inductor are controlled by parameters such as number of turns, inner hole diameter, strip width and spacing, substrate thickness, substrate permittivity and conductivity, and so on. Circular inductors perform slightly better than their square counterparts because the former exhibit smaller series resistance than other shapes. As a spiral inductor occupies substantial chip area, it can become a potential source and receptor of electromagnetic interference(EMI).

Guidelines to layout spiral inductor can be summarized as follows [7]-[11]. Spacing between strips should be minimized to reduce inductor's die area and increase its inductance and Q factor. However, narrow spacing incurs stronger capacitive coupling and higher loss, especially at high frequencies. Thicker substrate renders higher Q factor because the substrate capacitance is reduced. High Q factor is usually required of an inductor. For example, maximizing the Q factor of an inductor is closely related to minimizing the phase noise of an oscillator consisting of this inductor and other components. Two spiral inductors can be stacked and shunted together to form an inductor with larger strip cross section, hence smaller series resistance. Wider strip can also be used to achieve lower series resistance. However, inductor with wider strips occupies larger die area, induces larger parasitic capacitance, metallic and substrate losses. A hole is usually reserved in the middle of an inductor to reduce the effect of Eddy current and negative mutual inductance. Usually, the hole diameter is set equal to at least five times of strip width. The inductance strongly depends on the number of turns. The Q factor decreases with increasing turn number because the parasitic capacitance increases with die area and the series resistance increases with strip length.

At low frequencies, interconnections and passive components can be analyzed by using lumped circuit models. At high frequencies, they should be analyzed using transmission line theory or full-wave approaches. The Q factor of an inductor is defined as [12]

$$Q = 2\pi \frac{\text{net magnetic energy stored}}{\text{energy loss in one cycle}}$$

$$= 2\pi \frac{\text{peak magnetic energy} - \text{peak electric energy}}{\text{energy loss in one cycle}}$$

Since an inductor is intended to store magnetic energy, hence, its Q factor is defined to be proportional to the net magnetic energy stored which is equal to the difference between peak magnetic energy and its electric counterpart. If an inductor is modeled by a parallel RLC circuit, the Q factor can be reduced to

$$Q = \frac{R}{\omega L}\left[1 - \left(\frac{\omega}{\omega_o}\right)^2\right]$$

where $\omega_o = 1/\sqrt{LC}$ is the self-resonant frequency of the inductor, which is also the resonant frequency of the equivalent parallel RLC circuit. An inductor is defined to be at self-resonance when $Q = 0$ or its peak magnetic energy is equal to its peak electric energy. If the equivalent parallel RLC circuit is viewed as a lossy LC tank, the total stored energy will be the sum of average magnetic and electric energies, namely

$$Q_{\text{tank}} = 2\pi \left.\frac{\text{average magnetic energy} + \text{average electric energy}}{\text{energy loss in one cycle}}\right|_{\omega=\omega_o} = \frac{R}{\omega_o L}$$

From another point of view, Q factor of an inductor can be defined as the ratio of self-resonant frequency to its 3 dB bandwidth, $\Delta\omega$, namely, $Q = \omega_o/\Delta\omega$. The Q factor can also be defined as the ratio between the reactive and resistive powers fed into the inductor.

For a general two-port circuit, the net power fed into the circuit is $V_1 I_1^* + V_2 I_2^*$ which is equal to $P_d + jS_r$, where P_d is the power dissipated in the circuit and S_r is the reactive power stored in the circuit. By expressing the net power in terms of Y-parameters, we have

$$\begin{aligned} P_d + jS_r &= |V_1|^2 y_{11}' + |V_2|^2 y_{22}' + 2\text{Re}(V_1 V_2^*) y_{12}' \\ &\quad -j\left[|V_1|^2 y_{11}'' + |V_2|^2 y_{22}'' + 2\text{Re}(V_1 V_2^*) y_{12}''\right] \end{aligned}$$

where y_{nm}' and y_{nm}'' are the real and imaginary part, respectively, of y_{nm}. The Y-parameters are defined from the circuit point of view. However, they can be related to the S-parameters which are defined from the perspective of waves. If port two is shorted($V_2 = 0$) to emulate the scenario of measuring the inductor's Q factor, then $P_d + jS_r = |V_1|^2 y_{11}' - j|V_1|^2 y_{11}''$, and we have

$$Q = \frac{S_r}{P_d} = -\frac{y_{11}''}{y_{11}'}$$

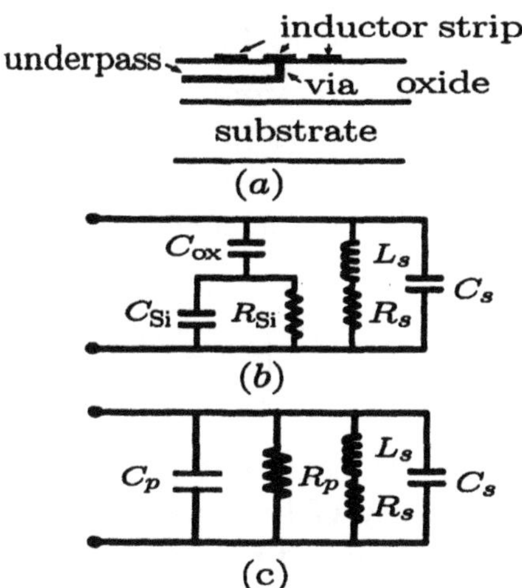

Figure 9.1. (a) Typical layout of spiral inductor on silicon substrate, (b) equivalent one-port lumped circuit model, (c) reduced circuit model where parasitics are reduced to two shunt elements, R_p and C_p.

Fig.9.1(a) shows the schematic of an inductor deposited on silicon substrate. Fig.9.1(b) shows an equivalent one-port circuit model of the spiral inductor with inductance L_s and with one terminal grounded. At high frequencies, the electric fields tend to penetrate into the silicon substrate which has relatively higher permittivity than the oxide layer. Extra loss is incurred and the inductance is reduced. The series resistance, R_s, of spiral loops increases exponentially with frequency due to skin effect. The series capacitance, C_s, accounts for the capacitance between the spiral strips and an underpass or air bridge, which is more significant than the capacitance between two spiral loops. Because the potential difference between the spiral strips and the underpass/air bridge is bigger than that between two spiral loops. The parasitics are modeled as elements C_{ox}, C_{Si} and R_{Si}, representing the oxide capacitance between spiral and substrate, the capacitance and Ohmic loss, respectively, of the silicon substrate. To further simplify the analysis, the equivalent circuit in Fig.9.1(b) is reduced to the parallel circuit shown in Fig.9.1(c) with

$$R_p = \frac{1}{\omega^2 C_{ox}^2 R_{Si}} + \frac{R_{Si}(C_{ox} + C_{Si})^2}{C_{ox}^2}$$

$$C_p = C_{ox} \frac{1 + \omega^2 (C_{ox} + C_{Si}) C_{Si} R_{Si}^2}{1 + \omega^2 (C_{ox} + C_{Si})^2 R_{Si}^2}$$

Q factor of the inductor can be factorized as

$$Q = \left(\frac{\omega L_s}{R_s}\right) \frac{R_p}{R_p + [(\omega L_s/R_s)^2 + 1]R_s} \left[1 - \frac{R_s^2(C_s + C_p)}{L_s} - \omega^2 L_s(C_s + C_p)\right]$$

where the first term accounts for the magnetic energy stored in the inductor and the Ohmic loss in the series resistance, the second term accounts for the substrate loss, and the last term accounts for the reduction due to the increase of peak electric energy with increasing frequency. Intuitively, the self-resonant frequency is defined by setting the last term to zero. In other words, the reduction of Q factor at high frequencies is a combined effect of substrate loss and self-resonance. Frequency dependence of the Q factor is determined by the parasitic resistance and capacitance. The Q factor increases as the frequency is raised from dc due to the increasing reactance of the inductor. Subsequent decrease of the Q factor from its peak is due to the parasitic related energy dissipation at higher frequencies.

A symmetric two-port physical circuit model can be built from the one-port circuit model shown in Fig.9.1(b) by adding another R_pC_p shunt branch to the right hand side of the inductor modeled by L_s, R_s and C_s. An equivalent π-network can be derived from which to extract L_s, R_s, R_p and C_p from simulated or measured Y-parameters as

$$Y_a = y_{11} + y_{12} = \frac{1}{R_p} + j\omega C_p$$

$$Y_c = -y_{12} = \frac{1}{R_s + j\omega L_s} + j\omega C_s$$

As reported in [12], C_s is insensitive to frequency, and can be approximated as a constant. The other four parameters can thus be determined as

$$L_s = \frac{1}{\omega}\frac{y_{12}'' + \omega C_s}{(y_{12}')^2 + (y_{12}'' + \omega C_s)^2}, \qquad R_s = \frac{-y_{12}'}{(y_{12}')^2 + (y_{12}'' + \omega C_s)^2}$$

$$R_p = \frac{1}{y_{11}' + y_{12}'}, \qquad C_p = \frac{1}{\omega}(y_{11}'' + y_{12}'') \tag{9.1}$$

A multiturn inductor can be designed with consecutive turns embedded in different layers and connected by vias. Modified helical inductor has been proposed in which only half a turn is embedded in each layer, and parasitic capacitance is reduced due to wider spacing between consecutive turns[4]. As a consequence, its self-resonant frequency is raised.

The Q factors of a two-turn planar inductor and three helical inductors with different inter-turn spacings are compared. Smaller inter-turn spacing incurs larger inductance. Trade-off between self-resonant frequency and Q factor is achieved by choosing an inter-turn spacing of two layers.

At low frequencies, most of the electric energy is stored within the oxide layer, hence C_p is dominated by C_{ox}. As frequency increases, the electric fields tend to penetrate into the silicon substrate, C_p is thus reduced to the series combination of oxide and substrate capacitances. In general, R_p is large since little conduction current flows in the silicon substrate. Drop of R_p signifies the increase of energy dissipation in the substrate.

The substrate loss is reduced if R_p in (9.1) approaches infinity. It can be achieved if R_{Si} approaches zero or infinity, which means the silicon substrate becomes an equivalent short or open circuit, respectively. Using high-resistivity silicon substrate or etching away part of the substrate renders the substrate a quasi-open circuit. Inserting a ground plane to shield the electric fields from penetrating into the substrate is equivalent to placing a small resistance in parallel with C_{Si} and R_{Si} in the physical circuit model of Fig.9.1(b).

In [13], it is shown that by shunting two metal layers together to increase the effective strip thickness, its series resistance is reduced and its Q factor is enhanced. Dual-level metal cascade inductor occupies about half of the die area required by conventional spiral inductors. However, parasitic capacitance between different layers of the spiral is larger in the former than in the latter. Hence, self-resonant frequency of the former is reduced more significantly.

Magnetic fluxes are enhanced in multilevel spiral inductors in which conventional underpasses are replaced by vias to link individual spiral coils. Thus, larger inductance can be achieved with the same die area and metal pitch. Since the required magnetic fluxes can be maintained at reduced wire length, the series resistance is reduced. The inductance of a multilevel spiral inductor is much larger than the sum of two single-level inductors due to the mutual inductance between stacked spiral coils, only at the expense of lower self-resonant frequency. Also, the maximum Q factor of multilevel spiral inductors is not as high as that of single-level inductors [14].

2. Challenging Issues

The Q factor of silicon-based inductor degrades at high frequencies due to energy dissipation in the substrate. Energy loss can be reduced by shielding the electric fields emanating from the inductor to the silicon

substrate. High-resistivity silicon substrates(150-200 Ω-cm) have been reported to have similar loss effects as low-loss semi-insulating GaAs substrates [15]. Etching a pit in the silicon substrate can also reduce substrate loss [16].

A solid ground may be inserted between the spiral and substrate to reduce the substrate loss. However, the ground itself contributes extra Ohmic loss and hence extra series resistance to the inductor. Polysilicon solid ground yields Q factor close to that of an inductor without any ground, indicating that polysilicon is resistive enough to impede the Eddy current. Magnetic fluxes are reduced when penetrating through the ground into the substrate due to skin effect. The inductance is thus decreased because the stored magnetic energy is reduced. To alleviate such effect, the ground should be much thinner than the skin depth at the frequencies of interest. With a solid ground of high conductivity, magnetic fluxes are reduced due to negative image current induced in the ground. Inductance of spiral without any ground is frequency insensitive because the magnetic fields in the substrate do not vary drastically with frequency. Inductance of spiral with a ground is more frequency sensitive because the induced image current and hence the associated reduction of magnetic energy is a function of frequency.

The ground may be patterned with slots orthogonal to the spiral strips to break the path of induced current. The slot should be sufficiently narrow so that electric fields cannot penetrate through the patterned ground into the substrate and incur Ohmic loss. It is suggested that the ground strips should be properly bundled yet not form a closed ring around the spiral to induce counteracting magnetic fluxes. The bundled strips are also suggested to be electrically connected to the ground on the top level to maintain their dc potential.

Adjacent inductors on a more conductive substrate(11 Ω-cm) may incur stronger coupling due to lower substrate impedance. Proper patterned ground is recommended as an isolation measure. Trade-off is required between the desired isolation level and the chip area required to implement the patterned ground.

Net effects of patterned ground on Q factor are still controversial. A patterned ground is intended to provide an alternative low-resistance return path and mitigate induced Eddy current in the substrate. However, a patterned ground inevitably introduces more parasitic capacitance between inductor and substrate.

Fig.9.2(a) shows a typical structure in the BiCMOS process with four layers of metallic strips deposited at interfaces between different oxide layers. An optional polysilicon layer can also be deposited between the substrate and metal layers. Metal layers are indexed from bottom to top.

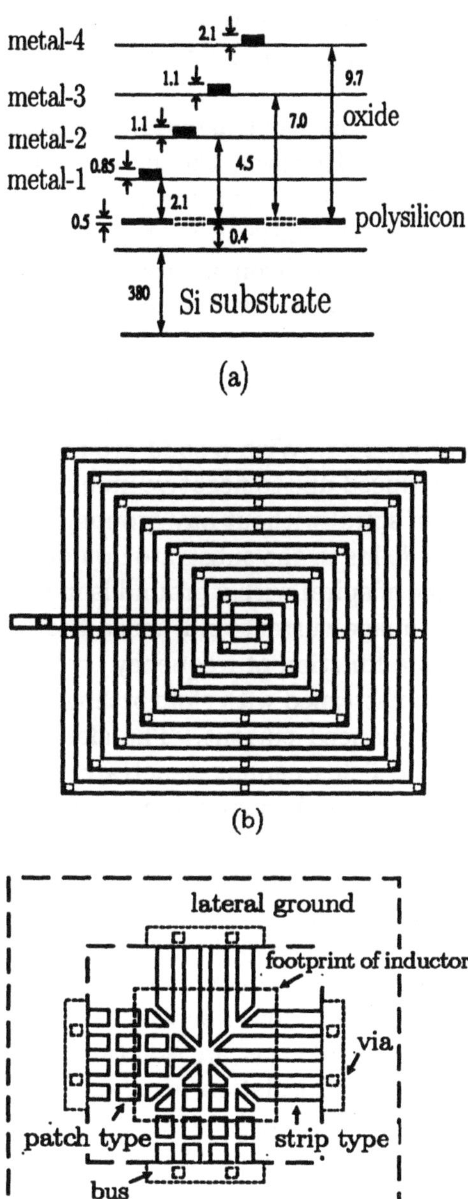

Figure 9.2. Typical four-layered 0.8 μm BiCMOS process with an optional polysilicon layer, (a) cross section(unit in μm), $\sigma_{\mathrm{Si}} = 5$ S/m(20 Ω-cm, BiCMOS) or $\sigma_{\mathrm{Si}} = 10^4$ S/m(10 mΩ-cm, CMOS), $\sigma_{\mathrm{Al}} = 4 \times 10^7$ S/m(50 mΩ per square, 0.5 μm thick), $\sigma_{\mathrm{PSi}} = 2 \times 10^5$ S/m(10 Ω per square, 0.5 μm thick), $\epsilon_{\mathrm{ox}} = 4\epsilon_o$, $\epsilon_{\mathrm{Si}} = 11.9\epsilon_o$, (b) top view of a seven-turn spiral inductor with optional vias shunting multiple metal layers, (c) schematic of patterned ground and lateral ground.

The oxide and metal layers above metal-n layer will not be fabricated if only metal-n layer and the layers below are used. The Q factor of inductors with patterned ground made in metal-1 layer is lower than those with patterned ground made in polysilicon layer. Because metal-1 layer is closer to the inductor layer than polysilicon layer is, parasitic capacitance in the former configuration is larger, hence its self-resonant frequency is lower.

In [12], a patterned ground is inserted between a spiral inductor and its silicon substrate to increase its inductance. However, the patterned ground introduces additional parasitic resistance and capacitance, hence affects the Q factor of the inductor. A patterned ground made of polysilicon(0.5 μm, 12 Ω per square) proves to increase the inductor's Q factor by 33% and claims an improvement of isolation between two adjacent inductors in the same substrate by 25 dB. In [17], it is argued that such an isolation improvement between adjacent inductors is not due to the patterned ground if the coupling mechanism is magnetic by nature, because the patterned ground is not expected to perturb the magnetic fields. At low frequencies, the mutual inductance between two adjacent inductors increases in proportion to frequency. At high frequencies, coupling decreases with frequency because C_p becomes an effective short circuit to bypass both inductors to the ground.

Frequency sensitivity of L_s and R_s increases as R_{Si} is decreased or as C_p is increased, where the dependence on R_{Si} is weaker. Frequency dependence of L_s and R_s for an inductor fabricated with metal-1 layer above a polysilicon patterned ground is stronger than those fabricated with metal-2 or metal-3 layer above the same polysilicon patterned ground because C_p is larger in the former case. However, it is not well explained why R_s of an inductor with patterned ground increases with frequency at low frequencies. The shape of patterned ground also affects the frequency dependence of L_s and R_s. The Q factor increases when the footprint of patterned ground is reduced. However, if the footprint is over-reduced, R_{Si} eventually becomes high enough to reduce the Q factor. The maximum Q factor can be slightly increased by using a polysilicon patterned ground, accompanied by C_p increase.

In [18], it is observed that the series resistance decreases when a ground is inserted under the spiral inductor, and the Q factor strongly depends on the ground layout. The Q factor of an inductor with patterned ground is lower than that without any ground at low frequencies, but vice versa at high frequencies because energy dissipation in substrate is reduced at high frequencies. It is also found that fingers of patterned ground beneath the spiral inductor do not significantly affect its Q factor. When a patterned ground is inserted, C_{Si} is decreased, but C_{ox} is

increased. Also, the series resistance, R_s, does not vary drastically with frequency. Hence, the Q factor with a patterned ground stays high at high frequencies. In contrast, the Q factor with a surrounding ground is lower than that without any ground for frequencies up to 5 GHz. This is associated with the fact that the series resistance of an inductor with a patterned ground does not increase as much as that with a surrounding ground at high frequencies.

In [19], a patterned ground is claimed to improve the maximum available gain of a transformer at high frequencies. The coils with patterned ground and bar ground have essentially the same inductance as that without any ground, and the Q factor is not always improved by inserting a patterned ground. Patterned ground serves as an electromagnetic shield between the inductor and substrate, especially at high frequencies. It also provides an additional return path for the induced ac currents through capacitive coupling. Such a return path introduces additional Ohmic loss, which partially cancels the improvement due to substrate shielding at low frequencies, leading to lower Q factor in comparison to the transformer without any patterned ground. However, relatively higher Q factor is obtained as frequency increases.

In [20], effects of a polysilicon ground on inductor performance are studied. The stored magnetic energy in inductors are affected by the substrate conductivity. A floating ground made of undoped polysilicon with sheet resistance of higher than 10 kΩ per square can create a quasi-open substrate to enhance the inductor's Q factor. The improvement is more obvious at low frequencies and around the self-resonant frequency. There is little improvement on either inductance or Q factor when a low-resistivity ground is used. With a low-resistivity polysilicon ground, the Q factor decreases because the parasitic capacitance, C_{ox}, is larger than that of an unshielded inductor. Around 7.8 GHz, the Q factor decrease of an inductor with floating high-resistivity polysilicon ground could be due to the parasitic capacitance associated with strips of the patterned ground.

Patterned ground is intended to reduce the substrate loss while does not significantly change the inductance below 1 GHz, only at the price of increasing parasitic capacitance. High-resistivity substrate improves the Q factor of an inductor by providing high R_p, while patterned ground improves the Q factor by providing a low-resistance path which is equivalent to lower R_p. Inserting a patterned ground in polysilicon or metal-1 layer does not improve the Q factor of an inductor. The degradation of Q factor due to a metal-1 patterned ground is worse than that due to a polysilicon patterned ground. Both polysilicon and metal-1 patterned grounds increase C_p significantly.

The n^+-diffusion layer turns out to be an effective layer to layout patterned ground due to small parasitic capacitance it may introduce. The inductor with an n^+-diffusion patterned ground demonstrates noticeable Q factor improvement up to 21% [21]. Although an n^+-diffusion patterned ground incurs higher Q factor than a polysilicon patterned ground, self-resonant frequency of the former is only a little higher than that of the latter. The n^+-diffusion patterned ground does not increase C_p much because it is farther away from the spiral inductor than the polysilicon is.

Heavily doped(10-20 mΩ-cm) substrates, also known as epi substrates, are routinely used in CMOS and BiCMOS processes while lightly doped(1-30 Ω-cm) substrates are commonly used in bipolar and some CMOS processes. Typical epi substrates consist of a lightly doped epitaxial layer grown on a heavily doped(10-20 mΩ-cm) bulk substrate. In [22], it is shown that the energy loss in lightly doped substrates is higher than that in epi substrates. The Q factor with epi substrates is higher than that with lightly doped substrates. An epi substrate incurs less loss as the electric fields only penetrate into a superficial layer with total volume considerably smaller than that of the entire lightly doped substrate. Eddy current induced by an inductor is negligible in heavily doped epi substrates up to several GHz, and degradation of Q factor for an inductor on epi substrates at high frequencies is due to larger substrate parasitic capacitance which incurs lower self-resonant frequency as compared to lightly doped substrates.

The inductor on an insulating quartz substrate has the highest Q factor because it has the lowest substrate loss and the smallest parasitic capacitance. With lightly doped substrates, the inductor's Q factor becomes higher if a thicker oxide is inserted between the spiral and the lossy substrate. Same type of inductors fabricated separately on lightly-doped and quartz substrates have similar series inductance and resistance, which implies that the Eddy current is still insignificant in lightly doped substrates. The inductor with a solid ground exhibits smaller inductance and larger series resistance owing to the Eddy current induced in the solid ground. A polysilicon patterned ground can be inserted between the inductor and an epi substrate to improve the Q factor by about 15%. Similar improvement is obtained with a source/drain diffusion patterned ground. Although the p^+ bulk has sheet resistance close to that of a solid ground, carriers in the p^+ bulk are distributed over larger volume. Hence, the effective ground appears to be farther away from the inductor. As a result, no significant Eddy current is induced.

3. Simulations and Discussions

A seven-turn spiral inductor as shown in Fig.9.2(b) is used as benchmark to study the effects of different parameters. The spiral may be fabricated in different metal layers with various features like ground, vias or bus as shown in Fig.9.2(c).

When applying (9.1) to extract the equivalent circuit parameters from Y-parameters, difficulties are encountered to solve for R_s, L_s and C_s simultaneously. In [12], C_s is assumed as a constant to derive R_s and L_s using (9.1). The value of C_s is difficult to estimate because the static potential is different at different spots on the spiral. Thus, conventional definition of capacitance is not applicable, especially at high frequencies. However, C_s is used mainly for describing the inductor characteristics using circuit concept. Hence, the sensitivity of R_s and L_s on C_s is studied first. Extracted inductances and resistances of the same spiral inductor with three different C_s, 0 fF, 18 fF and 36 fF, are compared. Define percentage deviation as

$$\frac{|L_s(C_s = 0/36 \text{ fF}) - L_s(C_s = 18 \text{ fF})|}{L_s(C_s = 18 \text{ fF})} \times 100\%$$

It is observed that the percentage deviation is less than 1% under 1 GHz.

Equivalent circuit parameters are useful for synthesizing microwave circuits. Usually, each distributed component of a microwave circuit is assigned a constant value over specific frequency range. Then, the microwave circuit can be synthesized using conventional circuit design approach to fulfill the specifications. Full-wave approaches are usually required to verify and fine-tune the circuit layout. The equivalent inductance of a distributed spiral inductor is approximately constant at low frequencies, and a spiral inductor is rarely used near its self-resonant frequency.

Fig.9.3 shows the effects of separation between inductor and substrate. The series resistance increases exponentially with frequency due to skin effect [23]. A metal-3(metal-4) inductor has lower series resistance than a metal-2(metal-3) inductor because the former is farther away from the substrate than the latter, and metal-4 layer is thicker than metal-2 and metal-3 layers. In a metal-n inductor, the spiral is fabricated in metal-n layer with underpass fabricated in metal-(n-1) layer. The Q factor is strongly affected by the series resistance. Smaller series resistance is assoiciated with higher Q factor. The inductor has the impedance of $Z_s = R_s + j\omega L_s$ as shown in Fig.9.1(b), and the inductance decreases along with the series resistance. As frequency increases, less electric fields tend to penetrate into the substrate. Hence, parasitic capacitance and resistance decrease with increasing frequency. Since metal-4 inductor

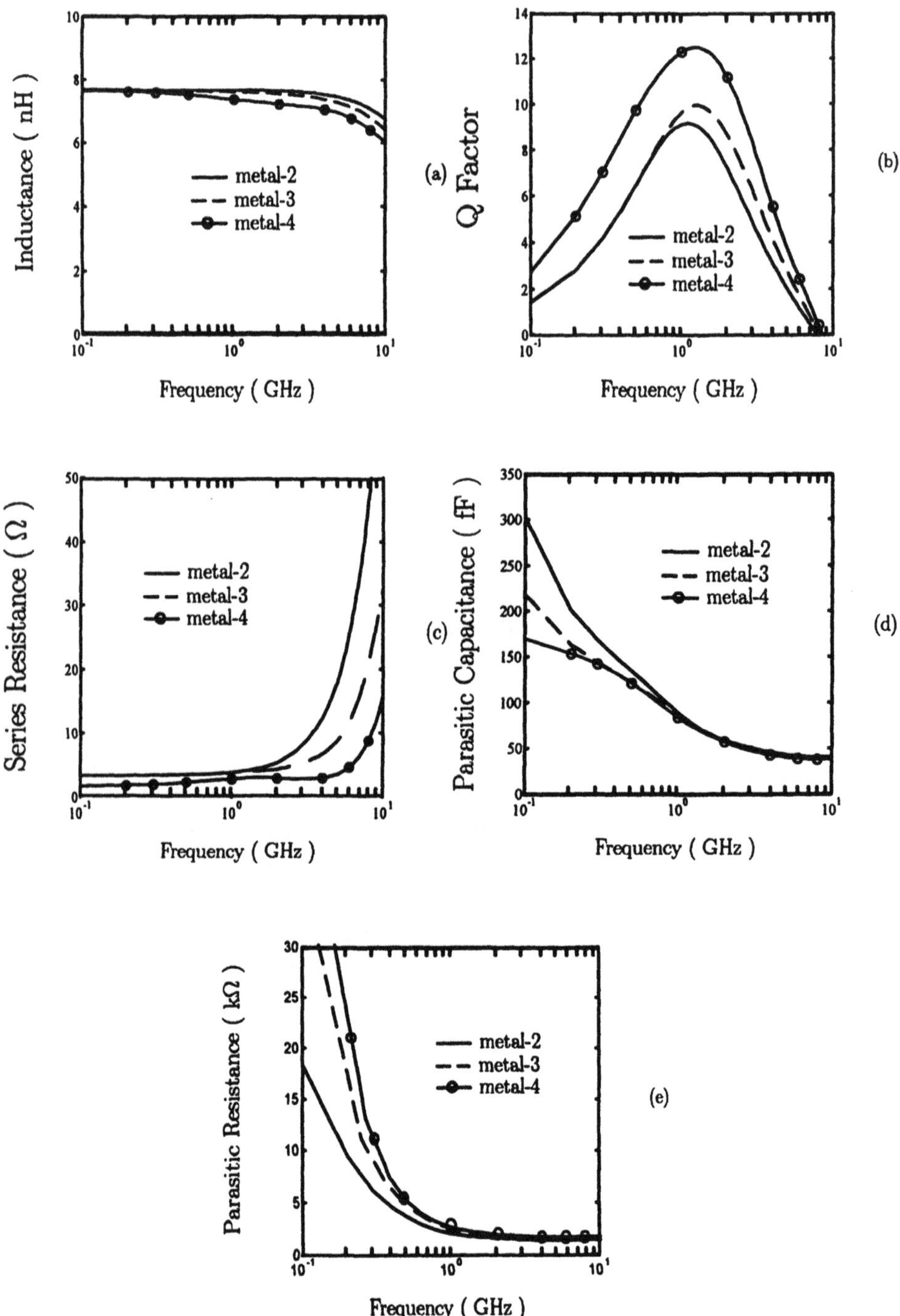

Figure 9.3. **Comparison of inductors fabricated in different metal layers, (a) inductance, (b) Q factor, (c) series resistance, (d) parasitic capacitance, (e) parasitic resistance.**

has the lowest indictance and parasitic capacitance, it has the highest self-resonant frequency among the three inductors.

The effects of a lateral ground made of the top metal layer are also analyzed, which is 100 μm wide and surrounds the inductor as shown in Fig.9.2(c). The parasitic capacitance increases slightly due to the coupling between lateral ground and inductor. The series resistance is also slightly increased. Due to higher resistance, the Q factor decreases obviously when the lateral ground is inserted, espectically at low frequencies. The inductance with a lateral ground is reduced from 7.7 nH to 7.3 nH because the induced current on the lateral ground contributes counteracting magnetic fluxes. The parasitic resistance is very close to that without any ground. The self-resonant frequency is only slightly reduced due to the balance of inductance decrease and parasitic capacitance increase.

The effects of vias connecting the lateral ground and patterned ground are also analyzed. Two and six vias, respectively, on each side of the surrounding ground are compared. All the results are very close except the parasitic capacitance. The parasitic capacitance of the inductor with six vias is higher than that with two vias. This could be due to the extra parasitic capacitance between inductor and vias.

Fig.9.4 shows the effects of inserting a polysilicon patterned ground between inductor and substrate. The patterned ground is connected to the lateral ground by vias. Strip width of the patterned ground is chosen to be 3 μm, 18 μm and 48 μm, respectively, while 2 μm slots are placed between ground strips. Results of an inductor with a solid ground are also presented for comparison. The inductances in all the cases are very close at low frequencies. The series resistance of inductors with a patterned ground decreases at high frequencies, and narrower strip incurs smaller series resistance. This implies that narrow strips are more effective to break the Eddy currents induced in the patterned ground. When strips become wider, the parasitic capacitance decreases and parasitic resistance increases. The Q factor of inductors with patterned ground increases obviously above 1 GHz due to lower series resistance, but the self-resonant frequency is reduced because the parasitic capacitance increases. The inductor with a patterned ground of 3 μm strips has the highest Q factor and the lowest self-resonant frequency among the three.

Fig.9.5 shows the effects of a polysilicon patterned ground which is not connected to the lateral ground. We also test a patterned ground with both strip width and spacing equal to 10 μm to study the effects of strip spacing. The parasitic capacitance and parasitic resistance of inductors with patterned grounds of different strip widths are very close

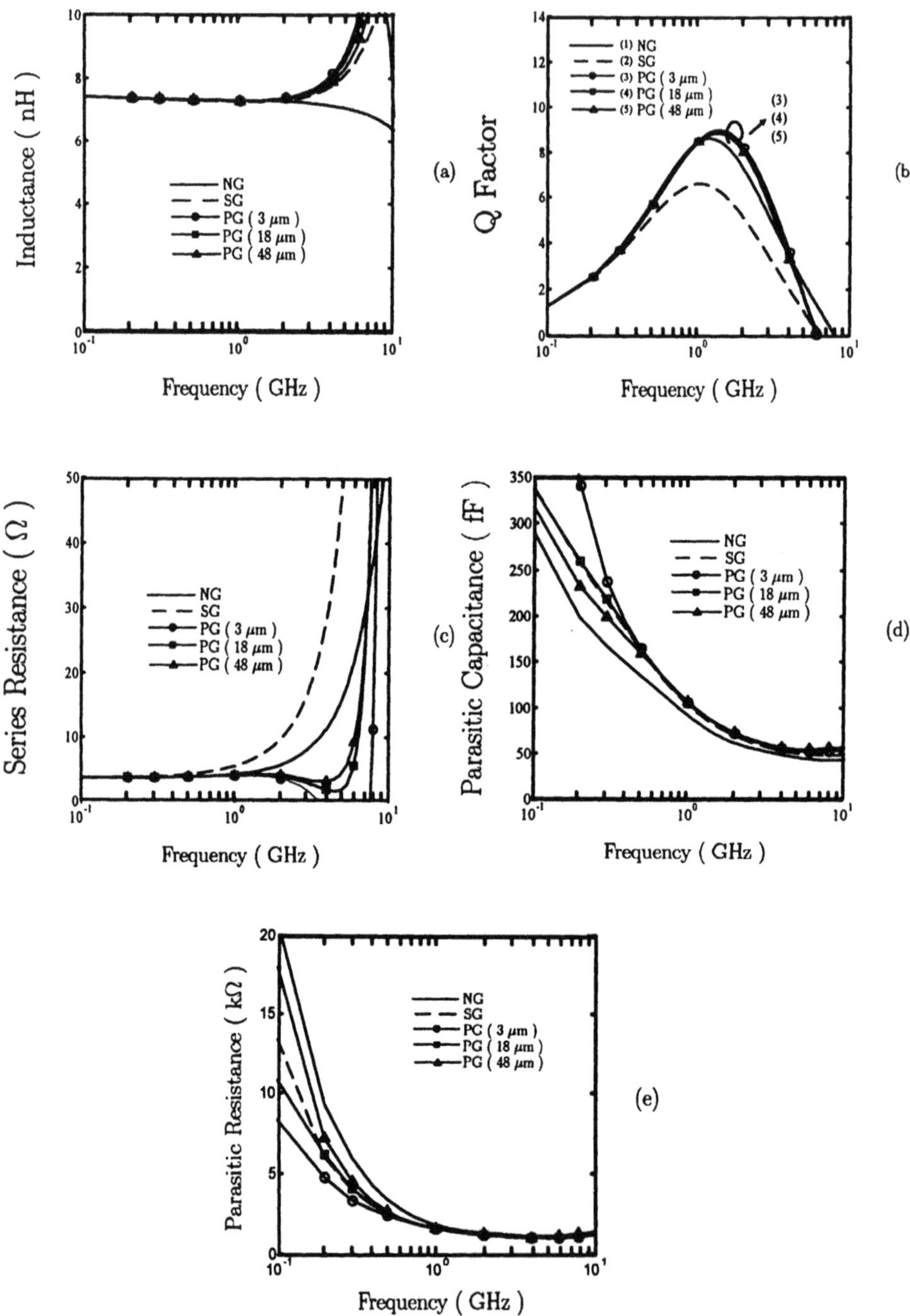

Figure 9.4. **Effects of inserting a polysilicon patterned ground between spiral and substrate, (a) inductance, (b) Q factor, (c) series resistance, (d) parasitic capacitance, (e) parasitic resistance.**

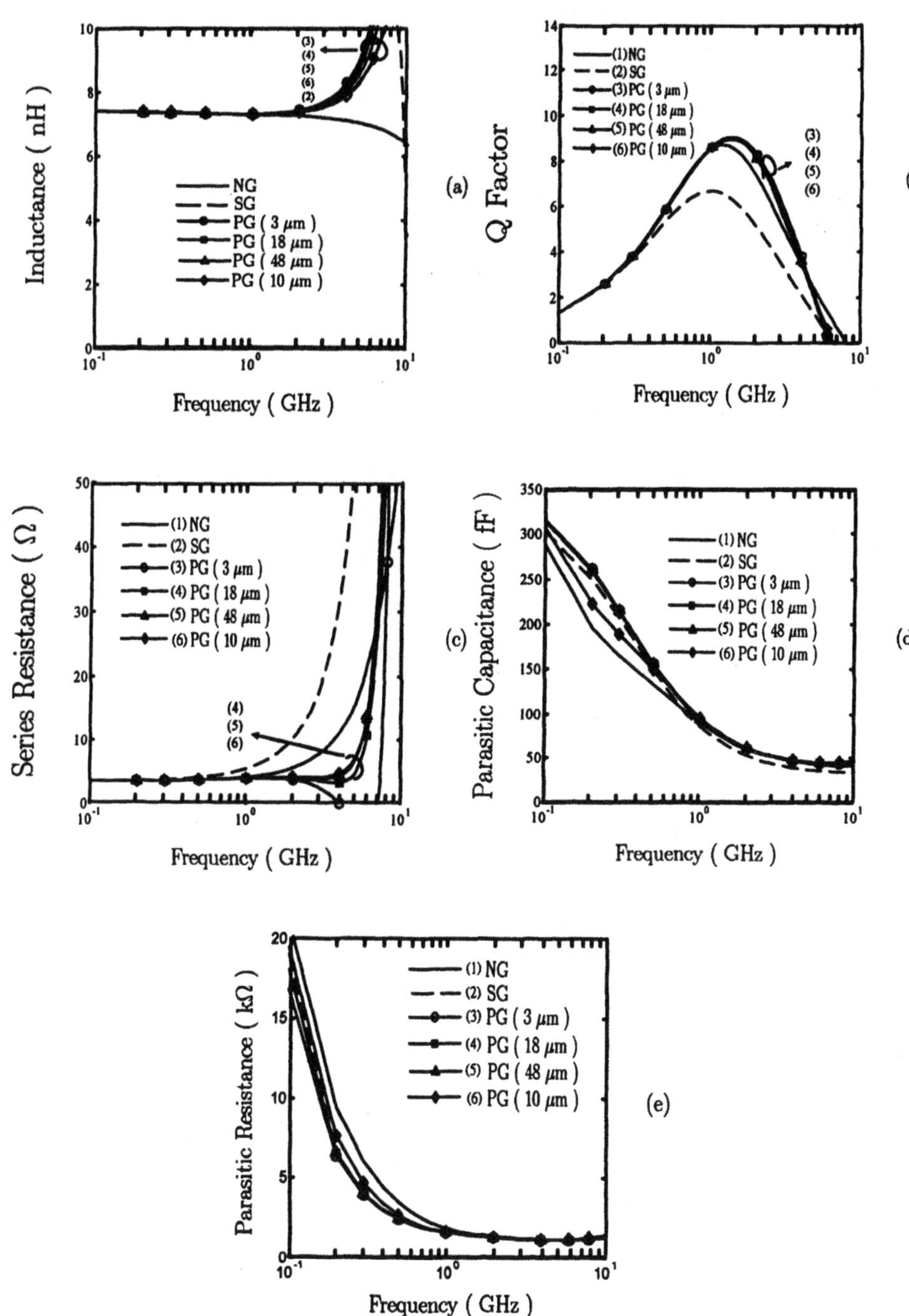

Figure 9.5. Effects of a polysilicon floating ground, (a) inductance, (b) Q factor, (c) series resistance, (d) parasitic capacitance, (e) parasitic resistance.

except the one with strip width and spacing equal to 10 μm, of which the parasitic capacitance is the smallest due to its relatively smaller ground area.

The effects of patterned grounds made of aluminum are also analyzed. The variation of parasitic capacitance and parasitic resistance are similar to that with a polysilicon ground. The inductance with a patterned ground of 10 μm strip width and 10 μm spacing increases slightly at low frequencies. The series resistance increases rapidly at high frequencies, and becomes larger than that without any ground. The series resistance increases and the Q factor decreases obviously with the increase of strip width in the patterned ground. The Q factor of the inductor with a patterned ground of 48 μm strips is even lower than that without any ground because stronger Eddy current can be induced on wider strips.

Fig.9.6 shows the effects of polysilicon patterned grounds which are laid out by segmenting all the strips into patches. The length of each patch is 18 μm while the strip widths are 3 μm, 18 μm and 48 μm, respectively. The length of patches made from 10 μm strips is 20 μm. The motivation of this patch ground is to further reduce Eddy current on strip ground. However, larger gaps may allow more electric fields to penetrate into the substrate. There is no obvious improvement in the Q factor, and all the other parameters are close to those without any ground.

The effects of aluminum patterned grounds with the same layouts are also analyzed. The series resistance increases with strip width due to stronger Eddy current induced on wider strips. The series resistance with a patterned ground increases slightly compared to that without any ground at high frequencies, thus the Q factor is decreased. The inductor without any ground has the highest Q factor and self-resonant frequency due to its lowest parasitic capacitance which increases only slightly with area of the patch ground.

Fig.9.7 shows the grounding effects of five different patterned grounds made of polysilicon. They are patterned grounds with or without vias connecting to lateral ground, patterned ground without bus and patterned ground with patches. Strip width and spacing of the patterned ground are 3 μm and 2 μm, respectively. Fig.9.7(d) shows that the parasitic capacitance of an inductor with its patterned ground connected to the lateral ground by vias increases, which may be contributed by the vias. Higher parasitic capacitance also results in lower self-resonant frequency. The inductor with a patch ground has the smallest parasitic capacitance due to large gaps which allow more electric fields to penetrate into the substrate. The inductor with the patterned ground connected by vias has the smallest series resistance at high frequencies,

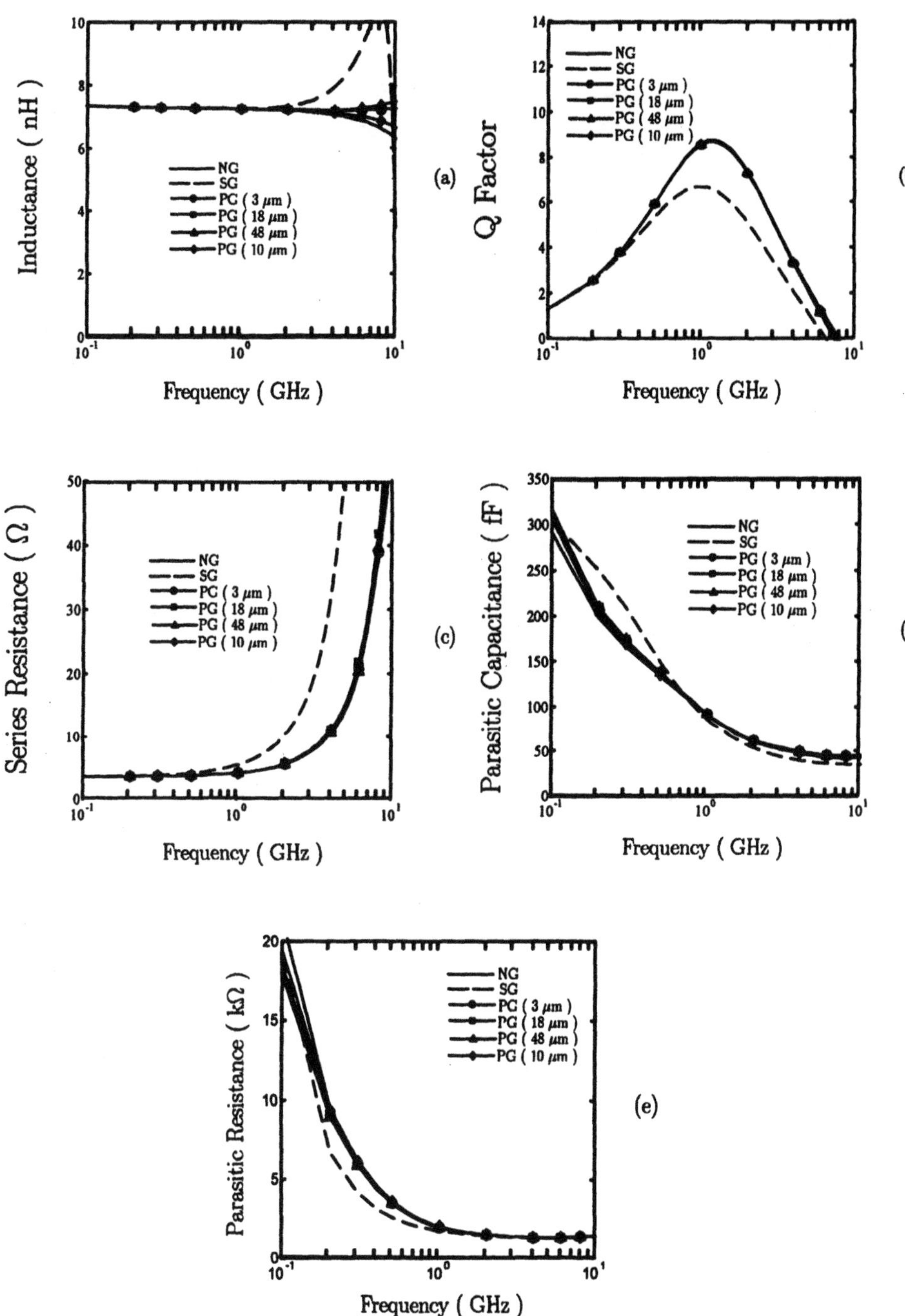

Figure 9.6. Effects of polysilicon patterned ground of patch type, (a) inductance, (b) Q factor, (c) series resistance, (d) parasitic capacitance, (e) parasitic resistance.

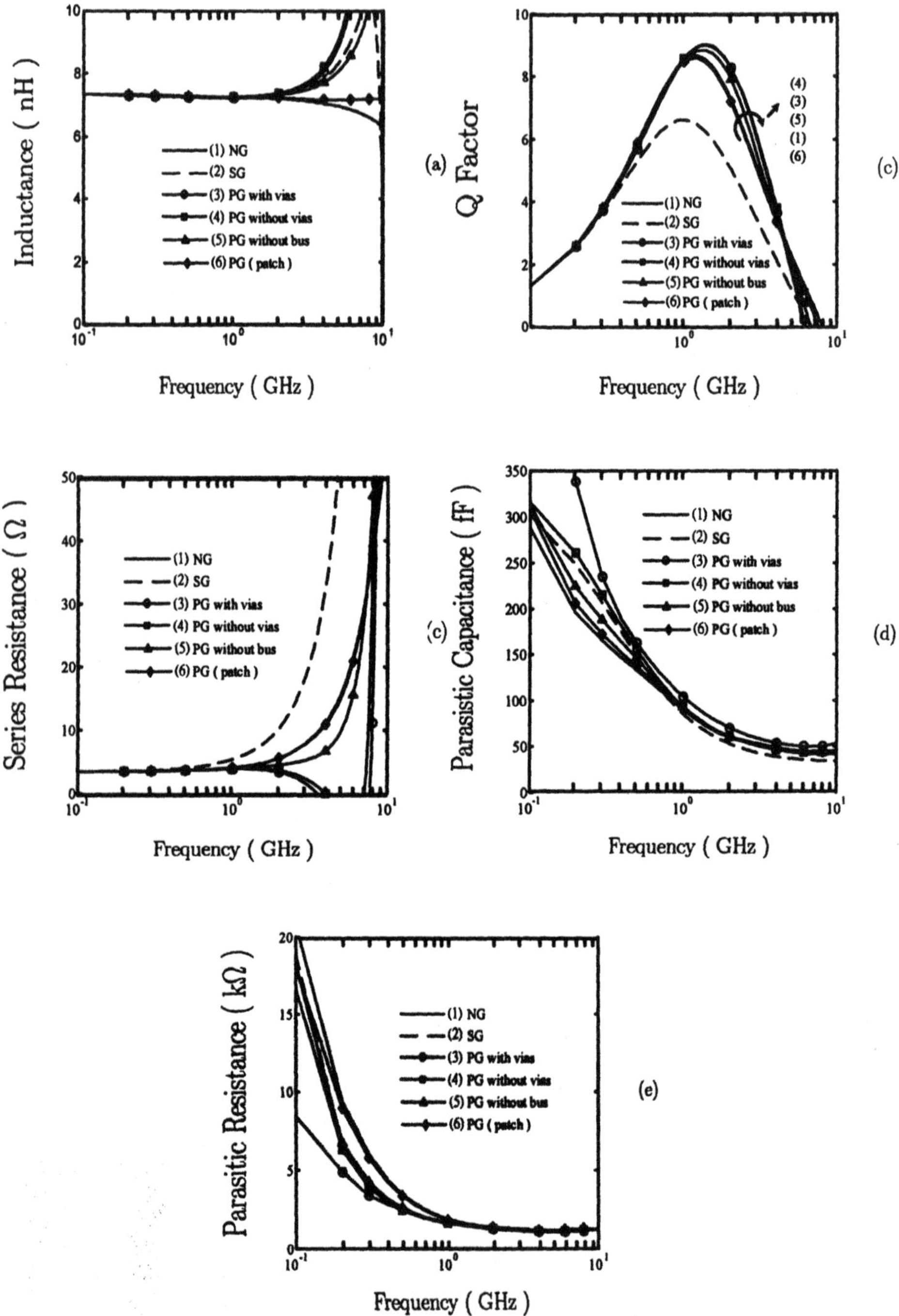

Figure 9.7. **Grounding effects of polysilicon patterned grounds, (a) inductance, (b) Q factor, (c) series resistance, (d) parasitic capacitance, (e) parasitic resistance.**

while that without vias has higher inductance. These two inductors have similar Q factors which are higher than those of the other configurations. However, self-resonant frequencies of the two inductors are lower than those of the others. The inductor with a patch ground has the lowest Q factor which is related to its high series resistance among all these configurations.

The effects of different patterned grounds made of aluminum are also analyzed. Both the strip width and spacing of the patterned ground are 10 μm. The inductor with the patterned ground connected by vias has the lowest series resistance and the highest parasitic capacitance, and results in lower self-resonant frequency. The inductor with patch ground is very similar to that without any patterned ground. This implies that patch ground has the least effects on electric fields due to its large gap area.

References

[1] R. R. Tummala and E. J. Rymaszewski, "*Microelectronics Packaging Handbook*," Van Nostrand Reinhold, 1989.

[2] W. Schwab and W. Menzel, "Compact bandpass filters with improved stop-band characteristics using planar multilayer structures," *IEEE MTT-S Int. Microwave Symp. Dig.*, vol.3, pp.1207-1209, June 1992.

[3] M. J. Tsai, C. Chen, N. G. Alexopoulos, and T. S. Horng, "Multiple arbitrary shape via-hole and air-bridge transitions in multilayered structures," *IEEE Trans. Microwave Theory Tech.*, vol.44, pp.2504-2511, Dec. 1996.

[4] A. Sutono, A. Pham, J. Laskar, and W. R. Smith, "Development of three dimensional ceramic-based MCM inductors for hybrid RF / microwave applications," *IEEE Radio Freq. Integ. Circuits Symp. Dig.*, pp.175-178, 1999.

[5] M. Nakatsugawa, A. Kanda, H. Okazaki, K. Nishikawa, and M. Muraguchi, "Line-loss and size reduction techniques for millimeter-wave RF front-end boards by using a polyimide / alumina-ceramic multilayer configuration," *IEEE Trans. Microwave Theory Tech.*, vol.45, no.12, pp.2308-2315, Dec. 1997.

[6] S. Consolazio, K. Nguyen, D. Biscan, K. Vu, A. Ferek, and A. Ramos, "Low temperature cofired ceramic(LTCC) for wireless applications," *IEEE MTT-S Int. Microwave Symp. Dig.*, pp.201-205, 1999.

[7] M. Park, S. Lee, H. K. Yu, and K. S. Nam, "The detailed analysis of high Q CMOS-compatible microwave spiral inductors in silicon technology," *IEEE Trans. Electron. Dev.*, vol.45, pp.1953-1959, Sept. 1998.

[8] M. Park, S. Lee, C. S. Kim, H. K. Yu, and K. S. Nam, "Optimization of high Q CMOS-compatible microwave inductors using silicon CMOS technology," *IEEE Radio Freq. Integ. Circuits Symp. Dig.*, pp.181-184, 1997.

[9] M. Park, S. Lee, H. K. Yu, J. G. Koo, and K. S. Nam, "High Q CMOS-compatible microwave inductors using double-metal interconnection silicon technology," *IEEE Microwave Guided Wave Lett.*, vol.7, pp.45-47, Feb. 1997.

[10] J. N. Burghartz, D. C. Edelstein, M. Soyuer, H. A. Ainspan, and K. A. Jenkins, "RF circuit design aspects of spiral inductors on silicon," *IEEE J. Solid-State Circuit*, vol.33, pp.2028-2034, Dec. 1998.

[11] J. R. Long and M. A. Copeland, "The modeling, characterization, and design of monolithic inductors for silicon RF IC's," *IEEE J. Solid-State Circuit*, vol.32, pp.357-369, Mar. 1997.

[12] C. P. Yue and S. S. Wong, "On-chip spiral inductors with patterned ground shields for Si-based RF IC's," *IEEE J. Solid-State Circuit*, vol.33, no.5, pp.743-752, May 1998.

[13] Y. J. E. Chen, D. Heo, J. Laskar, and T. Anderson, "Investigation of Q enhancement for inductors processed in BiCMOS technology," *IEEE*, pp.263-266, 1999.

[14] J. N. Burghartz, K. A. Jenkuns, and M. Soyuer, "Multilevel-spiral inductors using VLSI interconnect technology," *IEEE Electron. Dev. Lett.*, vol.17, no.9, pp.428-430, Sept. 1996.

[15] K. B. Ashby, I. A. Koukkias, W. C. Finley, J. J. Bastek, and S. Moinian, "High Q inductors for wireless applications in a complementary silicon bipolar process," *IEEE J. Solid-State Circuit*, vol.31, pp.4-9, Jan. 1996.

[16] J. Y. C. Chang, A. A. Abidi, and M.Gaitan, "Large suspended inductors on silicon and their use in a 2-mm CMOS RF amplifier," *IEEE Electron. Dev. Lett.*, vol.14, pp.246-248, May 1993.

[17] S. M. Yim, T. Chen, and K. K. O, "The effect of a ground shield on the characteristics and performance of spiral inductors," *IEEE J. Solid-State Circuit*, vol.37, no.2, Feb. 2002.

[18] "Effect of a ground shield of a silicon on-chip spiral inductor."

[19] K. Ng, B. Rejaei, and J. N. Burghartz, "Ground pattern for improved characteristics of spiral RF transformers on silicon," *IEEE*, pp.75-78, 2001.

[20] C. B. Sia, K. S. Yeo, W. L. Goh, T. N. Swe, C. Y. Ng, K. W. Chew, W. B. Loh, S. Chu, and L. Chan, "Effects of polysilicon shield on spiral inductors for silicon-based RF IC's," *IEEE*, pp.158-161, 2001.

[21] Y. E. Chen, D. Bien, D. Heo, and J. Laskar, "Q-enhancement of spiral inductor with N^+-diffusion patterned ground shields," *IEEE MTT-S Int. Microwave Symp. Dig.*, pp.1289-1292, 2001.

[22] C. P. Yue and S. S. Wong, "A study on substrate effects of silicon-based RF passive components," *IEEE MTT-S Int. Microwave Symp. Dig.*, pp.1625-1628, 1999.

[23] J. F. Kiang, "Integral equation solution to the skin efect problem in conductor strips of finite thickness," *IEEE Trans. Microwave Theory Tech.*, vol.39, no.3, Mar. 1991.

Chapter 10

GUIDANCE PROPERTIES OF STRIPS IN AN INHOMOGENEOUS LAYERED MEDIUM

Jean-Fu Kiang,[1] Chung-I G. Hsu,[2] and Ching-Her Lee[3]

[1] *Department of Electrical Engineering and*
Graduate Institute of Communication Engineering
National Taiwan University
Taipei, Taiwan, ROC

[2] *Department of Electrical Engineering*
Da-Yeh University
Changhua, Taiwan, ROC

[3] *Department of Electronic Engineering*
National Changhua University of Education
Changhua, Taiwan, ROC

Abstract In this Chapter, rigorous dyadic Green's formulation in the spectral domain is applied to study the guidance properties of waveguiding structures consisting of conducting strips and/or localized inhomogeneities in a layered medium. An integral equation in terms of surface current on the conducting strips and equivalent volume current associated with localized inhomogeneities is derived. Galerkin method is then applied to transform this mixed surface and volume integral equation into a matrix equation, from which the propagation constants are solved. Numerical results for exemplified structures are also presented.

Keywords: layered medium, inhomogeneous, integral equation, Galerkin method, propagation constant.

1. Introduction

Transmission lines are basic building blocks in modern electronic equipments, which serve as interconnections between modules on a board, between chips in a module, between gates in a chip, and between transistors in a gate [1]. In all these environments, interconnections are usually embedded in a layered medium. Each layer of the background medium is usually homogeneous. However, some practical dielectric materials contain inhomogeneities which may be manufacturing imperfections or designed features [2]-[5].

As technology evolves, devices tend to be more densely packed, thus making interconnections between devices closer to one another. This may cause strong crosstalk between transmission lines. To reduce the crosstalk between two edge-coupled microstrip lines, notch filled with air or low-permittivity material in the substrate was proposed [3]. Grooves can be carved in the substrate of microstrip line to create less dispersion as compared with conventional microstrip lines [4]. It is also found that both dielectric and conductor losses of the grooved microstrip line can be reduced as the groove width is increased. It has been experimentally verified that a notch created in the silicon substrate of a coplanar waveguide(CPW) with micromachining using CMOS process can reduce its insertion loss and dispersive characteristics [5]. In these cases, localized inhomogeneities are created in the layered medium.

Different approaches have been proposed to study transmission lines and dielectric waveguides with localized inhomogeneities. In [3], a notch is created in the middle of substrate to decouple two microstrip lines. Free-space Green's function is employed in the quasistatic analysis to solve for the free charge distributions on conducting strips and the equivalent bound charge distributions on dielectric interfaces. Method of lines technique is applied to analyze laterally unbounded strip lines with segmented layers [6]. This approach is capable of handling very narrow strips with good accuracy.

In [7], a coordinate transformation is proposed to transform the laterally open microstrip structure into a laterally closed one, which is then solved using conventional method of lines. This method is suitable for structures with segmented dielectric layers and/or finite ground planes. In [4], both finite difference time domain(FDTD) method and method of lines are applied to study the dispersion characteristics of grooved microstrip lines. Less dispersive results are observed as a groove is present.

In [8], finite element method with higher-order elements is applied to investigate slow wave transmission lines. If the depletion region due to Schottky contact is modeled as a localized inhomogeneity instead of a homogeneous layer, this approach yields more accurate solutions.

Fast advances in integrated circuits(IC) technology in the past three decades have significantly reduced the size of electronic devices and the propagation delays are also reduced in proportion. Such reductions are preferred in high-speed digital circuitries. Although the length of transmission lines can be reduced, there are limitations in scaling down the line thickness [1]. If the line thickness is over-reduced, the RC delays due to interconnection may start to dominate the IC chip performance. Besides, thinner interconnection may become unreliable due to electromigration. To overcome such drawbacks, optical interconnection like dielectric strip waveguide can be used [1].

In this Chapter, a rigorous spectral domain integral equation is formulated to analyze transmission line structures consisting of conducting strips embedded in a multilayered medium, where localized inhomogeneities in the dielectric may be present. The conducting strips can be of zero or finite thickness. The number of dielectric layers can be arbitrary. The localized inhomogeneity is modeled by an equivalent volume current embedded in a homogeneous layer. An electric type of dyadic Green's function for the multilayered medium is incorporated in the integral equation [9]-[14]. Subdomain basis functions are used to approximate the surface current densities on conducting strips and the volume current densities in the localized inhomogeneous regions. Galerkin method is then applied to derive a matrix eigenvalue equation for the propagation constants. Numerical results for exemplified structures are presented for demonstration.

2. Formulation

Fig.10.1 shows a waveguiding structure consisting of perfectly conducting strips and localized inhomogeneous dielectric regions embedded in a multilayered medium. The conducting strips are of zero thickness and the localized inhomogeneous dielectric regions are of arbitrary cross section. The thickness of layer (m) is h_m, and the dielectric layer is characterized by permittivity, ϵ_m, and conductivity, σ_m. The conductivity and permittivity can be combined to form a complex-valued permittivity, $\tilde{\epsilon}_m = \epsilon_m - j\sigma_m/\omega$. Although the layered medium is assumed isotropic, the localized inhomogeneous regions can be anisotropic in the formulation.

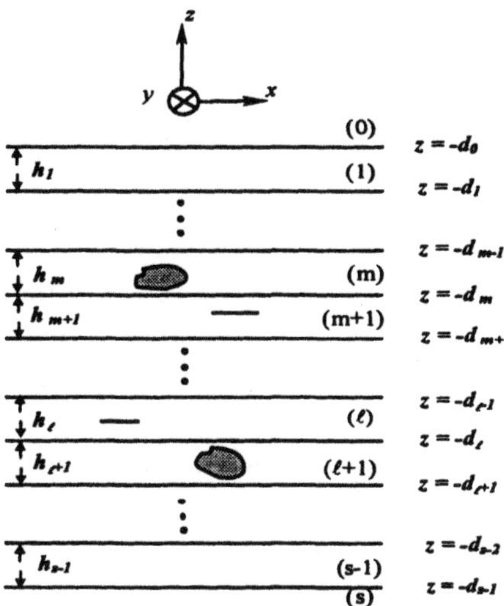

Figure 10.1. Perfectly conducting strips and inhomogeneous dielectric regions in a layered medium.

The electric field due to current distribution, $\bar{J}(\bar{r})$, in an unbounded homogeneous medium of ($\tilde{\epsilon}_m$, μ_0) can be expressed as [10], [12]

$$\bar{E}(\bar{r}) = \iiint\limits_V dv' \bar{\bar{G}}^{EJ}(\bar{r}, \bar{r}') \cdot \bar{J}(\bar{r}') \tag{10.1}$$

where

$$\bar{\bar{G}}^{EJ}(\bar{r}, \bar{r}') = \frac{1}{4\pi^2} \iint\limits_{-\infty}^{\infty} d\bar{k}_\rho e^{-j\bar{k}_\rho \cdot (\bar{\rho} - \bar{\rho}')} \tilde{\bar{\bar{G}}}^{\mathrm{PV}}(\bar{k}_\rho, z, z')$$
$$+ \hat{z}\hat{z} \frac{j}{\omega \epsilon_0 \tilde{\epsilon}_{rm}} \delta(\bar{r} - \bar{r}') \tag{10.2}$$

is the electric type of dyadic Green's function, $\bar{k}_\rho = \hat{x}k_x + \hat{y}k_y$ and $d\bar{k}_\rho = dk_x dk_y$. Substituting the first term of $\bar{\bar{G}}^{EJ}(\bar{r}, \bar{r}')$ into (10.1) results in a principal-value integral with a $\hat{z}$-directed pillbox as an exclusion volume. The second term of $\bar{\bar{G}}^{EJ}(\bar{r}, \bar{r}')$ is a source dyadic associated with the same pillbox [15]-[17]. The explicit form of the principal-value part in (10.2) is [9], [10]

$$\tilde{\bar{\bar{G}}}^{\mathrm{PV}}(\bar{k}_\rho, z, z') = -\frac{\omega \mu_0}{2k_{mz}} \left[\hat{v}^{\mathrm{TE}}(\bar{k}_\rho) \hat{v}^{\mathrm{TE}}(\bar{k}_\rho) \right.$$

$$+\hat{v}^{\rm TM}_{m\pm}(\bar{k}_\rho)\hat{v}^{\rm TM}_{m\pm}(\bar{k}_\rho)\Big] e^{-jk_{mz}|z-z'|} \qquad (10.3)$$

where the fields are decomposed into TE$_z$ and TM$_z$ parts, $k_m = k_0\sqrt{\tilde{\epsilon}_{rm}}$, $k_{mz} = \sqrt{k_m^2 - k_\rho^2}$, and

$$\hat{v}^{\rm TE}(\bar{k}_\rho) = \frac{\hat{x}k_y - \hat{y}k_x}{k_\rho}, \qquad \hat{v}^{\rm TM}_{m\pm}(\bar{k}_\rho) = \mp\frac{k_{mz}\bar{k}_\rho}{k_m k_\rho} + \hat{z}\frac{k_\rho}{k_m}$$

where the upper(lower) sign is for $z > z'(z < z')$. If the current distribution is restricted in layer (m), additional terms must be incorporated into (10.3) to take into account multiple reflections at the upper boundary, $z = -d_{m-1}$, and lower boundary, $z = -d_m$, to obtain [9]-[14]

$$\begin{aligned}
&\overline{\overline{G}}^{\rm PV}_{mm}(\bar{k}_\rho, z, z') \\
&= -\frac{\omega\mu_0}{2k_{mz}}\Big\{\Big[\hat{v}^{\rm TE}(\bar{k}_\rho)\hat{v}^{\rm TE}(\bar{k}_\rho) + \hat{v}^{\rm TM}_{m\pm}(\bar{k}_\rho)\hat{v}^{\rm TM}_{m\pm}(\bar{k}_\rho)\Big] e^{-jk_{mz}|z-z'|} \\
&+\frac{\hat{v}^{\rm TE}(\bar{k}_\rho)\hat{v}^{\rm TE}(\bar{k}_\rho)}{D^{\rm TE}_m}\Big[R^{\rm TE}_{\cap m}\Big(e^{-jk_{mz}(z_m+z'_m)} + R^{\rm TE}_{\cup m}e^{-jk_{mz}(2h_m+z_m-z'_m)}\Big) \\
&\quad +R^{\rm TE}_{\cup m}\Big(e^{-jk_{mz}(2h_m-z_m-z'_m)} + R^{\rm TE}_{\cap m}e^{-jk_{mz}(2h_m-z_m+z'_m)}\Big)\Big] \\
&+\frac{1}{D^{\rm TM}_m}\Big[R^{\rm TM}_{\cap m}e^{-jk_{mz}(z_m+z'_m)}\hat{v}^{\rm TM}_{m+}(\bar{k}_\rho)\hat{v}^{\rm TM}_{m-}(\bar{k}_\rho) \\
&\quad +R^{\rm TM}_{\cap m}R^{\rm TM}_{\cup m}e^{-jk_{mz}(2h_m+z_m-z'_m)}\hat{v}^{\rm TM}_{m+}(\bar{k}_\rho)\hat{v}^{\rm TM}_{m+}(\bar{k}_\rho) \\
&\quad +R^{\rm TM}_{\cup m}e^{-jk_{mz}(2h_m-z_m-z'_m)}\hat{v}^{\rm TM}_{m-}(\bar{k}_\rho)\hat{v}^{\rm TM}_{m+}(\bar{k}_\rho) \\
&\quad +R^{\rm TM}_{\cap m}R^{\rm TM}_{\cup m}e^{-jk_{mz}(2h_m-z_m+z'_m)}\hat{v}^{\rm TM}_{m-}(\bar{k}_\rho)\hat{v}^{\rm TM}_{m-}(\bar{k}_\rho)\Big]\Big\}
\end{aligned} \qquad (10.4)$$

where $z_m = z + d_m(z'_m = z' + d_m)$ is the local z coordinate of the observation(source) point in layer (m). Here, the subscript, mm, in $\overline{\overline{G}}^{\rm PV}_{mm}$ indicates that (10.4) is valid only when both the source and observation points are in layer (m). In (10.4), $R^\alpha_{\cup n}$ ($R^\alpha_{\cap n}$) with $\alpha =$ TE or TM is the upward(downward) reflection coefficient at the upper(lower) boundary of layer (m), defined as [9], [10]

$$R^\alpha_{\cup m} = \frac{R^\alpha_{m(m-1)} + R^\alpha_{\cup(m-1)}e^{-j2k_{(m-1)z}h_{m-1}}}{1 + R^\alpha_{m(m-1)}R^\alpha_{\cup(m-1)}e^{-j2k_{(m-1)z}h_{m-1}}}$$

$$R^\alpha_{\cap m} = \frac{R^\alpha_{m(m+1)} + R^\alpha_{\cap(m+1)}e^{-j2k_{(m+1)z}h_{m+1}}}{1 + R^\alpha_{m(m+1)}R^\alpha_{\cap(m+1)}e^{-j2k_{(m+1)z}h_{m+1}}}$$

where

$$R^{\rm TM}_{m(m\pm1)} = \frac{\tilde{\epsilon}_{r(m\pm1)}k_{mz} - \tilde{\epsilon}_{rm}k_{(m\pm1)z}}{\tilde{\epsilon}_{r(m\pm1)}k_{mz} + \tilde{\epsilon}_{rm}k_{(m\pm1)z}}, \qquad R^{\rm TE}_{m(m\pm1)} = \frac{k_{mz} - k_{(m\pm1)z}}{k_{mz} + k_{(m\pm1)z}}$$

are Fresnel reflection coefficients, and

$$D_m^\alpha = 1 - R_{\cup m}^\alpha R_{\cap m}^\alpha e^{-j2k_{mz}h_m}$$

Next, consider the case where the source and observation points are located in different layers. The fields in a source-free layer (ℓ) can be expressed as

$$\bar{E}_\ell(\bar{r}) = \frac{1}{4\pi^2} \iiint\limits_V d\bar{r}' \iint\limits_{-\infty}^{\infty} d\bar{k}_\rho e^{-j\bar{k}_\rho\cdot(\bar{\rho}-\bar{\rho}')} \left(\hat{x}\tilde{E}_{\ell x} + \hat{y}\tilde{E}_{\ell y} + \hat{z}\tilde{E}_{\ell z}\right)$$

$$\bar{H}_\ell(\bar{r}) = \frac{1}{4\pi^2} \iiint\limits_V d\bar{r}' \iint\limits_{-\infty}^{\infty} d\bar{k}_\rho e^{-j\bar{k}_\rho\cdot(\bar{\rho}-\bar{\rho}')} \left(\hat{x}\tilde{H}_{\ell x} + \hat{y}\tilde{H}_{\ell y} + \hat{z}\tilde{H}_{\ell z}\right)$$

where the transverse field components can be derived from the longitudinal components as [9], [10]

$$\tilde{E}_{\ell p} = \hat{p} \cdot \left(\frac{-j\bar{k}_\rho}{k_\rho^2} \frac{\partial \tilde{E}_{\ell z}}{\partial z} + \hat{z} \times \bar{k}_\rho \frac{\omega\mu_0}{k_\rho^2} \tilde{H}_{\ell z} \right), \qquad p = x, y$$

$$\tilde{H}_{\ell p} = \hat{p} \cdot \left(\frac{-j\bar{k}_\rho}{k_\rho^2} \frac{\partial \tilde{H}_{\ell z}}{\partial z} - \hat{z} \times \bar{k}_\rho \frac{\omega\epsilon_0\tilde{\epsilon}_{r\ell}}{k_\rho^2} \tilde{E}_{\ell z} \right), \qquad p = x, y \quad (10.5)$$

For $\ell > m$ ($z < z'$), the $\hat{z}$-directed field components can be expressed as

$$\tilde{E}_{\ell z}(\bar{k}_\rho) = A_\ell \left(e^{jk_{\ell z}z_\ell} + R_{\cap \ell}^{\rm TM} e^{-jk_{\ell z}z_\ell} \right)$$

$$\tilde{H}_{\ell z}(\bar{k}_\rho) = B_\ell \left(e^{jk_{\ell z}z_\ell} + R_{\cap \ell}^{\rm TE} e^{-jk_{\ell z}z_\ell} \right) \quad (10.6)$$

By substituting (10.6) into (10.5) and imposing the boundary conditions at $z = -d_{\ell-1}$, the fields in layer (ℓ) can be related to those in layer ($\ell - 1$). Similarly, for $\ell < m$ ($z > z'$), the $\hat{z}$-directed field components can be expressed as

$$\tilde{E}_{\ell z}(\bar{k}_\rho) = C_\ell \left[e^{jk_{\ell z}(h_\ell - z_\ell)} + R_{\cup \ell}^{\rm TM} e^{-jk_{\ell z}(h_\ell - z_\ell)} \right]$$

$$\tilde{H}_{\ell z}(\bar{k}_\rho) = D_\ell \left[e^{jk_{\ell z}(h_\ell - z_\ell)} + R_{\cup \ell}^{\rm TE} e^{-jk_{\ell z}(h_\ell - z_\ell)} \right] \quad (10.7)$$

By substituting (10.7) into (10.5) and imposing the boundary conditions at $z = -d_\ell$, the fields in layer (ℓ) can be related to those in layer ($\ell + 1$). This procedure can be extended to derive the spectral Green's functions $\bar{\bar{G}}_{\ell m}^{\rm PV}$ for $\ell \neq m$. The spectral Green's functions derived using the transmission line analogy can be found in [11], [18], [19].

For localized inhomogeneous regions in layer (m), define an equivalent volume current by using Ampere's law as

$$\nabla \times \bar{H} = j\omega\epsilon_0\bar{\bar{\epsilon}}_r(\bar{r}) \cdot \bar{E} + \bar{\bar{\sigma}}(\bar{r}) \cdot \bar{E} + \bar{J} = j\omega\epsilon_0\tilde{\epsilon}_{rm} + \bar{J}_{eq} + \bar{J}$$

where $\bar{\bar{\sigma}}$ and $\bar{\bar{\epsilon}}_r(\bar{r})$ are the conductivity and relative permittivity dyadics, respectively, of the localized inhomogeneous regions, and $\bar{J}_{eq}(\bar{r}) = \Delta\bar{\bar{\sigma}}(\bar{r})\cdot \bar{E}(\bar{r})$ is the equivalent volume(conduction and polarization) current pertinent to the inhomogeneous regions by $\Delta\bar{\bar{\sigma}}(\bar{r}) = \bar{\bar{\sigma}}(\bar{r}) - \sigma_m\bar{\bar{I}} + j\omega\epsilon_0[\bar{\bar{\epsilon}}_r(\bar{r}) - \epsilon_{rm}\bar{\bar{I}}]$. The fields can thus be regarded as being generated by the currents $\bar{J}$ and $\bar{J}_{eq}$ in the layered medium in the absence of localized inhomogeneous regions. When there are perfectly conducting strips in layer (ℓ) and localized inhomogeneous dielectric regions in layer (m), the electric field in the layered medium can be expressed as

$$\bar{E}(\bar{r}) = \iint_{S_\ell} d\bar{r}'\bar{\bar{G}}^{EJ}(\bar{r},\bar{r}')\cdot\bar{J}_s(\bar{r}') + \iiint_{V_m} d\bar{r}'\bar{\bar{G}}^{EJ}(\bar{r},\bar{r}')\cdot\bar{J}_{eq}(\bar{r}') \quad (10.8)$$

where $\bar{J}_s(\bar{r})$ is the surface current density on conducting strip, S_ℓ is the conducting strip surface in layer (ℓ), and V_m is the localized inhomogeneous dielectric region in layer (m). Substituting (10.2) into (10.8), we have [13], [14]

$$\bar{E}(\bar{r}) = \iint_{S_\ell} d\bar{r}'\bar{\bar{G}}^{EJ}(\bar{r},\bar{r}')\cdot\bar{J}_s(\bar{r}') + \iiint_{V_m} d\bar{r}'\bar{\bar{G}}^{EJ}(\bar{r},\bar{r}')\cdot\Delta\bar{\bar{\sigma}}(\bar{r}')\cdot\bar{E}(\bar{r}') \quad (10.9)$$

For a wave mode guided along the $\hat{y}$ direction, all the field and current components can be expressed as $P(\bar{r}) = p(\bar{r}_s)e^{-j\eta y}$, where $\bar{r}_s = \hat{x}x + \hat{z}z$ and η is the propagation constant of the eigenmode. Thus, (10.9) can be reduced to

$$\begin{aligned}\bar{E}(\bar{r}_s) - \hat{z}\frac{j}{\omega\epsilon_0\tilde{\epsilon}_{rm}}\Delta\sigma_z(\bar{r}_s)E_z(\bar{r}_s) &= \frac{1}{2\pi}\int_{\partial S_\ell} d\ell' \int_{-\infty}^{\infty} dk_x e^{-jk_x(x-x')}\tilde{\bar{\bar{G}}}^{EJ}(k_x,\eta,z,z')\cdot\bar{J}_s(\bar{r}'_s) \\ &+ \frac{1}{2\pi}\iint_{\partial V_m} ds' \int_{-\infty}^{\infty} dk_x e^{-jk_x(x-x')}\tilde{\bar{\bar{G}}}^{PV}(k_x,\eta,z,z')\cdot\Delta\bar{\bar{\sigma}}(\bar{r}'_s)\cdot\bar{E}(\bar{r}'_s)\end{aligned} \quad (10.10)$$

where ∂S_ℓ is the cross sectional contour of conducting strips in layer (ℓ) and ∂V_m is the cross sectional area of localized inhomogeneous dielectric

regions in layer (m). Eq.(10.10) is an integral equation in terms of the unknown surface current densities on conducting strips and the unknown electric fields in localized inhomogeneous dielectric regions.

Galerkin method is applied to solve this integral equation. The longitudinal(transversal) component of surface current densities on conducting strips can be expanded by a set of pulse(triangle) basis functions. Each component of electric fields over the inhomogeneous dielectric regions can be expanded by a set of pulse basis functions. Then, the same sets of basis functions are used as testing functions to construct a matrix equation from which both the phase constant and attenuation constant are obtained. Take the inlet in Fig.10.3 for example, there are one conducting strip and one localized dielectric region. First, divide the conducting strip along the $\hat{x}$ direction into K segments, with the x coordinates of the two ends of segment i denoted by x_{i-1} and x_i, respectively. Next, divide the cross sectional area, ∂V_m, of the localized region into N rectangular cells, where the nth cell is centered at (x_n, z_n) with cell width Δx_n and cell height Δz_n. Thus, the unknown surface current density on conducting strips and the electric fields in localized dielectric regions can be expressed as

$$\bar{J}_s(\bar{r}_s) = \hat{x}\sum_{i=1}^{K-1} a_{ix}g_i(x) + \hat{y}\sum_{i=1}^{K} a_{iy}p_i(x)$$

$$\bar{E}(\bar{r}_s) = \sum_{n=1}^{N}(\hat{x}b_{nx} + \hat{y}b_{ny} + \hat{z}b_{nz})P\left(\frac{x-x_n}{\Delta x_n}\right)P\left(\frac{z-z_n}{\Delta z_n}\right)$$

where a_{iq} with $q = x, y$ and b_{nq} with $q = x, y, z$ are the unknown coefficients, and

$$p_i(x) = \begin{cases} 1, & x_{i-1} \le x \le x_i \\ 0, & \text{elsewhere} \end{cases}$$

$$g_i(x) = \begin{cases} (x - x_{i-1})/(x_i - x_{i-1}), & x_{i-1} \le x \le x_i \\ (x_{i+1} - x)/(x_{i+1} - x_i), & x_i \le x \le x_{i+1} \\ 0, & \text{elsewhere} \end{cases}$$

$$P(t) = \begin{cases} 1, & |t| \le 1/2 \\ 0, & \text{otherwise} \end{cases} \qquad (10.11)$$

By using the basis functions given in (10.11) as testing functions in the Galerkin method, the integral equation (10.10) can be transformed

into a matrix equation as

$$\begin{bmatrix} [Z_{cc}] & [Z_{cd}] \\ [Z_{dc}] & [Z_{dd}] \end{bmatrix} \begin{bmatrix} [a] \\ [b] \end{bmatrix} = [0]$$

where the subscripts, c and d, denote conductor and localized dielectric, respectively, and

$$[Z_{cc}]_{ji} = \begin{bmatrix} Z^{xx}_{cc,ji} & Z^{xy}_{cc,ji} \\ Z^{yx}_{cc,ji} & Z^{yy}_{cc,ji} \end{bmatrix}, \qquad [Z_{cd}]_{jn} = \begin{bmatrix} Z^{xx}_{cd,jn} & Z^{xy}_{cd,jn} & Z^{xz}_{cd,jn} \\ Z^{yx}_{cd,jn} & Z^{yy}_{cd,jn} & Z^{yz}_{cd,jn} \end{bmatrix}$$

$$[Z_{dc}]_{mi} = \begin{bmatrix} Z^{xx}_{dc,mi} & Z^{xy}_{dc,mi} \\ Z^{yx}_{dc,mi} & Z^{yy}_{dc,mi} \\ Z^{zx}_{dc,mi} & Z^{zy}_{dc,mi} \end{bmatrix}$$

$$[Z_{dd}]_{mn} = \begin{bmatrix} Z^{xx}_{dd,mn} & Z^{xy}_{dd,mn} & Z^{xz}_{dd,mn} \\ Z^{yx}_{dd,mn} & Z^{yy}_{dd,mn} & Z^{yz}_{dd,mn} \\ Z^{zx}_{dd,mn} & Z^{zy}_{dd,mn} & Z^{zz}_{dd,mn} \end{bmatrix}$$

$$[a]_i = [a_{ix}, a_{iy}]^t, \qquad [b]_n = [b_{nx}, b_{ny}, b_{nz}]^t$$

The $Z^{\alpha\beta}_{dd,mn}$ element takes the following form

$$\begin{aligned} Z^{\alpha\beta}_{dd,mn} = & -\left(1 - \frac{j}{\omega\epsilon_o\tilde{\epsilon}_r}\Delta\sigma_{nz}\delta_{z\beta}\right)\delta_{\alpha\beta}\delta_{mn}\Delta A_m \\ & + \frac{\Delta\sigma_{n\beta}}{2\pi}\int_{-\infty}^{\infty} dk_x e^{-jk_x(x_m - x_n)}\frac{4}{k_x^2}\sin\left(\frac{k_x\Delta x_m}{2}\right) \\ & \sin\left(\frac{k_x\Delta x_n}{2}\right)\tilde{g}^{\mathrm{PV}}_{\alpha\beta,mn}(k_x,\eta) \end{aligned}$$

where $\alpha, \beta = x, y, z$, ΔA_m is the area of the mth rectangular cell in the dielectric region, and

$$\tilde{g}^{\mathrm{PV}}_{\alpha\beta,mn}(k_x,\eta) = \int dz P\left(\frac{z - z_m}{\Delta z_m}\right)\int dz' P\left(\frac{z' - z_n}{\Delta z_n}\right)\tilde{G}^{\mathrm{PV}}_{\alpha\beta}(k_x,\eta,z,z')$$

which can be integrated analytically since the variables z and z' in the spectral Green's functions appear only in the exponential terms. Other elements of the Z matrices can be derived in a similar manner. The propagation constant, η, is determined by solving $\det[Z] = 0$, using complex root searching algorithms such as secant method or Muller's method.

3. Results and Discussions

In this Section, numerical results are presented to demonstrate the applications of this method. Fig.10.2 shows the effective dielectric constant of two coupled microstrip lines. The results computed using this method are very close to those in [20]. In the presence of a rectangular region with lower dielectric constant than that of the substrate, the effective dielectric constant becomes lower, and coupling between these two lines can be reduced [3].

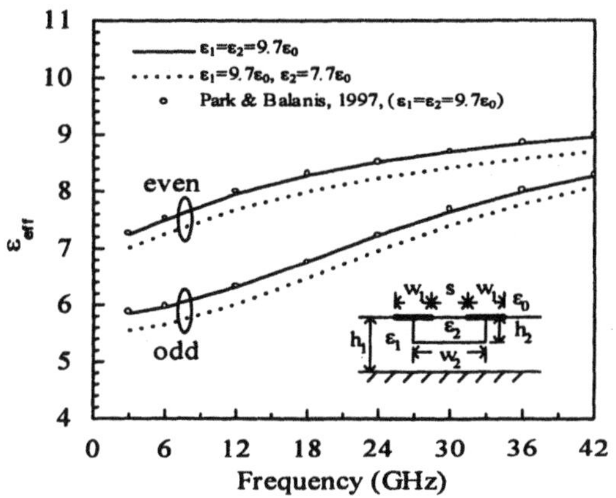

Figure 10.2. Effective dielectric constant of coupled microstrip lines, $w_1 = 1$ mm, $w_2 = 2$ mm, $s = 1$ mm, $h_1 = 1$ mm and $h_2 = 0.2$ mm.

Transmission lines fabricated using CMOS technology become practical [1]. Appropriate etching in the semiconductor substrate can make the guidance properties less frequency dependent [5]. The inlet in Fig.10.3 shows a transmission line with metal-insulator-semiconductor(MIS) configuration, where layer (2) is insulating silicon nitride and layer (1) is doped silicon. When dc bias voltage is applied on the conducting strip, a region depleted of charge carriers is formed in the lossy semiconductor substrate. This depletion region can be characterized by the same dielectric constant as that of the substrate, but with vanishing conductivity. The MIS line is known to exhibit slow wave phenomenon, namely, the phase constant is higher than that in the substrate [8], [21]. This phenomenon finds potential applications in the design of delay lines, phase shifters and tunable filters. Due to this depletion region, the slow wave factor, defined as λ_0/λ_g, and the attenuation constant are significantly reduced, especially at low frequencies.

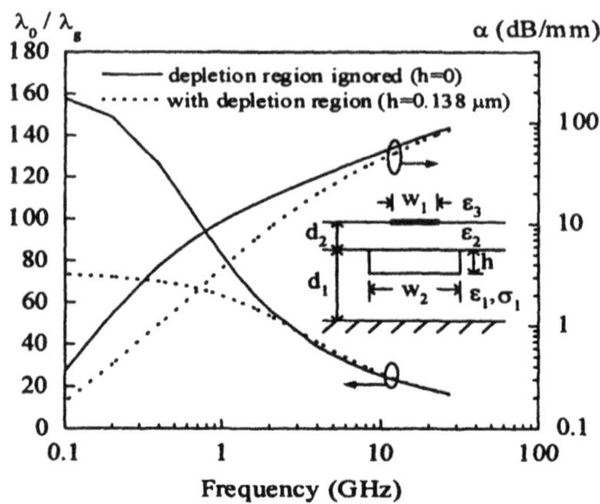

Figure 10.3. Slow wave factor and attenuation constant of an MIS line, $d_1 = 250$ mm, $d_2 = 0.01$ mm, $\epsilon_{r1} = 12$, $\epsilon_{r2} = 4$, $\epsilon_{r3} = 1$, $\sigma_1 = 1{,}000$ S/m, $w_1 = 160$ mm and $w_2 = 164$ mm.

Fig.10.4 shows a dielectric strip waveguide where there is no conducting strip, and the ground plane is assumed perfectly conducting. Wave modes in dielectric strip waveguide are hybrid in nature, and can be categorized into $E^x_{\delta pq}$ and $E^z_{\delta pq}$ modes. For the TE-like mode, $E^x_{\delta pq}$, its TE$_z$ portion is stronger than its TM$_z$ portion. For the TM-like mode, $E^z_{\delta pq}$, its TM$_z$ portion is stronger than its TE$_z$ portion [22]. The subscript pq denotes that the dominant electric field pattern has p and q extrema in $\hat{x}$ and $\hat{z}$ directions, respectively [23]. The subscript δ indicates that the cross section of dielectric strip is not rectangular. The phase constant of $E^z_{\delta 11}$ mode is higher than that of the TM$_0$ background mode. Hence, it is a bound mode. The phase constant of $E^x_{\delta 11}$ mode is lower than that of the TM$_0$ background mode. By exciting the latter, the former mode turns out to be a leaky mode with nonzero attenuation constant. Since the E_z component is larger than the E_x component in $E^z_{\delta 11}$ mode, and vice versa for $E^x_{\delta 11}$ mode, varying ϵ_{st} has stronger effect on the phase constant of $E^x_{\delta 11}$ mode than on that of $E^z_{\delta 11}$ mode.

Fig.10.5 shows the propagation constant of $E^x_{\delta 11}$ modes for coupled dielectric strip waveguides. As the separation between two dielectric strips is small, phase constant of the even $E^x_{\delta 11}$ mode is lower than that of the odd mode. This is opposite to the $E^z_{\delta 11}$ mode in [13]. This phenomenon can be explained as follows. The x, y and z components of electric fields are odd(even), even(odd), and even(odd) functions of x, respectively, for both even(odd) $E^x_{\delta 11}$ and $E^z_{\delta 11}$ modes. For the $E^z_{\delta 11}$

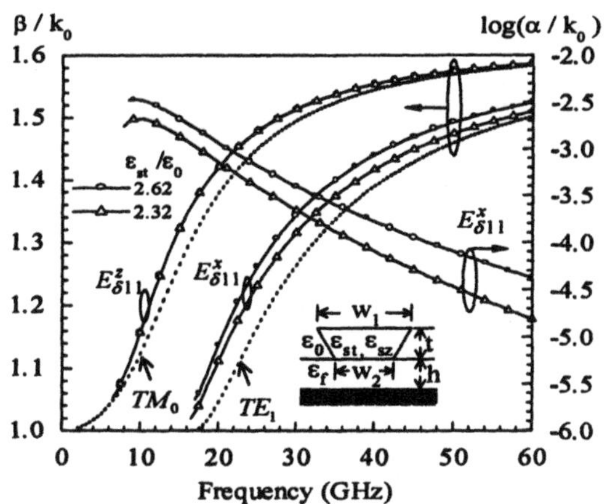

Figure 10.4. Propagation constant of a dielectric strip waveguide, $\epsilon_f = \epsilon_{sz} = 2.62\epsilon_o$, $w_1 = 9.75$ mm, $w_2 = 3.25$ mm, $h = 3.5$ mm and $t = 3.25$ mm.

modes, the E_z component has smoother variation along the $\hat{x}$ direction for the even mode than for the odd mode, thus rendering higher phase constant for the even mode. For the $E^{x}_{\delta 11}$ modes, the E_x component has smoother variation along the $\hat{x}$ direction for the odd mode than for the even mode, thus rendering higher phase constant for the odd mode.

Next, examine the oscillatory behavior on these curves. As stated in [13], the electric fields of leaky modes outside the dielectric strip core regions are the superposition of guided background surface waves that are excited by the two dielectric waveguides. At certain separations, the surface waves excited by each waveguide may add constructively, leading to maximum leakage for the even mode. For odd mode at the same separation, these surface wave modes add destructively to render minimum attenuation constant. Similarly, when the odd mode has maximum leakage at certain separation, the even mode exhibits cancellation effect.

Next, consider the strip line structure in Fig.10.6, where both the signal conductor and ground planes are lossy. Thus, the strip line mode is quasi-TEM. This approach takes into account the conductor loss exactly [14]. From the attenuation constants shown in Fig.10.6, it can be inferred that as the thickness of conducting strip is increased beyond three skin depths, the conductor loss approaches a constant at a given frequency. Notice that the abscissa is the conducting strip thickness normalized to skin depth at each frequency.

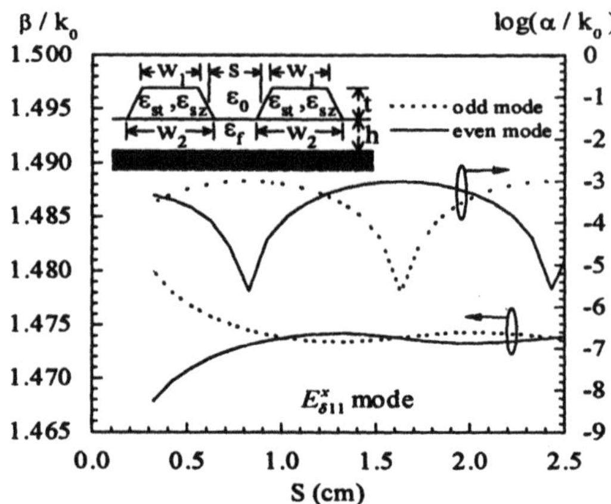

Figure 10.5. Propagation constant of coupled dielectric strip waveguides, $f = 40$ GHz, $w_1 = 3.25$ mm, $w_2 = 9.75$ mm, $t = 3.2$ mm, $h = 3.5$ mm, $\epsilon_{st} = \epsilon_f = 2.62\epsilon_o$ and $\epsilon_{sz} = 2.32\epsilon_o$.

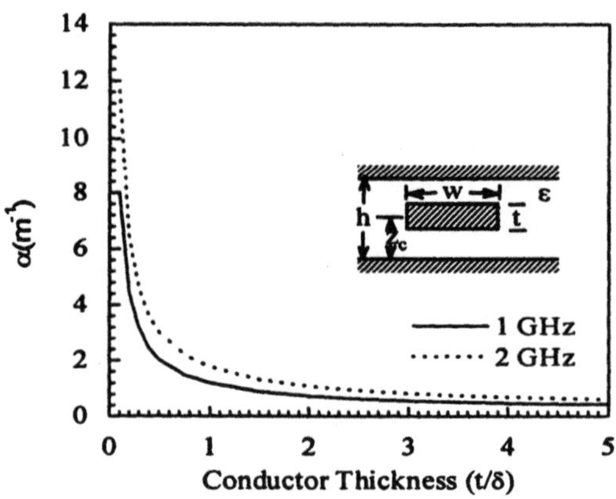

Figure 10.6. Attenuation constant of a stripline with $w = 100\mu$m, $h = 600\mu$m, $z_c = 300\mu$m, $\epsilon_r = 10$ and $\sigma = 5.92 \times 10^7$ S/m.

Fig.10.7 shows the effect of cross sectional shape on the conductor loss. With the cross sectional area fixed, the fish-eyed shape has relatively sharper corners and exhibits higher conductor loss. This phenomenon is more pronounced at high frequencies.

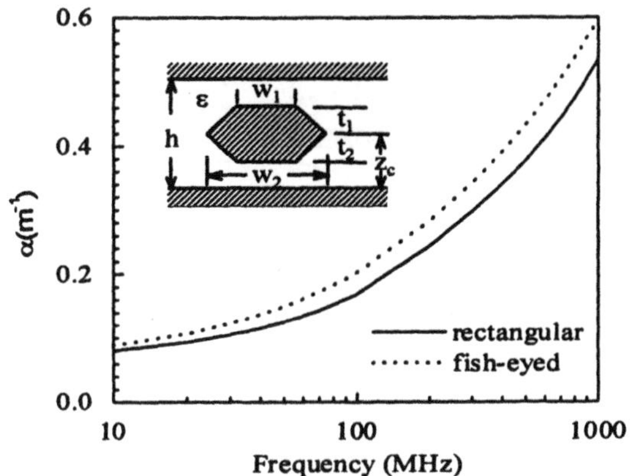

Figure 10.7. **Attenuation constant of strip lines with the same cross sectional area but different shapes, $t_1 = t_2 = 12.5\ \mu m$, $h = 600\ \mu m$, $z_c = 300\ \mu m$, $\epsilon_r = 10$, $\sigma = 5.92 \times 10^7$ S/m. With rectangular shape: $w_1 = w_2 = 100\ \mu m$, with fish-eyed shape: $w_1 = 75\ \mu m$, $w_2 = 125\ \mu m$.**

4. Conclusions

An integral equation formulation utilizing dyadic Green's function has been formulated to study the guidance properties of waveguiding structures consisting of conducting strips and/or localized inhomogeneities in a layered medium. The derivation of dyadic Green's function has been reviewed. Galerkin method has been used to solve the integral equation for the propagation constants. This approach is general and flexible for a large variety of waveguiding structures. Numerical results have been presented for some exemplified structures, including coupled microstrip lines on segmented substrate, MIS transmission lines with localized depletion region, single and coupled dielectric strip waveguides and strip lines.

References

[1] A. K. Goel, *High-Speed VLSI Interconnections: Modelling, Analysis, and Simulation*, New York: Wiley, 1994.

[2] J.-F. Kiang, "Microstrip lines on substrates with segmented or continuous permittivity profiles," *IEEE Trans. Microwave Theory Tech.*, vol.45, pp.229-235, Feb. 1997.

[3] S. He, A. Z. Elsherbeni, and C. E. Smith, "Decoupling between two conductor microstrip transmission line," *IEEE Trans. Microwave Theory Tech.*, vol.41, pp.53-61, Jan. 1993.

[4] W. Chamma, N. Gupta, and L. Shafai, "Dispersion characteristics of grooved microstrip line (GMSL)," *IEEE Trans. Microwave Theory Tech.*, vol.48, pp.611-615, Apr. 2000.

[5] V. Milanovic, M. Gaitan, E. D. Bowen, and M. E. Zaghloul, "Micromachined microwave transmission lines in CMOS technology," *IEEE Trans. Microwave Theory Tech.*, vol.45, pp.630-635, May 1997.

[6] K. Wu and R. G. Bosisio, "A technique for efficient analysis of planar integrated microwave circuits including segmented layers and miniature topologies," *IEEE Trans. Microwave Theory Tech.*, vol.42, pp.826-833, Mar. 1994.

[7] X. H. Yang and L. Shafai, "Full wave approach for analysis of open planar waveguides with finite width dielectric layers and ground planes," *IEEE Trans. Microwave Theory Tech.*, vol.42, pp.142-149, Jan. 1989.

[8] T.-K. Tzuang and T. Itoh, "Finite-element analysis of Schottky contact printed lines," *IEEE Trans. Microwave Theory Tech.*, vol.34, pp.1483-1489, Dec. 1986.

[9] J. A. Kong, *Electromagnetic Wave Theory*, New York: Wiley, 1986.

[10] W. C. Chew, *Waves and Fields in Inhomogeneous Media*, New York: Van Nostrand Reinhold, 1990.

[11] K. A. Michalski and J .R. Mosig, "Multilayered media Green's functions in integral equation formulations," *IEEE Trans. Antennas Propagat.*, vol.45, pp.508-519, Mar. 1997.

[12] J.-F. Kiang, S. M. Ali, and J. A. Kong, "Propagation properties of striplines periodically loaded with crossing strips," *IEEE Trans. Microwave Theory Tech.*, vol.37, pp.776-786, Apr. 1989.

[13] J.-F. Kiang, S. M. Ali, and J. A. Kong, "Integral equation solution to the guidance and leakage properties of coupled dielectric strip waveguides," *IEEE Trans. Microwave Theory Tech.*, vol.38, pp.193-203, Feb. 1990.

[14] J.-F. Kiang, "Integral equation solution to the skin effect problem in conductor strips of finite thickness," *IEEE Trans. Microwave Theory Tech.*, vol.39, pp.452-460, Mar. 1991.

[15] A. D. Yaghjian, "Electric dyadic Green's functions in the source region," *Proc. IEEE*, vol.68, pp.248-263, Feb. 1980.

[16] J. J. Wang, "A unified and consistent view on the singularityies of the electric dyadic Green's function in the source region," *IEEE Trans. Antennas Propagat.*, vol.30, pp.463-468, May 1982.

[17] R. Kastner, "On the singularity of the full spectral Green's dyad," *IEEE Trans. Antennas Propagat.*, vol.35, pp.1303-1305, Nov. 1987.

[18] K. A. Michalski, "Formulation of mixed-potential integral equations for arbitrarily shaped microstrip structures with uniaxial substrates," *J. Electromagn. Waves Appl.*, vol.7, no.7, pp.899-917, 1993.

[19] S. Barkeshli and P. H. Pathak, "On the dyadic Green's function for a planar multilayered dielectric/magnetic media," *IEEE Trans. Microwave Theory Tech.*, vol.40, pp.128-142, Jan. 1992.

[20] S.-O. Park and C. A. Balanis, "Closed-form asymptotic extraction method for coupled microstrip lines," *IEEE Microwave Guided Wave Lett.*, vol.7, pp.84-86, Mar. 1997.

[21] T.-C. Mu, H. Ogawa, and T. Itoh, "Characteristics of multiconductor, asymmetric, slow-wave microstrip transmission lines," *IEEE Trans. Microwave Theory Tech.*, vol.34, pp.1471-1477, Dec. 1986.

[22] A. A. Oliner, S.-T. Peng, T.-I. Hsu, and A. Sanchez, "Guidance and leakage properties of a class of open dielectric waveguides: Part II–New physical effects," *IEEE Trans. Microwave Theory Tech.*, vol.29, pp.855-869, Sep. 1981.

[23] C.-C. Su, "A surface integral equation method for homogeneous optical fibers and coupled image lines of arbitrary cross sections," *IEEE Trans. Microwave Theory Tech.*, vol.33, pp.1114-1119, Nov. 1985.

Chapter 11

MICROSTRIP LINES WITH A PERIODICALLY CORRUGATED GROUND PLANE

Jean-Fu Kiang, Hsiao-Lun Hsu, and Yuan-Shun Cheng
Department of Electrical Engineering and
Graduate Institute of Communication Engineering
National Taiwan University
Taipei, Taiwan, ROC

Abstract Mode matching technique combined with spectral domain representation is used to study the guidance properties of microstrip lines with periodically corrugated ground plane. Field components in the corrugated region are expanded by standing wave mode functions in the guidance direction and by Fourier integral in the transversal direction. Integral equations are formed in terms of surface currents on the microstrip. Dispersion relation is then obtained by applying method of moments to solve the integral equations. Various geometrical and electrical parameters affecting the pass-band/stop-band characteristics of the first stop-band are studied. Singularity distribution in the complex plane is also inspected.

Keywords: boundary condition, layered medium, microstrip line, spectral domain, mode matching, integral equation, singularity, method of moments, periodic structure, Floquet's theorem.

1. Introduction

Layout of high frequency circuit is affected by the frequency dispersion of microstrip lines, which have been widely discussed [1]-[4]. In a periodic waveguiding structure, stop-bands exist due to the interaction of different Floquet modes that have the same phase velocity. The interaction can be predicted by observing the intersection of dispersion curves of various Floquet modes. Such stop-bands can be used in filter design, or viewed as artifacts that may distort the guided signals. There

are many one-dimensional periodic waveguiding structures like nonuniform microstrip lines with periodic width-modulation [5], [6], strip lines and fin lines with periodic stubs [7], coplanar waveguide with periodically doped substrate [8], corrugated periodic structures in dielectric loaded waveguides [9], strip lines periodically load with crossing strips [10], periodic array of perforations on microstrip [11], and so on.

Dispersion characteristics of such structures can be studied with integral equations and method of moments, incorporating spectral domain technique and Floquet's theorem [5], [7], [10]. Approximate solution like wave amplitude transmission matrix method has also been used [6].

Different modes in the same periodic structure possess different field distributions, thus their stop-band properties are also different [12], [13]. Stop-band properties can be controlled by adjusting period and other geometrical parameters of the guiding structures. For a ferrite slab periodically load with metal strips, the stop-band properties can be adjusted by changing the bias magnetic field strength [14]. In [15], a microstrip line is put on a photoconductive substrate with periodic holes on the ground plane. Its stop-band properties can be controlled by shedding light on the ground plane side to make this structure function as tunable rejection filters.

Two-dimensional periodic structures have angle dependent stop-band properties. They can be applied in high-directivity surface wave antennas and space-frequency signal selectors [16]. These structures with proper geometrical parameters can be used to suppress leaky waves [17], [18]. Their dispersion relation can be obtained by using vector integral equation method combined with two-stage moments method [19].

To obtain wider stop-band, several structures with different stop-band properties can be combined. In [20], two-dimensional periodic square lattices consisting of circular holes of three different sizes are used to create wider stop-band. In [21], a filter of wide stop-band is proposed by connecting two periodic structures in parallel. Stop-bands of these two periodic structures have different center frequencies and bandwidths, wider stop-band is achieved by combining two adjacent stop-bands. In [22], a pattern is proposed that superposes various sinusoidal functions tuned at different frequencies to obtain wider stop-band.

In this Chapter, the effects of a periodically corrugated ground plane on uniform microstrip line is analyzed using a hybrid approach. This approach can be used to model the ground plane bulges caused by thermal process because the coefficients of thermal expansion of metal ground and substrate are different. Such periodic structures can also be made to function as filter. For example, orientation etchant in Si process can be used to create V-shaped or straight-walled grooves [23]. In the mod-

eling, each corrugation is approximated by a cascade of parallel-plate waveguides with different widths and heights, and the fields in each waveguide are expressed in terms of the standing wave modes in that waveguide. The fields outside of corrugations are expressed as a summation of Floquet modes. Reflection matrices are defined to facilitate the derivation. Integral equations are derived with surface currents on the strip as unknowns. Method of moments is then applied to convert the integral equations to a determinantal equation from which the dispersion relation can be obtained.

2. Formulation

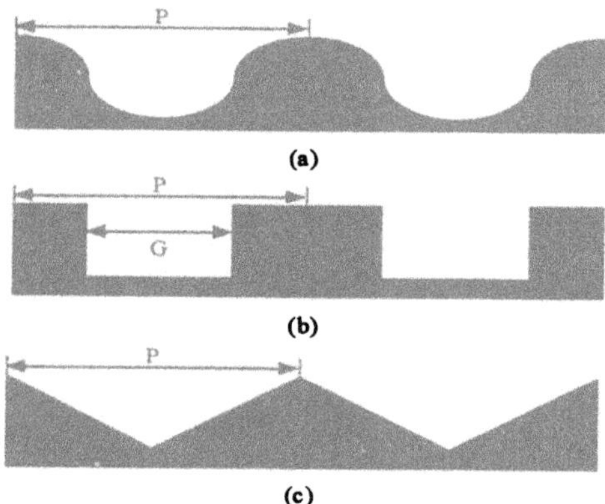

Figure 11.1. Three types of corrugated ground plane: (a) sinusoidal profile, (b) step profile, and (c) triangular profile.

Fig.11.1 shows the sideview of periodically corrugated ground plane where the orientation of corrugation is perpendicular to the microstrip. The period of corrugation is P, and each period of the ground plane is approximated by a set of connecting vertical and horizontal panels as shown in Fig.11.2. The corrugation region surrounded by the bulge ground plane is modeled as a multilayered medium where the ℓth layer starts at $y = c_\ell$, has a width w_ℓ in the $\hat{y}$ direction, and is bounded between $z = -d_{\ell-1}$ and $z = -d_\ell$. The z component of fields in each layer can thus be expanded as a Floquet series in the $\hat{y}$ direction and Fourier integral in the $\hat{x}$ direction as

$$E_{0z} = \sum_{n=-N}^{N} e^{ik_{yn}y} \int_{-\infty}^{\infty} dk_x e^{ik_x x + ik_{0zn} z_0} e_{0n}(k_x, k_y)$$

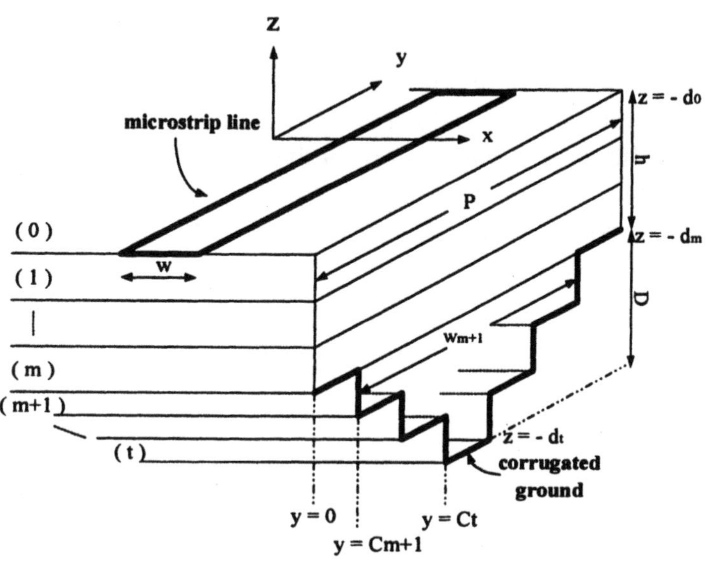

Figure 11.2. Geometrical configuration of a microstrip line with periodically corrugated ground plane.

$$H_{0z} = \sum_{n=-N}^{N} e^{ik_{yn}y} \int_{-\infty}^{\infty} dk_x e^{ik_x x + ik_{0zn}z_0} h_{0n}(k_x, k_y)$$

$$E_{\ell z} = \sum_{n=-N}^{N} e^{ik_{yn}y} \int_{-\infty}^{\infty} dk_x e^{ik_x x} \left[e_{\ell n}^{U}(k_x, k_y) e^{ik_{\ell zn}z_\ell} + e_{\ell n}^{D}(k_x, k_y) e^{-ik_{\ell zn}z_\ell} \right], \qquad 1 \le \ell \le m$$

$$H_{\ell z} = \sum_{n=-N}^{N} e^{ik_{yn}y} \int_{-\infty}^{\infty} dk_x e^{ik_x x} \left[h_{\ell n}^{U}(k_x, k_y) e^{ik_{\ell zn}z_\ell} + h_{\ell n}^{D}(k_x, k_y) e^{-ik_{\ell zn}z_\ell} \right], \qquad 1 \le \ell \le m$$

$$E_{\ell z} = \sum_{n=1}^{N_\ell} \sin\left[\alpha_{\ell n}(y - c_\ell)\right] \int_{-\infty}^{\infty} dk_x e^{ik_x x} \left[e_{\ell n}^{U}(k_x) e^{ik_{\ell zn}z_\ell} + e_{\ell n}^{D}(k_x) e^{-ik_{\ell zn}z_\ell} \right], \qquad m+1 \le \ell \le t$$

$$H_{\ell z} = \sum_{n=0}^{N_\ell} \cos\left[\alpha_{\ell n}(y - c_\ell)\right] \int_{-\infty}^{\infty} dk_x e^{ik_x x} \left[h_{\ell n}^{U}(k_x) e^{ik_{\ell zn}z_\ell} + h_{\ell n}^{D}(k_x) e^{-ik_{\ell zn}z_\ell} \right], \qquad m+1 \le \ell \le t$$

where $z_\ell = z + d_\ell$, $k_{yn} = k_y + 2n\pi/P$, $k_{\ell zn} = \sqrt{k_\ell^2 - k_x^2 - k_{yn}^2}$ with $0 \le \ell \le m$, $\alpha_{\ell n} = n\pi/w_\ell$, and $k_{\ell zn} = \sqrt{k_\ell^2 - k_x^2 - \alpha_{\ell n}^2}$ with $m+1 \le \ell \le t$. The tangential field components can be derived from these z components.

Boundary conditions at $z = -d_m$ will be elaborated first, followed by the fields above $z = -d_m$, finally the fields below $z = -d_m$. Define a reflection matrix to relate the upward field coefficients, $\{\bar{e}_\ell^U, \bar{h}_\ell^U\}$, to the downward field coefficients, $\{\bar{e}_\ell^D, \bar{h}_\ell^D\}$, as

$$\begin{bmatrix} \bar{e}_\ell^U \\ \bar{h}_\ell^U \end{bmatrix} = \begin{bmatrix} \bar{\bar{A}}_\ell & \bar{\bar{B}}_\ell \\ \bar{\bar{C}}_\ell & \bar{\bar{D}}_\ell \end{bmatrix} \cdot \begin{bmatrix} \bar{e}_\ell^D \\ \bar{h}_\ell^D \end{bmatrix} = \bar{\bar{R}}_{\cap\ell} \cdot \begin{bmatrix} \bar{e}_\ell^D \\ \bar{h}_\ell^D \end{bmatrix}, \qquad 1 \le \ell \le m$$

At $z = -d_m$, imposing the following continuity conditions

$$E_{ms} = \begin{cases} E_{(m+1)s}, & c_{m+1} \le y \le c_{m+1} + w_{m+1} \\ 0, & 0 \le y \le c_{m+1}, c_{m+1} + w_{m+1} \le y \le P \end{cases}$$

$$H_{ms} = H_{(m+1)s}, \qquad c_{m+1} \le y \le c_{m+1} + w_{m+1}$$

with $s = x, y$, we obtain

$$\bar{\bar{X}}_m \cdot \left(\bar{\bar{R}}_{\cap m} - \bar{\bar{J}}_m\right) \cdot \begin{bmatrix} \bar{e}_m^D \\ \bar{h}_m^D \end{bmatrix} = \bar{\bar{X}}_{m+1} \cdot \left(\bar{\bar{\bar{R}}}_{\cap(m+1)} - \bar{\bar{J}}_{m+1}\right) \cdot \begin{bmatrix} \bar{\bar{e}}_{m+1}^D \\ \bar{\bar{h}}_{m+1}^D \end{bmatrix}$$

$$\bar{\bar{Z}}_{m+1} \cdot \left(\bar{\bar{\bar{R}}}_{\cap(m+1)} + \bar{\bar{J}}_{m+1}\right) \cdot \begin{bmatrix} \bar{\bar{e}}_{m+1}^D \\ \bar{\bar{h}}_{m+1}^D \end{bmatrix} = \bar{\bar{Z}}_m \cdot \left(\bar{\bar{R}}_{\cap m} + \bar{\bar{J}}_m\right) \cdot \begin{bmatrix} \bar{e}_m^D \\ \bar{h}_m^D \end{bmatrix} \tag{11.1}$$

where

$$\bar{\bar{X}}_m = \begin{bmatrix} k_x \bar{\bar{K}}_{mz} \cdot \bar{\bar{K}}_{ms}^{-2} & \omega\mu_m \bar{\bar{K}}_y \cdot \bar{\bar{K}}_{ms}^{-2} \\ \bar{\bar{K}}_y \cdot \bar{\bar{K}}_{mz} \cdot \bar{\bar{K}}_{ms}^{-2} & -\omega\mu_m k_x \bar{\bar{K}}_{ms}^{-2} \end{bmatrix}$$

$$\bar{\bar{X}}_{m+1} = \begin{bmatrix} k_x \bar{\bar{U}}^{m(m+1)} \cdot \bar{\bar{K}}_{(m+1)z} \cdot \bar{\bar{\bar{K}}}_{(m+1)s}^{-2} & i\omega\mu_{m+1} \bar{\bar{U}}^{m(m+1)} \cdot \bar{\bar{\alpha}}_{m+1} \cdot \bar{\bar{\bar{K}}}_{(m+1)s}^{-2} \\ -\bar{\bar{\bar{U}}}^{m(m+1)} \cdot \bar{\bar{\alpha}}_{m+1}^t \cdot i\bar{\bar{K}}_{(m+1)z} \cdot \bar{\bar{\bar{K}}}_{(m+1)s}^{-2} & -\omega\mu_{m+1} k_x \bar{\bar{\bar{U}}}^{m(m+1)} \cdot \bar{\bar{\bar{K}}}_{(m+1)s}^{-2} \end{bmatrix}$$

$$\bar{\bar{Z}}_m = \begin{bmatrix} \omega\epsilon_m \bar{\bar{\bar{V}}}^{(m+1)m} \cdot \bar{\bar{K}}_y \cdot \bar{\bar{K}}_{ms}^{-2} & -k_x \bar{\bar{\bar{V}}}^{(m+1)m} \cdot \bar{\bar{K}}_{mz} \cdot \bar{\bar{K}}_{ms}^{-2} \\ \omega\epsilon_m k_x \bar{\bar{V}}^{(m+1)m} \cdot \bar{\bar{K}}_{ms}^{-2} & \bar{\bar{V}}^{(m+1)m} \cdot \bar{\bar{K}}_y \cdot \bar{\bar{K}}_{mz} \cdot \bar{\bar{K}}_{ms}^{-2} \end{bmatrix}$$

$$\bar{\bar{Z}}_{m+1} = \begin{bmatrix} -i\omega\epsilon_{m+1} \bar{\bar{\alpha}}_{m+1}^t \cdot \bar{\bar{K}}_{(m+1)s}^{-2} & -k_x \bar{\bar{\bar{K}}}_{(m+1)z} \cdot \bar{\bar{\bar{K}}}_{(m+1)s}^{-2} \\ \omega\epsilon_{m+1} k_x \bar{\bar{K}}_{(m+1)s}^{-2} & \bar{\bar{\alpha}}_{m+1} \cdot i\bar{\bar{\bar{K}}}_{(m+1)z} \cdot \bar{\bar{\bar{K}}}_{(m+1)s}^{-2} \end{bmatrix}$$

$$\bar{\bar{J}}_m = \begin{bmatrix} \bar{\bar{I}}_m & 0 \\ 0 & -\bar{\bar{I}}_m \end{bmatrix}, \qquad \bar{\bar{J}}_{m+1} = \begin{bmatrix} \bar{\bar{I}}_{m+1} & 0 \\ 0 & -\bar{\bar{\bar{I}}}_{m+1} \end{bmatrix}$$

and

$$\bar{e}_m^\alpha = \left[e_{m,-N}^\alpha, e_{m,-N+1}^\alpha, \cdots, e_{m,N-1}^\alpha, e_{m,N}^\alpha\right]^t, \qquad \alpha = U, D$$

$$\bar{h}_m^\alpha = \left[h_{m,-N}^\alpha, h_{m,-N+1}^\alpha, \cdots, h_{m,N-1}^\alpha, h_{m,N}^\alpha\right]^t, \qquad \alpha = U, D$$

$$\bar{e}_{m+1}^\alpha = \left[e_{m+1,1}^\alpha, e_{m+1,2}^\alpha, \cdots, e_{m+1,N_{m+1}}^\alpha\right]^t, \qquad \alpha = U, D$$

$$\bar{h}_{m+1}^\alpha = \left[h_{m+1,0}^\alpha, h_{m+1,1}^\alpha, \cdots, h_{m+1,N_{m+1}}^\alpha\right]^t, \qquad \alpha = U, D$$

$$\bar{\bar{K}}_{ms} = \text{diag.}\left\{k_{ms(-N)}, k_{ms(-N+1)}, \cdots, k_{ms(N-1)}, k_{msN}\right\}$$

$$\bar{\bar{K}}_{(m+1)s} = \text{diag.}\left\{k_{(m+1)s1}, k_{(m+1)s2}, \cdots, k_{(m+1)sN_{m+1}}\right\}$$

$$\bar{\bar{\bar{K}}}_{(m+1)s} = \text{diag.}\left\{k_{(m+1)s0}, k_{(m+1)s1}, \cdots, k_{(m+1)sN_{m+1}}\right\}$$

$$\bar{\bar{K}}_y = \text{diag.}\left\{k_{y(-N)}, k_{y(-N+1)}, \cdots, k_{y(N-1)}, k_{yN}\right\}$$

$$\bar{\bar{K}}_{mz} = \text{diag.}\left\{k_{mz(-N)}, k_{mz(-N+1)}, \cdots, k_{mz(N-1)}, k_{mzN}\right\}$$

$$\bar{\bar{K}}_{(m+1)z} = \text{diag.}\left\{k_{(m+1)z1}, k_{(m+1)z2}, \cdots, k_{(m+1)zN_{m+1}}\right\}$$

$$\bar{\bar{\bar{K}}}_{(m+1)z} = \text{diag.}\left\{k_{(m+1)z0}, k_{(m+1)z1}, \cdots, k_{(m+1)zN_{m+1}}\right\}$$

$$\bar{\bar{\alpha}}_{m+1} = \left[\ \bar{0} \ \mid \ \text{diag.}\left\{\alpha_{(m+1)1}, \alpha_{(m+1)2}, \cdots, \alpha_{(m+1)N_{m+1}}\right\}\right]$$

Here, $\bar{0}$ is a null vector of dimension N_{m+1}, $\bar{\bar{I}}_m$ is an identity matrix of order $2N_m + 1$, $\bar{\bar{I}}_{m+1}$ is an identity matrix of order N_{m+1}, $\bar{\bar{\bar{I}}}_{m+1}$ is an identity matrix of order $N_{m+1} + 1$. The scaled reflection matrix is

defined as

$$\tilde{\bar{\bar{R}}}_{\cap(m+1)} = \begin{bmatrix} \tilde{\bar{\bar{A}}}_{m+1} & \tilde{\bar{\bar{B}}}_{m+1} \\ \tilde{\bar{\bar{C}}}_{m+1} & \tilde{\bar{\bar{D}}}_{m+1} \end{bmatrix}$$

where

$$\begin{aligned}
\tilde{\bar{\bar{A}}}_{m+1} &= e^{i\bar{\bar{K}}_{(m+1)z}h_{m+1}} \cdot \bar{\bar{A}}_{m+1} \cdot e^{i\bar{\bar{K}}_{(m+1)z}h_{m+1}} \\
\tilde{\bar{\bar{B}}}_{m+1} &= e^{i\bar{\bar{K}}_{(m+1)z}h_{m+1}} \cdot \bar{\bar{B}}_{m+1} \cdot e^{i\tilde{\bar{\bar{K}}}_{(m+1)z}h_{m+1}} \\
\tilde{\bar{\bar{C}}}_{m+1} &= e^{i\tilde{\bar{\bar{K}}}_{(m+1)z}h_{m+1}} \cdot \bar{\bar{C}}_{m+1} \cdot e^{i\bar{\bar{K}}_{(m+1)z}h_{m+1}} \\
\tilde{\bar{\bar{D}}}_{m+1} &= e^{i\tilde{\bar{\bar{K}}}_{(m+1)z}h_{m+1}} \cdot \bar{\bar{D}}_{m+1} \cdot e^{i\tilde{\bar{\bar{K}}}_{(m+1)z}h_{m+1}}
\end{aligned}$$

and the scaled coefficients are defined as

$$\tilde{\bar{e}}^{D}_{m+1} = e^{-i\bar{\bar{K}}_{(m+1)z}h_{m+1}} \cdot \bar{e}^{D}_{m+1}, \qquad \tilde{\bar{h}}^{D}_{m+1} = e^{-i\tilde{\bar{\bar{K}}}_{(m+1)z}h_{m+1}} \cdot \bar{h}^{D}_{m+1}$$

The U and V matrices linking an open region with adjacent closed waveguide region are defined as

$$\begin{aligned}
\left\{\bar{\bar{U}}^{m(m+1)}\right\}_{nr} &= \frac{1}{P}\int_{c_{m+1}}^{c_{m+1}+w_{m+1}} dy e^{-ik_{yn}y} \sin\left[\alpha_{(m+1)r}(y - c_{m+1})\right] \\
\left\{\tilde{\bar{\bar{U}}}^{m(m+1)}\right\}_{nr} &= \frac{1}{P}\int_{c_{m+1}}^{c_{m+1}+w_{m+1}} dy e^{-ik_{yn}y} \cos\left[\alpha_{(m+1)r}(y - c_{m+1})\right] \\
\left\{\bar{\bar{V}}^{(m+1)m}\right\}_{nr} &= \frac{2}{w_{m+1}}\int_{c_{m+1}}^{c_{m+1}+w_{m+1}} dy \sin\left[\alpha_{(m+1)n}(y - c_{m+1})\right] e^{ik_{yr}y} \\
\left\{\tilde{\bar{\bar{V}}}^{(m+1)m}\right\}_{nr} &= \frac{\varepsilon_n}{w_{m+1}}\int_{c_{m+1}}^{c_{m+1}+w_{m+1}} dy \cos\left[\alpha_{(m+1)n}(y - c_{m+1})\right] e^{ik_{yr}y}
\end{aligned}$$

where $\varepsilon_n = 1$ when $n = 0$ and $\varepsilon_n = 2$ when $n > 0$. By solving (11.1), a recursive formula for the reflection matrix, $\bar{\bar{R}}_{\cap m}$, is derived in terms of $\tilde{\bar{\bar{R}}}_{\cap(m+1)}$ as

$$\begin{aligned}
\bar{\bar{R}}_{\cap m} = &\left\{\bar{\bar{X}}_m - \bar{\bar{X}}_{m+1} \cdot \left(\tilde{\bar{\bar{R}}}_{\cap(m+1)} - \bar{\bar{J}}_{m+1}\right)\right. \\
&\left. \cdot \left(\tilde{\bar{\bar{R}}}_{\cap(m+1)} + \bar{\bar{J}}_{m+1}\right)^{-1} \cdot \bar{\bar{Z}}^{-1}_{m+1} \cdot \bar{\bar{Z}}_m\right\}^{-1} \\
&\cdot \left\{\bar{\bar{X}}_m + \bar{\bar{X}}_{m+1} \cdot \left(\tilde{\bar{\bar{R}}}_{\cap(m+1)} - \bar{\bar{J}}_{m+1}\right)\right. \\
&\left. \cdot \left(\tilde{\bar{\bar{R}}}_{\cap(m+1)} + \bar{\bar{J}}_{m+1}\right)^{-1} \cdot \bar{\bar{Z}}^{-1}_{m+1} \cdot \bar{\bar{Z}}_m\right\} \cdot \bar{\bar{J}}_m
\end{aligned}$$

Next, impose the continuity condition of E_x E_y, H_x, and H_y at $z = -d_\ell$ with $\ell = 1, 2, \cdots, m-1$ to obtain another recursive formula for the reflection matrices as

$$\bar{\bar{R}}_{\cap\ell} = \left\{ \bar{\bar{X}}_\ell - \bar{\bar{X}}_{\ell+1} \cdot \left(\bar{\bar{\bar{R}}}_{\cap(\ell+1)} - \bar{\bar{J}}_{\ell+1} \right) \cdot \left(\bar{\bar{\bar{R}}}_{\cap(\ell+1)} + \bar{\bar{J}}_{\ell+1} \right)^{-1} \cdot \bar{\bar{Y}}_{\ell+1}^{-1} \cdot \bar{\bar{Y}}_\ell \right\}^{-1}$$
$$\cdot \left\{ \bar{\bar{X}}_\ell + \bar{\bar{X}}_{\ell+1} \cdot \left(\bar{\bar{\bar{R}}}_{\cap(\ell+1)} - \bar{\bar{J}}_{\ell+1} \right) \cdot \left(\bar{\bar{\bar{R}}}_{\cap(\ell+1)} + \bar{\bar{J}}_{\ell+1} \right)^{-1} \cdot \bar{\bar{Y}}_{\ell+1}^{-1} \cdot \bar{\bar{Y}}_\ell \right\} \cdot \bar{\bar{J}}_\ell, \qquad \ell = 1, 2, \cdots, m-1$$

Finally, imposing the continuity conditions at $z = -d_\ell$ in the corrugated region that

$$E_{\ell s} = \begin{cases} E_{(\ell+1)s}, & c_{\ell+1} \le y \le c_{\ell+1} + w_{\ell+1} \\ 0, & c_\ell \le y \le c_{\ell+1}, c_{\ell+1} + w_{\ell+1} \le y \le c_\ell + w_\ell \end{cases}$$
$$H_{\ell s} = H_{(\ell+1)s}, \qquad c_{\ell+1} \le y \le c_{\ell+1} + w_{\ell+1}$$

with $s = x, y$ and $\ell = m+1, \cdots, t-1$, we obtain a recursive formula for the reflection matrices in the corrugated region as

$$\bar{\bar{R}}_{\cap\ell} = \left\{ \bar{\bar{X}}_\ell - \bar{\bar{\bar{X}}}_{\ell+1} \cdot \left(\bar{\bar{\bar{R}}}_{\cap(\ell+1)} - \bar{\bar{J}}_{\ell+1} \right) \cdot \left(\bar{\bar{\bar{R}}}_{\cap(\ell+1)} + \bar{\bar{J}}_{\ell+1} \right)^{-1} \cdot \bar{\bar{Y}}_{\ell+1}^{-1} \cdot \bar{\bar{\bar{Y}}}_\ell \right\}^{-1}$$
$$\cdot \left\{ \bar{\bar{X}}_\ell + \bar{\bar{\bar{X}}}_{\ell+1} \cdot \left(\bar{\bar{\bar{R}}}_{\cap(\ell+1)} - \bar{\bar{J}}_{\ell+1} \right) \cdot \left(\bar{\bar{\bar{R}}}_{\cap(\ell+1)} + \bar{\bar{J}}_{\ell+1} \right)^{-1} \cdot \bar{\bar{Y}}_{\ell+1}^{-1} \cdot \bar{\bar{\bar{Y}}}_\ell \right\} \cdot \bar{\bar{J}}_\ell, \qquad \ell = m+1, \cdots, t-1$$

Notice that $\bar{\bar{\bar{R}}}_{\cap t} = \bar{\bar{J}}_t$ because the tangential electric fields vanish at $z = -d_t$.

The surface current can be expressed in terms of a Floquet series in the $\hat{y}$ direction and Fourier integral in the $\hat{x}$ direction as

$$J_\alpha = \sum_{n=-N}^{N} e^{ik_{yn}y} \int_{-\infty}^{\infty} dk_x e^{ik_x x} J_{\alpha n}(k_x, k_y), \qquad \alpha = x, y$$

Due to the presence of microstrip line, we have the conditions at $z = -d_0$ that $E_{0x} = E_{1x}$, $E_{0y} = E_{1y}$, $H_{0x} - H_{1x} = J_y$, and $H_{0y} - H_{1y} = -J_x$. Thus, we have

$$\begin{bmatrix} \bar{J}_x \\ \bar{J}_y \end{bmatrix} = \left\{ \bar{\bar{T}}_0 - \bar{\bar{T}}_1 \cdot \left(\bar{\bar{\tilde{R}}}_{\cap 1} + \bar{\bar{J}}_1 \right) \cdot \left(\bar{\bar{\tilde{R}}}_{\cap 1} - \bar{\bar{J}}_1 \right)^{-1} \right.$$

$$\left. \cdot \bar{\bar{X}}_1^{-1} \cdot \bar{\bar{X}}_0 \right\} \cdot \begin{bmatrix} \bar{e}_0 \\ \bar{h}_0 \end{bmatrix} = \bar{\bar{T}} \cdot \begin{bmatrix} \bar{e}_0 \\ \bar{h}_0 \end{bmatrix}$$

or

$$\begin{bmatrix} \bar{e}_0 \\ \bar{h}_0 \end{bmatrix} = \begin{bmatrix} \bar{\bar{S}}_{11} & \bar{\bar{S}}_{12} \\ \bar{\bar{S}}_{21} & \bar{\bar{S}}_{22} \end{bmatrix} \cdot \begin{bmatrix} \bar{J}_x \\ \bar{J}_y \end{bmatrix}$$

where

$$\bar{\bar{X}}_\ell = \begin{bmatrix} k_x \bar{\bar{K}}_{\ell z} \cdot \bar{\bar{K}}_{\ell s}^{-2} & \omega\mu_\ell \bar{\bar{K}}_y \cdot \bar{\bar{K}}_{\ell s}^{-2} \\ \bar{\bar{K}}_y \cdot \bar{\bar{K}}_{\ell z} \cdot \bar{\bar{K}}_{\ell s}^{-2} & -\omega\mu_\ell k_x \bar{\bar{K}}_{\ell s}^{-2} \end{bmatrix}, \qquad \ell = 0, 1$$

$$\bar{\bar{T}}_\ell = \begin{bmatrix} \omega\epsilon_\ell k_x \bar{\bar{K}}_{\ell s}^{-2} & \bar{\bar{K}}_y \cdot \bar{\bar{K}}_{\ell z} \cdot \bar{\bar{K}}_{\ell s}^{-2} \\ \omega\epsilon_\ell \bar{\bar{K}}_y \cdot \bar{\bar{K}}_{\ell s}^{-2} & -k_x \bar{\bar{K}}_{\ell z} \cdot \bar{\bar{K}}_{\ell s}^{-2} \end{bmatrix}, \qquad \ell = 0, 1$$

$$\bar{J}_\alpha = \left[J_{\alpha(-N)}, J_{\alpha(-N+1)}, \cdots, J_{\alpha(N-1)}, J_{\alpha N} \right]^t, \qquad \alpha = x, y$$

Finally, impose the boundary condition that the tangential electric fields vanish on the microstrip surface to obtain the following two integral equations

$$\int_{-\infty}^{\infty} dk_x e^{ik_x x} e^{i\bar{K}_y^t y} \left(\bar{\bar{G}}_0^{xx} \cdot \bar{J}_x + \bar{\bar{G}}_0^{xy} \cdot \bar{J}_y \right) = 0, \quad \bar{r}_s \text{ on } S \tag{11.2}$$

$$\int_{-\infty}^{\infty} dk_x e^{ik_x x} e^{i\bar{K}_y^t y} \left(\bar{\bar{G}}_0^{yx} \cdot \bar{J}_x + \bar{\bar{G}}_0^{yy} \cdot \bar{J}_y \right) = 0, \quad \bar{r}_s \text{ on } S \tag{11.3}$$

where

$$\bar{\bar{G}}_0^{xx}(k_x) = -k_x \bar{\bar{K}}_{0z} \cdot \bar{\bar{K}}_{0s}^{-2} \cdot \bar{\bar{S}}_{11} - \omega\mu_0 \bar{\bar{K}}_y \cdot \bar{\bar{K}}_{0s}^{-2} \cdot \bar{\bar{S}}_{21}$$

$$\bar{\bar{G}}_0^{xy}(k_x) = -k_x \bar{\bar{K}}_{0z} \cdot \bar{\bar{K}}_{0s}^{-2} \cdot \bar{\bar{S}}_{12} - \omega\mu_0 \bar{\bar{K}}_y \cdot \bar{\bar{K}}_{0s}^{-2} \cdot \bar{\bar{S}}_{22}$$

$$\bar{\bar{G}}_0^{yx}(k_x) = -\bar{\bar{K}}_y \cdot \bar{\bar{K}}_{0z} \cdot \bar{\bar{K}}_{0s}^{-2} \cdot \bar{\bar{S}}_{11} + \omega\mu_0 k_x \bar{\bar{K}}_{0s}^{-2} \cdot \bar{\bar{S}}_{21}$$

$$\bar{\bar{G}}_0^{yy}(k_x) = -\bar{\bar{K}}_y \cdot \bar{\bar{K}}_{0z} \cdot \bar{\bar{K}}_{0s}^{-2} \cdot \bar{\bar{S}}_{12} + \omega\mu_0 k_x \bar{\bar{K}}_{0s}^{-2} \cdot \bar{\bar{S}}_{22}$$

$$\bar{K}_y = \left[k_{y(-N)}, k_{y(-N+1)}, \cdots, k_{y(N-1)}, k_{yN} \right]^t$$

To solve the integral equations in (11.2) and (11.3), first choose a set of basis functions to expand the surface currents as

$$J_x(\bar{r}_s) = \sum_{r=-N_1}^{N_2} u_r e^{ik_{yr}y} f(x), \qquad |x| \leq w/2$$

$$J_y(\bar{r}_s) = \sum_{r=-N_1}^{N_2} v_r e^{ik_{yr}y} g(x), \qquad |x| \leq w/2$$

where $f(x) = \sin(2\pi x/w)$ and $g(x) = 1/\sqrt{(w/2)^2 - x^2}$. Fourier coefficients of the surface currents are $J_{xn}(k_x) = u_n\tilde{f}(k_x)$ and $J_{yn}(k_x) = v_n\tilde{g}(k_x)$ with

$$\tilde{\zeta}(k_x) = \frac{1}{2\pi}\int_{-w/2}^{w/2} dx e^{-ik_x x}\zeta(x), \qquad \zeta = f, g$$

Taking the inner product of $e^{-ik_{yn}y}f(x)$ with (11.2), and taking the inner product of $e^{-ik_{yn}y}$ with (11.3), we obtain the following determinantal equation

$$\det\begin{bmatrix} \int_{-\infty}^{\infty} dk_x \tilde{f}(-k_x)\bar{\bar{G}}_0^{xx}(k_x)\tilde{f}(k_x) & \int_{-\infty}^{\infty} dk_x \tilde{f}(-k_x)\bar{\bar{G}}_0^{xy}(k_x)\tilde{g}(k_x) \\ \int_{-\infty}^{\infty} dk_x \tilde{S}(-k_x)\bar{\bar{G}}_0^{yx}(k_x)\tilde{f}(k_x) & \int_{-\infty}^{\infty} dk_x \tilde{S}(-k_x)\bar{\bar{G}}_0^{yy}(k_x)\tilde{g}(k_x) \end{bmatrix} = 0$$

where

$$\tilde{S}(k_x) = \frac{a}{2\pi}\frac{\sin(k_x w/2)}{k_x w/2}$$

Dispersion relation is then obtained by solving the determinantal eqnation.

3. Distribution of Singularities

In computing the integrals in the determinantal equation, integration path in the complex k_x plane is determined by branch points, branch cuts, surface wave poles and leaky wave poles. The leaky wave modes are surface wave modes below their cutoff frequencies, and have complex propagation constants in general [24].

Consider a homogeneous dielectric slab backed by a flat ground plane. Branch points of the Green's function in the k_x plane are defined by imposing $k_{0z} = 0$, which implies that branch points of the mth space harmonic are at

$$k_{xm}^{(bp)} = \left(k_0^2 - k_{ym}^2\right)^{1/2} = \left[k_0^2 - (\beta_m + i\alpha)^2\right]^{1/2}$$

where $k_{ym} = \beta_m + i\alpha$ with $\alpha > 0$, and $\beta_m = \beta_0 + 2m\pi/P$. The Sommerfeld branch cuts are defined by imposing $\mathrm{Im}(k_{0z}) = 0$. Thus, branch cuts of the mth space harmonic are

$$k_{xm}^{(bc)} = \left(k_0^2 - k_{ym}^2 - k_{0z}'^2\right)^{1/2} = \left[(k_{xm}^{(bp)})^2 - k_{0z}'^2\right]^{1/2}$$

where $k_{0z} = k_{0z}' + ik_{0z}''$ with $k_{0z}' = \mathrm{Re}(k_{0z})$ and $k_{0z}'' = \mathrm{Im}(k_{0z})$. Notice that branch cuts emanate from the branch points at $k_{xm}^{(bp)}$.

Dyadic Green's function in the spectral domain may exhibit real and complex poles corresponding to real and complex modes, respectively, in the background medium. Let $k_{s(\mathrm{TM})}$ and $k_{s(\mathrm{TE})}$ be the wave number of a TM mode and a TE mode, respectively, in the background medium. The mth space harmonic of surface wave pole in the k_x plane is given by

$$k_{xm}^{(sw)} = \left(k_{sw}^2 - k_{ym}^2\right)^{1/2} = \left[k_{sw}^2 - (\beta_m + i\alpha)^2\right]^{1/2}$$

where k_{sw} is the k_s of specific TE or TM surface wave mode.

Similarly, the mth space harmonic of leaky wave pole in the k_x plane is given by

$$k_{xm}^{(\ell w)} = \left(k_{\ell w}^2 - k_{ym}^2\right)^{1/2} = \left[(\beta_{\ell w} + i\alpha_{\ell w})^2 - (\beta_m + i\alpha)^2\right]^{1/2}$$

where $k_{\ell w} = \beta_{\ell w} + i\alpha_{\ell w}$ is the propagation constant of specific TE or TM leaky wave mode.

Fig.11.3 shows the phase constant of surface wave modes in a slab waveguide backed by PEC ground plane. The cutoff frequencies are given by

$$f_c^n = \frac{n}{4t\sqrt{\epsilon_d\mu_d - \epsilon_0\mu_0}}$$

where $n = 0, 2, 4, \cdots$ for TM modes, and $n = 1, 3, 5, \cdots$ for TE modes. The dominant mode is TM_0 mode which propagates unattenuated at all frequencies.

Fig.11.4 shows a typical Brillouin diagram consisting of space harmonics of the fundamental guided mode. Dispersion curve of the first background surface wave mode is also shown for comparison. Guidance characteristics will be discussed in five frequency bands consecutively. In each frequency band, low-order space harmonics($n = -2, -1, 0, 1$) of singularities are inspected for their migrations in the k_x plane with increasing frequency.

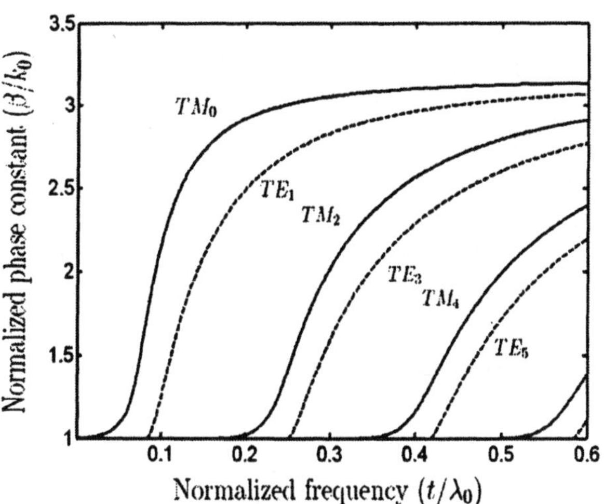

Figure 11.3. **Phase constant of surface wave modes in the background slab waveguide, t is slab thickness, λ_0 is the wavelength in free space.**

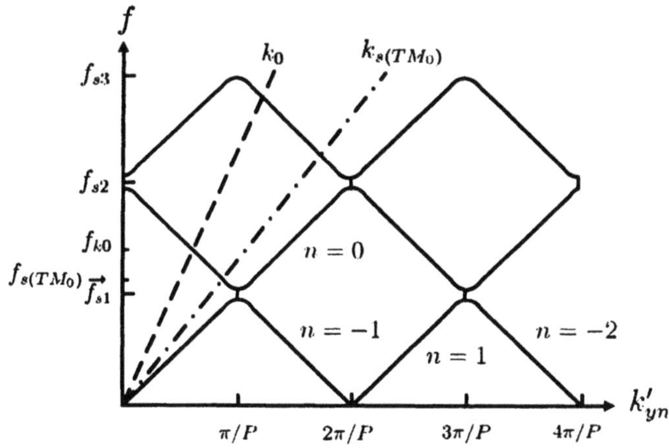

Figure 11.4. **Space harmonics of the fundamental guided mode accompanied by a background surface wave mode with $k_{y0} > k_{s(\mathrm{TM}_0)} > k_0$.**

Space harmonics of singularities in the k_x plane can be expressed as

$$k_{xn}^{(bp)} = \left[k_0^2 - (\beta + 2n\pi/P)^2 + \alpha^2 - 2i\,(\beta + 2n\pi/P)\,\alpha\right]^{1/2}$$

$$k_{xn}^{(bc)} = \left[k_0^2 - (\beta + 2n\pi/P)^2 + \alpha^2 - k_{0z}'^2 - 2i\,(\beta + 2n\pi/P)\,\alpha\right]^{1/2}$$

$$k_{xn}^{(sw)} = \left[k_{sw}^2 - (\beta + 2n\pi/P)^2 + \alpha^2 - 2i\,(\beta + 2n\pi/P)\,\alpha\right]^{1/2}$$

$$k_{xn}^{(\ell w)} = \Big[\beta_{\ell w}^2 - \alpha_{\ell w}^2 - (\beta + 2n\pi/P)^2 + \alpha^2 + 2i\beta_{\ell w}\alpha_{\ell w} - 2i\,(\beta + 2n\pi/P)\,\alpha\Big]^{1/2}$$

In all the demonstrations, branch points are marked by circles, surface wave poles are marked by squares, and leaky wave poles are marked by triangles. The zig-zag curves mark Sommerfeld branch cuts. Solid symbols denote singularities at the lower end of frequency band, and empty ones denote singularities at the upper end of frequency band.

In the case shown in Fig.11.4, $k_{y0} > k_{s(\mathrm{TM}_0)} > k_0$. The other two cases with $k_{s(\mathrm{TM}_0)} > k_{y0} > k_0$ and $k_{s(\mathrm{TM}_0)} > k_0 > k_{y0}$ can also be elaborated in a similar manner, and details are not presented here. Five consecutive frequency bands will be inspected separately. Here, f_{sn} is the center frequency of the nth stop-band which is the outcome of interaction between different space harmonics of the fundamental guided mode. The first background mode is also shown for reference. $f_{s(\mathrm{TM}_0)}$ is the frequency where the $n = -1$ space harmonic has the same phase constant with that of the TM_0 background mode, f_{k_0} is the frequency where the $n = -1$ space harmonic has the phase constant of k_0.

Fig.11.5(a) shows the distribution of singularities in the k_x plane in the frequency band of $0 < f < f_{s1}$. It is found that the branch points and surface wave poles of all the space harmonics lie on the imaginary axis, and the leaky wave poles lie in the first quadrant in the pass-band. As the frequency increases, the branch points and surface wave poles of the $n = 0$ and $n = 1$ space harmonics move upward, while those of the $n = -1$ and $n = -2$ space harmonics move downward. Notice that the propagation constant is a complex number in the first stop-band. Due to the nontrivial imaginary part of k_{y0}, the branch points, surface wave poles and leaky wave poles of the $n = 0$ and $n = 1$ space harmonics drift toward left, while those of the $n = -1$ and $n = -2$ space harmonics drift toward right.

Fig.11.5(b) shows the distribution of singularities in the k_x plane in the frequency band of $f_{s1} < f < f_{s(\mathrm{TM}_0)}$. As frequency increases, branch points, surface wave poles and leaky wave poles of the $n = 0$ and $n = 1$ space harmonics keep moving upward, while those of the $n = -1$ and

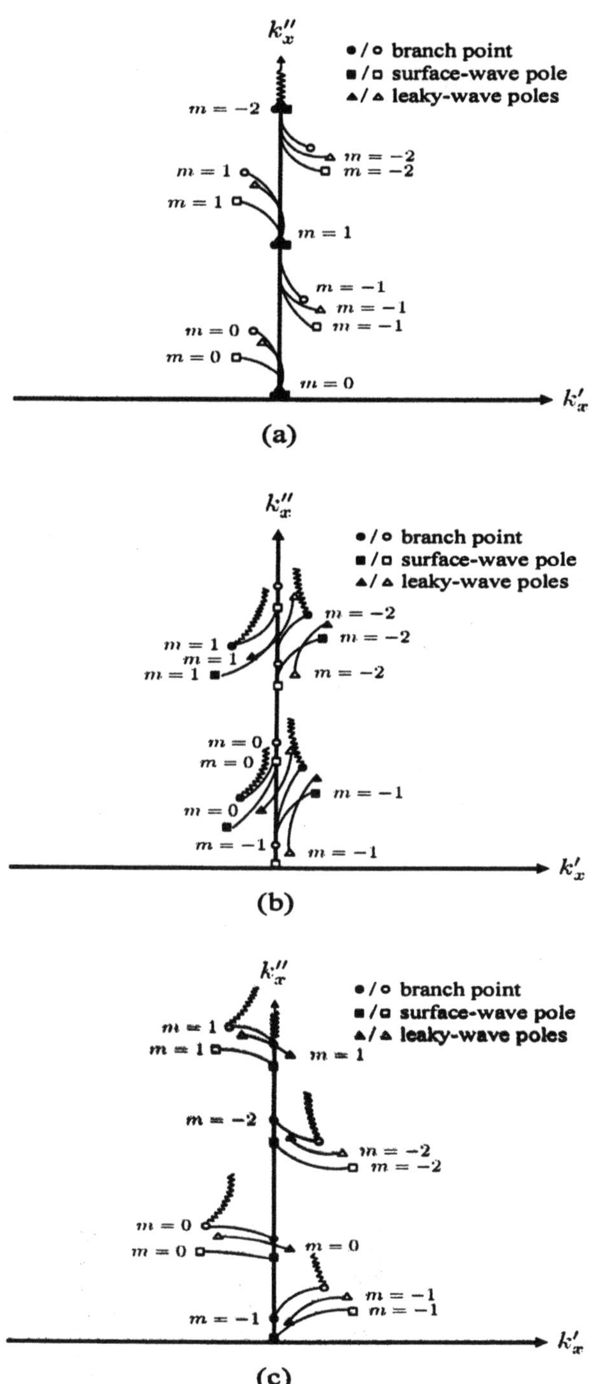

k_x''
●/○ branch point
■/□ surface-wave pole
▲/△ leaky-wave poles
$m = -2$
$m = -2$
$m = -2$
$m = 1$
$m = 1$
$m = 1$
$m = -1$
$m = -1$
$m = -1$
$m = 0$
$m = 0$
$m = 0$
k_x'
(a)
k_x''
●/○ branch point
■/□ surface-wave pole
▲/△ leaky-wave poles
$m = -2$
$m = -2$
$m = -2$
$m = 1$
$m = 1$
$m = 1$
$m = 0$
$m = 0$
$m = 0$
$m = -1$
$m = -1$
$m = -1$
k_x'
(b)
k_x''
●/○ branch point
■/□ surface-wave pole
▲/△ leaky-wave poles
$m = 1$
$m = 1$
$m = 1$
$m = -2$
$m = -2$
$m = -2$
$m = 0$
$m = 0$
$m = 0$
$m = -1$
$m = -1$
$m = -1$
k_x'
(c)

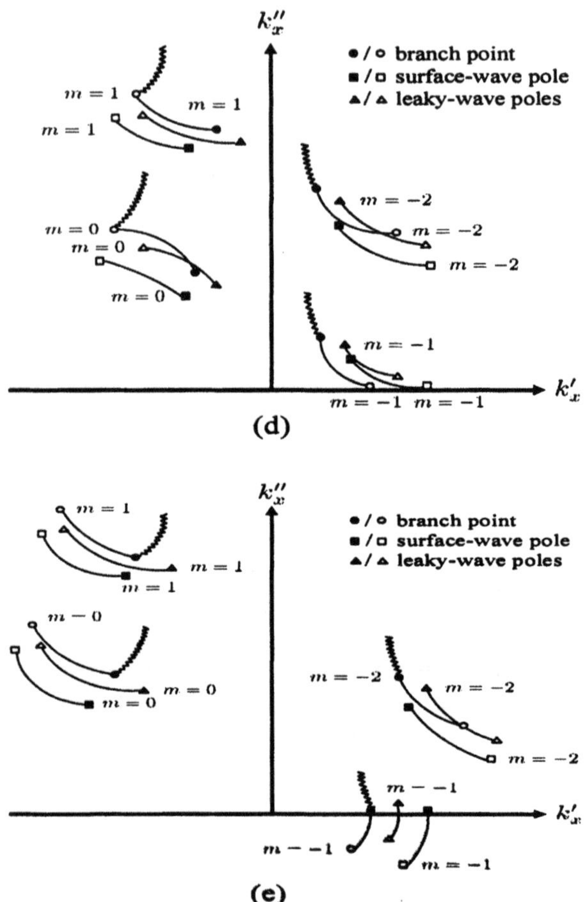

Figure 11.5. Singularities in the k_x plane, $k_{y0} > k_{s(\mathrm{TM}_0)} > k_0$, (a) $0 < f < f_{s1}$, (b) $f_{s1} < f < f_{s(\mathrm{TM}_0)}$, (c) $f_{s(\mathrm{TM}_0)} < f < f_{k0}$, (d) $f_{k0} < f < f_{s2}$, (e) $f_{s2} < f < f_{s3}$.

$n = -2$ space harmonics keep moving downward until they coincide at $f_{s(\mathrm{TM}_0)}$. Notice that in the first stop-band, branch points and surface wave poles of all the space harmonics move onto the imaginary axis and the leaky wave poles move to the first quadrant.

Up to $f = f_{s(\mathrm{TM}_0)}$, the structure does not radiate power into space or leak power into the background surface wave mode. Thus, the propagation constant is real except within the first stop-band.

Fig.11.5(c) shows the distribution of singularities in the k_x plane in the frequency band of $f_{s(\mathrm{TM}_0)} < f < f_{k0}$. When $f > f_{s(\mathrm{TM}_0)}$, part of the power leaks into TM_0 surface wave, and the propagation constant becomes a complex number. Branch points and surface wave poles of the $n = 0$ and $n = 1$ space harmonics drift left and upward from the imaginary axis to the second quadrant as frequency increases, while those of the $n = -1$ and $n = -2$ space harmonics drift right and downward to the first quadrant. Leaky wave poles of the $n = 0$ and $n = 1$ space harmonics move from the first quadrant across the imaginary axis to the second quadrant, while those of the $n = -1$ and $n = -2$ space hamonics move in the opposite sense and stay in the first quadrant.

Fig.11.5(d) shows the distribution of singularities in the k_x plane in the frequency band of $f_{k0} < f < f_{s2}$. Branch points, surface wave poles and leaky wave poles of the $n = 0$ and $n = 1$ space harmonics continue to move left and upward in the second quadrant, while those of the $n = -1$ and $n = -2$ space harmonics continue to move right and downward in the first quadrant. Notice that branch points and surface wave poles of the $n = -1$ space harmonic approach the real axis and cross it when $f > f_{s2}$.

Fig.11.5(e) shows the distribution of singularities in the k_x plane in the frequency band of $f_{s2} < f < f_{s3}$. Branch points, surface wave poles and leaky wave poles of the $n = 0$ and $n = 1$ space harmonics continue to move left and upward in the second quadrant, while those of the $n = -1$ and $n = -2$ space harmonics continue to move downward in the first quadrant. Notice that branch points and surface wave poles of the $n = -1$ space harmonic drift cross the real axis at certain frequencies.

Fig.11.6 shows the Brillouin diagram of the space harmonics associated with a higher-order guided mode. The dispersion curve of a background surface wave mode is also shown for reference. Dashed line indicates that the higher-order mode is no longer a guided mode when operating below its cutoff frequency.

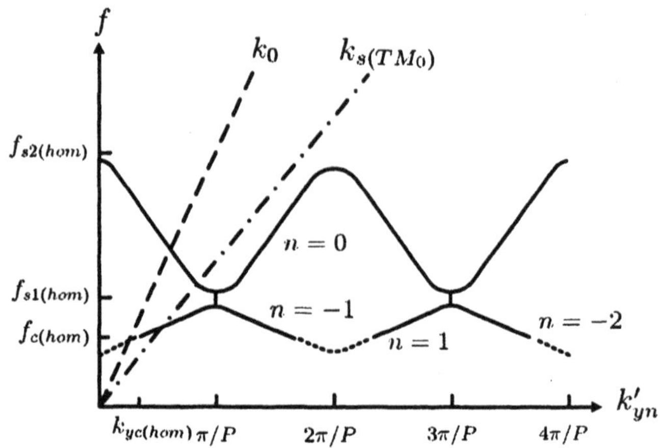

Figure 11.6. Space harmonics of a higher-order mode accompanied by a background surface wave mode.

4. Results and Discussions

Fig.11.7 shows the effective dielectric constant of a microstrip line with a periodically corrugated ground plane obtained by using this approach. Results from literatures are also shown for verification.

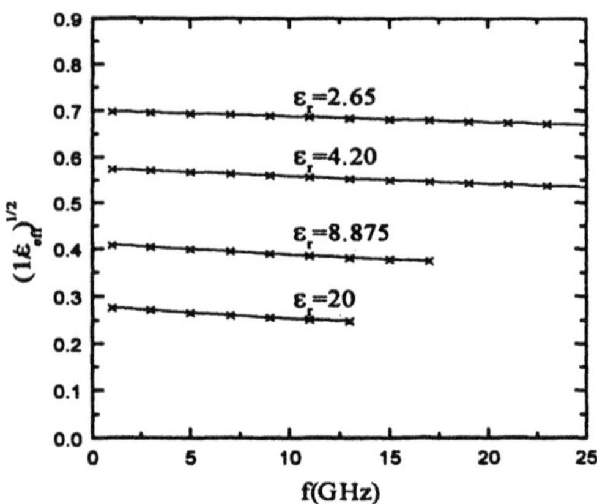

Figure 11.7. Effective dielectric constant of a microstrip line with corrugated ground of shallow step, P=1 cm, w=1.27 mm, h=1.27 mm, D/h=0.01, —— : this approach, × : data from [4].

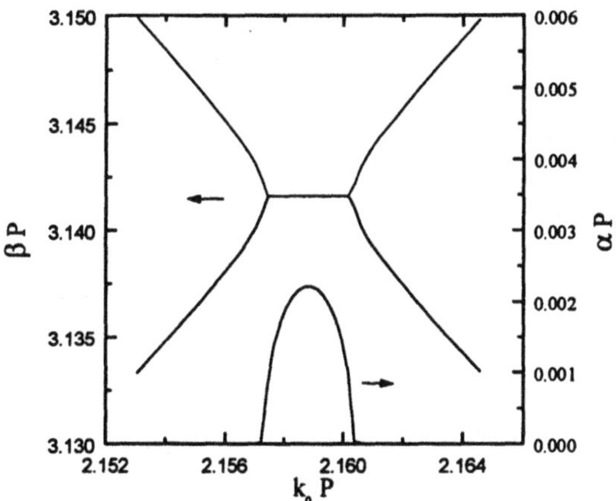

Figure 11.8. **Phase and attenuation constants near the first stop-band with corrugated ground of step profile, P=1 cm, w=1.27 mm, h=1.27 mm and D/h=0.01.**

A stop-band occurs where the Bragg condition, $k_y P = \pi$, is approximately satisfied. Since the corrugation depth(D) is only a small fraction of substrate thickness(h) and the frequencies are lower than the first stop-band, the dispersion curves are very close to those of a microstrip line with uniform ground plane [4]. A Pentium II 366MHz PC is used to run these simulations. It takes about 45 CPU minutes to obtain the propagation constant at one frequency using Muller's method.

Fig.11.8 shows the phase and attenuation constants of a microstrip line where the corrugated ground has a step profile. In the first stop-band, the phase constant equals π/P, and the attenuation constant is nonzero, reaching maximum near the center of stop-band.

Fig.11.9 shows the effects of profile depth on characteristics of the first stop-band with three different profiles. The normalized center frequency is defined as $k_c P = (2\pi f_c/c)P$ where f_c is the center frequency of the first stop-band, and c is the speed of light in free space. The normalized bandwidth is defined as $k_b P = (2\pi f_b/c)P$, where f_b is the bandwidth of the first stop-band. The normalized maximum attenuation constant is defined as $\alpha_{\max} P$ where $\alpha_{\max}$ is the maximum attenuation constant in the first stop-band. It is found that deeper corrugation incurs higher center frequency, wider stop-band width, and larger maximum attenuation constant.

The first stop-band occurs where reflected waves from discontinuities in all the periods add up in phase. Thus, power is enhanced in

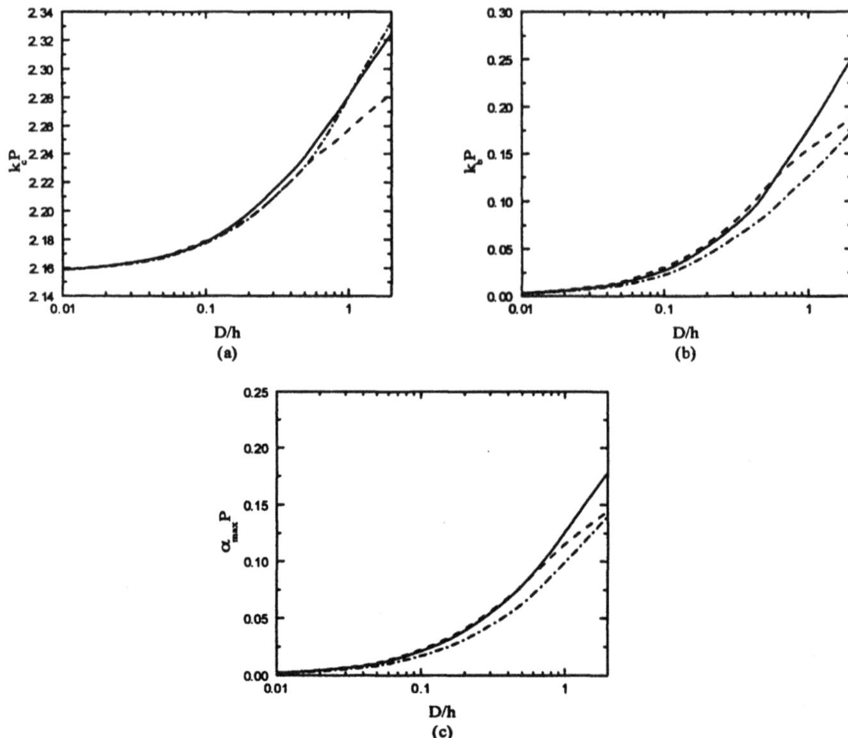

Figure 11.9. **Effects of profile depth on characteristics of the first stop-band, P=1 cm, w=1.27 mm, h=1.27 mm, $\epsilon_r = 2.65$, —— : sinusoidal profile, – – : step profile, – · : triangular profile, (a) normalized center frequency, (b) normalized bandwidth, (c) normalized maximum attenuation constant.**

the backward direction to establish the stop-band. Substituting $k_y = (2\pi f/c)\sqrt{\epsilon_{\text{eff}}}$ into the Bragg condition, it is found that with higher effective dielectric constant, ϵ_{eff}, Bragg condition is satisfied at lower frequency, hence the stop-band occurs at lower frequency. For a uniform microstrip line, deeper corrugation reduces the effective dielectric constant [2]. Microstrip line with deeper corrugation can be viewed as a microstrip line with thicker substrate, hence the center frequency becomes higher as shown in Fig.11.9(a). Deeper corrugation also implies stronger discontinuity. Thus, reflected wave becomes stronger, and results in higher maximum attenuation constant as shown in Fig.11.9(c). Notice that the variation with D/h of three different corrugation profiles are similar.

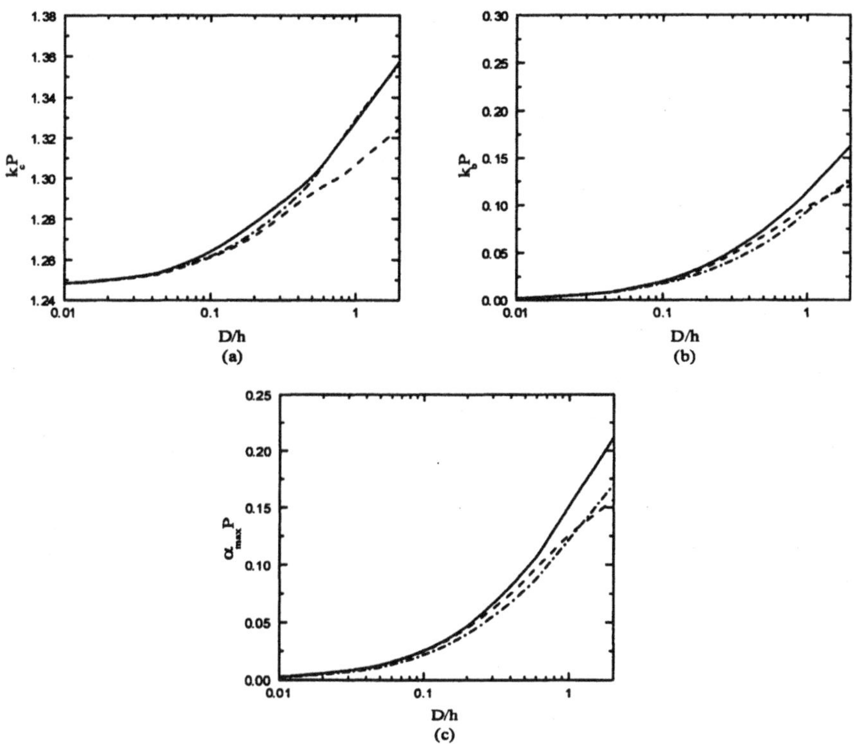

Figure 11.10. Effects of profile depth on characteristics of the first stop-band, parameters the same as in Fig.11.9 except $\epsilon_r = 8.875$.

Fig.11.10 shows the effects of profile depth on characteristics of the first stop-band, with the relative permittivity of substrate increased to 8.875. It is observed that increasing corrugation depth incurs higher center frequency, wider stop-band width and larger maximum attenuation constant, similar to those shown in Fig.11.9. It is also found that the center frequency and bandwidth of the stop-band in Figs.11.10(a) and 11.10(b), respectively, are lower and narrower than those in Figs.11.9(a) and 11.9(b) with the same corrugation depth. Maximum attenuation constant in the stop-band in Fig.11.10(c) is higher than that in Fig.11.9(c).

Next, study the effects of substrate dielectric constant on characteristics of the first stop-band. Results in Fig.11.11 show that higher substrate dielectric constant incurs lower center frequency, narrower bandwidth and larger maximum attenuation constant. For a uniform microstrip line, higher substrate dielectric constant implies higher effective dielectric constant [2]. Bragg condition implies that higher substrate

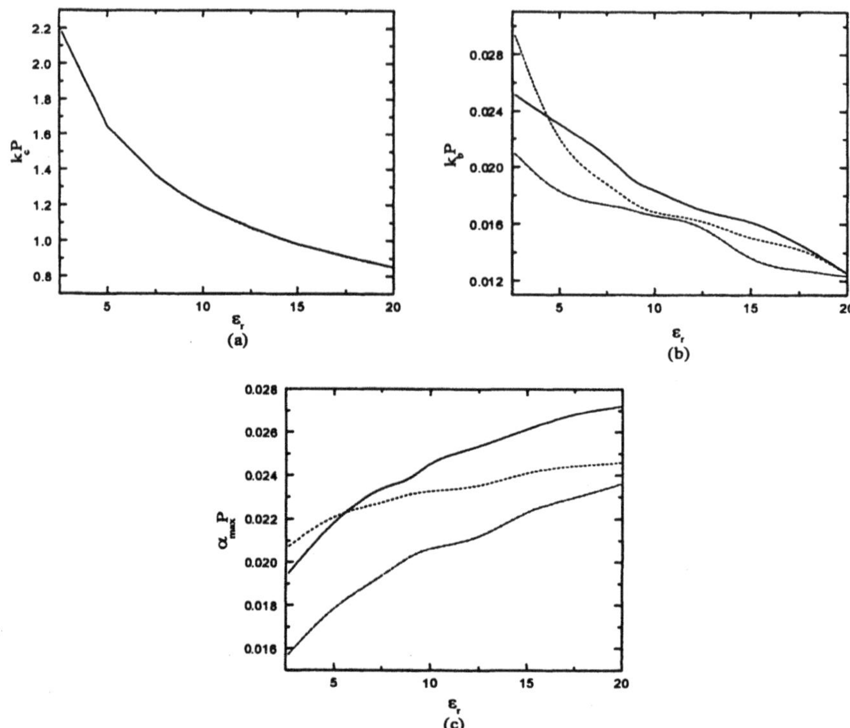

Figure 11.11. **Effects of substrate dielectric constant on characteristics of the first stop-band, P=1 cm, D=0.127 mm, h=1.27 mm, w = 1.27 mm, — : sinusoidal profile, – – : step profile, – · · : triangular profile, (a) normalized center frequency, (b) normalized bandwidth, (c) normalized maximum attenuation constant.**

dielectric constant incurs lower center frequency. Microstrip line with higher substrate dielectric constant tends to confine stronger field in the substrate between microstrip line and ground plane, hence the wave experiences stronger discontinuity due to corrugated ground. Thus, the reflected wave becomes stronger and results in larger maximum attenuation constant as shown in Fig.11.11(c).

Fig.11.12 shows the effects of profile depth on characteristics of the first stop-band, with a step profile of different corrugation widths(G). As discussed in Figs.11.9 and 11.10, deeper profile incurs higher center frequency, wider bandwidth and larger maximum attenuation constant. Comparing three different corrugation widths at any fixed profile depth in Fig.11.12(a), it is found that wider corrugation is equivalent to a uni-

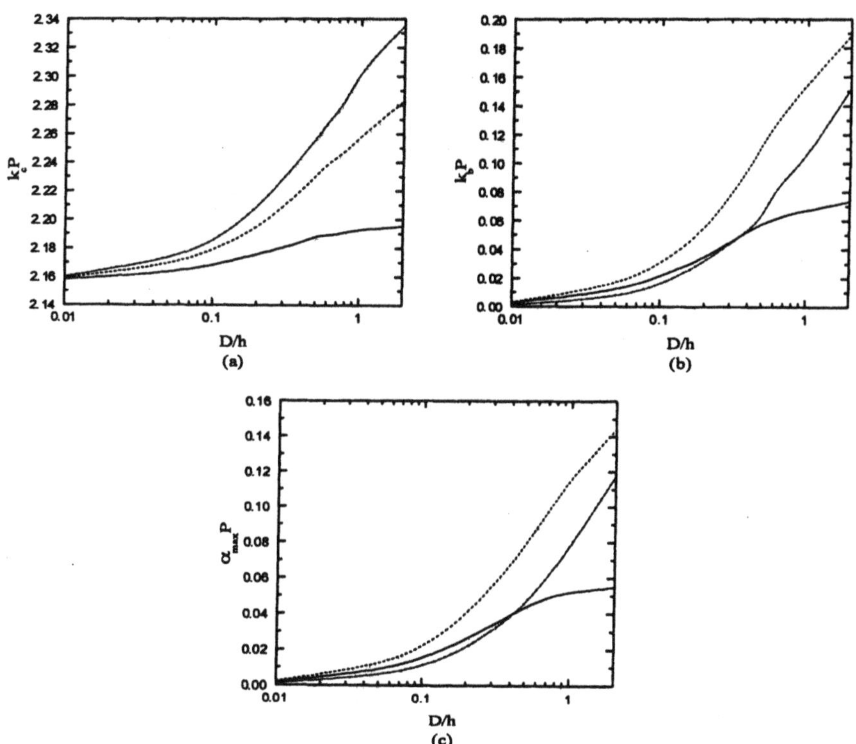

Figure 11.12. Effects of profile depth on characteristics of the first stop-band with a step profile of different widths, P=1 cm, w=1.27 mm, h=1.27 mm, $\epsilon_r = 2.65$, — : $G = 0.25P$, -- : $G = 0.5P$, - · : $G = 0.75P$, (a) normalized center frequency, (b) normalized bandwidth, (c) normalized maximum attenuation constant.

form microstrip line with thicker substrate, hence incurs lower effective dielectric constant and higher center frequency of the stop-band.

Fig.11.13 shows the effects of microstrip line width on chatacteristics of the first stop-band. It is found that wider microstrip line incurs lower center frequency. A wider uniform microstrip line possesses higher effective dielectric constant [2], thus implies lower center frequency by applying Bragg condition. The bandwidth and maximum attenuation constant increase with increasing microstrip width. As was discussed in Fig.11.11, wider microstrip line tends to confine stronger field in the substrate between microstrip line and ground plane. Thus, wider microstrip line incurs larger maximum attenuation constant as shown in Fig.11.13(c).

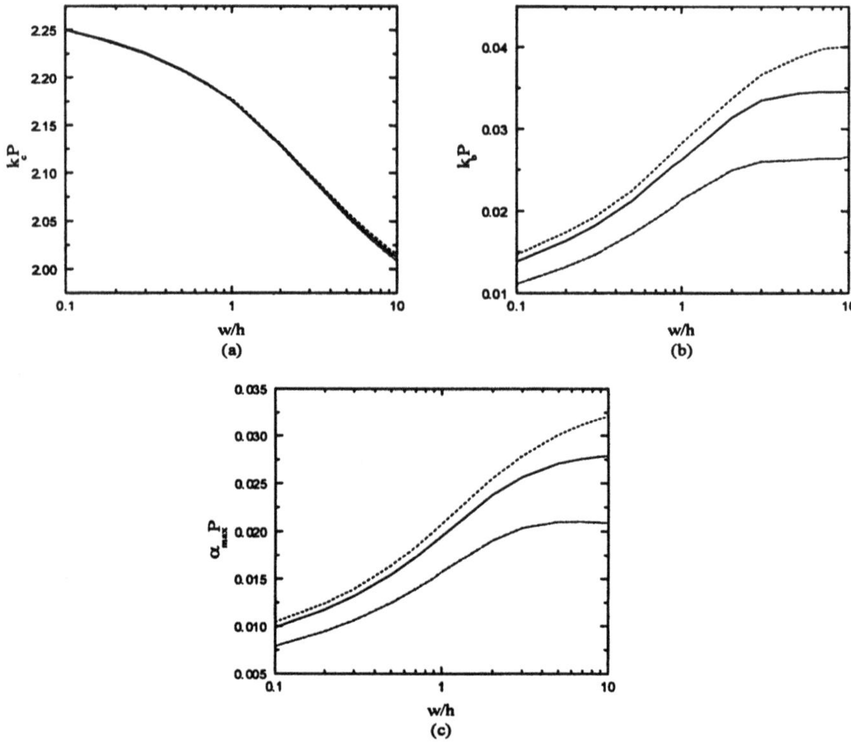

Figure 11.13. Effects of microstrip width on characteristics of the first stop-band, P=1 cm, D=0.127 mm, h=1.27 mm, ϵ_r = 2.65, — : sinusoidal profile, -- : step profile, - · : triangular profile, (a) normalized center frequency, (b) normalized bandwidth, (c) normalized maximum attenuation constant.

5. Conclusions

Mode matching technique and spectral domain representation have been used to develop integral equations to study guidance characteristics of a microstrip line with periodically corrugated ground plane. Three different corrugation profiles are chosen to demonstrate the effects of corrugation depth and substrate permittivity on characteristics of the first stop-band. Such ground planes can be used either to model a bulge ground or to design a transmission line filter. Distributions of the space harmonics of various singularities are also inspected to gain more understanding on the physical mechanisms.

References

[1] T. Leung and C. A. Balanis, "Attenuation distortion of transient signals in microstrip," *IEEE Trans. Microwave Theory Tech.*, vol.36, pp.765-769, Apr. 1988.

[2] E. J. Denlinger, "A frequency dependent solution for microstrip transmission lines," *IEEE Trans. Microwave Theory Tech.*, vol.19, pp.30-39, Jan. 1971.

[3] T. Itoh and R. Mittra, "Spectral-domain approach for calculating the dispersion characteristics of microstrip lines," *IEEE Trans. Microwave Theory Tech.*, vol.21, pp.496-499, July 1973.

[4] P. Bhartia and P. Pramanick, "A new microstrip dispersion model," *IEEE Trans. Microwave Theory Tech.*, vol.32, pp.1379-1384, Oct. 1984.

[5] F. J. Glandorf and I. Wolff, "A spectral-domain analysis of periodically nonuniform microstrip lines," *IEEE Trans. Microwave Theory Tech.*, vol.35, pp.336-343, Mar. 1987.

[6] N. V. Naur and A. K. Mallick, "An analysis of a width-modulated microstrip periodic structure," *IEEE Trans. Microwave Theory Tech.*, vol.32, pp.200-204, Feb. 1984.

[7] T. Kitazawa and R. Mittra, "An investigation of striplines and fin lines with periodic stubs," *IEEE Trans. Microwave Theory Tech.*, vol.32, pp.684-688, July 1984.

[8] Y. Fukuoka and T. Itoh, "Slow-wave coplanar waveguide on periodically doped semiconductor substrate," *IEEE Trans. Microwave Theory Tech.*, vol.31, pp.1013-1017, Dec. 1983.

[9] S. W. Chen, X. P. Liang, and K. A. Zaki, "Propagation in periodically loaded corrugated waveguides," *IEEE Trans. Magn.*, vol.25, pp.3055-3057, July 1989.

[10] J.-F. Kiang, S. M. Ali, and J. A. Kong, "Propagation properties of striplines periodically loaded with crossing strips," *IEEE Trans. Microwave Theory Tech.*, vol.37, pp.776-786, Apr. 1989.

[11] Q. Xue, K. M. Shum, and C. H. Chan, "Novel 1-D photonic bandgap microstrip transmission line," *IEEE MTT-S Int. Microwave Symp. Dig.*, pp.354-357, 2000.

[12] K. Ogusu, "Propagation properties of a planar dielectric waveguide with periodic metallic strips," *IEEE Trans. Microwave Theory Tech.*, vol.29, pp.16-21, Jan. 1981.

[13] F. J. Glandorf and I. Wolff, "A spectral-domain analysis of periodically nonuniform coupled microstrip lines," *IEEE Trans. Microwave Theory Tech.*, vol.36, pp.522-528, Mar. 1988.

[14] C. Surawatpunya, M. Tsutsumi, and N. Kumagai, "Bragg interaction of electromagnetic waves in a ferrite slab periodically loaded with metal strips," *IEEE Trans. Microwave Theory Tech.*, vol.32, pp.689-695, July 1984.

[15] D. Cadman, D. Hayes, R. Miles, and R. Kelsall, "Simulation results for a novel optically controlled photonic bandgap structure for microstrip lines," *IEEE MTT-S Int. Microwave Symp. Dig.*, pp.110-115, 2000.

[16] H. Y. D. Yang, "characteristics of guided and leaky waves on multilayer thin-film structures with planar material gratings," *IEEE Trans. Microwave Theory Tech.*, vol.45, pp.428-435, June 1997.

[17] R. B. Hwang and S. T. Peng, "Guidance characteristics of two-dimensionally periodic impedance surface," *IEEE Trans. Microwave Theory Tech.*, vol.47, pp.2503-2511, Dec. 1999.

[18] F. R. Yang, K. P. Ma, Y. Qian, and T. Itoh, "A uniplanar compact photonic-bandgap(uc-pbg) structure and its applications for microwave circuits," *IEEE Trans. Microwave Theory Tech.*, vol.47, pp.1509-1514, Aug. 1999.

[19] H. Y. D. Yang, "Theory of microstip lines on artificial periodic substrates," *IEEE Trans. Microwave Theory Tech.*, vol.47, pp.629-635, June 1997.

[20] I. Rumsey, M. P. May, and P. K. Kelly, "Photonic bandgap structure used as filters in microstrip circuits," *IEEE Microwave Guided Wave Lett.*, vol.8, pp.336-338, Oct. 1998.

[21] T. Kim, and C. Seo, "A novel photonic bandgap structure for low-pass filter of wide stopband," *IEEE Microwave Guided Wave Lett.*, vol.10, pp.13-15, Jan. 2000.

[22] M. A. G. Laso, T. Lopetegi, M. J. Erro, D. Benito, M. J. Garde, and M. Sorolla "Multiple-frequency-tuned photonic bandgap microstrip structures," *IEEE Microwave Guided Wave Lett.*, vol.10, pp.220-222, June 2000.

[23] K. E. Bean, "Anisotropic etching of Si," *IEEE Trans. Electron. Dev.*, vol.25, pp.1185-, 1978.

[24] M. Guglielmi and D. R. Jackson, "Low frequency location of the leaky-wave poles for a dielectric layer," *TEEE Trans. Microwave Theory Tech.*, vol.38, pp.1743-1746, Nov. 1990.

Chapter 12

THEORY AND APPLICATIONS OF ELECTROMAGNETIC BAND GAP STRUCTURES

Hung-Yu David Yang
Department of Electrical Engineering and Computer Science
University of Illinois at Chicago
Chicago, Illinois, USA

Abstract In recent years, there has been significant research interest in applications of electromagnetic(photonic) band gap structures. This Chapter provides an overview of recent research efforts on new applications of microwave band gap structures to integrated circuits and antennas. Results based on both rigorous numerical analysis and measurements are presented. The concept of modeless integrated substrates is reviewed. Examples of microwave circuit components and antennas on substrates without surface wave are given to demonstrate the reduction of surface wave loss. The characteristics of microstrip antennas and high-gain leaky wave antennas on periodic substrate are discussed. Basic properties of microstrip line on periodic substrates are also described.

Keywords: periodic structure, band gap, EBG, PBG, mode diagram, modeless substrate, DOVIE, EFIE, Green's function, Floquet's theorem.

1. Introduction

Recently, there have been significant research activities related to the development of artificial periodic materials or photonic crystals. Wave propagation in periodic structures has been an important subject in both microwave and optics for many decades. Artificial dielectrics made of arrays of periodic conductors have found applications in microwave lens [1]. Periodically loaded waveguides have found applications in a

variety of microwave devices such as traveling wave tubes, filters and surface waveguides [2]. Planar arrays of metallic elements on dielectric substrate have been used extensively in microwave frequency selective surfaces(FSS) as well as integrated antenna phased arrays [3]. Feedback lasers using X-ray diffraction from crystals and periodic dielectrics are examples of optical applications of periodic structures [4].

Then, the question comes to what is new about photonic crystals, or in a more general term, artificial periodic materials? It has been known for many decades in microwave theory that periodic loadings along the guided wave direction introduce stop-bands(or frequency band gaps) where electromagnetic wave propagation is prohibited. The new idea that spurs recent research activities in this old subject is to extend the band gap to two or three-dimensional periodic structures [5]. Why is this idea attracting worldwide research attention in physics and engineering community? If there indeed exists photonic(electromagnetic) band gap where photon(electromagnetic wave) propagation is prohibited, it will be dual to the electronic band gap that forms the basis of semiconductor industry. There have been tremendous expectations that optical and electromagnetic technologies may benefit from the photonic crystals in a similar way that the electronic technology benefits from the semiconductors [6], [7].

The properties of artificial periodic materials are scalable with wavelength and are applicable over a wide frequency range from radio up to optical frequencies. Although there have been many new concepts and applications proposed for optical applications in the past few years, the bottleneck is on accurate nanofabrication of three-dimensional photonic crystals [8], [9]. This is due to the fact that in order to create a photonic band gap(PBG), the dimension of a periodic cell is usually around one half wavelength in the medium. The fabrication of artificial materials with periodic elements in the nanometer range presents technological challenges. On the other hand, fabrication of microwave photonic crystals is relatively straightforward with printed circuit technology. Major issue in microwaves is to identify new technologies and new applications for the artificial periodic structures. Is there anything new for the scale-up microwave photonic crystals based upon well developed subject of periodic structures? A careful review reveals that features of two and three-dimensional electromagnetic band gaps(EBG) in microwave photonic crystal had received little attention, which means wave prohibition in the entire plane or space in a controllable way.

Microwave integrated circuits involve interaction between components and materials. The implementation of EBG materials in integrated circuits may result in many useful devices. It is noted that although the

geometry is similar, an EBG structure is very different from a frequency selective surface in that the EBG structure utilizes the stop-bands due to periodic elements along the energy propagation direction. Surface wave elimination and leaky wave generation are two unique characteristics of EBG structures identified recently with many potential applications [10].

Printed circuit elements placed on dielectric or semiconductor substrate are known to generate surface wave modes(or dielectric slab modes). It is often desirable to eliminate or minimize these surface waves. First, surface wave modes propagating laterally incur losses in the integrated circuits. Second, the crosstalk between devices printed on the substrate surface may be significant due to surface wave interaction. Third, the surface wave causes array scan blindness in a phased array. Recently, the use of periodic elements(metallic or dielectric) loading on integrated circuit substrates was suggested to eliminate surface wave modes within a frequency band gap [10]-[12]. The idea is to use planar arrays of elements on dielectric surface to create the Bragg diffraction condition for the surface wave in all planar directions. When the phase constant of a bound mode(a mode with fields decaying vertically away from the substrate) nearly satisfies the Bragg diffraction condition($\beta a = \pi$, where β is phase constant and a is period), this mode becomes a bound complex mode that is highly attenuating, namely, the mode becomes an evanescent mode that does not carry power. Leaky modes(fast waves with respect to free space) with small attenuation constant may exist in the same structure. In contrast to complex evanescent mode, a leaky mode will leak power into space, resulting in significant power loss from the circuits. Leaky mode structure may also be designed to function as high-gain and high-efficiency leaky wave antenna.

In this Chapter, fundamental principles of modeless(no surface wave or leaky wave) integrated circuit substrates are reviewed. A full-wave technique using integral equation and method of moments for the analysis of infinite arrays of microstrip elements is applied here to find the complex wave numbers of surface wave and leaky wave [13]. Examples of carefully designed mode diagrams for periodic metal patches are given. Measured results are also shown to confirm the design. Furthermore, numerical simulations and measurements are given to demonstrate the loss reduction of integrated circuit components on modeless substrate. The same modeless substrate supports leaky wave modes within certain frequency band, and such modes may result in patterns with high directivity if used for antenna application.

Printed metallic elements on materials are general features in microwave and millimeter wave integrated circuits. The use of periodic elements within the materials fits well in present integrated circuit tech-

nology. Electromagnetic wave theory and computational techniques are necessary for the understanding and characterization of circuits or antennas within artificial periodic media to devise new applications. Conventional analytic/numerical methods are suitable to fully periodic structures [14]-[17]. Periodic structures with localized sources or objects are important in many areas of engineering and science. Rigorous analysis on pertinent boundary-value problems involves radiating sources within planar periodic structures. The fields in any periodic structure with resonant antennas can be computed within certain degree of accuracy by using finite element or finite difference time domain(FDTD) methods if absorbing boundary conditions are used. However, in order to obtain exact numerical solutions and understand the fundamental physical mechanisms of the structure, integral equation formulation in conjunction with method of moments is preferred.

Since the pertinent problem is aperiodic, conventional integral equation and method of moments can not be applied directly. The analysis is carried out with a full-wave double vector integral equation(DOVIE) method developed for the analysis of antennas in an otherwise periodic structure [18]-[21]. The analysis begins with an analytic array scanning scheme to obtain a numerical Green's function. Then, the method of moments is used to obtain the fields and parameters of interest. This method is effective, stable and accurate, and applicable to general infinite periodic structures with anomalies. In this Chapter, we will discuss the computational and practical issues of a microstrip transmission line or a microstrip dipole antenna on artificial periodic substrate since they are the basic components in integrated circuits.

2. Theory of Modeless Substrates

Modeless substrate is a lossless substrate with no guided substrate mode in all planar directions within certain frequency range. Note that there are many useful modeless guided wave structures, including metallic waveguides below their cutoff frequency and parallel-plate waveguides with multilayered dielectric fillings. In many integrated circuit applications, particularly in the high-end of microwave bands, it is important to reduce or eliminate the substrate modes for loss and crosstalk reduction. A viable solution is to use planar periodic elements printed on or embedded within the substrate. Surface wave band gap structures with integrated circuit substrates and periodic implants were investigated earlier [10]-[11]. However, it is usually inconvenient to implant material elements into the substrate, and omnidirectional band gap is not viable unless the dielectric contrast is dramatic. On the other hand,

metallic elements can be easily printed on the substrate together with other circuit components. Also, the use of metallic elements renders the band gap edges relatively insensitive to azimuthal angle. Hence, surface wave band gap can be created in all angles [12]. We will focus on periodic elements made of rectangular metallic patches.

2.1 Formulation

To design a modeless substrate, we need to obtain the mode diagram numerically. An example of metallic patch array is shown in Fig.12.1. The method is identical to that for analyzing infinite arrays of printed patches. The problem is analyzed through the use of an electric field integral equation(EFIE) formulation in conjunction with the method of moments [22], [23]. Assuming that the metallic elements are of zero thickness, the electric field can be expressed as

$$\bar{E}(\bar{r}) = \iint_S \bar{\bar{G}}(\bar{r}, \bar{r}') \cdot \bar{J}(\bar{r}') d\bar{r}'$$

where the surface integration is only over one of the periodic elements, having current density, $\bar{J}(\bar{r}')$. The dyadic Green's function, giving the transversal field components in a layered medium, is expressed as

$$\bar{\bar{G}}(\bar{r}, \bar{r}') = \begin{bmatrix} G_{xx}(\bar{r}, \bar{r}') & G_{xy}(\bar{r}, \bar{r}') \\ G_{yx}(\bar{r}, \bar{r}') & G_{yy}(\bar{r}, \bar{r}') \end{bmatrix}$$

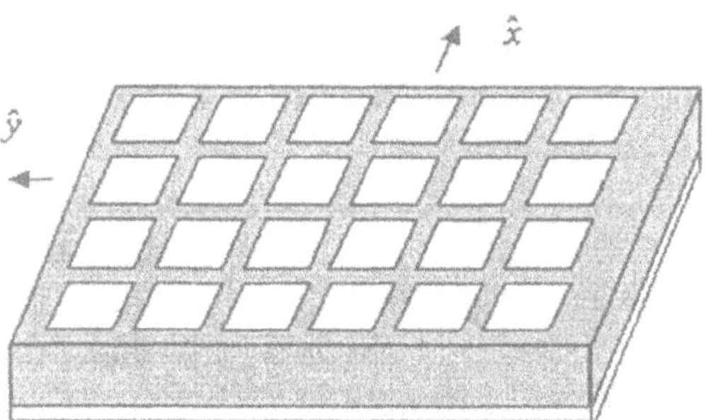

Figure 12.1. Planar periodic array of square patches on a grounded dielectric, array period is $a \times a$, patch size is $L \times L$.

Floquet's theorem is applied to reduce the problem to the fields within a unit cell with periods a and b, along the $\hat{x}$ and $\hat{y}$ directions, respectively. For planar periodic structures, components of the dyadic Green's

function can be expressed in terms of Floquet modes as [17]

$$G_{pq}(\bar{r},\bar{r}') = \frac{1}{ab}\sum_{m=-\infty}^{\infty}\sum_{n=-\infty}^{\infty}\tilde{G}_{pq}(k_{xm},k_{yn};z,z')e^{-jk_{xm}(x-x')-jk_{yn}(y-y')} \tag{12.1}$$

where p and q are either x or y, and the wave numbers are defined as $k_{xm} = k_{x0}+2m\pi/a$, $k_{yn} = k_{y0}+2n\pi/b$ with k_{x0} and k_{y0} the fundamental wave numbers(wave numbers of the zeroth order Floquet mode) in the $\hat{x}$ and $\hat{y}$ directions, respectively. $\tilde{G}_{pq}$ is the spectral Green's function, which is the same as that for a single(nonperiodic) element, and can be derived with the spectral matrix method [10].

A standard moments method is applied to determine the current distribution on the patch element in a unit cell. The matrix elements are in the form of a double infinite series as in (12.1), and are computed numerically. The propagation constant, $k_\rho = \beta - j\alpha$, in a given direction in the phase plane(namely, for a given ratio of $k_{\rho y}/k_{\rho x}$, or equivalently, a given angle ϕ where $k_{\rho x} = k_\rho \cos\phi$ and $k_{\rho y} = k_\rho \sin\phi$) is obtained by solving the matrix determinantal equation. If the propagation wave number is purely real, the associated mode is a bound mode. If the wave number is a complex number(in a lossless structure), the mode may either be an attenuating slow wave(an evanescent surface wave mode in the band gap) or a radiating leaky mode [24].

2.2 Discussions and Applications

An example of mode diagram for the planar patch array is shown in Fig.12.2 for wave propagation along the $\hat{x}$ direction. The substrate is 1.27 mm thick with dielectric constant of 10.2. The square patches are 6 mm long and the array spacing(period) is 8 mm. Dramatic variation of phase constant with frequency is due to patch resonance. At the edge of the Brillouin zone, the bound mode turns into a highly attenuating wave with complex wave number. It is observed that there also exists a weakly bound surface wave, which is not related to any particular patch mode. Fig.12.2 shows that there is only one proper leaky mode that radiates in the backward direction(the group velocity is negative). Other leaky modes(fast waves) are nonradiating. There are two frequency bands(from 7.4 to 9.6 GHz and from 12.45 to 13.9 GHz) where there is no bound or leaky modes. These frequency bands are true band gaps.

Measurement results for the case given in Fig.12.2 are shown in Fig.12.3, where a 10×10 patch array is fabricated and two probes are soldered at opposite edges of substrate. The transmission through substrate is mea-

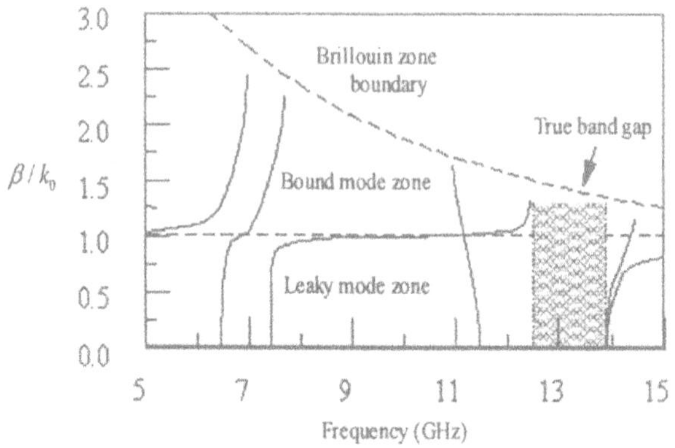

Figure 12.2. **Normalized phase constant(β/k_0) versus frequency for waves on the planar periodic patch array in Fig.12.1, propagation is in the $\hat{x}$ direction($\phi = 0$).**

sured by using an HP-8510C network analyzer. The transmission characteristics is pretty much as predicted in Fig.12.2. The results shown in Figs.12.2 and 12.3 are for the propagation in the $\hat{x}(\phi = 0)$ direction.

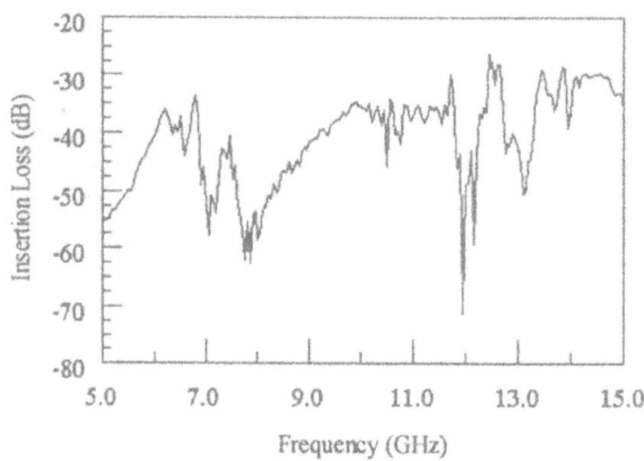

Figure 12.3. **Measured insertion loss across the dielectric slab described in Fig.12.2.**

Band gap that exists in one direction may not be present in other directions. For planar structures, one is interested in wave guidance in all possible planar directions(ϕ ranges from 0 to 360°). For square patches and lattices, guidance information with ϕ ranging from 0 to 45° is suffi-

cient due to symmetry. The existence of an omnidirectional band gap is demonstrated by the electromagnetic band diagram shown in Fig.12.4, which shows the frequency versus the boundary of irreducible reciprocal lattice, or the spatial boundary of Brillouin zone. The notations of Γ, X and M are adopted from the reciprocal lattice of natural crystal. The first sector of the plot(from Γ to X) with $0 \leq \beta_x a/\pi \leq 1$ and $\beta_y = 0$ corresponds to the principal axial direction. The third sector of the plot(from M to Γ) corresponds to the direction at $\phi = 45°$. The second sector(from X to M) shows the frequency versus $\beta_y a/\pi$ with $\beta_x a/\pi = 1$, which is basically a sampling of modes in various directions from $\phi = 0$ to $\phi = 45°$. Fig.12.4 indicates that there are indeed complete band gaps dictated by the band gap in the $\phi = 45°$ direction.

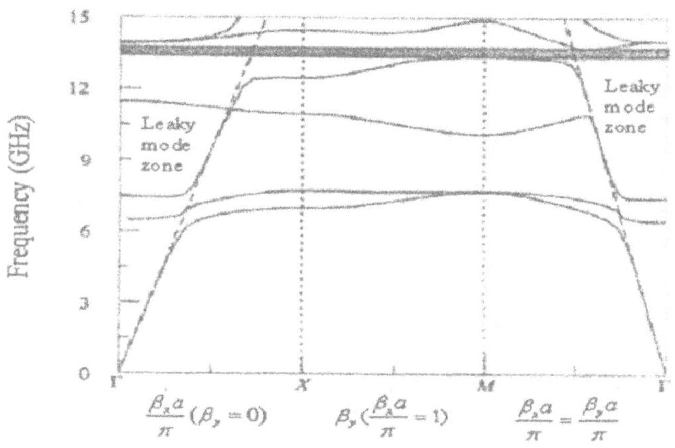

Figure 12.4. **Two-dimensional band diagram of guided wave modes associated with the structure in Fig.12.1. The horizontal axis indicates the normalized wave number in various directions. Γ, X and M are symmetric points in the Brillouin zone of a unit cell.**

The main purpose of using modeless substrate is to reduce the substrate loss which is the dominant loss factor in microwave integrated circuits. In order to demonstrate this loss reduction effect, an example of microstrip gap discontinuity on patch-loaded substrate is shown in Fig.12.5. The substrate is Duroid with thickness of 1.27 mm and dielectric constant of 10.2. The patch array is the same as that in Fig.12.1. Frequency variation of transmitted plus reflected power is shown for the cases with and without the patch array, respectively. It is found that the power loss is reduced significantly in the frequency band where only non-radiating leaky mode exists(8-10 GHz) as indicated in Fig.12.2. Since

the substrate is assumed lossless, the remaining power is radiated into space.

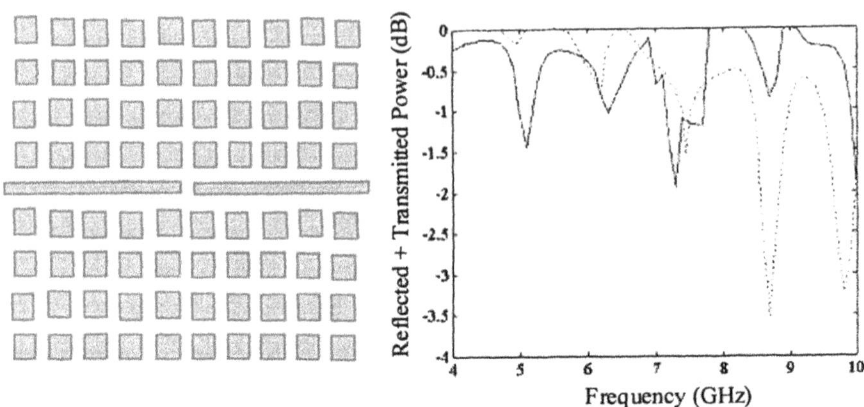

Figure 12.5. Power transmission and reflection of a two-port microstrip gap surrounded by periodic patches for surface wave loss reduction, —: with periodic patches, ···: without patch loading.

Another example is a design using the same substrate. The square patches are 3 mm long and the array spacing(period) is 8 mm in both principal directions(the same as in the previous case). The two-dimensional band diagram is shown in [12]. A complete band gap exists between 13 and 14 GHz. The measured results using this particular substrate is shown in Fig.12.6, where the structure is similar to that shown in Fig.12.5 except that the microstrip is a through line. It is observed that the insertion loss is noticeably reduced in the band gap(13-14 GHz). The sudden drop in transmission around 12 GHz is due to the stop-band effect of the periodically loaded transmission line.

In using patch array for mode elimination, the element shape in general is not critical, although planar symmetry is preferred. Recently, a high-impedance PBG substrate where the periodic element is consisted of a patch and a pin has been widely discussed [25]. From the field theory, the fundamental characteristics are not well understood. First of all, it is incorrect to describe it as a PBG structure since the element spacing is much smaller than the required length to satisfy the Bragg condition and the array is not necessarily periodic. Then, the question comes to why substrate mode elimination is possible. The basic concept of dielectric modes in planar substrate has been discussed in many microwave engineering books [26], [27]. The fundamental mode is a TM mode with vertical electric field component. It is well known and

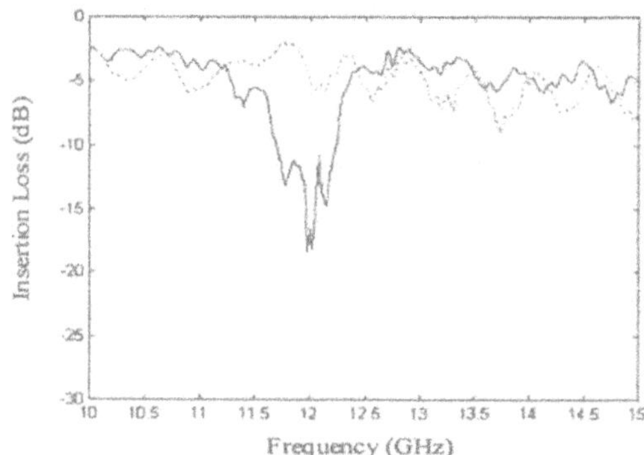

Figure 12.6. **Measured insertion loss through a microstrip line surrounded by patch arrays, patch size is 3 mm × 3 mm, other parameters are the same as used in Fig.12.2, —: with periodic patches, ···: without patch loading.**

has been extensively applied in microwave integrated circuits that randomly distributed vias in the substrate will prevent the propagation of the fundamental TM mode. This mode prohibition is similar to the case that the shorting pins prevent the fundamental mode in a parallel-plate waveguide.

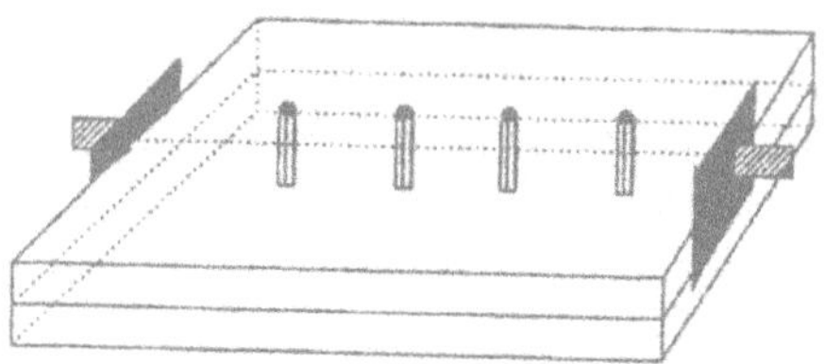

Figure 12.7. **Dielectric substrate with a linear array of four pins.**

Measurement has been carried out to verify this concept. The setup is shown in Figs.12.7 and 12.8. Two coaxial probes are mounted on opposite sides of the dielectric substrate. An HP-8510C network analyzer is used to measure the two-port S-parameters. The transmission coefficient, S_{21}, indicates the relative level of power transmission. Three sets of measurements are carried out with a homogeneous substrate, a substrate with a linear array of four pins, and a substrate with a 3×4 planar array of pins, respectively. The measured insertion loss for the three

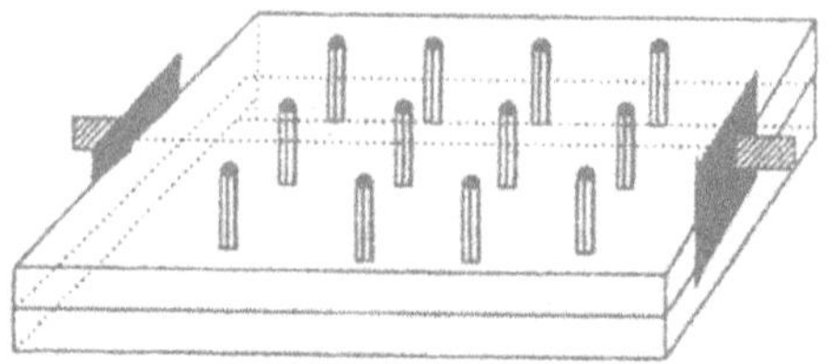

Figure 12.8. Dielectric substrate with a planar 3 × 4 array of pins.

cases are shown in Fig.12.9. It is seen that the pin arrays significantly reduce the transmission power at lower frequencies, which confirms the mechanism of mode elimination.

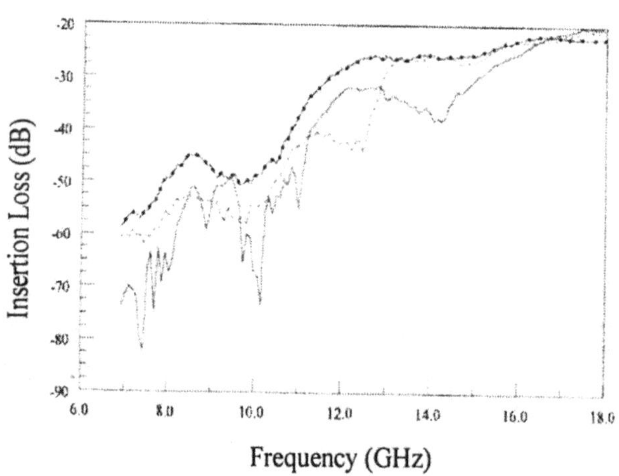

Figure 12.9. Measured insertion loss across the dielectric substrate, ···: four-pin array, —: 12-pin array, -*-: without pin.

3. Printed Antennas on EBG Substrates

Typical gain of printed antenna is about 6 dB. It has been demonstrated that using multiple periodic layers on top of a microstrip antenna can significantly enhance its gain [28], which is explained by leaky wave theory [29]. Those multiple superstrates can be treated as one-dimensional PBG overlay. Thick superstrates are too bulky for integrated circuit applications, while planar periodic structure is preferred. Recently, there are many research efforts on printed antennas in EBG

structures. Main applications include surface wave reduction(modeless effects) and leaky wave antennas. Experimental verifications of antenna gain increase due to surface wave reduction have been reported in [30]-[32].

Most prior research efforts on leaky wave antennas were limited to one-dimensional periodic structures, and leaky wave antennas on planar periodic structures have not been well understood. Fundamental investigation on leaky wave antenna design requires leaky mode analysis on planar periodic structures [33]. The basic characteristics of leaky wave antenna are determined by its radiation patterns. In the numerical computation of far-zone fields, the antenna can be replaced by a point source and reciprocity theorem is applied by interchanging the source and observation points. Instead of solving the antenna problem, we solve the boundary-value problem of TE or TM plane wave incident on the planar periodic structure as shown in Fig.12.1. Thus, the electric field at a given location within substrate will be the far-zone electric field due to an antenna at the same location. The numerical method is basically identical to the one for obtaining the mode diagram described in the previous Section. The far-zone fields can be used to determine the antenna directivity and gain patterns as a function of substrate characteristics including the periodic loading effects.

3.1 Leaky Wave on Planar Periodic Substrates

The modeless substrate shown in Fig.12.1 is also a structure of high-gain leaky wave antenna. As observed in Fig.12.2, there is a radiating leaky wave mode near 11 GHz, corresponding to a backward physical leaky mode [34]. The directivity patterns and the beam angles are computed for a small dipole embedded in the structure shown in Fig.12.1. Leaky wave antenna usually has high gain, and the beam angle is determined by the phase constant of the leaky mode [24], [35].

An example of radiation pattern using the parameters in Figs.12.1-12.3 is shown in Fig.12.10. The frequency is 12 GHz, the gain is 14 to 15 dB, phase constant of the leaky mode is small, and the beam angle is tilted toward the broadside. It is interesting to note that high gain is observed in the E-plane, only little radiation of lower than -20 dB in directivity is observed in the H-plane. The main beam points at about $\theta = 20°$.

The radiation patterns for the same structure but at 8 GHz(the modeless zone) are shown in Fig.12.11. The results without periodic patches are also shown for comparison. It is seen that due to surface wave elimination, there is less power radiation near the horizon, and the overall

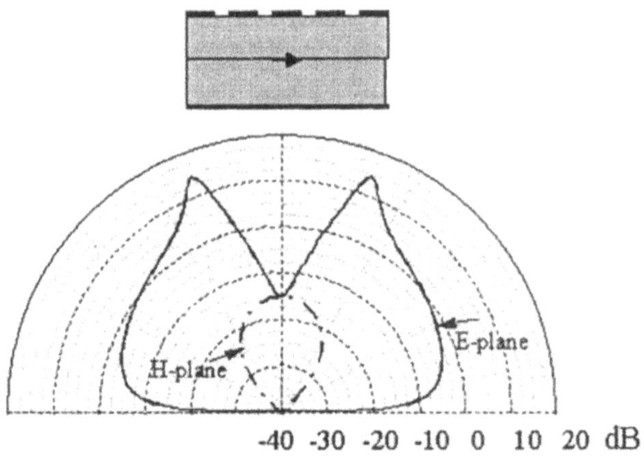

Figure 12.10. **Directivity patterns of a dipole under planar periodic patches depicted in Fig.12.1, $f = 12$ GHz, within the leaky mode zone.**

gain is higher. The beam angle and directivity as a function of frequency are shown in Fig.12.12. The directivity increases significantly within the leaky mode frequency band, which is greater than 10 dB as compared to 6 dB of conventional printed antenna structure. The beam angle is also tilted away from the broadside and is frequency sensitive, determined by the phase constant of leaky mode . Out of the leaky mode zone, the antenna has a broadside beam with about 6 dB in gain, which is typical of a printed antenna.

3.2 Dipole Antenna on Periodic Substrates

Periodic structures with localized sources or objects have important applications in many areas of engineering and science. In this Section, we first discuss a full-wave numerical method used to analyze antennas on an otherwise planar periodic structure. Fundamental properties of a microstrip dipole antenna on an artificial periodic substrate are studied as an example. The periodic elements are material blocks with holes carved in the substrate. The analysis is carried out with a full-wave double vector integral equation(DOVIE) method [18]-[21]. This method is effective, stable and accurate, and is applicable to general infinite periodic structures with anomalies.

There are two vector integral equations to be solved sequentially. The first integral equation formulation is used to find the electromagnetic fields in an artificial material structure with an infinite phased array of

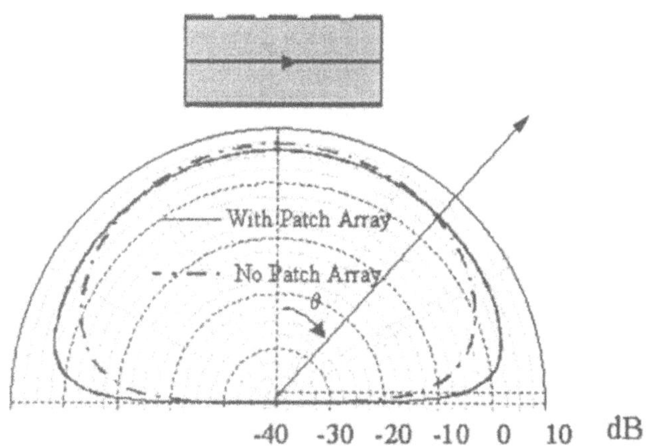

Figure 12.11. **E-plane radiation patterns of a dipole under planar periodic patches depicted in Figs.12.1-12.3, $f = 8$ GHz, within the modeless zone.**

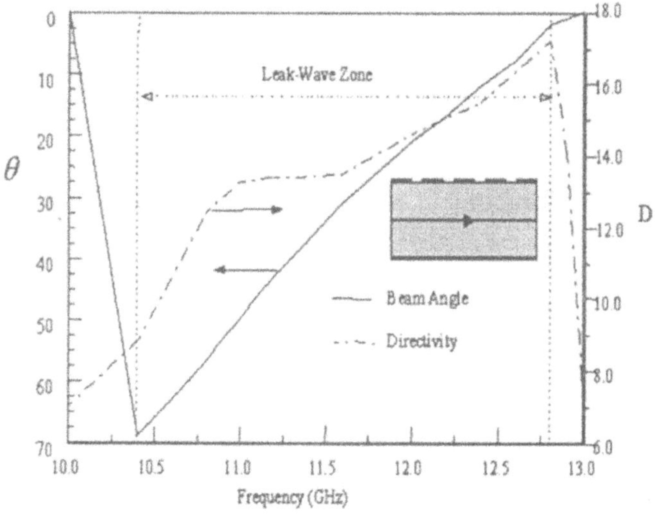

Figure 12.12. **Beam angle and directivity of a dipole under planar periodic patches depicted in Figs.12.1-12.3.**

δ sources. Array scanning or phase matching method transforms the fields of the infinite phased array to those of a single δ source while the otherwise periodic structures are unchanged. Those fields after trans-

formation are the dyadic Green's function for the second vector integral equation that is solved to find the conducting or displacement currents on the antennas or objects. This approach is a continuous method and its accuracy is similar to typical moments method for canonical boundary-value problems.

Input impedance and radiation efficiency of a gap-fed microstrip dipole are analyzed. For comparison, microstrip dipoles on both an artificial periodic substrate and an effective uniform substrate are analyzed. The geometry of a microstrip dipole antenna on an artificial periodic substrate is shown in Fig.12.13. The substrate, assumed of infinite extent, is a dielectric material with planar periodic blocks. For simplicity, it is also assumed that the microstrip is narrow with negligible transversal current. Although the dipole shown in Fig.12.13 is on top of the substrate, the theory is also applicable when the dipoles are embedded in the substrate.

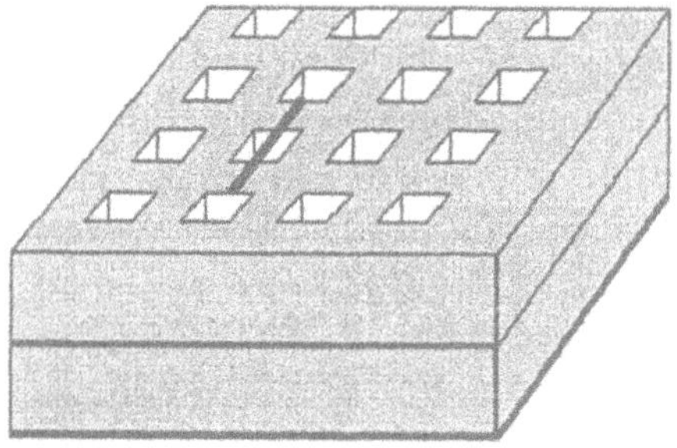

Figure 12.13. Microstrip dipole on an artifical periodic structure.

Although rectangular blocks are shown, the analysis is applicable to other irregular shapes of periodic elements. For practical purpose and without loss of generality, the periodic elements are assumed air filled. The periodic elements are placed on the x-y plane with spacing a along the x axis and b along the y axis, and a microstrip dipole is at the layer interface pointing in the $\hat{x}$ direction. The first step of analysis is to find the Green's function for the periodic structure, in which the dipole is replaced by a δ source.

Integral Equation Approach for Infinite Phased Arrays. Since the pertinent structure is aperiodic, conventional Floquet mode expansion method can not be applied directly. Instead of solving this problem directly, first consider the geometry where there is an infinite phased ar-

ray of Hertzian dipoles(δ sources) with progressive phase shifts, ϕ_x and ϕ_y, along the $\hat{x}$ and $\hat{y}$ directions, respectively. The dipole spacing and the material spacing are the same. This fictitious phased array problem can be solved using standard integral equation approach. With the use of Floquet's theorem and periodic boundary conditions, the problem is simplified to the modeling of electromagnetic waves within a unit cell.

A three-dimensional integral equation and moments method was discussed in details in [10]-[17]. Since all the three field and current components exist, it is necessary to deal with a full dyadic Green's function whose components are expressed in terms of Floquet modes(plane wave expansion). This spectral Green's function is valid for a multilayered structure without periodically implanted materials. Furthermore, define $\bar{E}^i(\bar{r})$ as the electric field due to the phased array of Hertzian dipoles without periodic material elements. If the periodic elements are air, we have the following electric field integral equation

$$\bar{E}(\bar{r}) = \bar{E}^i(\bar{r}) + j\omega\epsilon_0(1-\epsilon_r)\iiint_V \bar{\bar{G}}(\bar{r},\bar{r}')\cdot\bar{E}^e(\bar{r}')d\bar{r}' \tag{12.2}$$

The integration is carried only over the region of periodic elements. The electric fields within the material block in a unit cell are the unknowns to be solved by the method of moments. A finite-element moments method is applied to determine the electric fields within the blocks. This process converts the integral equation into a set of linear equations(a matrix equation). If the excitation term in (12.2) is removed, the problem becomes an eigenvalue problem. The eigenvalues are the progressive phase shifts, ϕ_x and ϕ_y, which determine the phase constant of the guided wave in the periodic structure in a specific planar($\hat{x}$ or $\hat{y}$) direction [10]. The electric field of the phased arrays can then be expressed as

$$\bar{E}^a(\bar{r},\phi_x,\phi_y) = \frac{1}{ab}\sum_{m=-\infty}^{\infty}\sum_{n=-\infty}^{\infty}\bar{\bar{E}}_{mn}(\phi_x,\phi_y,z)e^{-j(2m\pi/a+\phi_x)(x-x')} e^{-j(2n\pi/b+\phi_y)(y-y')}$$

Analytic Array Scanning for Aperiodic Problems. The moments method solutions of the EFIE give the displacement currents inside the periodic elements. Subsequently, these displacement currents are used in (12.2) to obtain the electromagnetic fields inside a unit cell. The field solutions for a δ source in an otherwise periodic structure are found from the infinite phased array solutions through analytic array scanning [36]. An example to demonstrate the theory of analytic array scanning is shown in Fig.12.14. If the electric field solution of case (a) in

Fig.12.14 is $\bar{E}^a(\bar{r},\phi)$, and the electric field solution of case (b) is $\bar{E}^b(\bar{r})$, then we have

$$\bar{E}^b(\bar{r}) = \frac{1}{2\pi}\int_{-\pi}^{\pi} \bar{E}^a(\bar{r},\phi)d\phi \tag{12.3}$$

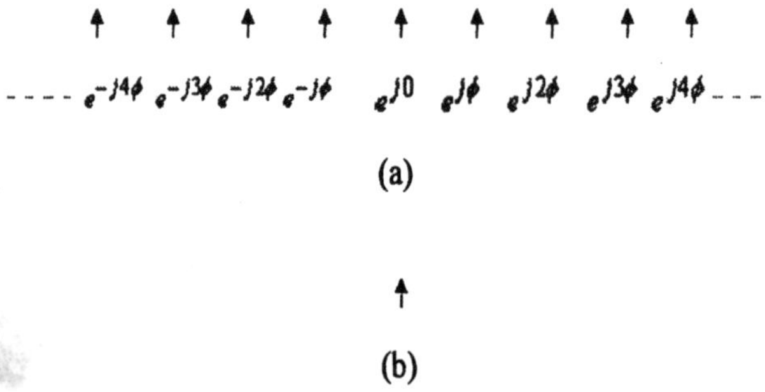

---- $e^{-j4\phi}$ $e^{-j3\phi}$ $e^{-j2\phi}$ $e^{-j\phi}$ e^{j0} $e^{j\phi}$ $e^{j2\phi}$ $e^{j3\phi}$ $e^{j4\phi}$ ---

(a)

(b)

Figure 12.14. **(a) Infinite phased array of δ sources, (b) a single δ source.**

The integration in (12.3) indicates the superposition of solutions for the periodic phased array. Performing the integration in (12.3) effectively cancels out all other Hertzian dipoles in the infinite array except the one with zero phase. As a result, the Green's function for planar periodic structure can be written in a general form as [37]

$$\bar{E}^a(\bar{r}) = \frac{1}{4\pi^2 ab}\sum_{m=-\infty}^{\infty}\sum_{n=-\infty}^{\infty}\iint_{-\pi}^{\pi} \tilde{\bar{E}}_{mn}(\phi_x,\phi_y,z)$$
$$e^{-j(2m\pi/a+\phi_x)(x-x')}e^{-j(2n\pi/b+\phi_y)(y-y')}d\phi_x d\phi_y$$

which is the Green's function due to an $\hat{x}$ directed δ source in the structure of Fig.12.13. Using proper change of variables, (12.3) can be rewritten as

$$\bar{E}^a(\bar{r}) = \frac{1}{2\pi^2}\iint_{-\infty}^{\infty} \tilde{\bar{E}}(\bar{k}_s,z)e^{-jk_x(x-x')}e^{-jk_y(y-y')}dk_x dk_y \tag{12.4}$$

where $\bar{k}_s = \hat{x}k_x + \hat{y}k_y$.

The Green's function for uniform dielectric substrate in the spectral domain formulation has the same form as (12.4). However, the spectral

components in homogeneous substrate have analytic form [38], while the spectral components in a periodic substrate need to be evaluated numerically through the volume integral equation solution described in this Section. Following the identical procedure as for the case of homogeneous substrate, we may apply standard moments method in conjunction with Galerkin method to solve for the current on the dipole excited by a voltage gap source. Current expansion using pulse functions in the transversal direction and triangular basis functions in the longitudinal direction can be applied to obtain the final impedance matrix equation as

$$[Z][I] = [V]$$

where $[I]$ is the vector containing the unknown amplitudes of basis functions on the dipole, $[V]$ is the excitation vector for the gap-fed dipole, and the matrix elements of $[Z]$ are in the form of

$$Z_{pq} = \frac{1}{4\pi^2 ab} \sum_{m=-\infty}^{\infty} \sum_{n=-\infty}^{\infty} \iint_{-\pi}^{\pi} \bar{\bar{E}}_{mn}(\phi_x, \phi_y) \cdot \hat{p} f_p(\bar{k}_s) f_q(\bar{k}_s) d\phi_x d\phi_y \quad (12.5)$$

where $f_q(\bar{k}_s)$ is the Fourier transform of the expansion function. Eq.(12.5) can be expressed in terms of continuous plane wave spectrum, similar to that for a printed antenna on homogeneous substrate. However, direct spectral domain integration will encounter infinite number of singularities, which are related to the surface waves in periodic substrate. Another approach is to deform the integration contours in (12.5) in the complex planes to avoid the singularities. The solution here is for a microstrip dipole, but this approach also applies to more complicated geometries.

Input Impedance of Dipole on Planar Periodic Substrates. The input impedance of a microstrip dipole on an artificial periodic substrate shown in Fig.12.13 is presented in Fig.12.15. The substrate is 1.27 mm thick and has $\epsilon_r = 10.2$. The periodic material blocks are 0.635 mm thick with $\epsilon_r = 4$, filling the unit cell completely. The resulting structure is in effect a homogeneous two-layered substrate that can be solved using conventional spectral domain method [39], [40]. Reasonable agreement in Fig.12.15 confirms the validity of this analysis.

Fig.12.16 shows the radiation efficiency as a function of frequency with a three-layered substrate. The radiation efficiency is defined as the ratio of the total radiated power to the total input power. The total radiated power is derived from the hemispherical integration of power density in the far zone. The input power is derived from the current and voltage over the dipole gap. The top layer is filled with a planar

Novel Technologies for Microwave and Millimeter – Wave Applications

Electronic Services <http://www.wkap.nl>

Library of Congress Cataloging-in-Publication

CIP info or:

Title: Novel Technologies for Microwave Millimeter-Wave Applications
Author (s): Jean-Fu Kiang

DOI 10.1007/978-1-4757-4156-8

Printed on acid-free paper.
www.springer.com/mycopy

Novel Technologies for Microwave and Millimeter – Wave Applications

by
Jean-Fu Kiang
National Taiwan University

Springer Science+Business Media, LLC

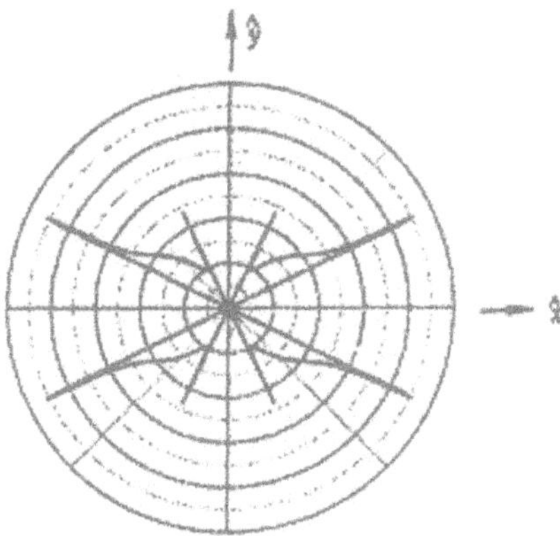

Figure 12.18. Surface wave pattern of a dipole on the planar periodic substrate depicted in Fig.12.15.

are phase constant and characteristic impedance. The fields are mostly confined underneath the microstrip line and hence, only the material in the vicinity of line affects its characteristics.

An example of microstrip line on periodic substrate is shown in Fig.12.19, where the line is parallel to one of the principal axes of array elements. The structure is basically a microstrip line loaded with periodic elements, where the pass and stop-band characteristics are well understood. There are two interesting features of the microstrip line mode that are different from those of conventional microstrip line. When $\beta_x a \simeq \pi$(the Bragg condition), there exists a band gap within which the propagation mode vanishes. Also, when $\beta_x \geq 2\pi/a - k_0$, the guided modes become leaky modes which are fast waves with complex propagation constants.

Notice that the surface wave band gap usually occurs at much higher frequencies than the band gap of microstrip line mode because the phase constant of a microstrip line mode is higher in the same structure. Since the phase constant increases with frequency, the Bragg condition is satisfied at lower frequencies for a microstrip line. At low frequencies, because the period and size of periodic elements are much smaller than a wavelength, the artificial substrate behaves like an effective homogeneous medium. As the frequency increases to near the band gap, the line impedance increases drastically and the microstrip line becomes open circuited at the edge of band gap. When the frequency moves just out of the band gap, the microstrip line behaves like a short circuit.

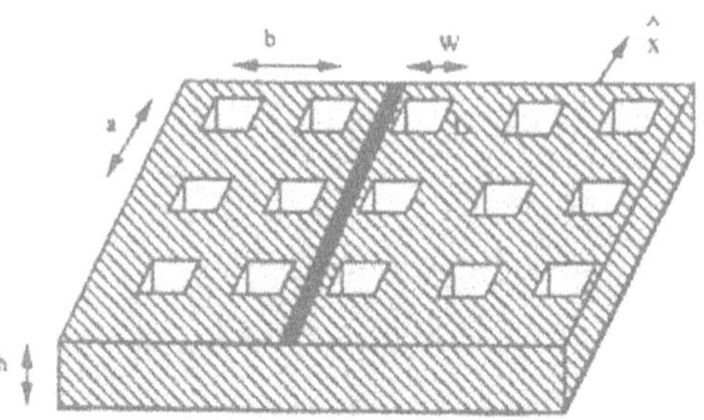

Figure 12.19. Microstrip line on an artificial periodic substrate.

This phenomenon is similar to the waveguide modes near their cutoff frequencies.

An example of mode diagram of a microstrip line on a three-layered planar periodic substrate is shown in Fig.12.20. The initial design has a normalized phase constant(β_x/k_0) of about 2 at 5-6 GHz. With the chosen element spacing of 14 mm, one can estimate the band gap center frequency from the Bragg condition($\beta_x a \simeq \pi$) as 5.36 GHz. Both the propagation constant and line impedance obtained by using full-wave method are shown in Fig.12.20. The band gap occurs in the frequency range of 4.9-5.9 GHz.

Transmission lines loaded with periodic elements are frequently used in microwave filters. An example is a microstrip line on a two-layered structure periodically loaded with metallic strips. There exists a transmission band gap satisfying the Brag condition, $\beta_L a \simeq \pi$, where β_L is the phase constant of the periodically loaded transmission line, and a is the period. The phase constant is much higher than that of the surface wave. Therefore, the surface wave band gap occurs at much higher frequencies than the transmission line band gap.

There are several interesting subjects related to periodically loaded microstrip lines that have not been fully explored. Leaky wave antenna using wide microstrip lines to generate higher-order modes had been explored recently with success [42]. Leaky mode of a periodically loaded microstrip line can be solved using full-wave spectral domain method. Proper leaky mode can be used to design high-gain antenna with specified beam angle. Another interesting subject is the characteristics of a microstrip line on planar periodic substrate with the line tilted from the principal axis of periodicity. The transmission band gap is a function of the relative location of elements as well as line orientation. However,

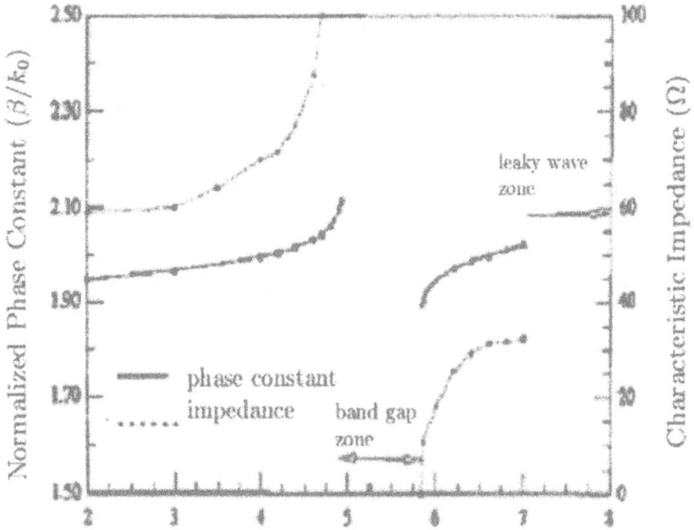

Figure 12.20. **Mode diagram of a microstrip line on a three-layered planar periodic substrate, with a periodic middle layer sandwiched by two homogeneous layers. Top and bottom layers have $\epsilon_r = 9.8$ and 0.635 mm in thickness, strip width is 3 mm, air blocks are 6.5 mm $\times$ 6.5 mm $\times$ 4 mm, period is 14 mm.**

the phase constant and line impedance cannot be obtained rigorously by continuous methods.

5. Conclusions

This Chapter reviews recent progress on microwave applications of electromagnetic band gap structures. For practical purposes, the emphasis is on antennas and transmission lines on two-dimensional planar periodic substrates. New microwave applications of planar periodic structures that have been explored include surface wave mode elimination for antenna gain enhancement and circuit loss reduction, high-gain leaky wave antennas and transmission line filters.

The issues of using periodic metallic patches on an integrated circuit substrate to eliminate surface wave propagation are discussed. It is shown to be fairly easy to eliminate surface wave modes completely in a band gap by using periodic metallic patches. The design principles to obtain modeless substrate in all directions are described, where all surface wave modes and leaky wave modes are eliminated. Both theory and measurement confirm the possibility of completely modeless substrate. Examples are also given to demonstrate the applications of modeless substrate. Microstrip gap discontinuity and through microstrip line on a substrate with a planar array of patches are considered. It is found in both theory and measurement that the loss is reduced significantly as compared to the structure without patch array. Modeless substrates using pin arrays are also discussed. Surface wave mode is suppressed not by the band gap, but by mode disturbance.

Leaky wave generation on planar periodic substrate is also discussed. Physical leaky modes associated with backward leaky modes on the mode diagram can be used to design high-gain leaky wave antennas. Several examples are given and their properties are discussed. The beam angle can be determined by the phase constant of leaky mode.

Design of microstrip antenna on planar periodic substrates requires accurate numerical models. We discussed the full-wave DOVIE method which is capable of analyzing sources on otherwise periodic structures. The procedure consists of two stages of moment methods. Examples of microstrip dipole on planar periodic substrate are given to demonstrate the validity of this numerical method. The properties of a microstrip dipole on such a structure are also discussed.

Finally, examples of microstrip line on planar periodic substrate are given to demonstrate the pass and stop-band characteristics. The transmission properties are similar to those of periodic waveguides. The microstrip line may also support leaky modes. High-gain antenna may be designed using periodic microstrip line structure and the research is still ongoing.

References

[1] W. E. Kock, "Metallic delay lenses," *Bell Syst. Tech. J.*, vol.27, pp.58-82, 1948.

[2] R. E. Collin, *Field Theory of Guided Waves*, IEEE Press, 1991.

[3] T. K. Wu, *Frequency Selective Surface and Grid Array*, ed., John Wiley, 1995.

[4] P. Sheng, *Scattering and Localization of Classical Waves in Random Media*, ed., Singapore: World Scientific, 1990.

[5] E. Yablonovitch, "Photonic band-gap structures," *J. Opt. Soc. Am. B*, vol.10, no.2, pp.283-294, Feb. 1993.

[6] "Photonic band-gap structure," *J. Opt. Soc. Am. B*, vol.10, no.2, special issue, Feb. 1993.

[7] "Photonic band-gap structure," *J. Modern Opt.*, special issue, Feb. 1994.

[8] "Science base for nanolithography," Dept. of Defense BAA for Multidiscipl. Res. Prog. Univ. Res. Init., Mar. 1998.

[9] "Nonascale modeling and simulation," NSF Prog. Solicit., NSF 00-36, Jan. 11, 2000.

[10] H. Y. D. Yang, "Characteristics of guided and leaky waves on multilayer thin-film structures with planar material gratings," *IEEE Trans. Microwave Theory Tech.*, vol.45, pp.428-435, Mar. 1997.

[11] H. Y. D. Yang, "Surface-wave elimination in integrated circuits with periodic substrates," *Electromag.*, vol.20, no.2, pp.188-193, Mar./Apr. 2000.

[12] H. Y. D. Yang, R. Kim, and D. R. Jackson, "On the design of modeless integrated circuit substrate using planar periodic patches," *IEEE Trans. Microwave Theory Tech.*, vol.48, no.1, pp.233-239, Dec. 2000.

[13] R. J. Mittra, C. H. Chan, and T. A. Cwik, "Techniques for analyzing frequency selective surfaces-a review," *Proc. IEEE*, vol.76, no.12, pp.1593-1614, Dec. 1988.

[14] S. D. Gedney, J. F. Lee, and R. Mittra, "A combined FEM/MoM approach to analyze the plane wave diffraction by arbitrary gratings," *IEEE Trans. Microwave Theory tech.*, vol.40, no.2, pp.363-370, Feb. 1992.

[15] W. P. Pinello, R. Lee, and A. C. Cangellaris, "Finite element modeling of electromagnetic wave interactions with periodic dielectric structures," *IEEE Trans. Microwave Theory Tech.*, vol.42, no.12, pp.2294-2301, Dec. 1994.

[16] E. W. Lucas and T. P. Fontana, "A 3-D hybrid finite element/boundary element method for the unified radiation and scattering analysis of general infinite periodic arrays," *IEEE Trans. Antennas Propagat.*, vol.43, no.2, pp.145-153, Feb. 1995.

[17] H. Y. D. Yang, R. Diaz, and N. G. Alexopoulos, "Reflection and transmission of waves from multilayer structures with planar periodic material blocks," *J. Opt. Soc. Am. B*, vol.14, no.10, pp.2513-1521, Oct. 1997.

[18] H. Y. Yang, "A double-vector integral equation method for antenna interaction with photonic band-gap materials," *IEEE AP-S Int. Symp. Dig.*, pp.792-795, Montreal, Canada, July 1997.

[19] H. Y. D. Yang, "Field computational method for source interaction with periodic structure," *PIERS*, Taipei, Taiwan, Mar. 1999.

[20] H. Y. D. Yang, "An analytic array scanning method for field computation of sources within a periodic structure," *IEEE AP-S Int. Symp. Dig.*, Orlando FL, July 15, 1999.

[21] H. Y. D. Yang, "Analysis of Microstrip Dipoles on Planar Artificial Periodic Dielectric Structures," *IEEE AP-S Int. Symp. Dig.*, pp.1800-1803, Salt Lake City, July 2000.

[22] D. M. Pozar and D. H. Schaubert, "Analysis of infinite array of rectangular microstrip patches with idealized probe feeds," *IEEE Trans. Antennas propagat.*, vol.32, no.10, pp.1101-1107, Oct. 1984.

[23] H. Y. Yang and J. A. Castaneda, "Infinite phased arrays of microstrip antennas on generalized anisotropic substrates," *Electromag.*, vol.11, no.1, pp.107-124, 1991.

[24] S. T. Peng, T. Tamir, and H. Bertoni, "Theory of periodic dielectric waveguides," *IEEE Trans. Microwave Theory Tech.*, vol.23, no.1, pp.123-133, Jan. 1975.

[25] D. Sievenpiper, L. Zhang, R. F. Broas, N. G. Alexopoulos, and E. Yablonovitch, "High-impedance electromagnetic surfaces with a forbidden frequency band," *IEEE Trans. Microwave Theory Tech.*, vol.47, no.11, pp.2059-2074, Nov. 1999.

[26] D. M. Pozar, *Microwave Engineering*, Wiley, 1998.

[27] R. S. Elliott, *An Introduction to Guided-Waves and Microwave Circuits*, Prentice Hall, 1993.

[28] H. Y. Yang and N. G. Alexopoulos, "Gain enhancement methods for printed circuit antennas through multiple superstrates," *IEEE Trans. Antennas Propagat.*, vol.35, no.7, pp.860-863, July 1987.

[29] D. R. Jackson, A. A. Oliner, and A. Ip, "Leaky-wave propagation and radiation for a narrow beam multiplayer dielectric structure," *IEEE Trans. Antennas Propagat.*, vol.41, no.3, pp.344-348, Mar. 1993.

[30] E. R. Brown, C. D. Parker, and E. Yablonovitch, "Radiation properties of a planar antenna on a photonic-crystal substrate," *J. Opt. Soc. Am. B*, vol.10, no.2, pp.404-407, Feb. 1993.

[31] R. Coccioli, F. R. Yang, K. P. Ma, and T. Itoh, "Aperture-coupled patch antenna on a UC-PBG substrate," *IEEE Trans. Microwave Theory Tech.*, vol.47, no.11, pp.2123-2130, Nov. 1999.

[32] Qian, R. Coccioli, D. Sievenpiper, V. Radisic, E. Yablonovitch, and T. Itoh, "Microstrip patch antenna using novel photonic band-gap structures," *Microwave J.*, vol.42, no.1, pp.66-76, Jan. 1999.

[33] H. Y. D. Yang, N. G. Alexopoulos, and E. Yablonovitch, "Photonic band-gap materials for high-gain printed circuit antennas," *IEEE Trans. Antennas Propagat.*, vol.45, no.1, pp.186-187, Jan. 1997.

[34] R. E. Collin and F. J. Zucker, *Antenna Theory*, McGraw-Hill, 1969.

[35] H. Y. D. Yang and D. R. Jackson, "Theory of Line source radiation from a metal-strip grating dielectric-slab structure," *IEEE Trans. Antennas Propagat.*, vol.48, no.4, pp.556-564, Apr. 2000.

[36] R. A. Sigelmann and A. Ishimaru, "Radiation from aperiodic structures excited by an aperiodic source," *IEEE Trans. Antennas Propagat.*, vol.13, no.4, pp.354-366, May 1965.

[37] H. Y. D. Yang, "Theory of antenna radiation from photonic band-gap materials," *Electromag.*, vol.19, no.3, pp.255-276, May/June 1999.

[38] P. B. Katehi and N. G. Alexopoulos, "On the modeling of electromagnetically coupled microstrip antennas-the printed dipoles," *IEEE Trans. Antennas Propagat.*, vol.32, no.11, pp.1179-1186, Nov. 1984.

[39] R. Jackson and D. Pozar, "Full-wave analysis of microstrip open-end and gap discontinuities," *IEEE Trans. Microwave Theory Tech.*, vol.33, no.10, pp.1036-1042, Oct. 1985.

[40] H. Y. Yang and N. G. Alexopoulos, "A dynamic model for microstrip-slotline transition and related structures," *IEEE Trans. Microwave Theory Tech.*, vol.36, no.2, pp.286-293, Feb. 1988.

[41] H. Y. D. Yang, "Theory of a microstrip line on artificial periodic substrates," *IEEE Trans. Microwave Theory Tech.*, vol.47, no.5, pp.629-635, May 1999.

[42] C. N. Hu and C.-K. Tzuang, "Microstrip leaky-wave antenna array," *IEEE Trans. Antennas Propagat.*, vol.45, no.11, pp.1698-1699, Nov. 1997.

Chapter 13

FINITE ELEMENT METHODS FOR MICROWAVE ENGINEERING

Jin-Fa Lee
ElectroScience Laboratory
Department of Electrical Engineering
The Ohio State University
Columbus, Ohio, USA

Abstract This Chapter reviews recent advancements in finite element methods, particularly the H(curl)-conforming vector finite element methods(FEMs), for solving three-dimensional time-harmonic Maxwell's equations. Advances in six major technical areas have made commercial field simulation softwares, based upon vector FEMs, widely accepted in microwave engineering community for analyzing and designing RF/microwave devices. They are automatic mesh generation, H(curl)-conforming vector finite elements, fast and efficient matrix solution techniques, accurate mesh truncation methods, a posteriori error estimate and adaptive mesh refinement, and fast frequency sweep algorithms. Research and development activities in these areas are also briefly reviewed.

Keywords: finite element method, adaptive mesh refinement, matrix solution technique.

1. Introduction

This Chapter briefly reviews recent advancements in finite element methods(FEM) for solving three-dimensional time-harmonic Maxwell's equations. Appropriate references are also provided for further studies. Breakthroughs of finite element methods in recent years have made it possible to design RF/microwave devices using field simulation tools. Commercial softwares, based upon the use of FEMs, are now commonly used to analyze and design complicated RF/microwave circuits with success. At the core of this increasingly accepted FEM technology are the development of great user friendly graphical interfaces, affordable yet powerful computers, and the research and development efforts in six major technical areas which will be reviewed herein. These six areas are automatic mesh generation, H(curl)-conforming vector finite elements, accurate mesh truncation methods, fast and efficient matrix solution techniques, a posteriori error estimate and adaptive mesh refinements, and fast frequency sweep algorithms.

2. Automatic Mesh Generation

Mesh generation is known to play a critical role in solving electromagnetic problems using finite element methods. Stable mesh generation of complicated structures with extreme ratios between the smallest and largest feature sizes is a necessary step to achieve good convergence and accurate numerical solutions with conventional H(curl)-conforming vector FEMs.

For the purpose of automating the mesh generation process, triangles and tetrahedrals have overwhelming advantages over other types of elements because they are the simplexes in two and three dimensions, respectively. Furthermore, for many computational problems, use of unstructured grids offers many advantages over structured grids, such as permitting the tessellation of geometrically complicated domains or allowing the mesh density to be adapted according to the solution behavior.

Delaunay tessellation is a convenient and proven way to automatically discretize an arbitrary geometry into a group of triangles and tetrahedrals in two and three dimensions, respectively. Many commercial FEM packages have already successfully employed the Delaunay tessellation technique. Practically, there are two general approaches of implementing the Delaunay tessellation: the Bowyer-Watson algorithm [1] and the advancing front technique [2], [3]. The Bowyer-Watson algorithm is based upon incremental point insertion into a pre-existing Delaunay mesh, and subsequently requires initial meshing of a convex hull that is large enough to enclose all the vertices and grid nodes. On the other hand, the advancing front technique constructs the tessellation one tetrahedron at a time, beginning at the boundaries of domain, thus advancing or sweeping a front throughout the domain. The advancing front technique offers some advantages over the Bowyer-Watson algorithm. Tessellation of the convex hull of vertices and grid nodes is not needed. Thus, the meshing can be efficiently restricted to the domain to be meshed. It also means that grid nodes outside of the meshed domain never affect the meshing procedure. Thus, the advancing front technique can be used to construct a constrained Delaunay tessellation where the tetrahedral mesh conforms to the prescribed boundary triangulation.

Although the Delaunay tessellation can be used to generate an FEM mesh automatically, it does not guarantee that bad elements can be avoided. In particular, slivery tetrahedrals are frequently encountered in three-dimensional problems. Smoothing and swapping approaches are commonly employed to improve the mesh quality once a tetrahedral mesh is created. The smoothing approach moves the grid nodes around to improve the mesh quality while maintaining the mesh topology. Improvements in mesh quality for a tetrahedral mesh can also be achieved by local modifications based upon simple edge-face swapping primitives. Fig.13.1 shows two meshes generated using modified Delaunay tessellation, a combination of the Bowyer-Watson algorithm and a series of swapping and smoothing operations afterwards. In general, use of either the Bowyer-Watson algorithm or the advancing front technique in conjunction with smoothing and swapping operations do result in good-quality tetrahedral meshes even for complicated geometrical domains.

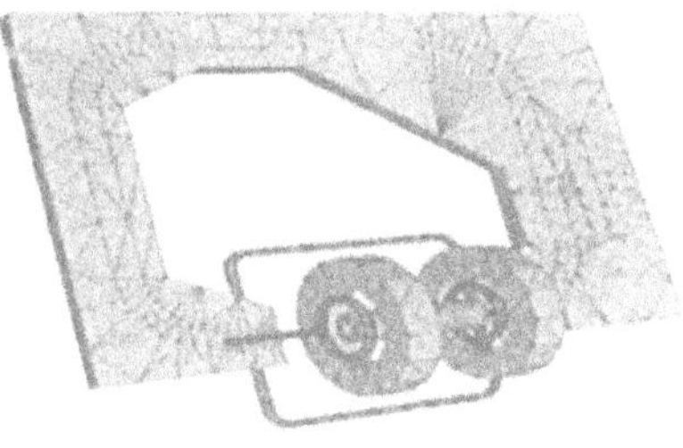

Figure 13.1. Sample meshes generated by using modified Delaunay tessellation algorithm.

3. E^p(curl) Vector Finite Elements

In 1980, Nedelec defined a set of vector finite element basis functions having the properties that they are complete to order p in the range space of the curl operator, that they conform to tangential but not normal continuity, and that they are unisolvent [4]. In this Chapter, we shall refer finite elements based on these functions the Nedelec elements, and denote the function space spanned by these elements as E^p(curl). The Nedelec elements have demonstrated their importance in the solution of electromagnetic field problems [5]. In particular, references show that these elements eliminate the problem of spurious modes that plague conventional node-based approximations of the vector wave equation derived from Maxwell's equations [5]-[7].

Numerous authors have derived alternative forms to the Nedelec bases, both for the low-order edge elements and for the high-order tangential vector elements. The goal of these constructions has often been to make the Nedelec bases interpolatory or hierarchical. Hiptmair recently provided a general abstract framework for systematic construction of these vector bases [8]. However, the overall performance of these elements varies drastically when matrix solution speed is considered. Researchers have found that the convergence of iterative matrix solution algorithms such as conjugate gradient algorithm varies considerably when the system of equations is created using different tangential vector finite elements. In [9], Sun, Lee and Cendes present a systematic approach to construct high-order E^p(curl) elements for multilevel finite element solution of electromagnetic wave problems where the multilevel corresponds to the order of basis functions. When these bases are used in the multilevel solutions, the resulting p-multilevel ILU preconditioned conjugate gradient method(MPCG) provides an optimal rate of convergence.

Fig.13.2 shows that use of high-order E^p(curl) elements results in much better numerical solutions with less computer resources for a rectangular cavity($8 \times 10 \times 16$ mm^3). Notice that higher-order basis functions are, in general, much more preferred when the numerical solutions are smooth. However, situation may not be clear when the numerical solutions involve field singularities. Since most electromagnetic problems have singular or nearly singular solutions, the hp-adaptive version of finite element method is suggested as the only viable technique to reach exponential rate of convergence [10].

4. Accurate Mesh Truncation Methods

In modeling unbounded electromagnetic problems, it is necessary to employ mesh truncation methods to confine the problem domain to a finite region. This has been a very active research area and many interesting approaches have been proposed. Here, three approaches are listed which I personally feel have the best potential: hybrid FEM and

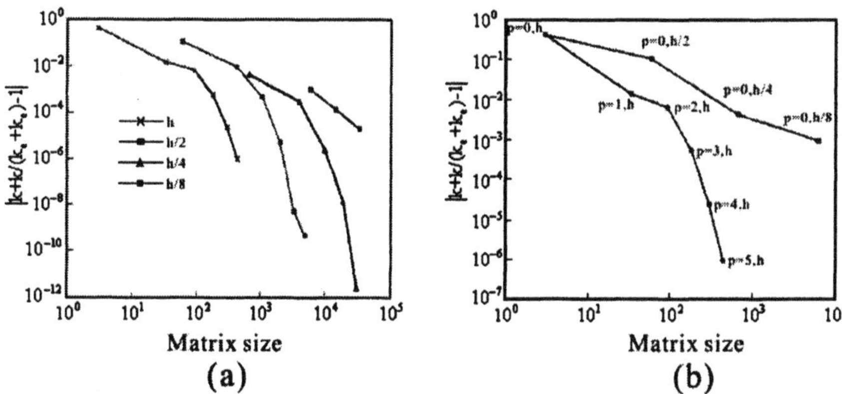

Figure 13.2. Lowest eigenvalue of a rectangular cavity of dimensions $8 \times 10 \times 16$ mm^3, (a) convergence with respect to *p*-refinement, and (b) comparison of *p* and *h*-refinements.

boundary-element method(BEM), Robin absorbing boundary conditions and perfectly matched layers(PMLs).

The hybrid FEM and BEM approach has been benchmarked as the exact mesh truncation method. In the computational electromagnetic community, Jin and Volakis have accomplished significant works on this subject [11], [12]. There are two major drawbacks. Since the truncation boundary is handled through BEM, dense and often nonsymmetric matrix blocks are created. For different configurations and applications such as the presence of ground planes and layered PCB, appropriate Green's functions need to be derived and implemented.

The Robin absorbing boundary conditions can be viewed as a combination of classical first-order absorbing boundary condition(ABC)/ impedance boundary condition(IBC) and integral equation/method of moments(MoM) [13]. The idea is to express the impedance at every vertex point on the truncation boundary through evaluation derived from integral equation. In this way, impedance on the truncation boundary will vary from location to location and can be conformed to the local wave impedance and subsequently results in very little reflection. Since the exact impedance at each vertex point is not known a priori, this procedure is carried out in an adaptive or iterative manner. Remarkable results have been obtained using this approach. However, the iterative process is more expensive and convergence is not guaranteed for all electromagnetic problems.

Perfectly matched layers were introduced into FDTD computation by Berenger in 1994 [14]. Since then, extensive research and development efforts have been spent to address various aspects of this approach. Among them, the anisotropic PML approach is most suitable for FEM implementation because it fulfills the PML property using anisotropic material tensors without splitting the Maxwell's equations [15]. Recently, Bardi and Cendes have automated the PMLs for unbounded scattering problems and used them together with the h-version adaptive mesh refinement such that the tetrahedral in the PML regions can be adjusted according to the numerical solutions [16]. In this way, extra overhead of the PMLs can be kept minimum. Very impressive results have been obtained using this approach. The major drawback is that occasionally, nonphysical resonances in the PML regions can contaminate the solution within the problem domain.

5. Fast and Efficient Matrix Solution Techniques

Three major advancements in recent years for solving sparse matrices are worth mentioning here. They are recursive bisection graph partitioning algorithm which greatly reduces the fill-ins during LU factorization, serial and parallel multi-frontal methods for sparse LU factorization, and p-multigrid conjugate gradient(pMGCG) method.

Traditional graph partitioning algorithms execute k-way partitioning of a graph such that the number of edges that are cut by partitioning is minimized, and each partition has an equal number of vertices. In [17], Karypis and Kumar present a multilevel recursive bisection graph partitioning algorithm. Their works result in a public domain high-quality graph partitioning package, METIS, which can be used to reorder a sparse matrix. Numerical experiments show that the METIS in general produces superior reordering that has much less fill-ins than other reordering algorithms.

The multi-frontal method of Duff and Reid is designed with regular memory access in the innermost loops [18]. Its kernel is one or more steps of LU factorization within each square, dense frontal matrix defined by the nonzero pattern of pivot row and column. These steps of LU factorization compute a submatrix of updated terms that are held within the frontal matrix until they are assembled into the frontal matrix of its parent in the assembly tree. The assembly tree controls the parallelism across multiple frontal matrices, while dense matrix operations provide parallelism and opportunity to use the highly efficient basic linear algebra subroutines(BLAS) on specific computers.

New hierarchical curl-conforming elements allow easy computation of a preconditioner to eliminate or at least weaken the indefiniteness of the resulting system matrix, and thus reduce the condition number of the system matrix [9]. When these bases are used in the multilevel solutions, where the multilevels correspond to orders of basis functions, the resulting p-multigrid conjugate gradient method provides an optimal rate of convergence [9], [19]. Here, p refers to the order of elements. Fig.13.3 shows the comparison of the new pMGCG method to conventional incomplete Choleski conjugate gradient(ICCG) method in solving a matrix equation resulted from employing the $p = 2$ curl-conforming elements for a passive microwave component. It is observed that the number of iterations is drastically reduced. Numerical experiments show that the number of iterations needed by the pMGCG is basically constant, regardless of the order of basis or matrix.

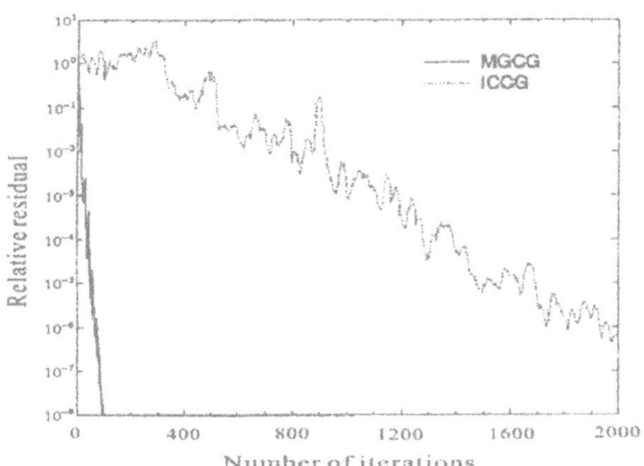

Figure 13.3. Comparison of matrix convergence behaviors using ICCG and pMGCG methods.

5.1 pMGCG as a Schwarz Method

We apply the Schwarz method, often used in domain decomposition area, to form an efficient preconditioner for the conjugate gradient algorithm with p-type of inite elements($p = 2$ in current case). Schwarz introduced the earliest domain decomposition method in 1870. Though not originally intended as a numerical method, classical alternating Schwarz method has been used intensively to solve elliptic boundary-value problems in domains that are union of two subdomains by alternatingly solving the same elliptic boundary problem restricted to each individual subdomain. However, our work focuses on the use of p-type finite elements, and treats each p group as a domain. Thus, the Schwarz method employed here is a nonoverlapping Schwarz method, or more specifically, the Schur complement method. In particular, we use a multiplicative Schwarz preconditioner which can be viewed as a nonoverlapping block Gauss-Seidel preconditioner. Even without conjugate gradient acceleration, the multiplicative method can take far less iterations than the additive version [20]. Conventionally, multilevel methods are associated with a nested grid that includes multilevel of grids [21]. In this work, the multilevel algorithm employs a single grid but multilevel of basis functions. We may think of the approach presented here as a p-refinement multilevel method instead of the more traditional h-refinement multilevel method, where h refers to the element size. We employ the Schur factorization to obtain an approximate inverse of the system matrix and treat it as a preconditioner in the conjugate gradient method. The resulting procedure is the MPCG method.

5.2 Multilevel Preconditioned Conjugate Gradient Method

Reordering unknowns from the $p = 1$ group(vertex gradients and cotree edge elements) to the $p = 2$ group, the system matrix, $\bar{\bar{A}}$, is partitioned into a 2×2 block matrix as

$$\bar{\bar{A}} = \begin{bmatrix} \bar{\bar{A}}_{11} & \bar{\bar{A}}_{12} \\ \bar{\bar{A}}_{21} & \bar{\bar{A}}_{22} \end{bmatrix} \tag{13.1}$$

Since the vector basis functions are of very different nature, diagonal entries in the system matrix will vary drastically. Therefore, it is always a good idea to diagonally scale the system matrix first before applying matrix solution techniques. The system matrix, $\bar{\bar{A}}$, in (13.1) is the matrix after scaling. Thus, all the diagonal entries are unity. The Schur factorization process begins by factorizing matrix $\bar{\bar{A}}$ into a product form as

$$\bar{\bar{A}} = \begin{bmatrix} \bar{\bar{I}} & 0 \\ \bar{\bar{A}}_{21} \cdot \bar{\bar{A}}_{11}^{-1} & \bar{\bar{I}} \end{bmatrix} \cdot$$

$$\begin{bmatrix} \bar{\bar{A}}_{11} & 0 \\ 0 & \bar{\bar{A}}_{22} - \bar{\bar{A}}_{21} \cdot \bar{\bar{A}}_{11}^{-1} \cdot \bar{\bar{A}}_{12} \end{bmatrix} \cdot \begin{bmatrix} \bar{\bar{I}} & \bar{\bar{A}}_{11}^{-1} \cdot \bar{\bar{A}}_{12} \\ 0 & \bar{\bar{I}} \end{bmatrix}$$

Next, apply two incomplete Choleski factorizations to $\bar{\bar{A}}_{11}$ and $\bar{\bar{A}}_{22}$ with threshold values of 10^{-5} and 10^{-2}, respectively, namely,

$$\bar{\bar{A}}_{11} = \bar{\bar{C}}_{11}^t \cdot \bar{\bar{C}}_{11} + \bar{\bar{E}}_1, \qquad \text{with } ||E_1|| \leq 10^{-5}$$
$$\bar{\bar{A}}_{22} = \bar{\bar{C}}_{22}^t \cdot \bar{\bar{C}}_{22} + \bar{\bar{E}}_2, \qquad \text{with } ||E_2|| \leq 10^{-2}$$

which means that we drop the entries in the factorization process that are smaller than 10^{-5} (10^{-2}) in $\bar{\bar{A}}_{11}$ ($\bar{\bar{A}}_{22}$). With $\bar{\bar{C}}_{11}$ and $\bar{\bar{C}}_{22}$ obtained, the preconditioner takes the form of

$$\bar{\bar{M}} = \begin{bmatrix} \bar{\bar{I}} & 0 \\ \bar{\bar{A}}_{21} \cdot (\bar{\bar{C}}_{11}^t \cdot \bar{\bar{C}}_{11})^{-1} & \bar{\bar{I}} \end{bmatrix} \cdot$$
$$\begin{bmatrix} \bar{\bar{C}}_{11}^t \cdot \bar{\bar{C}}_{11} & 0 \\ 0 & \bar{\bar{C}}_{22}^t \cdot \bar{\bar{C}}_{22} \end{bmatrix} \cdot \begin{bmatrix} \bar{\bar{I}} & (\bar{\bar{C}}_{11}^t \cdot \bar{\bar{C}}_{11})^{-1} \cdot \bar{\bar{A}}_{12} \\ 0 & \bar{\bar{I}} \end{bmatrix}$$

Its inverse is readily available as

$$\bar{\bar{M}}^{-1} = \begin{bmatrix} \bar{\bar{I}} & -(\bar{\bar{C}}_{11}^t \cdot \bar{\bar{C}}_{11})^{-1} \cdot \bar{\bar{A}}_{12} \\ 0 & \bar{\bar{I}} \end{bmatrix} \cdot$$
$$\begin{bmatrix} (\bar{\bar{C}}_{11}^t \cdot \bar{\bar{C}}_{11})^{-1} & 0 \\ 0 & (\bar{\bar{C}}_{22}^t \cdot \bar{\bar{C}}_{22})^{-1} \end{bmatrix} \cdot \begin{bmatrix} \bar{\bar{I}} & 0 \\ -\bar{\bar{A}}_{21} \cdot (\bar{\bar{C}}_{11}^t \cdot \bar{\bar{C}}_{11})^{-1} & \bar{\bar{I}} \end{bmatrix}$$

which can be used as the preconditioner in the preconditioned conjugate gradient method to constitute the MPCG method that is adopted in the current work. Its performance to solve the matrix equation derived from Maxwell's equations using vector finite elements($p = 2$) is truly remarkable as listed in Tables 13.1 and 13.2.

Splitting of edge elements using the tree-cotree approach is implemented. It is possible just to employ the edge elements in its entirety and still apply the MPCG method to solve the resulting matrix equations. The convergence is in general as good as in the current approach. However, there are two major advantages to advocate splitting. Firstly, the splitting allows us to employ incomplete Choleski factorization instead of complete factorization generally used for pure edge elements.

Table 13.1. Computational performance of MPCG method for a microstrip band-pass filter using the h-version adaptive mesh refinement, N: number of unknowns, NZ: total number of nonzero entries in the matrix, $NZ(C_{11}/C_{22})$: total number of nonzero entries in the C_{11}/C_{22} matrix, $NZ(\mathrm{PC}) = NZ(C_{11})+NZ(C_{22})$, NI: number of CG iterations to achieve the relative residual of 10^{-2}, τ_1: CPU time needed to construct preconditioner, τ_2: CPU time for solving the matrix equation.

N	NZ	$NZ(C_{11})$	$NZ(C_{22})$	$NZ(\mathrm{PC})$	NI	τ_1	τ_2
349,536	8,608,239	3,853,213	3,359,532	7,212,745	8	239	72
467,416	11,542,797	5,258,523	4,490,901	9,749,424	13	366	161
626,886	15,533,111	7,254,245	6,018,860	13,273,105	13	582	209
837,254	20,764,633	9,744,901	8,035,521	17,780,422	15	826	305
1,115,976	27,669,388	12,759,915	10,697,658	23,457,573	12	1,222	558
1,484,224	36,739,841	16,574,881	14,213,080	30,787,961	18	1,668	853

Secondly, by using pure edge elements for the $p = 1$ block, failure to converge is observed when very small elements are present in the problem domain, whereas such convergence problem is not present in our approach. This is related to the large round-off errors for poor conditioned matrices mainly due to low-frequency instability and Gaussian elimination without extensive pivoting.

Table 13.2. Computational performance of MPCG method for a bow-tie antenna using the h-version adaptive mesh refinement, notations are the same as in Table 13.1.

N	NZ	$NZ(C_{11})$	$NZ(C_{22})$	$NZ(\mathrm{PC})$	NI	τ_1	τ_2
139,860	3,426,490	4,218,775	2,013,728	6,232,503	16	573	70
177,774	4,326,729	4,644,427	2,434,750	7,079,177	8	525	47
233,754	5,674,294	5,953,185	3,016,030	8,969,215	9	778	66
309,616	7,516,031	6,986,562	3,775,765	10,762,327	8	819	78
412,942	10,035,904	8,855,653	4,797,907	13,653,560	7	1,106	90

5.3 Numerical Results

Table 13.1 lists the computational performance of MPCG method during the adaptive mesh refinement process when applying to solve matrix equation for a microstrip band-pass filter. It is shown that the size of preconditioner is always smaller than that of the original matrix and the number of iterations is kept very small(nearly constant), regardless of the matrix size.

The second example is a bow-tie antenna operating at 5 GHz. The mesh is adaptively refined, and the performance of MPCG method for solving the matrix equation is listed in Table 13.2. In this case, there are more nonzero entries in the preconditioner than in the original matrix. Also, it takes much longer CPU time to compute the preconditioner than in the first example with the same number of unknowns. However, a few key features remain in this example. The size of preconditioner is still less than twice the original matrix size, even in the worst case. The number of CG iterations is even slightly reduced. With 412,000 unknowns for the bow-tie antenna, it takes about 20 minutes to solve the matrix equation using a Pentium III/550 MHz PC Linux with 1 GB RAM.

6. Adaptive Mesh Refnements

6.1 Bilinear Form

To study a multiport microwave device, we apply the Galerkin method to the vector wave equation derived from Maxwell's equations. By assuming a two-port device and no Γ_∞ present in the domain, the resulting bilinear form can be written as

$$B(\bar{v},\bar{u}) = a(\bar{v},\bar{u}) - k^2(\bar{v},\bar{u})_\Omega$$

Weak solution of the multiport microwave device is a vector function $\bar{u} \in H_0(\text{curl})$ such that

$$a(\bar{v},\bar{u}) - k^2(\bar{v},\bar{u})_\Omega = -jk\eta\langle\bar{v},\bar{J}_1^p\rangle_{\Gamma_1^p}, \qquad \forall\bar{v} \in H_0(\text{curl}) \tag{13.2}$$

In the discretized version or the FEM process, finite dimensional approximation is obtained through seeking $\bar{u}^h \in V^h \subset H_0(\text{curl})$ such that

$$a(\bar{v}^h,\bar{u}^h) - k^2(\bar{v}^h,\bar{u}^h)_{\Omega^h} = -jk\eta\langle\bar{v}^h,\bar{J}_1^p\rangle_{\Gamma_1^p}, \qquad \forall\bar{v}^h \in V^h$$

where

$$a(\bar{v}^h,\bar{u}^h) = \int_{\Omega^h} \nabla\times\bar{v}^h \cdot \frac{1}{\mu_r}\nabla\times\bar{u}^h d\Omega$$

$$(\bar{v}^h,\bar{u}^h)_{\Omega^h} = \int_{\Omega^h} \bar{h}\cdot\epsilon_r\bar{u}^h d\Omega$$

$$\langle\bar{v}^h,\bar{u}^h\rangle_\Gamma = \int_\Gamma \bar{v}^h\cdot(\hat{n}\times\hat{n}\times\bar{u}^h)d\Gamma$$

In current implementation, the finite dimensional subspace is formed by $p = 2$ hierarchical Nedelec curl-conforming elements of the first kind. Assume that the port mesh is sufficiently fine so that interpolation error of the input current is negligible.

Theorem 1. If $\bar{e}^h = \bar{u} - \bar{u}^h$ is the error in the finite element solution, then we have $a(\bar{e}^h, \bar{u}^h) - k^2(\bar{e}^h, \bar{u}^h)_{\Omega^h} = 0$.

6.2 Error Functional

In modeling multiport microwave devices, we are usually interested in the S-parameters. Once the electromagnetic fields have been solved, S-parameters of the devices can usually be computed through certain surface integrals. In particular, the reflection coefficient, S_{11}, can be expressed as

$$S_{11} = 1 - \langle \bar{u}, \bar{J}_1^p \rangle_{\Gamma_1^p} \tag{13.3}$$

where Γ_1^p is the input port, $\bar{u}$ is the electric field, and $\bar{J}_1^p$ is the surface current of the fundamental mode(input waveform) flowing into the computational domain. Consequently, the error functional can be derived from (13.3) as

$$\mathrm{err}(\bar{u}^h) = S_{11}^h - S_{11} = \langle \bar{u}, \bar{J}_1^p \rangle_{\Gamma_1^p} - \langle \bar{u}^h, \bar{J}_1^p \rangle_{\Gamma_1^p} = \langle \bar{e}^h, \bar{J}_1^p \rangle_{\Gamma_1^p}$$

Notice that $\bar{e}^h = \bar{u} - \bar{u}^h \in H_0(\mathrm{curl})$, and with (13.2), we have

$$\mathrm{err}(\bar{u}^h) = \frac{-1}{jk\eta}\left[a(\bar{u}, \bar{e}^h) - k^2(\bar{u}, \bar{e}^h)_\Omega\right]$$

By applying Theorem 1, the error functional can be further expressed as

$$\mathrm{err}(\bar{u}^h) = \frac{-1}{jk\eta}\left[a(\bar{e}^h, \bar{e}^h) - k^2(\bar{e}^h, \bar{e}^h)_\Omega\right] \tag{13.4}$$

which will be the basis for deriving the refinement index for each tetrahedron in the FEM mesh. Eq.(13.4) indicates that the error functional is the reaction of the error field, $\bar{e}^h = \bar{u} - \bar{u}^h$.

6.3 Refinement Index

To find a computable refinement index, we begin with

$$\begin{aligned} & a(\bar{e}^h, \bar{e}^h) - k^2(\bar{e}^h, \bar{e}^h)_{\Omega^h} \\ & = \sum_K \int_K \left(\nabla \times \bar{e}^h \cdot \frac{1}{\mu_r} \nabla \times \bar{e}^h - k^2 \bar{e}^h \cdot \epsilon_r \bar{e}^h\right) dK \\ & = \sum_K \int_K \bar{e}^h \cdot \left(\nabla \times \frac{1}{\mu_r} \nabla \times \bar{e}^h - k^2 \epsilon_r \bar{e}^h\right) dK \\ & \quad + \sum_\Delta \int_\Delta \bar{e}^h \cdot \left(\frac{1}{\mu_r} \nabla \times \bar{e}^h \times \hat{n}\right)_\Delta d\Delta \end{aligned}$$

where K indicates tetrahedral element in the mesh, Δ indicates triangular face, and $(\hat{n} \times \bar{v})_\Delta$ denotes the jump of $\hat{n} \times \bar{v}$ across Δ. To further reduce the expression, notice that

$$\begin{aligned}
&\nabla \times \frac{1}{\mu_r} \nabla \times \bar{e}^h - k^2 \epsilon_r \bar{e}^h \Big|_K \\
&= -\left(\nabla \times \frac{1}{\mu_r} \nabla \times \bar{u}^h - k^2 \epsilon_r \bar{u}^h \right)\Big|_K = jk\eta \bar{j}_K \\
&\left[\frac{1}{\mu_r} \nabla \times \bar{e}^h \times \hat{n} \right]_\Delta = -\left[\frac{1}{\mu_r} \nabla \times \bar{u}^h \times \hat{n} \right]_\Delta = jk\eta \bar{j}_\Delta
\end{aligned}$$

where $\bar{j}_K$ and $\bar{j}_\Delta$ are volume and surface residual current densities for element K and triangular face Δ, respectively. Note that in current implementation, all the triangles that are on the port boundaries are shared with the prism elements in the PMA regions. In this way, all the triangles in the tetrahedral mesh can be viewed as internal faces and the jump of tangential magnetic fields can be computed accordingly across all the triangles.

However, the only exception is the triangles on perfect electric conductors(PECs). Since the exact currents on these PECs are unknown, there is no way of knowing what the residual surface currents are on these surfaces. Nonetheless, we can still define a volume residual index, α_K, based on an estimate of the reaction of element K as

$$\begin{aligned}
\alpha_K = \frac{\omega}{4\pi} \Bigg\{ &\left| \int_K \rho_K(\bar{r}) \int_K \frac{e^{-jkR}}{\epsilon R} \rho_K(\bar{r'}) dK' dK \right| \\
&+ \left| \int_k \bar{j}_K(\bar{r}) \cdot \int_K \frac{\mu e^{-jkR}}{R} \bar{j}_K(\bar{r'}) dK' dK \right| \Bigg\}
\end{aligned} \tag{13.5}$$

where $R = |\bar{r} - \bar{r'}|$ and the charge density, ρ_K, is defined by $\nabla \cdot \bar{j}_K = -j\omega\rho_K$. Similarly, a surface residual index, α_Δ, can be calculated for each triangular face, Δ, as

$$\begin{aligned}
\alpha_\Delta = \frac{\omega}{4\pi} \Bigg\{ &\left| \int_\Delta \rho_\Delta(\bar{r}) \int_\Delta \frac{e^{-jkR}}{\epsilon_{\text{avg}} R} \rho_\Delta(\bar{r'}) d\Delta' d\Delta \right| \\
&+ \left| \int_\Delta \bar{j}_\Delta(\bar{r}) \cdot \int_\Delta \frac{\mu e^{-jkR}}{R} \bar{j}_\Delta(\bar{r'}) d\Delta' d\Delta \right| \Bigg\}
\end{aligned} \tag{13.6}$$

Through the volume and surface residual indices, a refinement index for element K is defined as

$$\theta_K = \alpha_K + \sum_{i=0}^{3} \frac{\alpha_K}{\alpha_K + \alpha_i} \alpha_{\Delta_i}$$

where Δ_i is the ith triangular facet, and α_i is the volume residual index of the ith neighbor of element K. As can be seen from (13.5) and (13.6), computation of volume and surface residual indices involve significant approximations. Consequently, they may not represent accurate a posteriori error estimate. However, through various numerical experiments, it is found that they do give good information about the ranking of actions in tetrahedra. Moreover, from the physical insights of error functional, the current refinement index consists of the contributions from residual charges and residual currents. Performances of this new approach for various numerical examples are very promising. Fig.13.4 shows the final tetrahedral mesh resulted from applying the new h-version adaptive mesh refinement procedure to a high-temperature superconductor(HTS) band-pass filter centered at 4 GHz. The associated frequency response is shown in Fig.13.5. Note that very fine elements are created around the microstrip lines automatically during the process. Since this filter has a very high Q factor, it is very difficult to simulate its frequency response satisfactorily without the adaptive process.

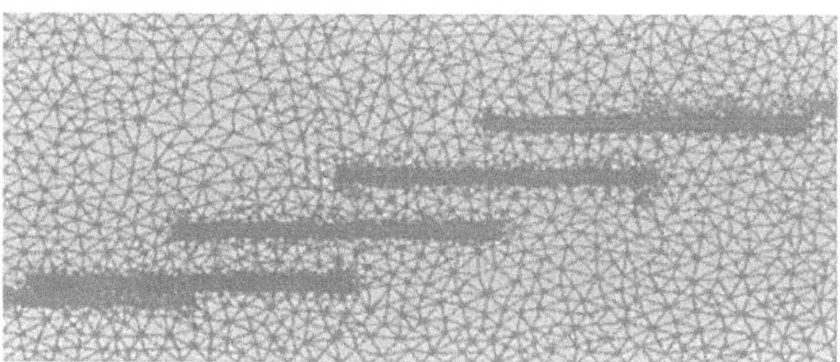

Figure 13.4. **Final mesh of a microstrip band-pass filter through h-version adaptive mesh refinement.**

7. Fast Frequency Sweep Algorithms

An order reduction technique called Galerkin asymptotic waveform evaluation(GAWE) or multipoint Galerkin asymptotic waveform evaluation(MGAWE), if multiple expansion points are considered simultaneously, can be used to reduce the matrices resulted from the application of E^p(curl) elements to a smaller space while still accurately approximating the characteristics of original system matrix equation [22]. The resulting solution procedure of using the MGAWE to solve the FEM equations allows for wideband frequency simulations with significant reduction in the total computation time. For example, frequency response of the HTS band-pass filter shown in Fig.13.5 is obtained using seven moments at the center frequency of 4.25 GHz.

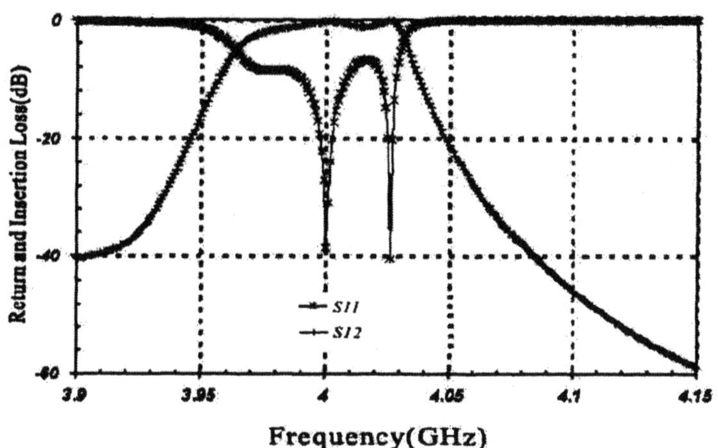

Figure 13.5. Frequency response of the band-pass filter depicted in Fig.13.4.

8. Conclusions

Many new advancements have been made in the past few years. Consequently, it is becoming a common practice to use field simulation packages to analyze and design RF/microwave devices. In this Chapter, the author briefly reviews some of the major technical contributions, based upon personal experiences. It is foreseen that progress in the FEMs for computational electromagnetics in the near future will evolve in a much faster pace, and more robust, efficient and versatile software tools will become available. Moreover, significant research and development efforts have also been launched not only on analyses but also on syntheses.

References

[1] J.-F. Lee and R. Dyczij-Edlinger, "Automatic mesh generation using a modified Delaunay tessellation," *IEEE Antennas Propagat. Mag.*, vol.39, pp.34-45, Feb. 1997.

[2] D. J. Mavriplis, "Unstructured mesh generation and adaptivity," *NASA ICASE*, Rep.195069, Apr. 1995.

[3] M. L. Merriam, "An efficient advancing front algorithm for Delaunay triangulation," *AIAA* 91-0792, Jan. 1991.

[4] J. C. Nedelec, "Mixed finite elements in R3," *Numer. Math.*, vol.35, pp.315-341, 1980.

[5] Z. J. Cendes, "Vector finite elements for electromagnetic field computation," *IEEE Trans. Magn.*, vol.27, pp.3953-3966, 1991.

[6] A. Bossavit and J. C. Verite, "A mixed FEM-BIEM method to solve three-dimensional Eddy current problem," *IEEE Trans. Magn.*, vol.18, pp.431-435, Mar. 1982.

[7] J.-F. Lee, D.-K. Sun, and Z. J. Cendes, "Tangential vector finite elements for electromagnetic field computation," *IEEE Trans. Magn.*, vol.27, pp.4032-4035, Sept. 1991.

[8] R. Hiptmair, "Canonical construction of finite elements," *Math. Comput.*, vol.68, pp.1325-1346, 1999.

[9] D.-K. Sun, J.-F. Lee, and Z. J. Cendes, "Construction of nearly orthogonal Nedelec bases for rapid convergence with multilevel preconditioned solvers," *SIAM J. Sci. Comput.*, vol.23, pp.1053-1076, 2001.

[10] L. Vardapetyan, "*hp*-adaptive finite element method for electromagnetics with applications to waveguiding structures," Ph. D. thesis, Univ. Texas, Austin, 1999.

[11] J. M. Jin and J. L. Volakis, "A finite element-boundary integral formulation for scattering by three-dimensional cavity-backed apertures," *IEEE Trans. Antennas Propagat.*, vol.39, pp.97-104, Jan. 1991.

[12] J. M. Jin, J. L. Volakis, and J. D. Collins, "A finite element-boundary integral method for scattering and radiation by two- and three-dimensional structures," *IEEE Antennas Propagat. Mag.*, vol.33, pp.22-32, June 1991.

[13] S. Alfonzetti, G. Borzi, and N. Salerno, "Iteratively-improved robin boundary conditions for the finite element solution of scattering problems in unbounded domains," *Int. J. Numer. Meth. Engng.*, vol.42, pp.601-629, June 1998.

[14] J. P. Berenger, "A perfectly matched layer for the absorption of electromagnetic waves," *J. Comput. Phys.*, vol.114, pp.185-200, 1994.

[15] Z. S. Sacks, D. M. Kingsland, R. Lee, and J.-F. Lee, "A perfectly matched anisotropic absorber for use as an absorbing boundary

condition," *IEEE Trans. Antennas Propagat.*, vol.43, pp.1460-1463, Dec. 1995.

[16] Bardi and Cendes, "New directions in HFSS for designing microwave devices," *Microwave J.*, Aug. 1998.

[17] G. Karypis and V. Kumar, "METIS 4.0: Unstructured graph partitioning and sparse matrix ordering system," Tech. Rep., http://www.cs.umn.edu/ metis, Computer Sci. Dept., Univ. Minnesota, 1998.

[18] S. Duff and J. K. Reid, "The multifrontal solution of indefinite sparse symmetric linear equations," *ACM Trans. Math. Softw.*, vol.9, pp.302-325, 1983.

[19] G. Peng, R. Dyczij-Edlinger, and J.-F. Lee, "Hierarchical methods for solving matrix equations from TVFEMs for microwave components," *IEEE Trans. Magn.*, vol.35, pp.1474-1477, May 1999.

[20] J. H. Bramble, J. E. Pasciak, J. Wang, and J. Xu, "Convergence estimates for product iterative methods with applications to domain decompositions and multigrid," *Math. Comput.*, vol.57, pp.1-21, 1991.

[21] R. Hiptmair, "Multigrid method for Maxwell's equations," *SIAM J. Numer. Anal.*, vol.36, pp.204-225, 1998.

[22] R. D. Slone, R. Lee, and J.-F. Lee, "Multipoint Galerkin asymptotic waveform evaluation for model order reduction of frequency domain FEM electromagnetic radiation problems," *IEEE Trans. Antennas Propagat.*, vol.49, pp.1504-1513, Oct. 2001.

[23] J. P. Webb and B. Forghani, "Hierarchical scalar and vector tetrahedra," *IEEE Trans. Magn.*, vol.29, pp.1495-1498, 1993.

[24] S. Savage and A. F. Peterson, "Higher-order vector finite elements for tetrahedral cells," *IEEE Trans. Microwave Theory Tech.*, vol.44, pp.874-879, June 1996.

[25] P. Webb, "Hierarchical vector basis functions of arbitrary order for triangular and tetrahedral finite elements," *IEEE Trans. Antennas Propagat.*, vol.47, pp.1244-1253.

[26] D. K. Sun, Z. J. Cendes, and J. F. Lee, "Adaptive mesh refinement, h-version, for solving multiport microwave devices in three dimensions," *IEEE Trans. Magn.*, special issue for COMPUMAG-Sapporo, Japan, 1999.

Chapter 14

DIELECTRIC RESONATOR ANTENNAS ON PRINTED CIRCUIT BOARDS

Jean-Fu Kiang and Ching-I Cho
Department of Electrical Engineering and
Graduate Institute of Communication Engineering
National Taiwan University
Taipei, Taiwan, ROC

Abstract In this Chapter, radiation properties of cylindrical dielectric resonator antennas are briefly reviewed. The input impedance and radiation pattern with various geometrical aspect ratios, dielectric constants and feeding structures are studied. Three feeding structures are considered: coaxial-fed probe, coaxial-fed conformal microstrip and CPW-fed conformal microstrip. The last feeding structure can be easily implemented and integrated with MMIC while retaining the advantages of DR antennas such as low cost, small size and free of conductor loss.

Keywords: dielectric resonator, printed circuit board, resonant frequency, input impedance, radiation pattern, coplanar waveguide, conformal microstrip, probe.

1. Review of Dielectric Resonators

Due to fast evolution of VLSI technologies, microwave circuits can be significantly reduced in size and cost, making them competitive to commercial applications. For example, microstrip lines and strip lines have replaced bulky waveguides and rigid coaxial lines as interconnections in many microwave systems. Resonators are key components to build oscillators, tuners, and so on. Low loss and temperature-stable dielectric resonators can be used to replace waveguide filters in applications such as satellite communication where extremely low loss and high temperature stability are required. Analysis and design of dielectric resonators have been discussed extensively in literatures [1]-[32]. The size of a dielectric resonator can be considerably smaller than that of a resonant cavity operating at the same frequency, provided that the relative dielectric constant of the former is substantially higher than unity.

In general, resonant modes in a dielectric resonator can be categorized into transverse electric(TE), transverse magnetic(TM) and hybrid electromagnetic(HEM) modes [2], [3]. Understanding these modes is helpful not only in selecting proper modes for a specific application, but also in avoiding undesired modes which may degrade performance for certain applications. When the dielectric constant is around 40 ϵ_0, more than 95% of the stored electric energy and more than 60% of the stored magnetic energy are confined within the resonator. Rest of the energy is distributed in the air around resonator, and decays rapidly away from its surface.

Other than radiation applications, dielectric resonators are usually shielded in a metallic enclosure to prevent radiation loss [2]. Its resonant frequency is altered when the metallic enclosure is push closer to the dielectric resonator, and the frequency deviation can be estimated by using perturbation technique. The resonant frequency decreases(increases) when a resonant cavity is protruded inward where the dominant stored energy is electric(magnetic).

For radiation applications, a resonator stores an equal amount of average electric and magnetic energies at resonant frequencies. Thus, input impedance of the driving probe is purely real since the reactance is proportional to the difference between average electric and magnetic stored energies [2].

The Q factor is an important parameter which relates the total stored energy in a resonator to its energy dissipation as [4]

$$Q = \frac{\omega_0 W_0}{P_d}$$

where W_0 is the total stored energy, P_d is the power dissipation, and ω_0 is the resonant frequency. The Q factor of certain dielectric resonators can be as high as 10,000. At lower frequencies, the Q factor is usually between 50 and 500. The resonator's bandwidth, $\Delta\omega$, is related to its Q factor by [4].

$$Q = \frac{\omega_0}{\Delta\omega}$$

As a hollow waveguide section forms a resonant cavity, a section of dielectric waveguide can form a dielectric resonator. Properties of a dielectric resonator are closely related to its field distributions which are similar to those in two-dimensional dielectric waveguides. A dielectric rod waveguide is capable of guiding modes with specific transverse wave numbers [33]. For a cylindrical dielectric rod, the separation-of-variables technique can be used to solve the scalar wave equation for the field distributions.

Dielectric resonators of different shapes like rectangular [5]-[7], cylindrical [8], [9], spherical [10]-[15], ring [26], [17] and others have been proposed. Even for a dielectric resonator with simple geometrical shape, an exact solution to Maxwell's equations is considerably more complicated than that for hollow metallic cavities. Usually, resonant frequencies of resonant modes can only be obtained by applying complicated numerical techniques. However, resonant frequencies of an isolated dielectric resonator can be estimated by enclosing the resonator with perfect magnetic conductors(PMC). Empirical formulas relating resonant frequencies to geometrical parameters and dielectric constant can also become available by regression analysis on numerical results.

By applying numerical techniques to solve for the fields in a dielectric resonator, sophisticated Green's function or a large number of unknowns may be required. It is of practical interest to find an approximate solution of the electromagnetic fields in a dielectric resonator to estimate certain parameters with reasonable accuracy. A first-order model is proposed by enclosing the dielectric resonator by perfect magnetic conductors to form a cavity. The resonant frequencies are then obtained using similar techniques as for analyzing waveguide sections. Such a model renders better prediction when the dielectric constant is higher.

However, the estimated frequencies are 20% or more off the measured results. Thus, a second-order model is proposed by Cohn [18]. The cylindrical PMC walls are retained, but the end caps are removed and replaced by two air-filled hollow waveguides extending to infinity, which operate below cutoff. The fields in these waveguides decay exponentially away from the interface with the dielectric resonator. Two parallel metal plates may also be placed to truncate the air-filled PMC waveguides.

The second-order model is suitable for describing the fields of an isolated dielectric resonator which may be used as an antenna, and the radiation efficiency is related to its Q factor. For example, the measured Q factor of an isolated resonator operating at $TE_{01\delta}$ mode is about 50. If an unloaded Q factor of about 5,000 is required, a typical solution is to place the dielectric resonator on top of a substrate. The entire fixture is then placed within a metallic enclosure which prevents external fields from interfering the dielectric resonator, and also reduces its radiation.

In the Cohn's model, fields of the dielectric resonators are zero outside of PMC walls. In reality, the tangential fields outside of the resonator surface should be equal to that on the inner side of that surface, and decreases outward. This implies that part of the stored energy sticks around the dielectric resonator, which is entirely neglected in the previous two models. To further improve Cohn's second-order model, the same electric and magnetic fields inside the dielectric resonator are retained. However, PMC walls are removed, and the field distributions outside of the dielectric resonator are so postulated that the tangential electric fields are continuous across the dielectric surface. The overall electric and magnetic fields are then used to determine the resonant frequency by using perturbation technique.

In practical applications, dielectric resonator is preferred to operate in a given frequency band, with its field distributions relevant to a specific mode. However, resonant frequencies of certain undesired modes may stay in proximity. Hence, it is of practical importance to determine the resonant frequencies and field distributions of these modes in order to design proper excitation mechanism. Resonant frequencies can also be adjusted by tuning objects near the resonator, such as metallic cavity walls, metallic tuning screws or dielectric tuning rods [19]. It will be more convenient for tuning purpose if resonant frequencies and relevant field distributions can be accurately predicted.

Integral equation technique is useful to analyze dielectric resonators. There are two typical integral equation formulations: volume integral equation and surface integral equation. In the volume integral equation, the dielectric contrast between resonator and background medium is represented by equivalent polarization currents. In the surface integral equation, the dielectric resonator is replaced by equivalent electric and magnetic surface currents on its surface [20]. Either of these formulations requires proper Green's function which is usually a dyadic. The integral equations are usually solved numerically by applying method of moments(MoM) with suitable basis and testing functions. The resonant frequencies are then obtained by searching zeros of the matrix determinant.

The surface integral equation approach proves useful in computing the field distributions, resonant frequency and Q factor associated with specific resonant modes. It is more computationally efficient than the volume integral equation approach when the resonator is made of homogeneous material. The volume integral equation approach can be implemented in a similar manner, and is more suitable to study resonators with inhomogeneous interior. Both volume and surface integral equation approaches can be extended to include the effects of complete or partial metallic enclosures, multiple resonators, tuning screws, coupling with microstrip lines, and so on.

Low-temperature cofired ceramics(LTCC) is an advanced ceramic material to fabricate resonators. LTCC has very low loss and controllable temperature stability in mechanical and electrical attributes [34]. Recently, ceramics with dielectric constant of $100\epsilon_0$, good temperature stability and low dielectric losses became available [35]. In describing ceramic materials, conductivity, σ, is used to account for the Ohmic loss, and the polarization loss can be described by appending an imaginary part to the dielectric constant, namely $\epsilon = \epsilon' - j\epsilon''$. By reorganizing the Ampere's law as $\nabla \times \bar{H} = [j\omega\epsilon' + (\sigma + \omega\epsilon'')]\bar{E}$, the loss tangent can be defined as the ratio of imaginary part to real part of the effective dielectric constant as

$$\tan\delta = \frac{\epsilon''}{\epsilon'} + \frac{\sigma}{\omega\epsilon'}$$

where the first part on the right hand side dominates at microwave frequencies. When a resonant cavity is uniformly filled with lossy dielectric, the inverse of loss tangent is equal to the Q factor of that cavity.

To integrate dielectric resonators with microwave circuits, it is helpful to know the coupling between resonators and microwave circuits like transmission lines or waveguides, and the mutual coupling between adjacent dielectric resonators. For example, Fig.14.1(a) shows the coupling of the commonly used $\mathrm{TE}_{01\delta}$ mode of a cylindrical dielectric resonator with a microstrip line [2]. The electric field distribution of $\mathrm{TE}_{01\delta}$ mode inside the resonator looks like concentric circles about the cylindrical axis, resembling the field distribution of a magnetic dipole. The coupling between microstrip line and resonator is significant when the equivalent magnetic dipole is close to and perpendicular to the microstrip line so that the magnetic fluxes of resonator link with those of microstrip line. The equivalent circuit of these coupled components is shown in Fig.14.1(b), which is further simplified to that in Fig.14.1(c). The coupling coefficient, κ, is defined as the ratio between power losses external and internal, respectively, to the system. From Fig.14.1(c), we have $\kappa = Z(\omega_0)/2Z_0$ where $Z(\omega_0)$ is the equivalent impedance at the resonant frequency and Z_0 is the characteristic impedance of transmission line. At the resonant frequency, the equivalent impedance, Z, becomes real. The external quality factor, Q_e, can be related to the internal quality factor, Q_0, by $Q_e = Q_0/\kappa$.

Almost any practical filter type can be implemented using waveguides or microstrip lines. More versatility can be provided by incorporating dielectric resonators in distributed filter designs. A typical Chebyshev

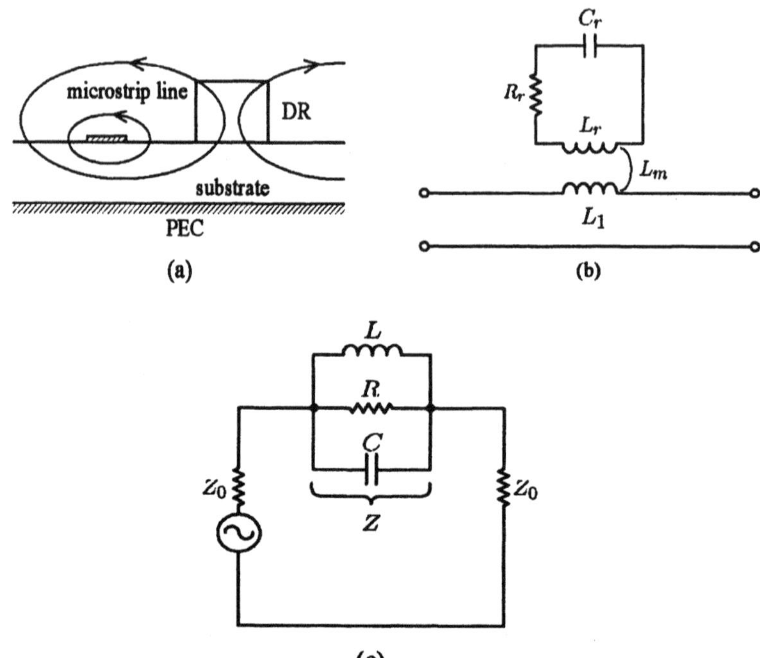

Figure 14.1. Field coupling between dielectric resonator and microstrip line, (a) geometrical configuration, (b) equivalent circuit, (c) equivalent circuit observed from the driving source, $L = L_m/L_r$, $R = L_r/(\omega_0^2 L_m)$, $R = Q_0\omega_0 L_m^2/L_r$.

design procedure starts from specifying the filter type, center frequency and bandwidth. Other types of Chebyshev filters can be transformed from the low-pass prototype with the following attenuation specifications [36]

$$L_A(\omega') = \begin{cases} 10\log_{10}\left[1 + \epsilon\cos^2\left(n\cos^{-1}(\omega'/\omega_1')\right)\right], & \omega' < \omega_1' \\ \\ 10\log_{10}\left[1 + \epsilon\cosh^2\left(n\cosh^{-1}(\omega'/\omega_1')\right)\right], & \omega' > \omega_1' \end{cases}$$

where $\epsilon = 10^{L_{Ar}/10} - 1$, with L_{Ar} the pass-band ripple in dB, ω' is the operating frequency of the low-pass prototype, ω_1' is the cutoff frequency, n is the order of Chebyshev polynomial, which is equal to the number of reactive elements in hardware implementation. The frequency response and equivalent circuits are shown in Fig.14.2.

Band-pass filter can be derived from the low-pass prototype by the following frequency transformation

$$\frac{\omega'}{\omega_1'} = \frac{1}{\xi}\left(\frac{\omega}{\omega_0} - \frac{\omega_0}{\omega}\right)$$

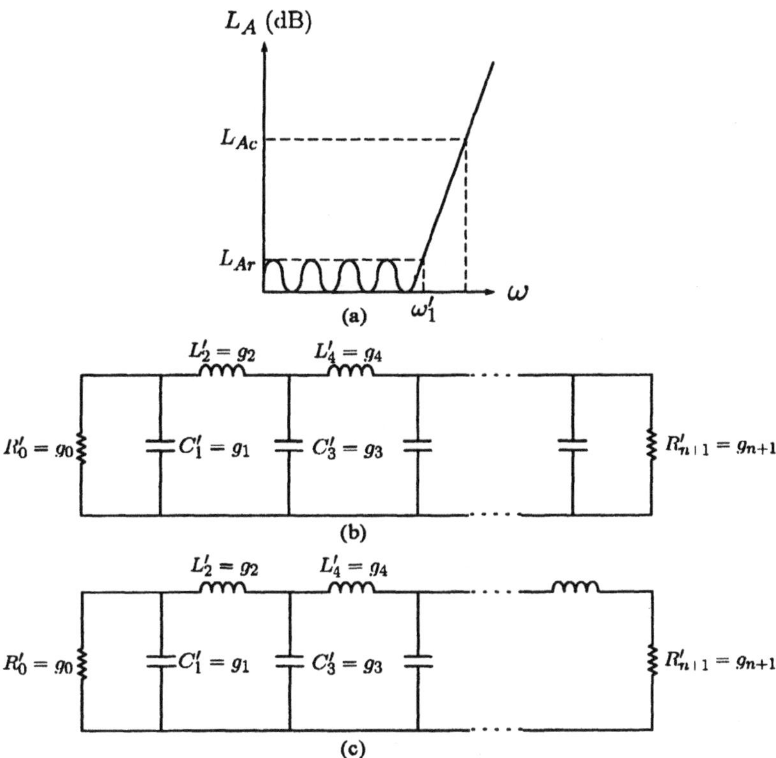

Figure 14.2. Prototype of an nth order Chebyshev low-pass filter, (a) specification of frequency response, (b) equivalent circuit when n is odd, (c) equivalent circuit when n is even.

where ξ is the fractional bandwidth defined as $\xi = (\omega_2 - \omega_1)/\omega_0$, ω_1 and ω_2 are the lower and upper corner frequencies, respectively, of the band-pass filter, and ω_0 is the center frequency with $\omega_0 = \sqrt{\omega_1 \omega_2}$. Fig.14.3 shows a ladder-type of equivalent circuit used to implement such a band-pass filter. The ladder-type circuit can be further transformed to other circuit configurations more suitable for implementation.

Microwave oscillators are widely used in radars, communication devices, navigation equipments, and so on. General requirements on oscillators include low noise, compact size, low cost, high efficiency, temperature stability and reliability. Before 1960s, massive klystrons or magnetron tubes requiring huge power supplies are used as microwave sources. In the late 1960s, microwave oscillators using Gunn and impact avalanche transit time(IMPATT) diodes were invented. In the 1970s, bipolar transistor oscillators and GaAs MESFET extended the operat-

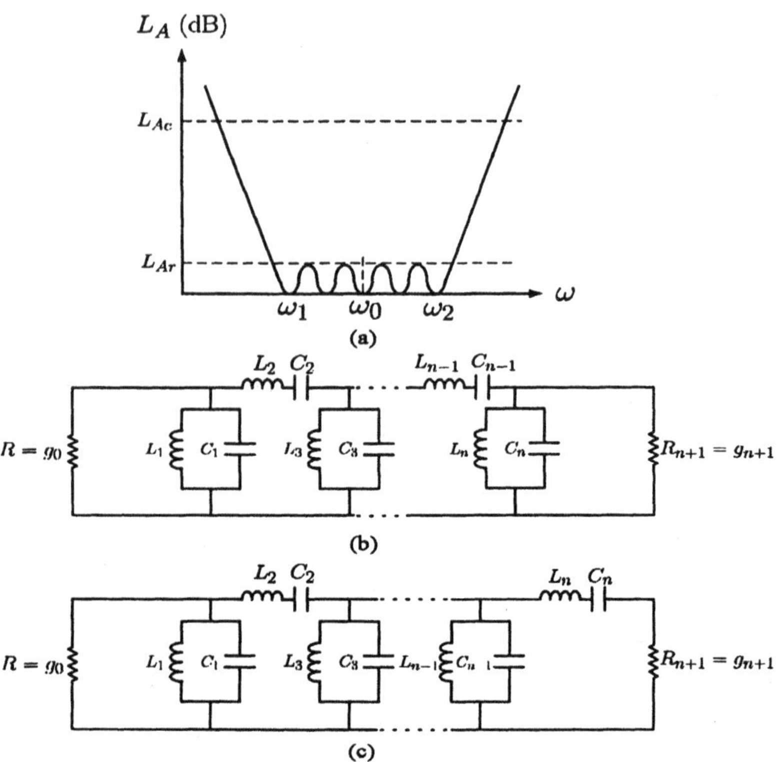

Figure 14.3. **Transformed nth order Chebyshev band-pass filter, (a) specification of frequency response, (b) equivalent circuit when n is odd, (c) equivalent circuit when n is even.**

ing frequency to the millimeter wave band. Due to their high Q factor, small size, and excellent integrability with microwave integrated circuits, dielectric resonators can be used as frequency-selective elements to implement stable transistor oscillator for single tone or narrow band applications [21]. Transistor oscillators can be implemented using either bipolar or GaAs FET. GaAs FET oscillators can operate up to 60 GHz, while oscillators using bipolar transistors can not operate above X-band. Typically, bipolar oscillator has 6 to 10 dB less frequency modulation(FM) noise close to the carrier as compared to GaAs FET oscillator.

Dielectric resonator can be used in two different ways. In the dielectrically stabilized oscillator, the dielectric resonator is used as a passive stabilization element properly coupled to a free-running transistor oscillator [37]. Passive stabilization mechanism is possible only for those free-running oscillators whose oscillation frequency is sensitive to load impedance variation. In the stable transistor dielectric resonator oscillator, the dielectric resonator is used as a circuit element in the feedback or matching network to determine the oscillation frequency [38].

In recent years, the frequency range of interest has gradually progressed to the 100-300 GHz band. As the conductor loss increases with increasing frequency, radiation efficiency of conventional metallic antennas drops significantly. Hence, most of the antennas in microwave band cannot be directly scaled down with increasing frequency. In addition, traditional microwave antennas with miniaturized size are expensive to manufacture. Thus, dielectric resonator antennas found potential applications in the millimeter wave band and above.

The use of low-loss dielectric resonators as radiating elements was proposed in 1983 [1]. Dielectric resonator can be placed on top of a substrate and driven by either probe or microstrip line through slot [22]. In designing miniaturized microwave filters and oscillators, high-permittivity materials($\epsilon_r \simeq$ 25-100) are preferred. In contrast, dielectric resonator antennas usually require low-permittivity materials($\epsilon_r \leq 10$) in order to achieve high radiation efficiency.

Many research endeavors have been devoted to the study of dielectric resonator antennas in the past two decades. It is found that dielectric resonators of rectangular [5]-[7], cylindrical [8], [9], spherical [10]-[15], ring [26], [17] shapes can be designed to radiate efficiently through proper feeding structures. Different types of feeding structures such as coaxial probe [1], microstrip-fed aperture [22], microstrip line [23] and coplanar waveguide [24] have been proposed. Bandwidth enhancement techniques have also been studied. By stacking a parasitic dielectric resonator on top of a directly-fed dielectric resonator, antenna with more than 25 % of bandwidth for SWR < 2 is achieved [25]. Bandwidth widening of a CPW coupled dielectric resonator antenna was reported using a similar stacked configuration [26]. In these two cases, attention was paid to the broadside $\text{HEM}_{11\delta}$ mode. For an annular ring dielectric resonator antenna operating in the end-fire $\text{TM}_{01\delta}$ mode, it is shown that the bandwidth can be increased by introducing an air gap between dielectric resonator and ground plane [26].

Similarly, bandwidth can also be increased for hemispherical dielectric resonator antenna with a hollow air gap [13] or with dielectric coating [14]. In [17], it is shown that inserting air gaps between dielectric resonators or between the driven dielectric resonator and ground plane of a stacked annular ring dielectric resonator antenna significantly increases its impedance bandwidth. Dielectric resonator antennas with circular polarization have been reported [6], [27], [28]. In designing array of dielectric resonator antennas, it is observed that the mutual impedance between two hemispherical dielectric resonator antennas may be significant [15]. Finite dielectric resonator arrays have also been presented [7], [9], [29]. In addition to applications in the millimeter wave band, dielectric resonator antennas with high permittivity($\epsilon_r > 80$) have been proposed as low-profile antennas in the lower microwave frequencies(1-10 GHz) [30]-[32].

2. Resonant Modes with PMC Walls

Fig.14.4 shows the configuration of a cylindrical dielectric resonantor placed above an infinite PEC ground plane. A first-order approximate solution for the fields can be obtained by enclosing the dielectric cylinder with perfect magnetic conductors(PMC). Hence, the fields can be categorized into transverse electric(TE_z) and transverse magnetic(TM_z) modes with the potential functions of

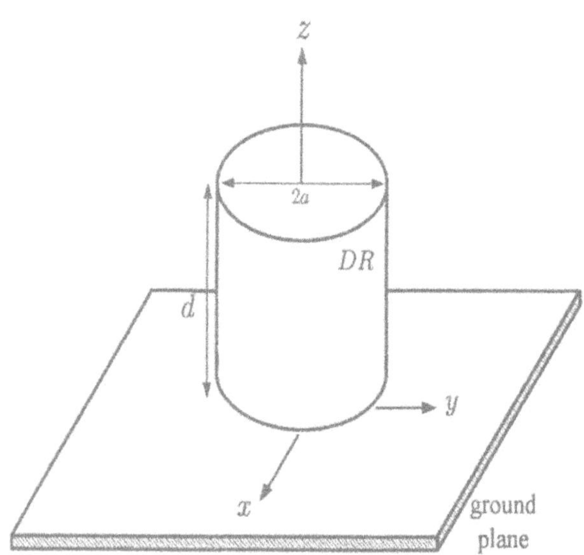

Figure 14.4. Geometry of a dielectric cylinder above an infinite PEC ground plane.

$$\psi^{\rm TE}_{npm} = J_n\left(\frac{\chi_{np}}{a}\rho\right)\left\{\begin{array}{c}\sin n\phi\\ \cos n\phi\end{array}\right\}\sin\left[\frac{(2m+1)\pi z}{2d}\right]$$

$$n = 0,1,2,\cdots, p = 1,2,3,\cdots, m = 0,1,2,\cdots$$

$$\psi^{\rm TM}_{npm} = J_n\left(\frac{\chi'_{np}}{a}\rho\right)\left\{\begin{array}{c}\sin n\phi\\ \cos n\phi\end{array}\right\}\cos\left[\frac{(2m+1)\pi z}{2d}\right]$$

$$n = 1,2,3,\cdots, p = 1,2,3,\cdots, m = 0,1,2,\cdots$$

where $J_n(\cdot)$ is the nth order Bessel function of the first kind, a and d are the radius and length of dielectric cylinder, respectively, χ_{np} and χ'_{np} are zeros of $J_n(\cdot)$ and $J'_n(\cdot)$, respectively, namely, $J_n(\chi_{np}) = 0$ and $J'_n(\chi'_{np}) = 0$. The separation relation leads to an expression for resonant frequency of the npm mode as

$$f^{\alpha}_{npm} = \frac{1}{2\pi a\sqrt{\mu\epsilon}}\sqrt{\beta^2 + \left[\frac{\pi a}{2d}(2m+1)\right]^2}$$

where $\beta = \chi_{np}$ with $\alpha =$ TE, and $\beta = \chi'_{np}$ with $\alpha =$ TM. TM_{110} mode is the dominant mode with the lowest resonant frequency where $n = 1$, $p = 1$, $m = 0$ and $\chi'_{11} = 1.841$.

To calculate the far field, first apply the equivalence principle, $\bar{M} = \bar{E} \times \hat{n}$, on the dielectric surface to derive the magnetic surface current, where $\hat{n}$ is a unit normal vector on the dielectric surface. Next, transform these magnetic surface currents to the spherical coordinates to facilitate the derivation of far field [1]. The electric vector potentials are then expressed as [33]

$$F_\alpha(r,\theta,\phi) = \frac{e^{-jk_0 r}}{4\pi r}\iiint \rho' d\rho' d\phi' dz' M_\alpha(\theta,\phi;\rho',\phi',z') \\ e^{jk_0[\rho'\sin\theta\cos(\phi-\phi')+z'\cos\theta]}, \qquad \alpha = \theta,\phi$$

where k_0 is the free-space wave number. The explicit forms of F_θ and F_ϕ are

$$\begin{aligned}
F_\theta(r,\theta,\phi) &= (-1)^m(2m+1)C_{1n} \\
&\left\{ n\left[I_{n(n-1)}(\chi'_{np}) + I_{n(n+1)}(\chi'_{np})\right] + \frac{\chi'_{np}}{2a}\left[I_{(n-1)(n-1)}(\chi'_{np})\right.\right. \\
&\qquad \left. -I_{(n-1)(n+1)}(\chi'_{np}) - I_{(n+1)(n-1)}(\chi'_{np}) + I_{(n+1)(n+1)}(\chi'_{np})\right] \\
&\quad +2nj^{n+1}J_n(\chi'_{np})k_0\sin\theta J_n(k_0 a\sin\theta)D_{1m} \\
&\quad \left. +\frac{\chi'^2_{np}}{a}J_n(\chi'_{np})\left[j^{n-1}J_{n-1}(k_0 a\sin\theta) - j^{n+1}J_{n+1}(k_0 a\sin\theta)\right]D_{1m}\right\} \\
F_\phi(r,\theta,\phi) &= (-1)^m C_{2n} \\
&\left\{ I_{n(n-1)}(\chi'_{np}) - I_{n(n+1)}(\chi'_{np}) + \frac{\chi'_{np}}{2a}\left[I_{(n-1)(n-1)}(\chi'_{np})\right.\right. \\
&\qquad \left. +I_{(n-1)(n+1)}(\chi'_{np}) - I_{(n+1)(n-1)}(\chi'_{np}) - I_{(n+1)(n+1)}(\chi'_{np})\right] \\
&\quad \left. +\frac{\chi'^2_{np}}{a}J_n(\chi'_{np})\left[j^{n-1}J_{n-1}(k_0 a\sin\theta) + j^{n+1}J_{n+1}(k_0 a\sin\theta)\right]D_{1m}\right\}
\end{aligned}$$

where

$$
\begin{aligned}
C_{1n} &= -j\frac{\pi^2}{\omega\epsilon d}\frac{e^{-jk_0 r}}{4\pi r}\cos(k_0 d\cos\theta)\sin n\phi \\
C_{2n} &= -j\frac{\pi^2}{\omega\epsilon d}\frac{e^{-jk_0 r}}{4\pi r}\cos(k_0 d\cos\theta)\cos n\phi \\
D_{1m} &= \left[\frac{(2m+1)^2\pi^2}{4d^2} - k_0^2\cos^2\theta\right]^{-1} \\
I_{nm}(\alpha) &= j^m \int_0^a J_n\left(\frac{\alpha}{a}\rho'\right) J_m(k_0\rho'\sin\theta)\rho' d\rho'
\end{aligned}
\tag{14.1}
$$

The far fields can then be derived from $\bar{F}$ [33].

3. Radiation Properties of Dielectric Resonantor Antennas

Fig.14.5 shows a dielectric resonator on top of an infinite ground plane, driven by a current probe. The current probe is extended from the inner conductor of coaxial cable beneath the ground plane. First, we study characteristics of the dielectric resonator without feeding probes. An approximate model assuming perfect magnetic conductor walls predicts the resonant frequencies and radiation patterns reasonably well compared with measured results. However, the predictions become less accurate when the permittivity of dielectric resonator is increased. More rigorous full-wave model is needed to analyze the near-field properties like input impedance which are more sensitive to field distributions.

Simulations are conducted using the HFSS package which is based on full-wave approach. Resonant frequencies of cylindrical dielectric resonator with different permittivities enclosed by PMC walls are calculated and compared with those obtained by using the PMC model. The simulation results shown in Fig.14.6 match well with those of PMC model. It is shown that the resonant frequencies decrease with increasing permittivity.

Fig.14.7 shows the resonant frequencies of the same dielectric resonator which is placed on top of an infinity ground plane and exposed to free space. The resonant frequencies are slightly higher than those in Fig.14.6. Hybrid HEM modes are also observed, which do not exist in the PMC model.

Fig.14.8 shows the radiation patterns of a dielectric resonator with different aspect ratios. Symmetry of the field patterns is related to the feed probe location at $\theta = 0^o$ as shown in Fig.14.5. The field is essentially omnidirectional on the x-z plane for cylinders with higher a/d ratios. A dip

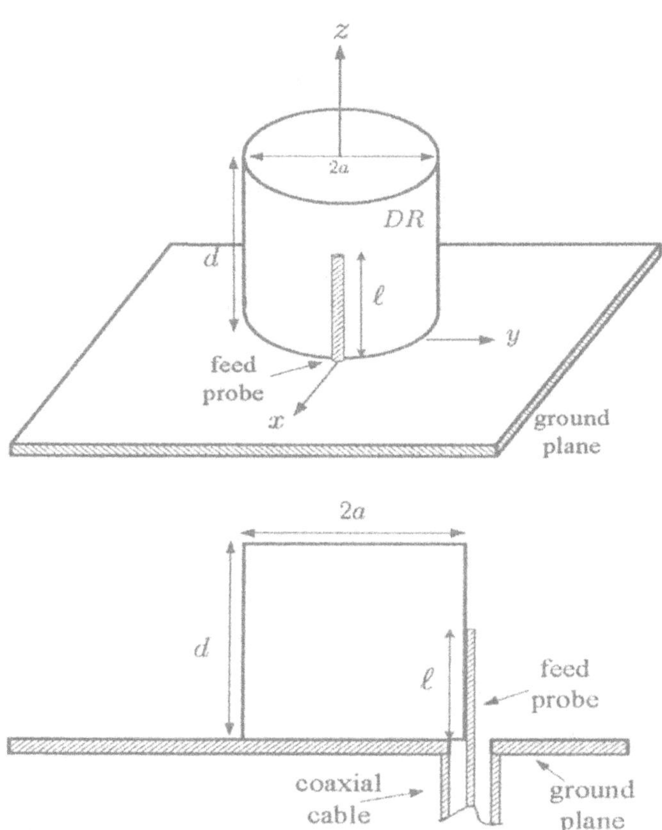

Figure 14.5. Geometry of dielectric resonator antenna above PEC ground plane.

appears in the axial($\hat{z}$) direction for smaller a/d ratios. Fig.14.9 shows the simulated input impedance compared with the measured results in [1].

Various parameters are checked to verify the validity of our simulation results as listed in Table 14.1. Polygonal tube with 12 sides are used to approximate the cylindrical tube and probe. The symbol PMC indicates that the dielectric resonator is enclosed by PMC walls, PML means that perfectly matching layer is used as absorbing boundary condition(ABC) to model the dielectric resonator exposed to free space, Im = 0 indicates that the resonant frequencies are defined as the frequencies at which the probe input impedance becomes purely real. The symbol c-c means that the polygonal tube and probe are attached to each other corner to corner in the model, the symbol s-s means that they are attached side by side,

Table 14.1. Resonant frequencies of a dielectric resonator above a PEC ground plane, $a/d = 0.5$, $\ell = 1.54$ cm, probe radius is 1 mm, coaxial cable impedance is 50 Ω, dielectric resonator is approximated by a 12-side polygon, $a_1 = 1.27$ cm, $a_2 = 1.283$ cm.

ϵ_r	(1)HFSS [1]-theory $a = a_1$ PMC	(2)HFSS $a = a_1$ PML	(3)HFSS $a = a_2$ PMC	(4)HFSS $a = a_2$ PML	(5)[1]-measure (Im = 0) $a = a_2$	(6)dev. (5)-(4)
15.2	1.93 GHz	2.19 GHz	1.95 GHz	2.18 GHz	2 GHz	−0.18 GHz (−8.3 %)
8.9	2.52 GHz	2.78 GHz	2.54 GHz	2.76 GHz	2.62 GHz	−0.14 GHz (−5.1 %)
6.6	2.93 GHz	3.14 GHz	2.95 GHz	3.12 GHz	2.95 GHz	−0.17 GHz (−5.4 %)
4.5	3.55 GHz	3.63 GHz	3.53 GHz	3.62 GHz	3.45 GHz	−0.17 GHz (−4.6 %)

ϵ_r	(7)HFSS (c-c) (Im=0) $a = a_1$	(8)dev. (7)-(2)	(9)HFSS (s-s) (Im=0) $a = a_1$	(10)dev. (9)-(2)	(11)HFSS (c-c) (Im = 0) $a = a_2$	(12)dev. (11)-(5)
15.2	2.14 GHz	−0.05 GHz (−2.2 %)	2.09 GHz	−0.1 GHz (−4.6 %)	2.12 GHz	−0.12 GHz (−5.6 %)
8.9	2.61 GHz	−0.17 GHz (−6.1 %)	2.6 GHz	−0.18 GHz (−6.5 %)	2.59 GHz	−0.03 GHz (−1.1 %)
6.6	2.98 GHz	−0.16 GHz (−5.1 %)	2.83 GHz	−0.31 GHz (−9.8 %)	2.96GHz	−0.01 GHz (−0.33 %)
4.5	3.56 GHz	−0.07 GHz (−1.9 %)	3.44 GHz	−0.19 GHz (−5.2 %)	3.52 GHz	−0.07 GHz (−1.9 %)

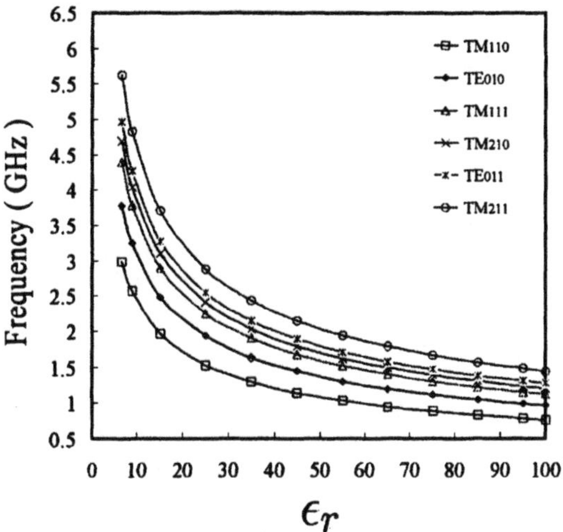

Figure 14.6. Resonant frequencies of the first few resonant modes of a dielectric resonator on top of an infinite PEC ground plane, the dielectric resonator is enclosed by PMC on sidewalls and top surface, $a = 1.27$ cm, $d = 2.54$ cm.

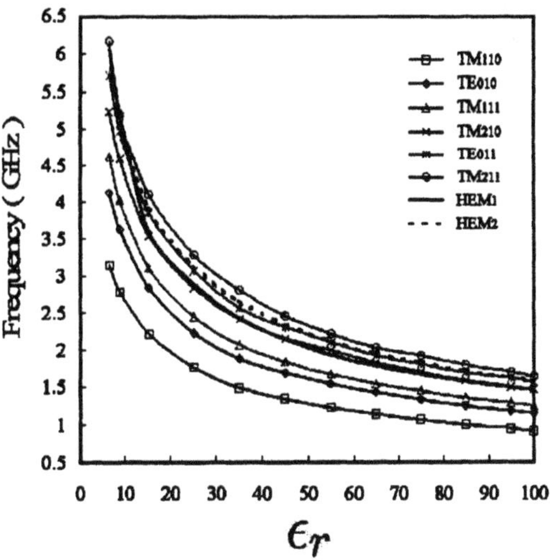

Figure 14.7. Resonant frequencies of the first few resonant modes of a dielectric resonator on top of an infinite PEC ground plane, $a = 1.27$ cm, $d = 2.54$ cm.

dev. is the difference of results between different approaches. Deviation of the corner-to-corner model is smaller than that of the side-by-side

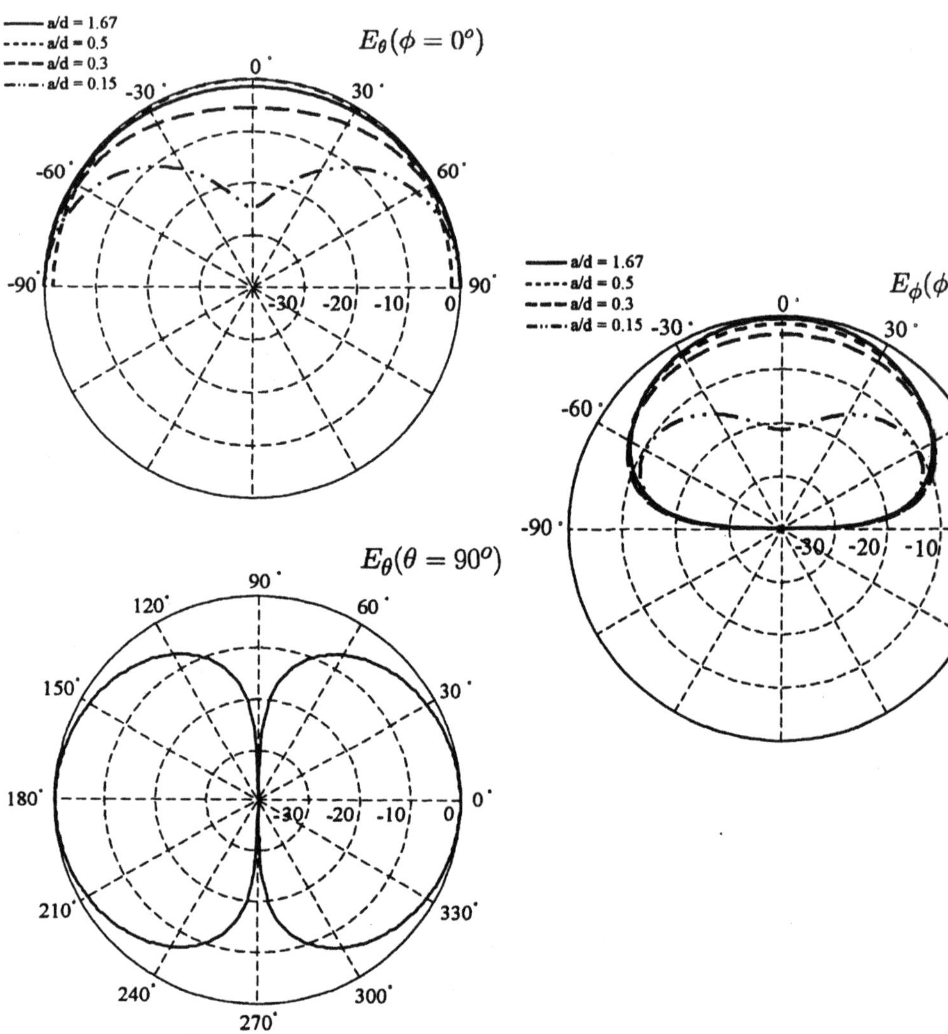

Figure 14.8. Radiation patterns of dielectric resonator on top of an infinite PEC ground plane, at the resonant frequency of TM_{110} mode, $\epsilon_r = 8.9$, (a) E_θ on x-z plane, (b) E_ϕ on y-z plane, (c)E_θ on x-y plane.

model because the separation between probe and dielectric resonator axis in the former model is closer to the physical configuration than the latter model. Polygonal tube of 12 sides is used to approximate the cylindrical tube and probe in the following discussions. Column (6) indicates that the resonant frequencies predicted by using HFSS match reasonably well with the measurements in [1].

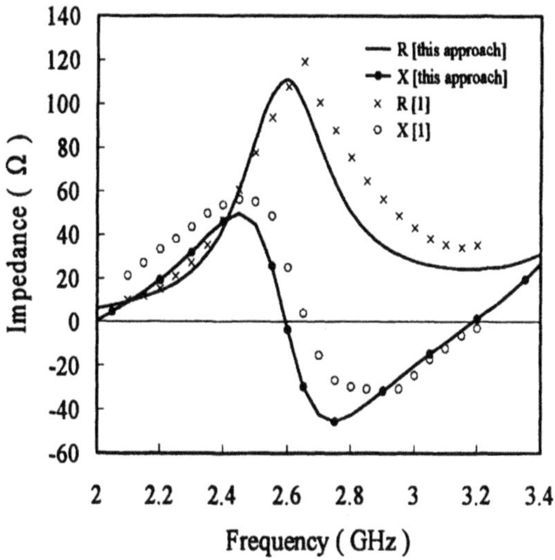

Figure 14.9. Simulated and measured impedance of dielectric resonator driven by coaxial probe, $\epsilon_r = 8.9$, $a = 1.283$ cm, $d = 2.566$ cm, $\ell = 1.54$ cm, probe radius is 1 mm, coaxial cable impedance is 50 Ω.

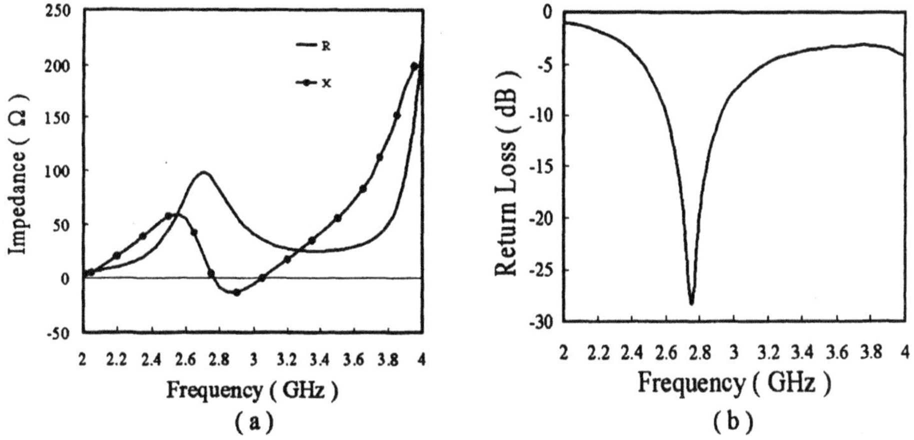

Figure 14.10. Feeding characteristics of dielectric resonator above a finite PEC ground plane of 15 cm × 15 cm, driven by coaxial probe, $\epsilon_r = 8.9$, $a = 1.283$ cm, $d = 2.566$ cm, coaxial cable impedance is 100 Ω, (a) input impedance, (b) return loss.

Input resistance at the first resonant frequency is about 110 Ω as shown in Fig.14.9. Hence, a fictitious coaxial cable of 100 Ω is chosen to achieve a better impedance match. As shown in Fig.14.10, the return loss is significantly reduced at the resonant frequency of TM_{110} mode.

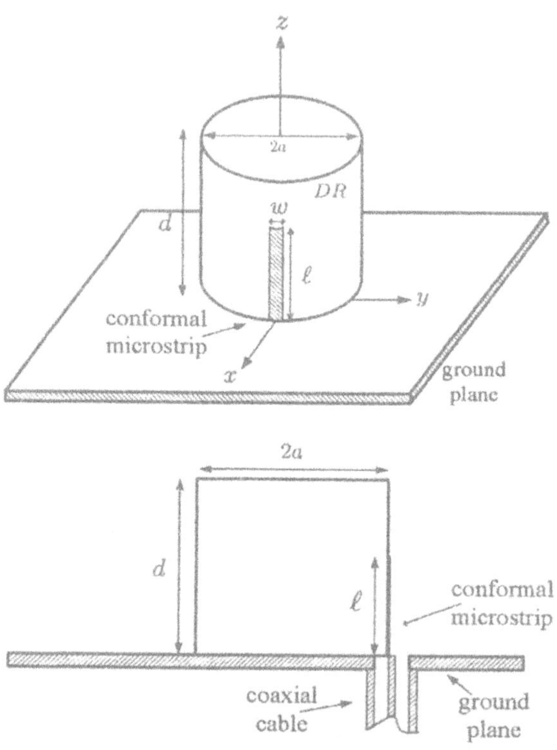

Figure 14.11. **Geometry of a dielectric resonator above PEC ground plane driven by conformal microstrip.**

Fig.14.11 shows a dielectric resonator on top of a finite ground plane, driven by a conformal microstrip which is connected to the inner conductor of coaxial cable beneath the ground plane. This configuration reserves the same advantage of coaxial-probe feed while avoiding the disadvantage of drilling a hole in the dielectric resonator. Furthermore, it allows easy integration and trimming without scratching the dielectric.

Fig.14.12 shows the input impedance and return loss of the dielectric resonator driven by conformal microstrip, a coaxial cable of 100 Ω impedance is used as the feed. At the first and third resonant frequencies, the input resistances are about 170 Ω. The return loss drops below −10 dB at the first and third resonant frequencies where the dominant modes are TM_{110} and TM_{111} mode, respectively. The return loss can be further reduced by using proper impedance transformer.

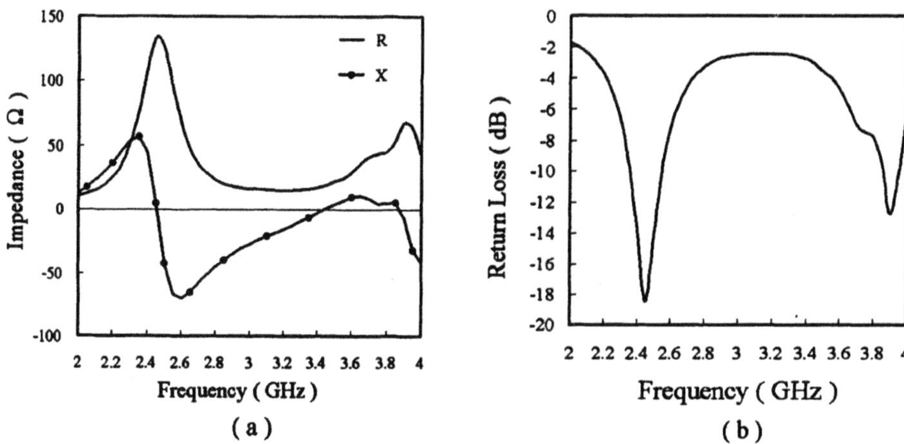

Figure 14.12. Feeding characteristics of dielectric resonator driven by conformal microstrip, $\epsilon_r = 8.9$, $a = 1.283$ cm, $d = 2.566$ cm, $\ell = 1.54$ cm, $w = 0.628$ cm, (a) input impedance, (b) return loss.

Fig.14.13 shows a dielectric resonator on top of a finite ground plane, driven by conformal microstrip fed by CPW. The conformal microstrip is extended from the signal line of CPW. This configuration reserves the advantages of coaxial-fed probe, and is easier to implement.

Fig.14.14 shows its input impedance and return loss, driven by conformal microstrip fed by a CPW of 100 Ω impedance. At the first and third resonant frequencies, the input resistances are about 170 Ω and 100 Ω, respectively. The return loss is significantly reduced at the first and third resonant frequencies where the dominant modes are TM_{110} and TM_{111} mode, respectively. Since the input resistance at the third resonant frequency is closer to 100 Ω, the return loss is much lower than that at the first one. When the conformal microstrip becomes wider, the input resistance increases at the first resonant frequency, but decreases at the third one.

Fig.14.15 shows the radiation patterns of the dielectric resonator driven by probe fed by a coaxial cable of 100 Ω. Since the ground plane is of finite extent, radiation in the backward direction is observed.

Fig.14.16 shows the radiation patterns of the dielectric resonator driven by conformal microstrip at the resonant frequency of TM_{111} mode. The main beam points at 300° on the x-z plane accompanied by several side lobes. Pattern on the y-z plane is almost symmetrical with respect to the ground plane. The E_θ pattern on the x-y plane becomes more omnidirectional compared with that in Fig.14.15(c).

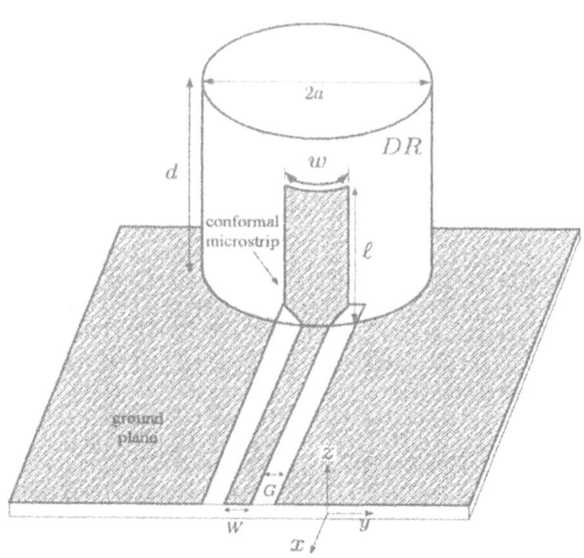

Figure 14.13. **Geometry of a dielectric resonator above PEC ground plane driven by conformal microstrip fed by CPW.**

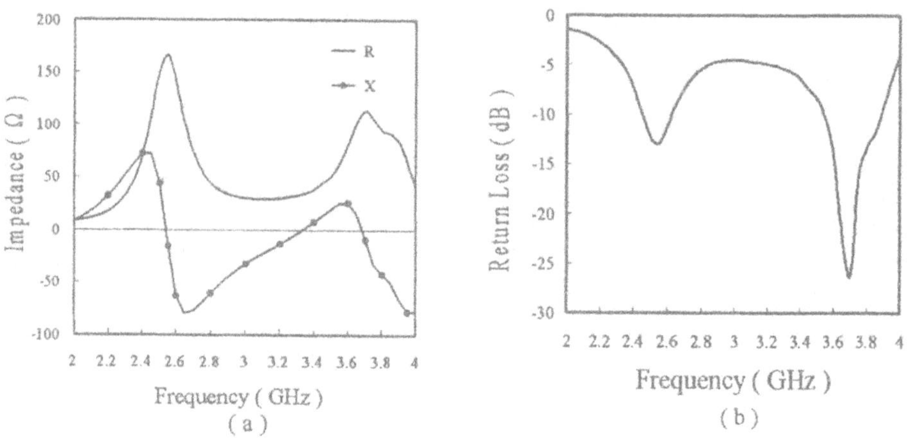

Figure 14.14. **Feeding characteristics of dielectric resonator driven by conformal microstrip fed by a CPW of 100 Ω, $\epsilon_r = 8.9$, $a = 1.283$ cm, $d = 2.566$ cm, $\ell = 1.54$ cm, $w = 3$ mm, $W = 3$ mm, $G = 2.8$ mm, (a) input impedance, (b) return loss.**

Fig.14.17 shows the radiation patterns at the resonant frequency of TM_{110} mode of the dielectric resonator driven by conformal microstrip fed by a CPW of 100 Ω. Compared with Fig.14.15, the radiation patterns

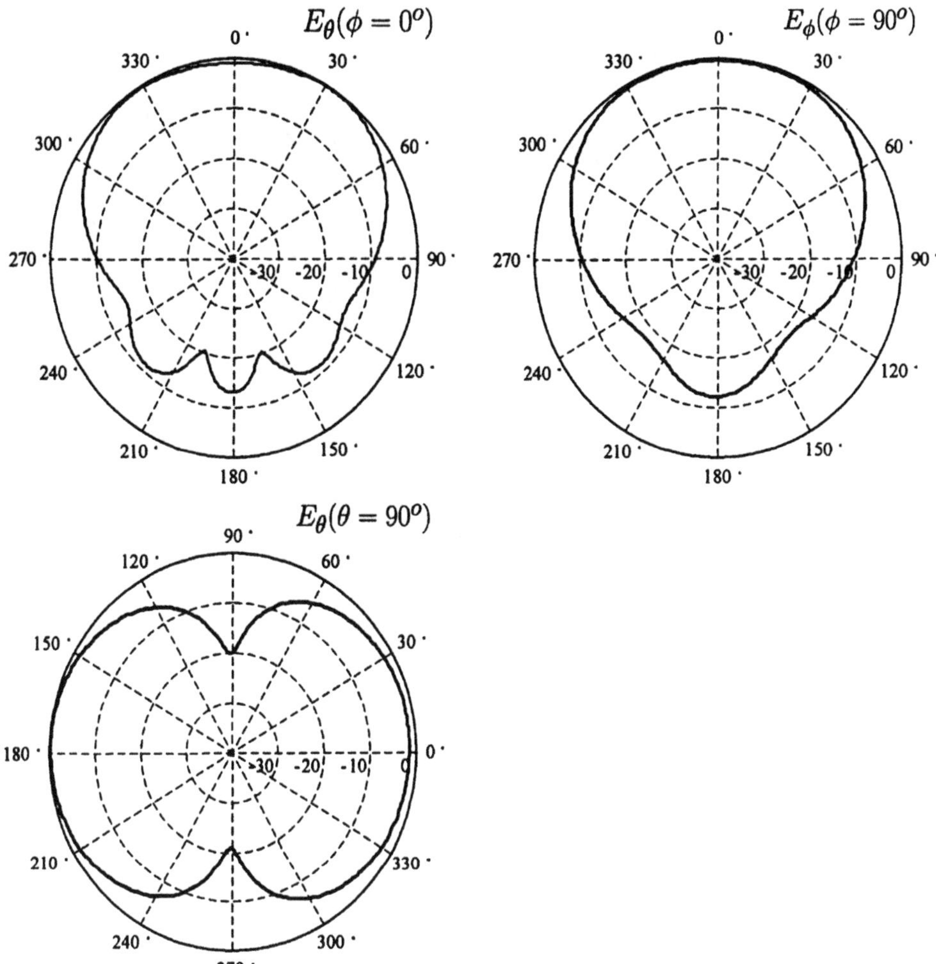

Figure 14.15. **Radiation patterns of dielectric resonator on top of a finite PEC ground plane of 15 cm × 15 cm, driven by coaxial probe, $\epsilon_r = 8.9$, $a = 1.283$ cm, $d = 2.566$ cm, coaxial cable impedance is 100 Ω, at the resonant frequency of TM_{110} mode($f = 2.75$ GHz), (a) E_θ on x-z plane, (b) E_ϕ on y-z plane, (c) E_θ on x-y plane.**

are similar, the E_θ pattern appears to rotates 45° on the x-z plane, the E_ϕ pattern on the y-z plane decreases below the ground plane, and two nulls appear in the E_θ pattern on the x-y plane.

4. Conclusions

Feeding and radiation characteristics of cylindrical dielectric resonator antennas have been analyzed. The input impedance and radiation pat-

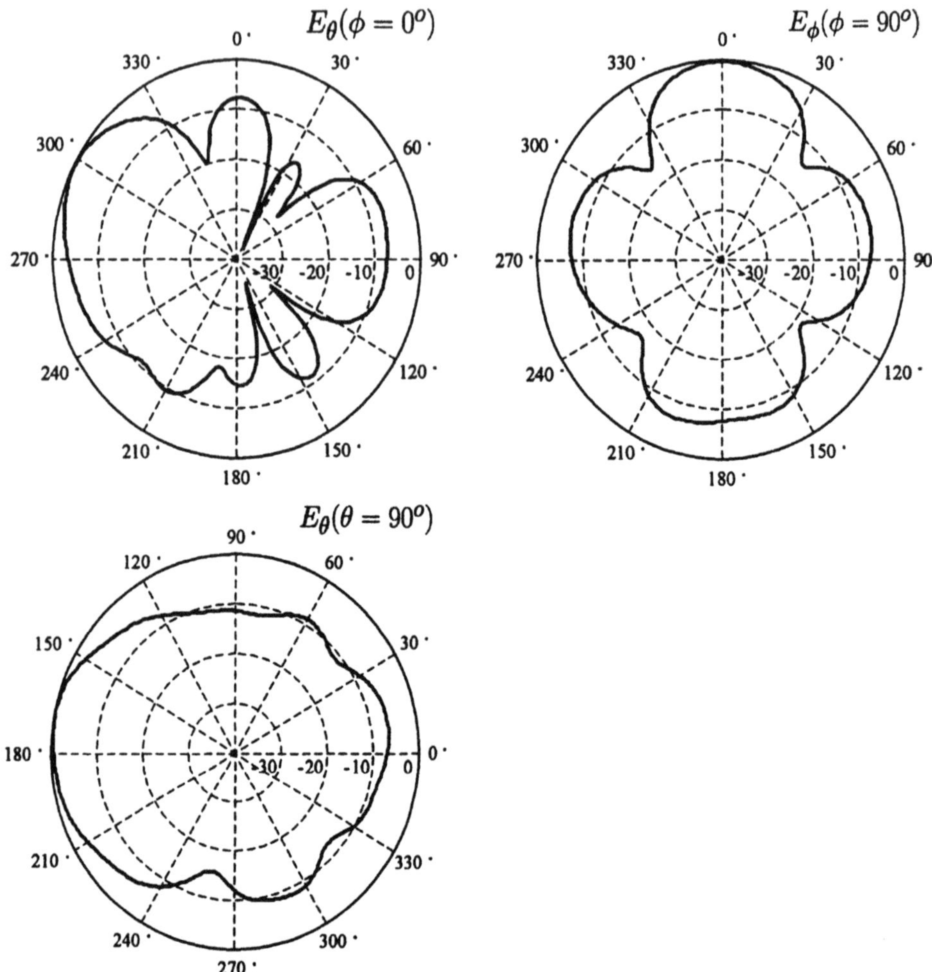

Figure 14.16. **Radiation patterns of dielectric resonator on top of a finite PEC ground plane of 15 cm × 15 cm, driven by conformal microstrip, $\epsilon_r = 8.9$, $a = 1.283$ cm, $d = 2.566$ cm, $\ell = 1.54$ cm, $w = 0.628$ cm, at the resonant frequency of TM_{111} mode($f = 3.9$ GHz), (a) E_θ on x-z plane, (b) E_ϕ on y-z plane, (c) E_θ on x-y plane.**

terns with various geometrical aspect ratios, dielectric constants and feeding configurations are simulated. Dielectric resonator above PEC ground plane driven by conformal microstrip fed by CPW is proposed. This configuration can be easily implemented and integrated with MMIC while taking advantages of dielectric resonator antennas in their low cost, small size and free of conductor loss. This antenna is shown to be capable of radiating efficiently in the direction normal to its ground plane while

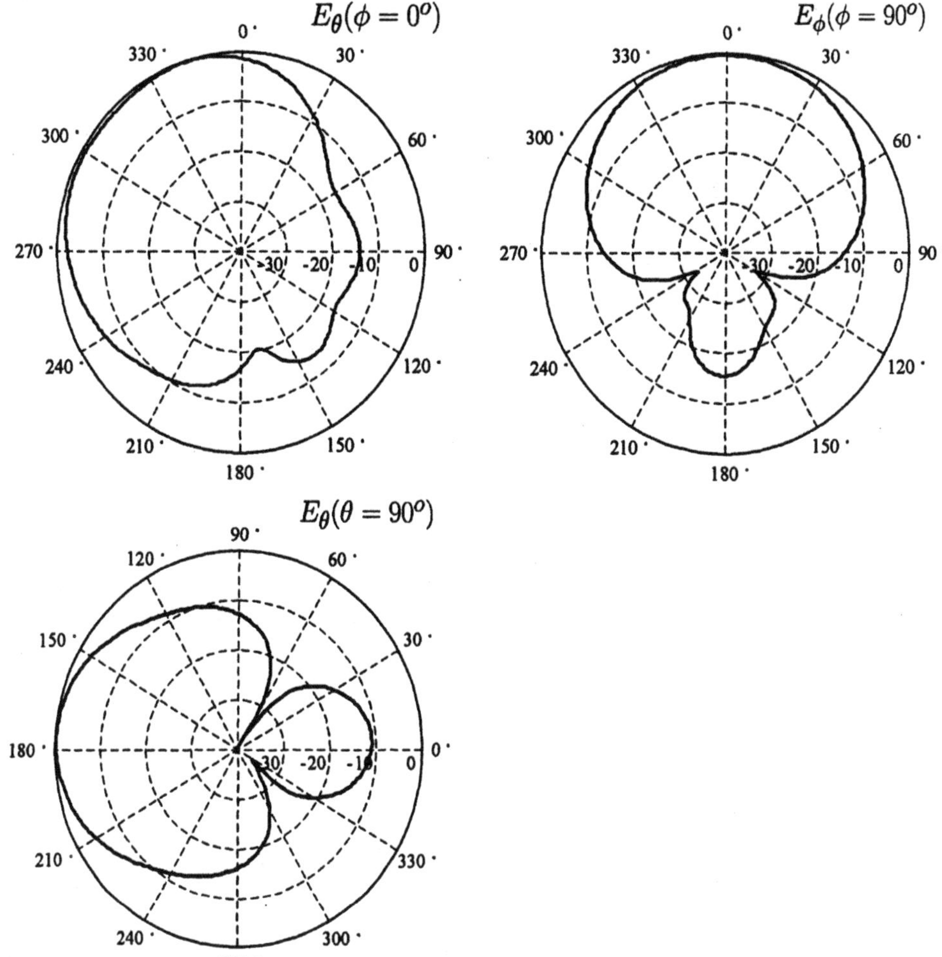

Figure 14.17. Radiation patterns of dielectric resonator on top of a finite PEC ground plane of 15 cm × 15 cm, driven by conformal microstrip fed by a CPW of 100 Ω, ϵ_r = 8.9, a = 1.283 cm, d = 2.566 cm, ℓ = 1.54 cm, w = 3 mm, W = 3 mm, G = 2.8 mm, at the resonant frequency of TM_{110} mode(f = 2.55 GHz), (a) E_θ on x-z plane, (b) E_ϕ on y-z plane, (c) E_θ on x-y plane.

retaining many desirable features suitable for applications in millimeter wave frequencies.

References

[1] S. A. Long, M. W. McAllister, and L. C. Shen, "The resonant cylindrical dielectric cavity antenna," *IEEE Trans. Antennas Propagat.*, vol.31, pp.406-412, May 1983.

[2] D. Kajfez and P. Guillon, *Dielectric Resonators*, Artech House, 1986

[3] A. W. Snyder, "Asymptotic expressions for eignfunctions and eigenvalues of a dielectric or optical waveguide," *IEEE Trans. Microwave Theory Tech.*, vol.17, pp.1130-1138, Dec. 1969.

[4] A. P. S. Khanna and Y. Garault, "Determination of loaded, unloaded and external quality factors of a dielectric resonator coupled to a microstrip line," *IEEE Trans. Microwave Theory Tech.*, vol.31, pp.261-264, Mar. 1983.

[5] M. W. McAllister, S. A. Long, and G. L. Conway, "Rectangular dielectric resonator antenna," *Electron. Lett.*, vol.19, pp.218-219, 1983.

[6] M. B. Oliver, Y. M. M. Anter, R. K. Mongia, and A. Ittipiboon, "Circularly polarised rectangular dielectric resonator antenna," *Electron. Lett.*, vol.31, pp.418-419, Mar. 1995.

[7] G. D. Loss and Y. M. M. Antar, "A new aperture-coupled rectangular dielectric resonator antenna array," *Microwave Opt. Technol. Lett.*, vol.7, pp.642-644, 1994.

[8] M. Jaworski and M. W. Pospieszalski, "An accurate solution of the cylindrical dielectric resonator problem," *IEEE Trans. Microwave Theory Tech.*, vol.27, pp.639-643, July 1979.

[9] K. Y. Chow, K. W. Leung, K. M. Luk, and E. K. N. Yung, "Cylindrical dielectric resonator antenna array," *Electron. Lett.*, vol.31, pp.1536-1537, Aug. 1995.

[10] M. W. McAllister and S. A. Long, "Resonant hemispherical dielectric antenna," *Electron. Lett.*, vol.20, pp.657-659, 1984.

[11] R. K. Mongia, "Half-split dielectric resonator placed on metallic plane for antenna applications," *Electron. Lett.*, vol.25, pp.426-464, Mar. 1989.

[12] K. W. Leung, K. M. Luk, and E. K. N. Yung, "Spherical cap dielectric resonator antenna using aperture coupling," *Electron. Lett.*, vol.30, pp.1366-1367, Aug. 1994.

[13] K. W. Wong, N. C. Chen, and H. T. Chen, "Analysis of a hemispherical dielectric resonator antenna with an air-gap," *IEEE Microwave Guided Wave Lett.*, vol.3, pp.355-357, Oct. 1993.

[14] N. C. Chen, H. C. Su, K. L. Wong, and K. W. Leung,"Analysis of a broadband slot-coupled dielectric-coated hemispherical dielectric resonator antenna," *Microwave Opt. Technol. Lett.*, vol.8, pp.13-16, 1995.

[15] K. M. Luk, W. K. Leung, and K. W. Leung, "Mutual impedance of hemispherical dielectric resonator antennas," *IEEE Trans. Antennas Propagat.*, vol.42, pp.1652-1654, Dec. 1994.

[16] S. M. Shum and K. M. Luk, "Characteristics of dielectric ring resonator antenna with an air-gap," *Electron. Lett.*, vol.30, pp.277-278, Feb. 1994.

[17] S. M. Shum and K. M. Luk, "Stacked annular ring dielectric resonator antenna excited by axi-symmetric coaxial probe," *IEEE Trans. Antennas Propagat.*, vol.43, pp.889-892, Aug. 1995.

[18] S. B. Cohn, "Microwave bandpass filters containing high-Q dielectric resonators," *IEEE Trans. Microwave Theory Tech.*, vol.16, pp.218-227, Apr. 1968.

[19] U. S. Hong and R. H. Jansen, "Numerical analysis of shielded dielectric resonators including substrate support disc and tuning post," *Electron. Lett.*, vol.18, no.23, pp.1000-1002, Nov. 1982.

[20] A. W. Glisson, D. Kajfez, and J. James, "Evaluation of modes in dielectric resonators using a surface integral equation formulation," *IEEE Trans. Microwave Theory Tech.*, vol.31, pp.1023-1029, Dec. 1983.

[21] J. K. Plourde and C. L. Ren, "Application of dielectric resonators in microwave components," *IEEE Trans. Microwave Theory Tech.*, vol.29, pp.754-769, Aug. 1981.

[22] J. T. H. st. Martin, Y. M. M. Anter, A. A. Kishk, A. Ittipiboon, and M. Cuhaci, "Dielectric resonator antenna using aperture coupling," *Electron. Lett.*, vol.26, pp.2015-2016, Nov. 1990.

[23] R. A. Kranenburg and S. A. Long, "Microstrip transmission line excitation of dielectric resonator antenna," *Electron. Lett.*, vol.24, pp.1156-1157, Aug. 1988.

[24] R. A. Kranenburg, S. A. Long, and J. T. Williams, "Coplanar wave guide excitation of dielectric resonator antennas," *IEEE Trans. Antennas Propagat.*, vol.39, pp.119-122, Jan. 1991.

[25] A. A. Kishk, B. Ahn, and D. Kajfez, "Broadband stacked dielectric resonator antenna," *Electron. Lett.*, vol.25, pp.1232-1233, Aug. 1989.

[26] R. N. Simons and R. Q. Lee, "Effect of parasitic dielectric resonators on CPW/aperture-coupled dielectric resonator antennas," *Proc. Inst. Elect. Eng.*, pt.H, vol.140, pp.336-338, Oct. 1993.

[27] M. Haneishi and H. Takazaawa, "Broadband circularly polarised planar array composed of a pair of dielectric resonator antennnas," *Electron. Lett.*, vol.21, pp.437-438, Aug. 1985.

[28] R. K. Mongia, A. Ittipiboon, M. Cuhaci, and D. Roscoe, "Circularly polarised dirlectric resonator antenna," *Electron. Lett.*, vol.30, pp.1361-1363, Aug. 1994.

[29] A. Petosa, R. K. Mongia, A. Ittibipoon, and J. S. Wight, "Design of microstrip-fed series array of dielectric resonator antennas," *Electron. Lett.*, vol.31, pp.1306-1307, Aug. 1995.

[30] R. K. Mongia, A. Ittibipoon, and M. Cuhaci, "Low profile dielectric resonator antennas using a very high permittivity material," *Electron. Lett.*, vol.30, pp.1362-1363, Aug. 1994.

[31] K. W. Leung, K. M. Luk, E. K. N. Yung, and S. Lai, "Characteristics of a low-profile circular disk DR antenna with very high permittivity," *Electron. Lett.*, vol.31, pp.417-418, Mar. 1995.

[32] J. van Bladel, "On the resonances of a dielectric resonator of a dielectric resonator of very high permittivity," *IEEE Trans. Microwave Theory Tech.*, vol.23, pp.199-208, 1975.

[33] R. F. Harrington, *Time-Harmonic Electromagnetic Fields*, New York: McGraw-Hill, 1961.

[34] A. Bailey, W. Foley, M. Hageman, C. Murray, A. Piloto, K. Sparks, and K. Zaki, "Miniature LTCC filters for digital receivers," *IEEE MTT-S Int. Microwave Symp. Dig.*, vol.2, pp.999-1002, June 1997.

[35] Y. Rong, K. A. Zaki, M. Hageman, D. Stevens, and J. Gipprich, "Low-temperature cofired ceramic (LTCC) ridge waveguide bandpass chip filters," *IEEE Trans. Microwave Theory Tech.*, vol.47, pp.2317-2324, Dec. 1999.

[36] G. L. Mattheai, L. Young, and E. Jones, *Microwave Filter Impedance Matching Networks and Coupling Structures*, New York: McGraw-Hill, 1965.

[37] H. Abe, Y. Yoichiro, A. Higashisaka, and H. Takamizawa, "A highly stabilized low-noise GaAs FET integrated oscillator with a dielectric resonator in the C band," *IEEE Trans. Microwave Theory Tech.*, vol.26, pp.156-162, Mar. 1978.

[38] A. Podcameni and L. Bermudez, "Large signal design of GaAs FET oscillators using input dielectric resonators," *IEEE Trans. Microwave Theory Tech.*, vol.31, pp.358-261, Apr. 1983.

Chapter 15

MULTIBAND AND WIDEBAND PATCH ANTENNAS

Kin-Lu Wong
Department of Electrical Engineering
National Sun Yat-Sen University
Kaohsiung, Taiwan, ROC

Abstract This Chapter reviews recent advances in designs of multiband and wideband patch antennas. Many promising patch antenna designs for dual band operations of orthogonal and same polarizations are described. Designs for dual band, dual polarized and circularly polarized operations are also presented. Some novel patch antenna designs for wireless communication are shown, and some challenges of patch antenna designs are also addressed.

Keywords: patch antennas, multiband, wideband, circular polarization, dual polarized.

1. Introduction

General patch antennas are fabricated by printing conducting patch on a grounded dielectric substrate. They are flat in appearance and have low profile. They also show attractive features of light weight, easy fabrication and conformability to mounting hosts [1]. However, with dielectric substrate, patch antennas usually have narrow impedance bandwidth which greatly limits their practical applications. To overcome this disadvantage, a variety of novel dual band and multiband techniques have been developed recently [2]. For dual band operations, designs with orthogonal polarizations [3]-[9] and same polarizations [10]-[17] have been reported. In addition, some novel dual band designs with dual polarized [18] and circular polarization(CP) [19]-[21] operations have also been devised. The former designs capable of dual band and dual polarized operations can find applications in 900 and 1,800 MHz cellular systems to alleviate multipath fading problem. The latter designs for generating dual band CP operations are suitable for 1,227 and 1,575 MHz dual band global positioning system(GPS). In addition, some interesting triple band patch antenna designs suitable for practical applications such as in the 900, 1,800, and 2,450 MHz bands have also been reported [22], [23]. These novel dual band and triple band designs are described in Sections 2 and 3.

There are also vast advances in wideband design techniques for patch antennas. By using foam or air gap to replace the dielectric substrate for patch antenna, much wider impedance bandwidths have been demonstrated. It has been reported that, by using U-slotted radiating patch [24], [25] or E-shaped radiating patch [26], [27], a probe-fed patch antenna with an air gap can easily achieve an impedance bandwidth(2:1 voltage standing wave ratio) of wider than 25%, with good radiation characteristics across the bandwidth. With the use of other feeding methods such as slot-coupled feed, capacitively coupled feed, or three-dimensional microstrip transition feed, an even wider impedance bandwidth(greater than 40%) with good radiation characteristics is possible for patch antennas with single radiating patch. For wideband CP operations, great progresses have also been achieved recently [28], [29]. In these designs, CP bandwidths with 3 dB of axial ratio can be greater than 40%. These wideband patch antennas will be described in Section 4.

In Section 5, some practical applications of patch antennas for wireless communication such as cellular systems of GSM(global system for mobile communication, 890-960 MHz), DCS(digital communication system, 1,710-1,880 MHz), PCS(personal communication system, 1,850-1,990 MHz) and UMTS(universal mobile telecommunication system, 1,920-2,170 MHz) and wireless local area network(WLAN) in the 2.4 GHz band(2,400-2,484 MHz) and 5.2 GHz band(5,150-5,350 MHz) are introduced. They include GSM/DCS dual band internal mobile phone antennas [30], cellular system base station antennas [31], WLAN antennas [32], and so on. Finally, some challenges of patch antenna designs for wireless communication are addressed in Section 6.

2. Dual Band Design Techniques

2.1 Orthogonal Polarization

Fig.15.1 shows a probe-fed patch antenna design for dual band operation with orthogonal polarizations using two feeds(points A and B) or single feed(point C). In this design, the feeds at points A and B are used to excite the fundamental TM_{01} and TM_{10} modes, respectively,

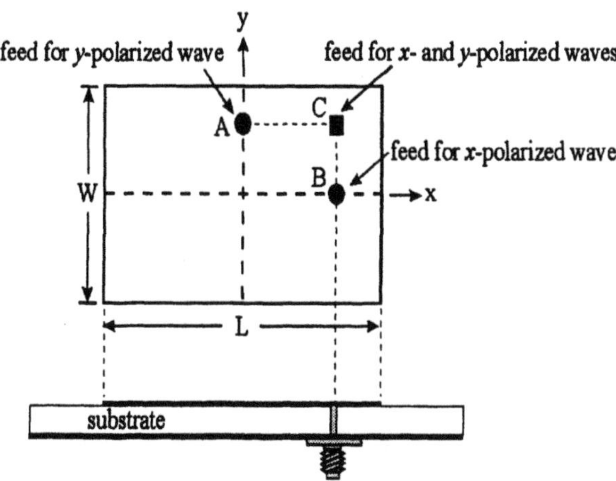

Figure 15.1. Geometry of probe-fed rectangular patch antenna for dual band operation with orthogonal polarizations.

whose resonant frequencies can be approximated as(more accurate for thin substrate)

$$f_{01} = \frac{c}{2W\sqrt{\epsilon_r}}, \qquad f_{10} = \frac{c}{2L\sqrt{\epsilon_r}}$$

where L and W are the length and width, respectively, of the rectangular patch, c is the speed of light in free space, and ϵ_r is the relative permittivity of substrate. Notice that these two resonant modes are of orthogonal polarization.

By using a single feed at point C as shown in Fig.15.1, effective excitation of the TM_{01} and TM_{10} modes can also be achieved [3]. In this case, single-feed dual band operation is obtained. Instead of using the probe feed method, similar dual band operation can also be obtained by using slot-coupled feed method as shown in Fig.15.2 [4].

Fig.15.3 presents some other plausible radiating patches for the dual band patch antenna shown in Fig.15.1. In Fig.15.3(a), a cross slot is embedded in the center of a rectangular patch to yield dual band operation [2], [5]. By inserting four narrow slits at the four edges of a rectangular patch [Fig.15.3(b)] or embedding a circular hole in a rectangular patch [6] [Fig.15.3(c)] or circular patch [7] [Fig.15.3(d)], dual band operations have also been realized. Designs using notched square patch [8] [Fig.15.3(e)] and slit-loaded triangular patch [9] are also promising in achieving single-feed dual band operations with orthogonal polarizations.

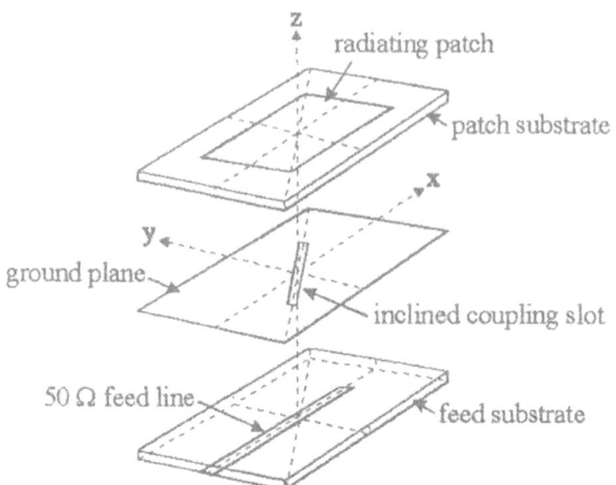

Figure 15.2. Exploded view of slot-coupled rectangular patch antenna for dual band operation with orthogonal polarizations.

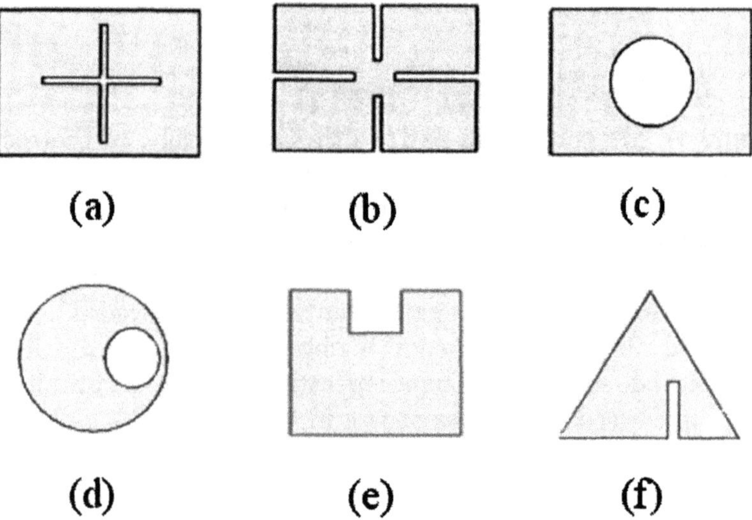

Figure 15.3. Plausible radiating patches for the dual band patch antenna shown in Fig.15.1.

2.2 Same Polarization

Many single-feed dual band antennas with same polarizations can be found in the literatures [2], [10]-[17]. Fig.15.4 shows the geometry of a probe-fed rectangular patch antenna with a pair of bent slots for dual band operation with same polarizations [10]. In this design, a pair of properly bent slots are placed close to the nonradiating edges of patch. It is found that, with the presence of bent slots, a new resonant mode with resonant frequency between those of the TM_{10} and TM_{20} modes is excited. When the bend angle is within 15°-30°, null-current point of this new resonant mode is moved toward the radiating edge closer to the bent slot, and the excited surface current distribution on the central portion of patch becomes uniformly distributed, which makes the radiation efficiency of this new mode close to that of TM_{10} mode. In addition, owing to their similar surface current distributions on the central portion of patch, this new mode and the TM_{10} mode are expected to have the same polarization planes and similar radiation patterns, and both can be excited using single probe feed with good impedance matching. In this design, dual band operation with the frequency ratio between 1.29 and 1.60 has been obtained [10].

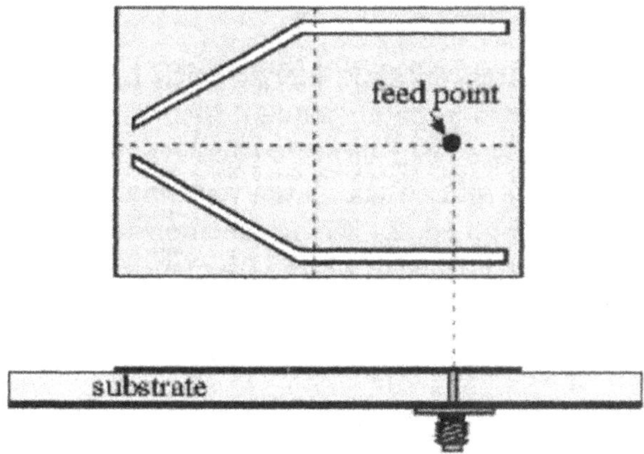

Figure 15.4. Geometry of probe-fed rectangular patch antenna for dual band operation with same polarizations.

Fig.15.5 shows other promising radiating patches for the dual band patch antenna shown in Fig.15.4. One can use a pair of narrow slots embedded close to the radiating edges [11] [Fig.15.5(a)] or a pair of step slots embedded close to the nonradiating edges [12] [Fig.15.5(b)]. A circular patch can also be used [13]. In Fig.15.5(c), a pair of arc-shaped slots are embedded in a circular patch, and similar dual band operation with same polarizations has been obtained.

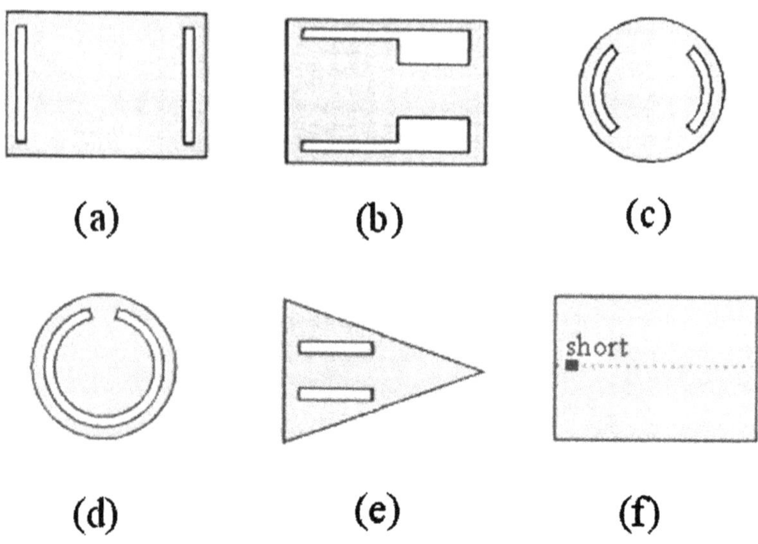

Figure 15.5. Promising radiating patches for the dual band patch antenna shown in Fig.15.4.

Recently, circular patch antenna loaded with an open ring slot [14] [Fig.15.5(d)] and triangular patch loaded with a pair of narrow slots [15], [16] [Fig.15.5(e)] have been shown to achieve dual band operations. By attaching a shorting pin to the patch antenna, compact dual band antennas can also be designed [2], [17]. Such shorted patch antennas can achieve dual band operations with a large frequency ratio in the ranges of 2.0-3.2 GHz, 2.5-3.8 GHz and 2.5-4.9 GHz for rectangular patch, circular patch, and triangular patch, respectively.

2.3 Dual Polarization Operations

To mitigate multipath fading problem and enhance system performance, slot-coupled, dual band, dual polarized patch antenna for 900/ 1,800 MHz cellular systems has been proposed [18]. Fig.15.6 shows an antenna geometry with four major factors to be considered. The first two factors are dimensions of the rectangular ring patch and rectangular patch for the 900 and 1,800 MHz bands, respectively. The two patches are printed on the same substrate, and the rectangular patch is placed within the rectangular ring patch to obtain a compact structure. The third factor is the arrangement of coupling slots on the ground plane, and the fourth factor is the feed network design. The coupling slots and feed network are printed on separate sides of the feed substrate.

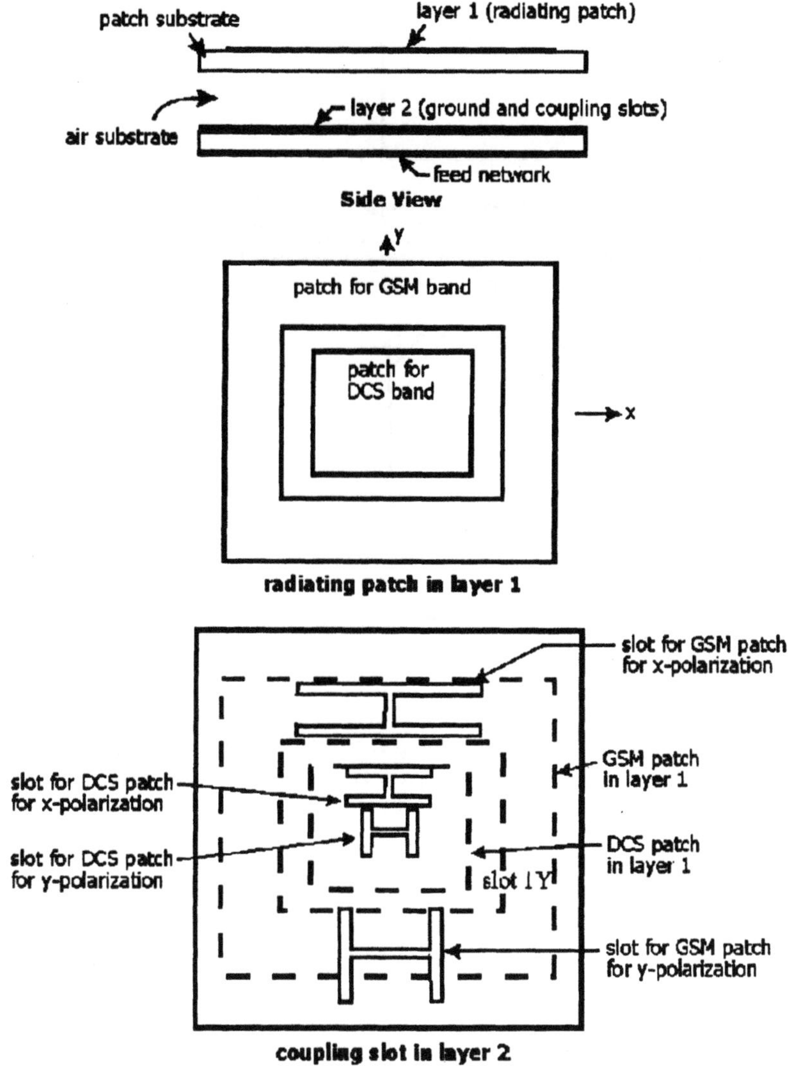

Figure 15.6. Geometry of slot-coupled, dual band, dual polarized patch antenna for 900/1,800 MHz cellular systems.

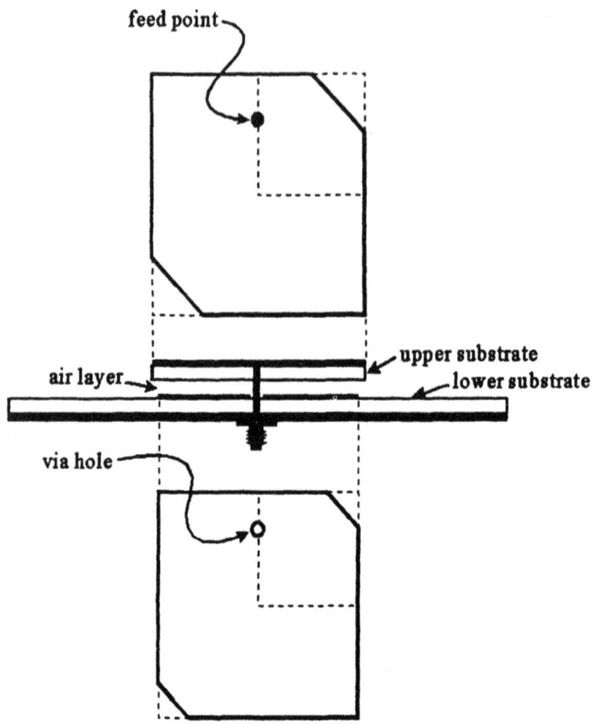

Figure 15.7. Geometry of probe-fed dual band circularly polarized patch antenna for GPS in the 1,227 and 1,575 MHz bands.

In this design, the rectangular ring patch is designed for the 900 MHz band, and is slot-coupled by using two orthogonal H-shaped coupling slots to obtain dual linear polarizations. On the other hand, the rectangular patch is designed for the 1,800 MHz band. All the four H-shaped coupling slots have the same width. In order to achieve high degree of port isolation, the two slots for the 900 MHz band are so arranged that the central arm of one H slot is parallel to the microstrip feed line of the second H slot. The same arrangement applies to the H slots for the 1,800 MHz band. High degree of isolation between ports 1 and 2 is obtained, and good dual polarized performance has been obtained for both the 900 and 1,800 MHz bands.

2.4 Circular Polarization Operations

Several dual band CP designs with patch antenna have also been reported recently [2], [19]-[21]. A promising design for GPS operations in the 1,227 and 1,575 MHz bands is shown in Fig.15.7[19]. This dual band

CP patch antenna is achieved by stacking two corner-truncated square microstrip patches. The CP bandwidths, defined by 3 dB of axial ratio, are about 15 MHz(1.2%) and 17 MHz(1.1%) at 1,227 and 1,575 MHz, respectively. Good CP radiation patterns and antenna gain have also been obtained. In Fig.15.8, two other promising single-layered patch antenna designs for dual band CP operations are shown. In Fig.15.8(a), the dual band CP operation is obtained by embedding two pairs of arc-shaped slots of proper lengths close to the perimeter of a circular patch and extending one of the arc-shaped slots with a narrow slot. Two separate CP bands are centred at 1,561 and 2,335 MHz, with CP bandwidths of about 1.3 and 1.1%, respectively [20]. In Fig.15.8(b), a probe-fed square patch antenna with a center slot and four bent slits also shows the capability of dual band CP operations [21].

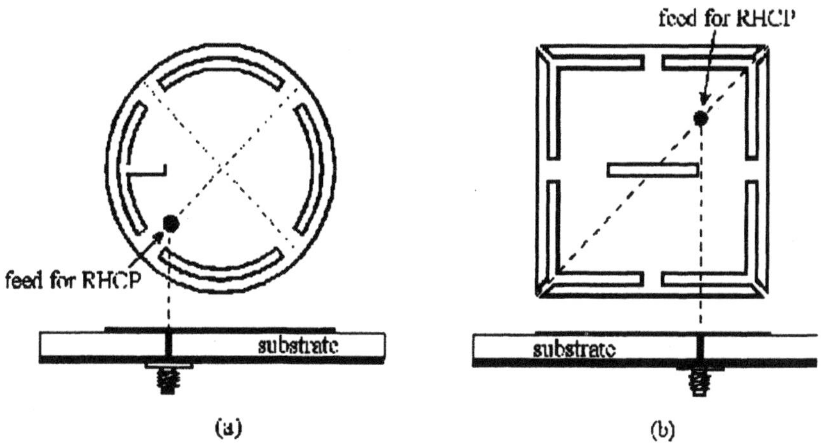

Figure 15.8. **Promising single-layered patch antenna designs for dual band circularly polarized operations, (a) circular patch, (b) square patch.**

3. Triple Band Design Techniques

Fig.15.9 shows two triple band shorted patch antennas for cellular and WLAN operations in the 900, 1,800 and 2,450 MHz bands. In Fig.15.9(a), feed point 1 drives the larger patch for the 900/1,800 MHz cellular bands, and feed point 2 drives the smaller patch for the 2,450 MHz WLAN band. Some related designs have also been studied in [22]. In Fig.15.9(b), single-feed design for 900/1,800/2,450 MHz bands is shown, in which a meandered rectangular patch with a shorting pin is devised [23]. Such meandering leads to a more compact size of patch and strongly affects the first three resonant frequencies of antenna. By

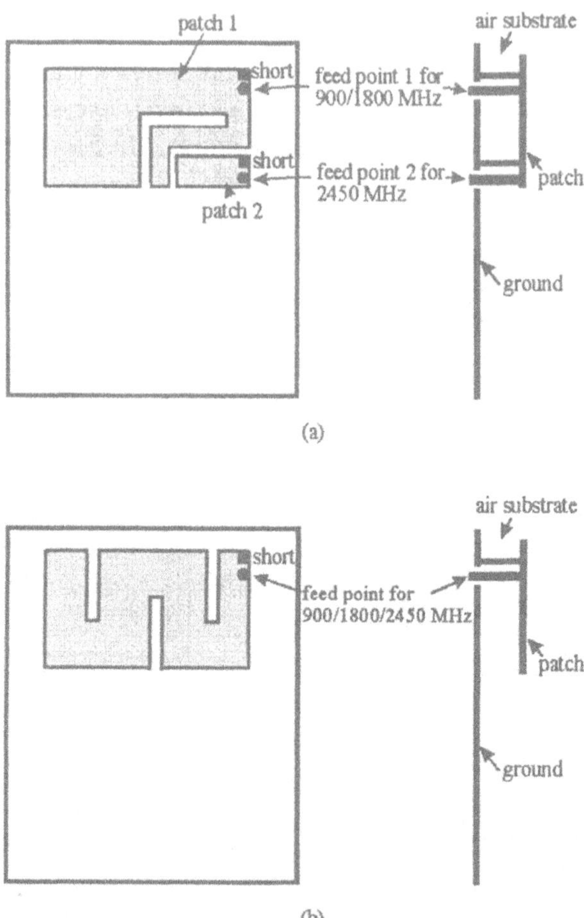

Figure 15.9. **Geometries of 900/1,800/2,450 MHz triple band shorted patch antennas, (a) two-feed design, (b) single-feed design.**

choosing proper dimensions of the linear slits, the first three resonant frequencies can be adjusted to be around 900, 1,800 and 2,450 MHz, respectively.

4. Wideband Design Techniques

4.1 Linear Polarization Operations

Patch antennas with air gap are less expensive to fabricate and can provide wideband operations. Fig.15.10 shows two promising wideband patch antennas with air gap. Fig.15.10(a) shows a patch in which a U-shaped slot is embedded [24], [25]. In Fig.15.10(b), the radiating patch has an E shape [26], [27]. Reactance due to the long probe in the air gap is high and impedance matching over a wide band is difficult to achieve. However, such high reactance is compensated for by the U slot or slit. For the two designs shown here, impedance bandwidths wider than 25% can be obtained. More relevant wideband patch antennas are described in [2].

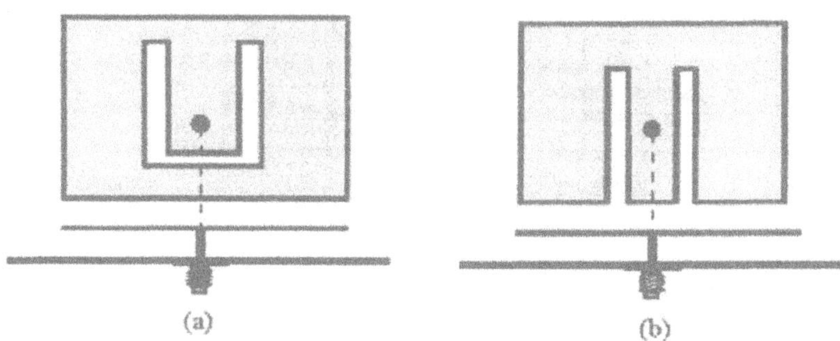

Figure 15.10. Geometries of wideband probe-fed rectangular patch antenna with air gap, (a) U-slotted patch, (b) E-patch.

4.2 Circular Polarization Operations

Wideband CP operation is another subject in patch antenna designs. Fig.15.11 shows two promising wideband circularly polarized patch antennas. In Fig.15.11(a), wideband CP operation is achieved by using two gap-coupled probe feeds placed in orthogonal directions [28], and the design shown in Fig.15.11(b) uses two capacitively coupled feeds [29]. In both designs, a Wilkinson power divider is used to provide equal input powers and 90° phase shift to these two feeds. For the former design with a thick air gap, measurement results show that 65% of impedance bandwidth(2:1 VSWR) and 46% of 3 dB axial ratio CP bandwidth are obtained [28]. Within the CP bandwidth, the gain bandwidth, defined to be within 3 dB of the peak antenna gain(about 6 dBi) is 44.6%. For the latter, impedance bandwidth(2:1 VSWR) of about 49% and 3 dB axial ratio CP bandwidth of about 35% have been achieved [29]. In this design, the 1 dB gain bandwidth is 28%, and the antenna gain is about 7 dBi.

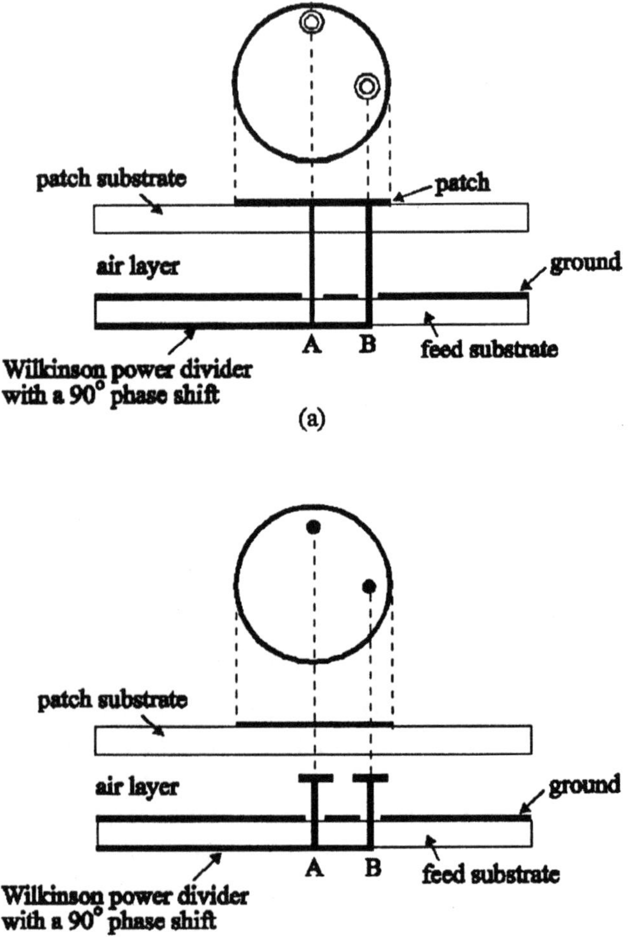

Figure 15.11. Geometries of wideband circularly polarized patch antennas, (a) with two gap-coupled probe feeds, (b) with two capacitively coupled feeds.

5. Practical Applications in Wireless Communication

5.1 Antennas for Dual Band Internal Mobile Phone

Applications of patch antennas in mobile phones as GSM/DCS internal antennas have proved successful [30]. Fig.15.12 shows the geometry of dual band planar inverted-F patch antenna(PIFA) in 900/1,800 MHz as an mobile phone internal antenna. In this design, an inverted-L slit is embedded in the top patch to separate it into two subpatches. The larger subpatch has a longer resonant path associated with the resonant mode in the 900 MHz band. The smaller subpatch, on the other hand, has a shorter resonant path associated with the 1,800 MHz band operation. These PIFA designs are usually compact and can be integrated within the mobile phone housing, thus named concealed or internal antennas.

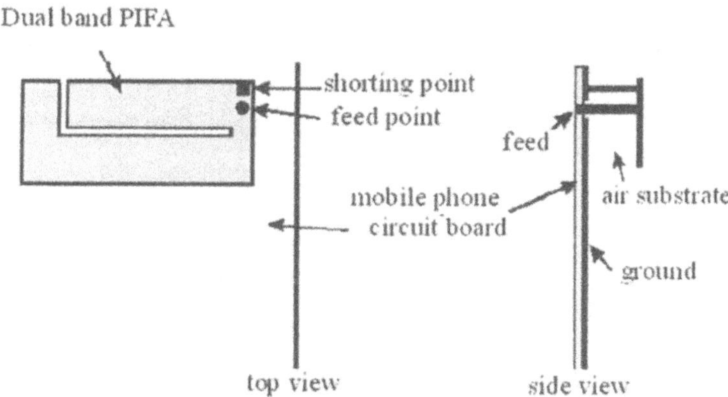

Figure 15.12. Geometry of dual band planar inverted-F patch antenna for 900/1,800 MHz mobile phone.

Compared to conventional whip or rod antennas used for mobile phones, PIFAs are less likely to break or tangle with other objects. Moreover, PIFAs usually render lower specific absorption rate(SAR) [33], [34] than conventional whip antennas. This implies that electromagnetic energy absorbed by the user's head is reduced. These advantages have led to many novel dual band PIFA designs for applications in mobile phones.

5.2 Base Station Antennas in Cellular Systems

Patch antennas are also considered for base stations in cellular systems. Fig.15.13 shows a coplanar probe-fed patch antenna with a U-shaped ground plane for GSM/DCS/PCS triple band operation [31]. Notice that the probe is connected to the center of a patch edge with a flare angle of about 170°. Impedance matching of this antenna can be improved by tuning the flare angle. There are two vertical ground planes. The first one is primarily used to hold the coplanar probe feed, and the second one has significant effects on improving the impedance matching of two desired resonant modes to cover the GSM, DCS and PCS bands. From the measurement results, the lower band has an impedance bandwidth(1.5:1 VSWR) of 388 MHz(nearly 50%) at the center frequency of 787 MHz, and covers the GSM band [31]. The upper band has a bandwidth of 575 MHz(about 30%) at the center frequency of 1,950 MHz, and covers both the DCS and PCS bands. Good radiation characteristics have also been observed.

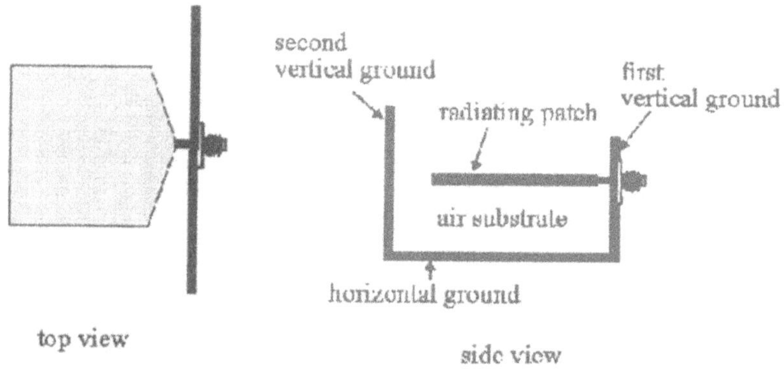

Figure 15.13. **Geometry of coplanar probe-fed patch antenna with a U-shaped ground plane for GSM/DCS/PCS triple band operation.**

5.3 WLAN Antennas

Owing to its low profile and compact structure, shorted patch antenna and PIFA are also very attractive for WLAN operations in the 2.4 and 5.2 GHz bands. Fig.15.14 shows a probe-fed, dual band PIFA for WLAN systems [32]. Top patch of the antenna is 18×27 mm^2, and is printed on an FR4 substrate of thickness 0.8 mm and relative permittivity of 4.4. The FR4 substrate supports the top patch, and an air gap of thickness 3.2 mm lies between the FR4 substrate and ground plane. The first two resonant frequencies of the antenna are 2.4 and 5.2 GHz, respectively, which are mainly controlled by the shorting pin attached to the patch's centerline and the four slits cut at nonradiating edges of the patch.

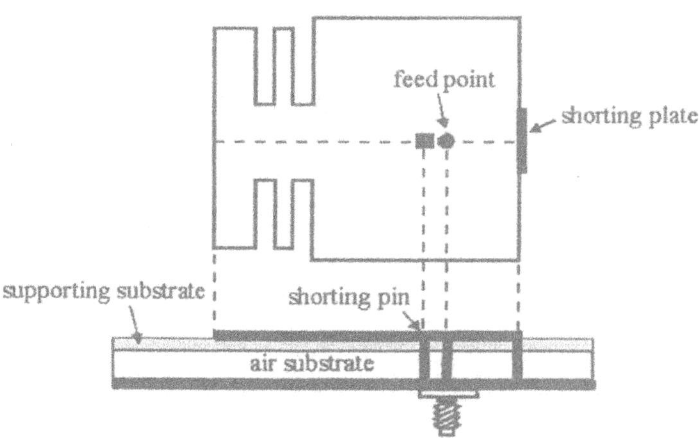

Figure 15.14. **Geometry of probe-fed planar inverted-F patch antenna for WLAN operations in 2.4 and 5.2 GHz bands.**

With a ground plane of 35×35 mm^2, measurement results show that two resonant modes at about 2.4 and 5.2 GHz are excited with good impedance matching [32]. Impedance bandwidth(2.5:1 VSWR) for the lower band reaches 90 MHz, 3.7% with respect to 2.45 GHz, and covers the WLAN band(2,400-2,484 MHz). On the other hand, the upper band has a bandwidth of 277 MHz, 5.3% with respect to 5.25 GHz, and covers the WLAN band(5,150-5,350 MHz). Good radiation characteristics across the 2.4 and 5.2 GHz bands have also been obtained.

6. Design Challenges for Wireless Communication

With the rapid progress in wireless communication, competition persists to reduce the size of portable communication devices. This leads to miniaturization requirement on patch antenna designs. Designing compact patch antennas capable of dual band, multiband, dual polarized or CP operations thus become an important issue for practical applications of patch antennas. In general, when the antenna size is reduced at a fixed operating frequency, the antenna gain is also decreased. Thus, relevant designs to achieve gain enhancement becomes a challenge in patch antenna designs [2].

Cost is another major considerations for practical applications of patch antennas. The most effective way to reduce cost is to use inexpensive substrates such as FR4. However, FR4 substrate has high loss, especially at higher frequencies, which usually leads to degradation in antenna performance. Thus, new dielectric materials with low cost and low loss are desirable.

Recently, there are public concerns about potential hazards caused by radiation from the mobile phone antennas [33], [34]. Some studies have also indicated that the mobile phone users may be exposed to harmful electromagnetic radiation from mobile phone antennas. For this reason, patch antennas having significantly reduced SAR value are demanded for use in cellular systems.

References

[1] K. L. Wong, *Design of Nonplanar Microstrip Antennas and Transmission Lines*, New York: John Wiley, 1999.

[2] K. L. Wong, *Compact and Broadband Microstrip Antennas*, New York: John Wiley, 2002.

[3] J. S. Chen and K. L. Wong, "A single-layer dual-frequency rectangular microstrip patch antenna using a single probe feed," *Microwave Opt. Technol. Lett.*, vol.11, pp.83-84, Feb. 5, 1996.

[4] Y. M. M. Antar, A. I. Ittipiboon, and A. K. Bhattacharyya, "A dual-frequency antenna using a single patch and an inclined slot," *Microwave Opt. Technol. Lett.*, vol.8, pp.309-311, Apr. 20, 1995.

[5] K. L. Wong and K. P. Yang, "Small dual-frequency microstrip antenna with cross slot," *Electron. Lett.*, vol.33, pp.1916-1917, Nov. 6, 1997.

[6] H. D. Chen, "A dual-frequency rectangular microstrip antenna with a circular slot," *Microwave Opt. Technol. Lett.*, vol.18, pp.130-132, June 5, 1998.

[7] J. H. Lu and K. L. Wong, "Compact dual-frequency circular microstrip antenna with an offset circular slot," *Microwave Opt. Technol. Lett.*, vol.22, pp.254-256, Aug. 20, 1999.

[8] H. Nakano and K. Vichien, "Dual-frequency square patch antenna with rectangular notch," *Electron. Lett.*, vol.25, pp.1067-1068, Aug. 3, 1989.

[9] K. L. Wong, S. T. Fang, and J. H. Lu, "Dual-frequency equilateral-triangular microstrip antenna with a slit," *Microwave Opt. Technol. Lett.*, vol.19, pp.348-350, Dec. 5, 1998.

[10] K. L. Wong and J. Y. Sze, "Dual-frequency slotted rectangular microstrip antenna," *Electron. Lett.*, vol.34, pp.1368-1370, July 9, 1998.

[11] S. Maci and G. B. Gentili, "Dual-frequency patch antennas," *IEEE Antennas Propagat. Mag.*, vol.39, pp.13-20, Dec. 1997.

[12] J. H. Lu, "Single-feed dual-frequency rectangular microstrip antenna with pair of step-slots," *Electron. Lett.*, vol.35, pp.354-355, Mar. 4, 1999.

[13] K. B. Hsieh and K. L. Wong, "Inset-microstrip-line-fed dual-frequency circular microstrip antenna and its application to a two-element dual-frequency microstrip array," *Proc. Inst. Elect. Eng.*, pt.H, vol.147, pp.359-361, Oct. 1999.

[14] J. Y. Jan and K. L. Wong, "Single-feed dual-frequency circular microstrip antenna with an open-ring slot," *Microwave Opt. Technol. Lett.*, vol.22, pp.157-160, Aug. 5, 1999.

[15] S. T. Fang and K. L. Wong, "A dual-frequency equilateral-triangular microstrip antenna with a pair of narrow slots," *Microwave Opt. Technol. Lett.*, vol.23, pp.82-84, Oct. 20, 1999.

[16] J. H. Lu, C. L. Tang, and K. L. Wong, "Novel dual-frequency and broadband designs of single-feed slot-loaded equilateral-triangular microstrip antennas," *IEEE Trans. Antennas Propagat.*, vol.48, pp.1048-1054, July 2000.

[17] K. L. Wong and W. S. Chen, "Compact microstrip antenna with dual-frequency operation," *Electron. Lett.*, vol.33, pp.646-647, Apr. 10, 1997.

[18] T. W. Chiou and K. L. Wong, "A compact dual-polarized aperture-coupled patch antenna for GSM 900/1800-MHz systems," *Proc. Asia-Pacific Microwave Conf.*, pp.95-98, 2001.

[19] C. M. Su and K. L. Wong, "A dual-band GPS microstrip antenna," *Microwave Opt. Technol. Lett.*, vol.33, May 20, 2002.

[20] K. B. Hsieh, M. H. Chen, and K. L. Wong, "Single-feed dual-band circularly polarized microstrip antenna," *Electron. Lett.*, vol.34, pp.1170-1171, June 11, 1998.

[21] K. P. Yang and K. L. Wong, "Dual-band circularly-polarized square microstrip antenna," *IEEE Trans. Antennas Propagat.*, vol.49, pp.377-382, Mar. 2001.

[22] C. T. P. Song, P. S. Hall, H. Ghafouri-Shiraz, and D. Wake, "Triple band planar inverted F antennas for handheld devices," *Electron. Lett.*, vol.36, pp.112-114, Jan. 20, 2000.

[23] W. P. Dou and Y. W. M. Chia, "Novel meandered planar inverted-F antenna for triple-frequency operation," *Microwave Opt. Technol. Lett.*, vol.27, pp.58-60, Oct. 5, 2000.

[24] T. Huynh and K. F. Lee, "Single-layer single-patch wideband microstrip antenna," *Electron. Lett.*, vol.31, pp.1310-1311, Aug. 3, 1995.

[25] K. L. Wong and W. H. Hsu, "Broadband triangular microstrip antenna with U-shaped slot," *Electron. Lett.*, vol.33, pp.2085-2087, Dec. 4, 1997.

[26] K. L. Wong and W. H. Hsu, "A broadband rectangular patch antenna with a pair of wide slits," *IEEE Trans. Antennas Propagat.*, vol.49, pp.1345-1347, Sept. 2001.

[27] B. L. Ooi and Q. Shen, "A novel E-shaped broadband microstrip patch antenna," *Microwave Opt. Technol. Lett.*, vol.27, pp.348-252, Dec. 5, 2000.

[28] T. W. Chiou and K. L. Wong, "Single-layer wideband probe-fed circularly polarized microstrip antenna," *Microwave Opt. Technol. Lett.*, vol.25, pp.74-76, Apr. 5, 2000.

[29] K. L. Wong and T.W. Chiou, "A broadband single-patch circularly polarized microstrip antenna with dual capacitively-coupled feeds," *IEEE Trans. Antennas Propagat.*, vol.49, pp.41-44, Jan. 2000.

[30] I. A. Korisch, "Planar dual frequency band antenna," US patent no.5926139, July 20, 1999.

[31] F. S. Chang and K. L. Wong, "Broadband dual-frequency coplanar probe-fed patch antenna for GSM/DCS/PCS operations," *Microwave Opt. Technol. Lett.*, vol.33, June 5, 2002.

[32] H. C. Tung and K. L. Wong, "A shorted microstrip antenna for 2.4/5.2-GHz dual-band operation," *Microwave Opt. Technol. Lett.*, vol.30, pp.401-402, Sept. 20, 2001.

[33] M. A. Jensen and Y. Rahmat-Samii, "EM Interaction of handset antennas and a human in personal communications," *Proc. IEEE*, vol.83, pp.7-17, Jan 1995.

[34] G. Lazzi, J. Johnson, S. S. Pattnaik, and O. P. Gandhi, "Experimental study on compact, high-gain, low SAR single- and dual-band patch antenna for cellular telephones," *IEEE AP-S Int. Symp. Dig.*, pp.130-133, 1998.

Chapter 16

PHASED ANTENNA ARRAY BASED ON NONLINEAR DELAY LINE TECHNOLOGY

Chia-Chan Chang, Cheng Liang, and Neville C. Luhmann, Jr.
Department of Electrical and Computer Engineering
University of California, Davis
California, USA

Abstract True time delay(TTD) technology is necessary when beam scanning capability over broad bandwidth is required. The use of transmission lines periodically loaded with diodes, which is referred to as nonlinear delay line(NDL) technology, has been studied at UC-Davis and has been successfully demonstrated in phased antenna arrays, both in hybrid and monolithic embodiments.

Keywords: phased antenna array(PAA), true time delay(TTD).

1. Introduction

Demand for high-performance microwave and millimeter wave phased antenna array(PAA) systems has increased significantly during the last decade. For many applications, coverage over a broad range of frequencies is required. However, a particularly severe limitation results from the use of phase shifters. Since phase shift is frequency dependent, the direction of beam pointing will consequently change when frequency is varied. Such distortion is referred to as beam squinting [1]. It is therefore necessary to implement true time delay(TTD) devices instead of phase shifters.

As shown in Fig.16.1, a variety of approaches have been reported to realize TTD, including digital delay lines [2], optical delay lines [3], [4], dielectric delay lines [5], varactor loaded delay lines [6], [7] and piezoelectric delay lines [8].

This Chapter will concentrate on a nonlinear delay line(NDL) which is realized by high-impedance transmission line periodically loaded with

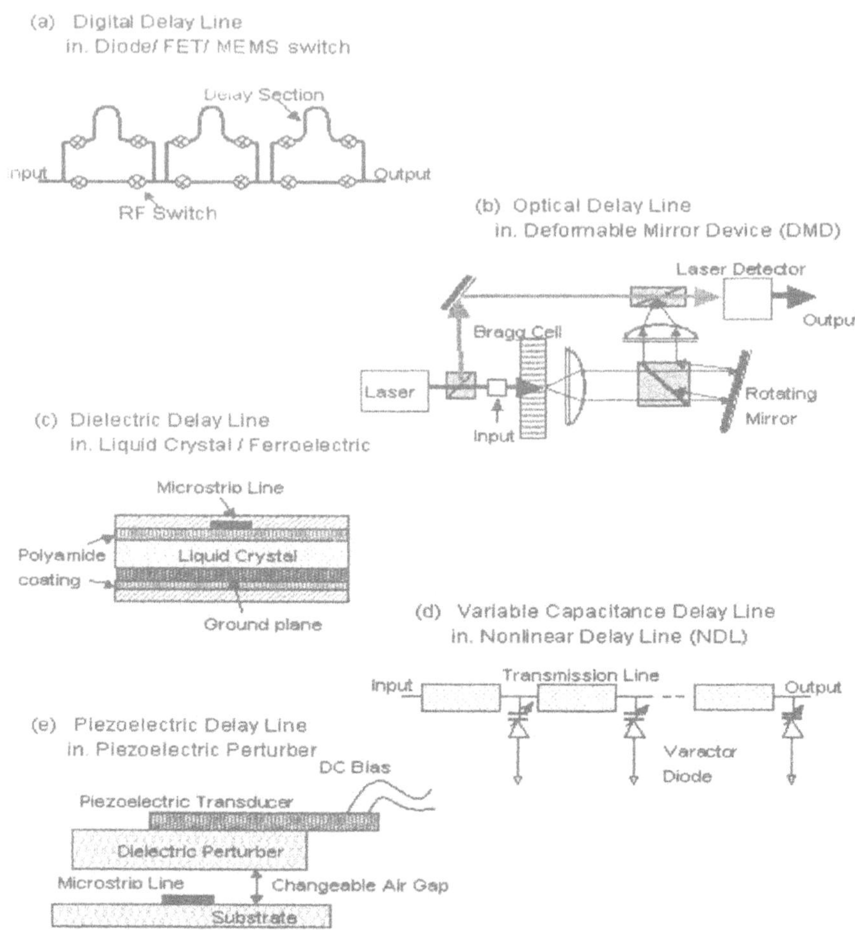

Figure 16.1. **Typical true time delay lines: (a) digital delay line [2], (b) optical delay line [3], [4], (c) dielectric delay line [5], (d) nonlinear delay line [6], [7], (e) piezoelectric delay line [8].**

varactor diodes. The capacitance of varactor diodes changes when different reverse dc bias levels are applied, which results in a change of propagation speed along the transmission line.

Nonlinearity and dispersion characteristics of such structure were first reported in 1977 by D. Jager [9], and have continued to receive attention. Previously, such nonlinear transmission lines(NLTLs) have only been utilized for their large-signal properties in the nonlinear regime and have been employed as short pulse generators and frequency multipliers [10]-

[12]. For short pulse generation, NLTLs are operated in the region near Bragg cutoff frequency.

Nonlinear delay line(NDL) employs the same structure as NLTL even though it is operated in the linear regime(namely, small-signal regime). The concept was first proposed by W. M. Zhang during the course of his Ph.D. thesis research in NLTL-based short pulse generation [13]. At low RF voltage levels(small signals), nonlinearity of the diodes is sufficiently weak that the varactors can be treated as linear elements. The diode capacitance depends only on the dc bias voltage. For frequencies far below the Bragg cutoff frequency, the group velocity is essentially frequency independent with negligible dispersion. Hence, the line provides time delay directly controllable by the reverse dc bias level.

Comparing with other TTD techniques, this NDL approach offers attractive features which are summarized in this paragraph. NDLs with low insertion loss can be realized. A hybrid NDL has demonstrated less than 6 dB loss up to 5 GHz with 160 ps time delay [7]. A similar monolithic NDL has also been reported by UCSB researchers with less than 4 dB loss up to 20 GHz [14]. The NDL time delay can be simply controlled by varying external dc bias level with response time potentially less than 100 ps. NDL is an analog device, and thus its delay precision is limited only by the precision of bias control circuitry. Hybrid NDLs can be built inexpensively by employing commercial beam-lead diodes and standard printed circuit boards. NDL can also be fabricated monolithically with low cost. Monolithic NDL is compatible with existing GaAs MMIC technology. It can also achieve high operating frequency and wide frequency bandwidth due to its extremely small size. Size of an NDL in either hybrid or monolithic form is very compact, and its weight is negligible as compared to many other TTD components.

This Chapter provides a brief overview of progress in NDL-related works with emphasis on those performed at UC-Davis. It will begin with NDL theory and analysis. Design and fabrication information for nonlinear delay lines, both in hybrid and monolithic embodiments, will be provided later. Moreover, experimental results of NDL-based PAAs will be presented. Finally, the challenges on NDL design and current development activities will be summarized.

2. Nonlinear Delay Line Theory and Design Considerations

2.1 Nonlinear Delay Line Theory

By utilizing a first order approximation, a high-impedance transmission line loaded with varactor diodes can be approximated by lumped

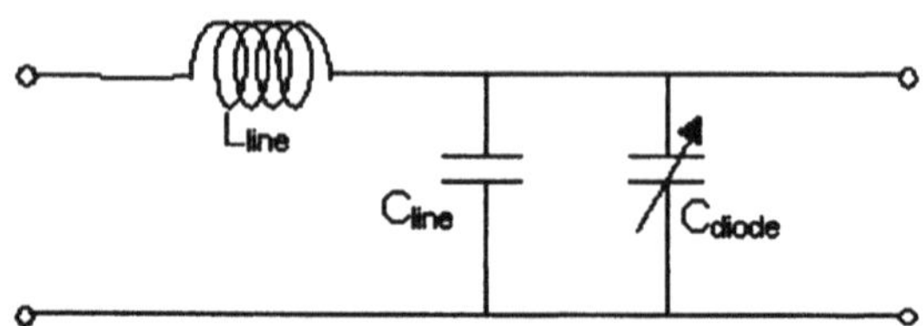

Figure 16.2. Equivalent lumped circuit model of an NLTL section.

LC sections as shown in Fig.16.2. Here, ℓ is the length of section, L_{line} is series line inductance, and C_{line} is shunt line capacitance, both are approximated from the transmission line model. Finally, $C_{\text{diode}}(V)$ is the shunt diode capacitance which arises from the varactor diodes regularly spaced along the line. The diode capacitance depends upon both applied dc bias voltage and ac voltage of the guided wave.

Unloaded NLTL is usually designed to have a high impedance, Z_i, which is determined by the line capacitance, C_{line}, and line inductance, L_{line}, as

$$Z_i = \sqrt{\frac{L_{\text{line}}}{C_{\text{line}}}}$$

The characteristic impedance of loaded line, Z_L, is given by

$$Z_L = \sqrt{\frac{L_{\text{line}}}{C_{\text{line}} + C_{\text{diode}}(V)}}$$

Notice that the characteristic impedance of this periodically loaded NLTL is also a function of dc bias. The unloaded characteristic impedance, Z_i, is usually chosen to have a higher value in order to make the loaded line impedance, Z_L, 50 Ω when the varactors have maximum capacitance of $C_{\text{diode,max}}$.

Referring to the equivalent circuit model, the periodically loaded transmission line will act as an intrinsic low-pass filter, with a ladder network cutoff frequency(Bragg cutoff frequency) of

$$f_{\text{Bragg}} = \frac{1}{\pi\sqrt{L_{\text{line}}\left[C_{\text{line}} + C_{\text{diode}}(V)\right]\ell}} \tag{16.1}$$

At lower frequencies, impedance of the shunt capacitance is significantly higher than that of the transmission line, and the wavelength is much longer than the length of each transmission line section.

Assuming traveling wave solutions for current and voltage along the transmission line, and then applying continuity conditions at each node, one can derive a general dispersion relation for this system as [6]

$$\cos(\beta\ell) - \cos(k\ell) + \frac{\omega C Z_0}{2}\sin(k\ell) = 0$$

where Z_0 and k are impedance and wave number, respectively, of the transmission line, and β is the NDL small-signal wave number. The phase velocity can be derived from (16.1) as

$$v_p = \frac{\omega}{\beta} = \frac{\ell}{\cos^{-1}\left[\cos(\omega\ell/v) - (\omega C Z_0/2)\sin(\omega\ell/v)\right]}$$

and the group velocity is calculated as

$$v_g = \frac{\partial\omega}{\partial\beta} = \frac{\ell\sin(\beta\ell)}{(\ell/v)\sin(\omega\ell/v) + [\omega C Z_0\ell/(2v)]\cos(\omega\ell/v)} \qquad (16.2)$$

Fig.16.3 displays the calculated group velocity and phase velocity of an NDL. In this calculation, the NDL has been designed to have a cutoff frequency of 100 GHz, unloaded line impedance of 90 Ω, and loaded line impedance of 50 Ω. For frequencies far below the Bragg cutoff frequency(dc to 20 GHz), the group velocity is essentially frequency independent. For this particular transmission line design, the group velocity varies by less than 1% over the frequency range from dc to 20 GHz [13].

As indicated above, the varactor diode capacitance is a function of dc bias voltage. By changing this dc voltage, group velocity of this nonlinear transmission line can also be controlled, hence resulting in the desired true time delay. However, this distributed *LC* equivalent circuit model is valid as long as the frequency is much lower than the Bragg cutoff frequency [13]. In other words, for frequencies significantly lower than the Bragg cutoff frequency, group velocity of this nonlinear transmission line is essentially frequency independent, and only dependent upon the dc bias level.

Further simplifying (16.2) under the premise of small signal and low frequency(far below f_{Bragg}), group velocity in the equivalent *LC* circuit model becomes

$$v_g = \frac{\ell}{\sqrt{L_{\text{line}}\left[C_{\text{line}} + C_{\text{diode}}(V)\right]}}$$

and the propagation time is given by

$$t = \frac{\ell}{v_g} = \sqrt{L_{\text{line}}\left[C_{\text{line}} + C_{\text{diode}}(V)\right]}$$

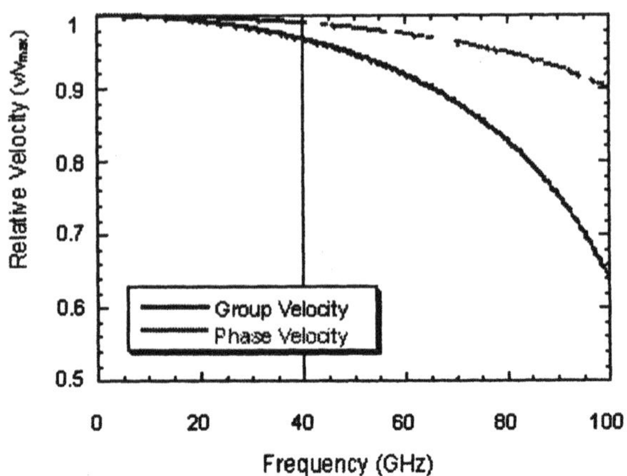

Figure 16.3. **Comparison of calculated group velocity and phase velocity.**

Hence, the delay time along each NLTL section is a function of transmission line length, varactor diode capacitance and dc bias voltage as

$$\Delta t = \sqrt{L_{\text{line}}\left[C_{\text{line}} + C_{\text{diode}}(V_1)\right]} - \sqrt{L_{\text{line}}\left[C_{\text{line}} + C_{\text{diode}}(V_2)\right]}$$

The maximum delay time, Δt_{max}, is then determined by the maximum and minimum capacitances of the varactor diodes as

$$\Delta t_{\text{max}} = \sqrt{L_{\text{line}}\left[C_{\text{line}} + C_{\text{diode,max}}\right]} - \sqrt{L_{\text{line}}\left[C_{\text{line}} + C_{\text{diode,min}}\right]} \quad (16.3)$$

It is observed that to achieve the longest delay time with the shortest(lowest loss) transmission line, the highest possible ratio of diode capacitances, $C_{\text{diode,max}}/C_{\text{diode,min}}$, is required. In general, increasing the unloaded line impedance, Z_i(increasing line inductance given a fixed line capacitance), will decrease the Bragg cutoff frequency, but will increase the time delay per section. Alternatively, decreasing line capacitance has the effect of increasing the cutoff frequency but decreasing the time delay.

In NDL applications, varactor diodes are always reverse biased. At low frequencies(under 20 GHz), NDL circuit can be designed to have very low insertion loss(< 3 dB) over a very wide frequency band.

2.2 Choice of Varactor Diode

Varactor diode has a wide range of junction capacitance with variation in reverse dc bias level and has high cutoff frequency. The capacitance ranges from a few pF to over 100 pF in commercial beam-lead products. As mentioned previously, the change of transmission line characteristics is primarily due to capacitance change of the varactor diodes. Therefore, diode performance becomes a key issue for NDL design. There are several important parameters. The first is the diode capacitance ratio, $C_{\text{diode,max}}/C_{\text{diode,min}}$, within the applied dc bias range. According to (16.3), higher $C_{\text{diode,max}}/C_{\text{diode,min}}$ ratio results in longer time delay. To increase this ratio, commercial varactor diodes are usually fabricated on substrates which have hyperabrupt or even superhyperabrupt doping profiles. Another parameter is the diode dynamic cutoff frequency, f_c, which is defined as

$$f_c = \frac{1}{2\pi R_s}\left(\frac{1}{C_{\text{diode,min}}} - \frac{1}{C_{\text{diode,max}}}\right)$$

where R_s is the series resistance of diode.

Dynamic cutoff frequency is required to be significantly higher than the Bragg cutoff frequency, not only because of the high operating frequency, but also to ensure that the series resistance is reduced. It is suggested that the diode cutoff frequency should be approximately five times higher than the Bragg cutoff frequency [15]. When the diode cutoff frequency is near the Bragg cutoff frequency(R_s is increased), loss increases mainly due to the series resistance of diode.

Physical size of the diode determines the section length, ℓ. Eq.(16.1) indicates that the Bragg cutoff frequency is increased when the section length is reduced. In order to achieve high operating frequency, it is necessary to fabricate the NDL devices by monolithic technologies to efficiently reduce their size.

2.3 Choice of Interconnection Transmission Line

Examples of coplanar transmission lines are coplanar waveguide(CPW) and coplanar strip line(CPS). One distinct advantage of coplanar lines is that mounting of planar components in shunt or series configuration is much easier, since drilling of holes through substrate for ground connections is not necessary. Hence, they are attractive for NDL design.

At low frequencies, both CPW and CPS support quasi-TEM mode, with its group velocity equal to phase velocity, and both are frequency independent. CPWs have relatively low loss at low impedance levels of 50 Ω, while CPS lines have low loss at higher impedance levels of greater than 90 Ω [16]. In this NDL design, CPW structure is chosen because back-to-back diodes can double the usable capacitance with a fixed section number comparing with the CPS structure.

3. Proof-of-Principle Nonlinear Delay Line

3.1 Hybrid Nonlinear Delay Line Fabrication

Hybrid NDL technologies developed at UC-Davis have already undergone two generations. Hybrid NDLs basically use printed circuit boards(PCBs) upon which commercial beam-lead diodes are mounted. Dimensions of hybrid NDLs are not only limited by physical diode size, but also by fabrication capability. Fig.16.4 shows a ten-section hybrid NDL consisting of a 10-mil printed circuit board(RT/Duroid 6010.5) loaded with commercial varactor diodes(MA/COM MA46580 beam-lead GaAs tuning diodes) as nonlinear elements. Silver epoxy (Epoxy Technology Inc., EPO-TEK-H20E) is used as electrically conductive adhesive for diode mounting [17]. Measured insertion loss and time delay variations with reverse dc bias levels are shown in Fig.16.5 [7].

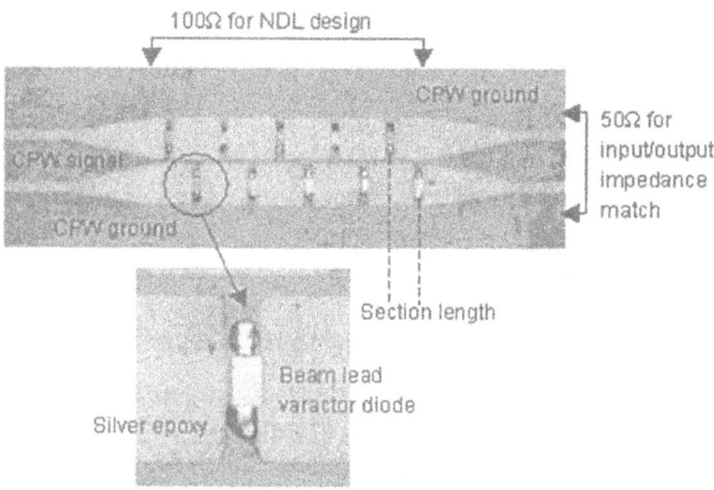

Figure 16.4. A ten-section hybrid NDL with silver epoxy mounted diode.

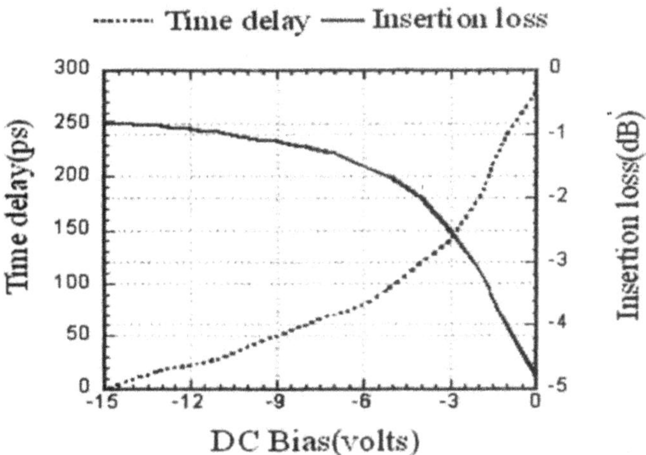

Figure 16.5. **Measured insertion loss and time delay as a function of bias voltage [7].**

3.2 Monolithic Nonlinear Delay Line Fabrication

Monolithic fabrication technology can significantly increase the Bragg cutoff frequency and reduce the total loss. A monolithic NDL development effort has been initiated, where nonlinear delay lines are fabricated on a 600 μm thick molecular beam epitaxy(MBE) GaAs wafer. As mentioned in the previous Section, the purpose of varactor design for NDL is to maximize the cutoff frequency, the $C_{\text{diode,max}}/C_{\text{diode,min}}$ ratio, and the breakdown voltage. All of these characteristics are related to the active layer doping profile, which means that these parameters can be calculated and optimized as a function of the doping profile.

Fig.16.6 shows the cross section of an MBE growth wafer. To acquire wide capacitance variation, active n-layer of the varactor is designed with a hyperabrupt doping profile from $3 \times 10^{17}\text{cm}^{-3}$ to $2.24 \times 10^{15}\text{cm}^{-3}$, namely, the net doping concentration of epitaxy decreases with distance from the interface between semiconductor and metal. An n^{+}-layer with thickness of 2.5 μm and doping of $4 \times 10^{18}\text{cm}^{-3}$(the highest possible substrate doping) is chosen to minimize the series resistance and thereby maximize the cutoff frequency. In addition, a very pure, undoped buffer layer is grown between the semi-insulating substrate and n^{+}-layer. It ensures low series resistance and prevents impurities in the substrate from diffusing into the epitaxial layer during wafer processing [18].

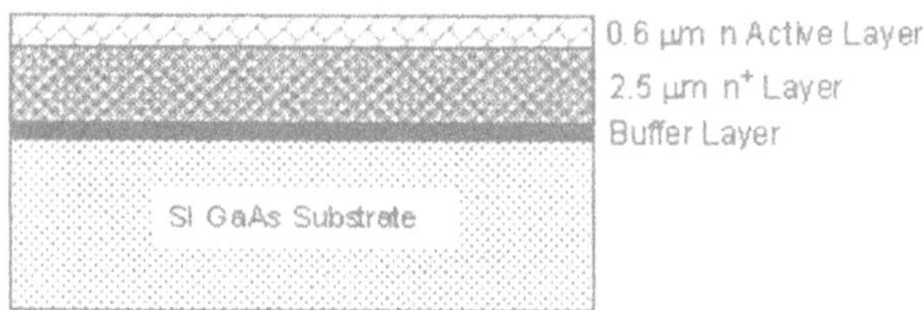

Figure 16.6. Cross section of an MBE growth wafer.

Further investigation on varactor performance has been conducted by utilizing the two-dimensional device simulator, Atlas, from Silvaco International Company. Fig.16.7 shows the varactor structure being simulated. In addition to the shape of different regions, critical dimensions of the device are also shown.

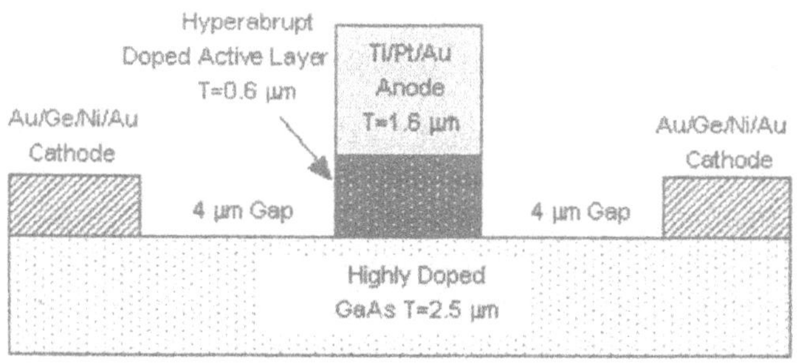

Figure 16.7. Varactor model in Silvaco simulation.

Since the program is limited to two-dimensional problems, only the cross section is required to build a device model. Junction capacitance and current are obtained by multiplying the associated program output with the anode length. In this example, the anode length is 10 μm. Using dc and ac analysis of the Atlas simulator, the varactor's C-V characteristics are extracted as shown in Fig.16.8.

The process is designed based on standard industry practice while considering specific facilities available in the GaAs foundry, Global Communication Semiconductor, Inc., for NDL fabrication. Fig.16.9 shows the wafer cross sections and top views after each fabrication step. Pho-

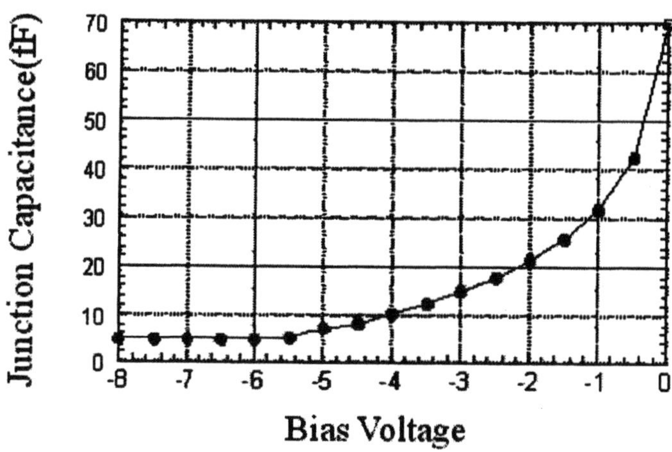

Figure 16.8. *C-V* characteristics extracted from the Silvaco simulation.

tographs of the fabricated nonlinear delay line are shown in Fig.16.10.

In this fabrication, an Au/Ge/Au/Ni/Au(0.18 μm) recipe has been applied for the Ohmic contact. After metal deposition, standard lift-off processing is used to remove metal from the Ohmic contact region. Before Au/Ge/Au/Ni/Au can be used as Ohmic contact, the wafer surface is heated until metallization alloys into GaAs by employing rapid thermal annealing for three minutes at 360° C. The Ohmic contact is found to have a contact resistance of $2 \times 10^{-6}\Omega\text{-cm}^2$. The $H_3PO_4/H_2O_2/H_2O$ wet etchant system has been employed to perform active layer etching for diode anode. Mesa etching is used to create an isolation, which will reduce the possibility to damage the varactor's active region and reduce leakage current among them.

Silicon nitride(Si_3N_4) is used most often as dielectric in GaAs processing, which provides a better diffusion barrier than silicon dioxide [19]. After the mesa etching, 0.3 μm silicon nitride deposition is performed in order to planarize and passivate the wafer. In addition, this nitride layer serves as an isolation layer between n^+ layer and metal interconnection on the mesa step. Reactive ion beam etching(RIE) is employed to open these vias for connection between circuit metal and diode anode and cathode. Finally, TiPtAu alloy of 1.6 μm thickness is deposited to form diode anode and circuit metal. TiPtAu is a standard material for GaAs Schottky contacts, which has a barrier of around 0.8 V [18].

Measured *C-V* and *I-V* characteristics of varactor exhibit 90 fF of capacitance at zero bias and 20 fF to 30 fF near breakdown region(6-

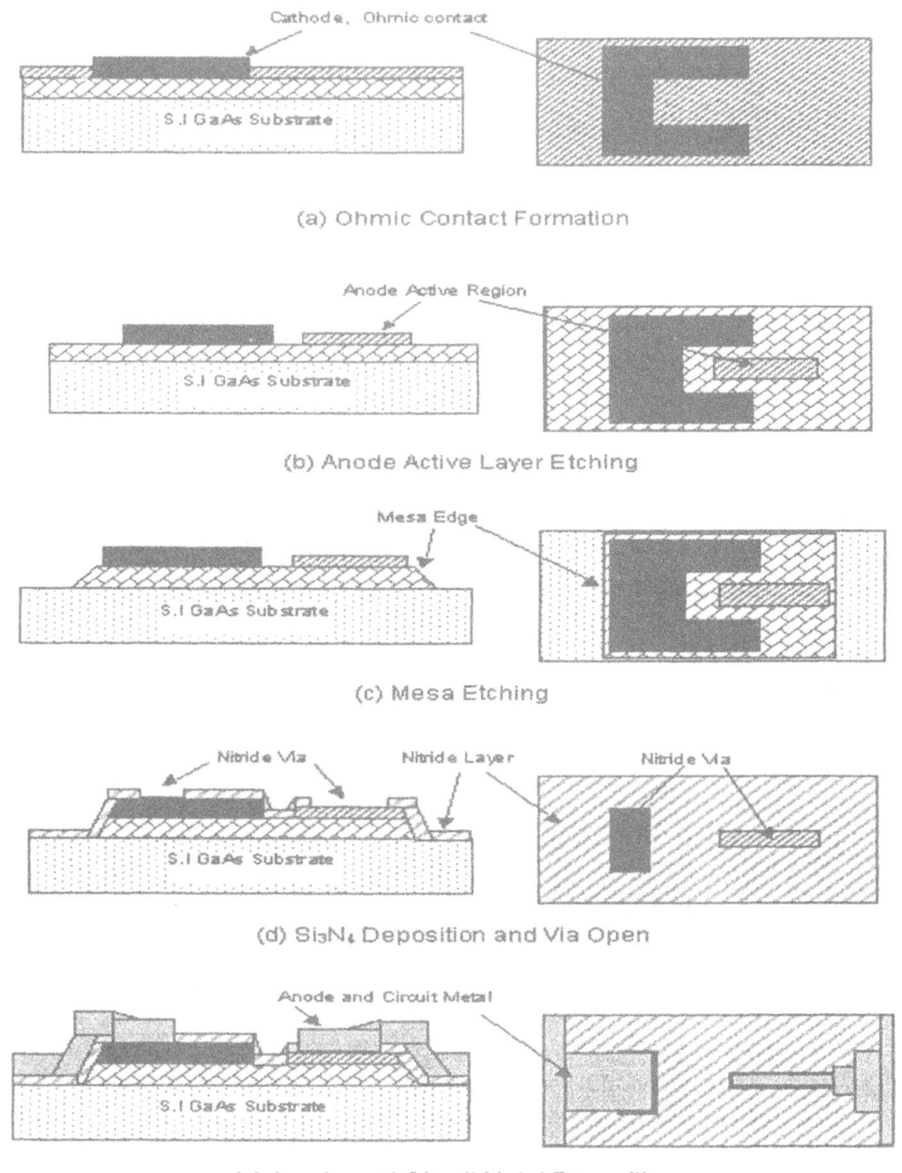

Figure 16.9. **Fabrication steps of monolithic nonlinear delay line.**

7 V), which yields a maximum achievable $C_{\text{diode,max}}/C_{\text{diode,min}}$ ratio of 4.5. By employing these measured varactor parameters in the Agilent-

Figure 16.10. (a) Eight parallel NDLs with each NDL consisting of 80 sections, (b) single section with a pair of Schottky varactors.

ADS simulator, the predicted insertion loss and propagation time versus frequency and bias voltage are shown in Fig.16.11.

Insertion loss and time delay of the delay line are measured using an HP8510C vector network analyzer and shown in Fig.16.12. Two-port TRL calibration has been performed before measurement. Due to the monolithic configuration of NDL, a probe station and Cascade MicroTech ACP-50 probe system have been utilized in the frequency domain measurement. Measured data from 0 to -4V show that the insertion loss and time delay performance are close to the predictions by Agilent-ADS simulation. However, for frequencies in excess of 12 GHz, the NDL shows an extra loss and the time delay becomes unpredictable. Remainder of the monolithic NDLs are tested only over the bias range from 0 to -4V to avoid possible early breakdown damage which is caused by yield problems and nonuniformity in fabrication. Early breakdown of one or more varactors will result in short circuits, and the NDL becomes extremely lossy and exhibits an input return loss near 0 dB. Reduction of bias range causes reduction in both usable frequency band and usable delay time range. The delay time exhibits ripples over the operating frequencies in both measured and simulated data. It is caused by the mismatch of loaded delay line, and further optimization is required to eliminate or minimize ripples [18].

In summary, the measured data indicate that the NDLs will provide about 70 ps of time delay and a maximum loss of about 14 dB from 2 GHz up to 12 GHz. It should be mentioned that a similar structure serving as $0°$-$360°$ monolithic phase shifter has also been studied and

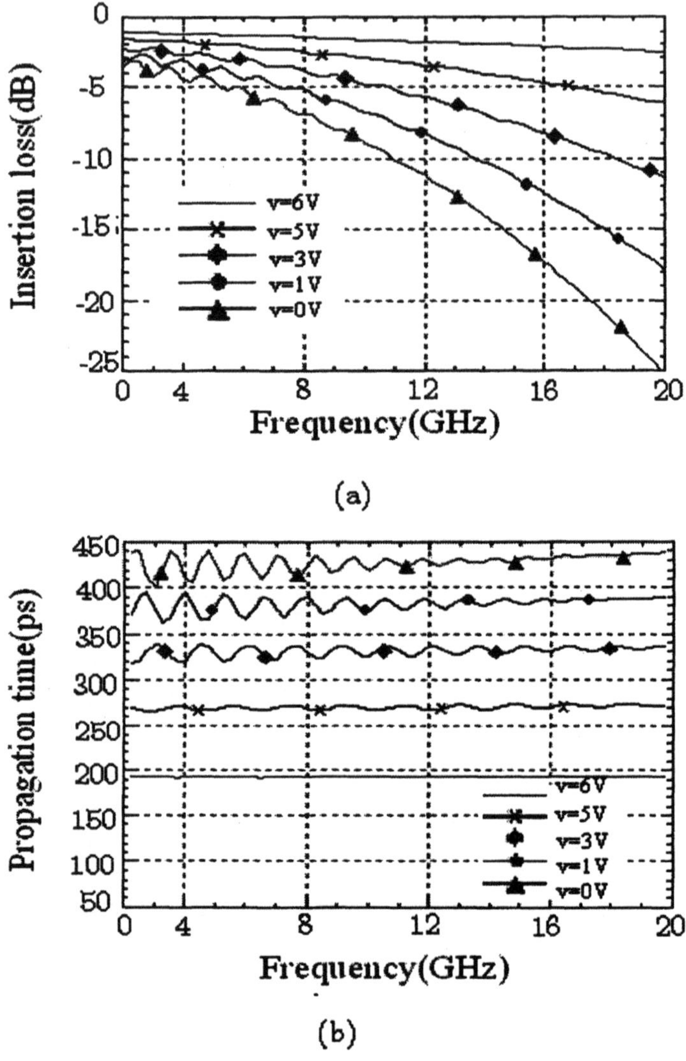

Figure 16.11. Simulated (a) insertion loss and (b) propagation time.

reported by researchers at UC-Santa Barbara, with a performance of less than 4 dB loss up to 20 GHz [14].

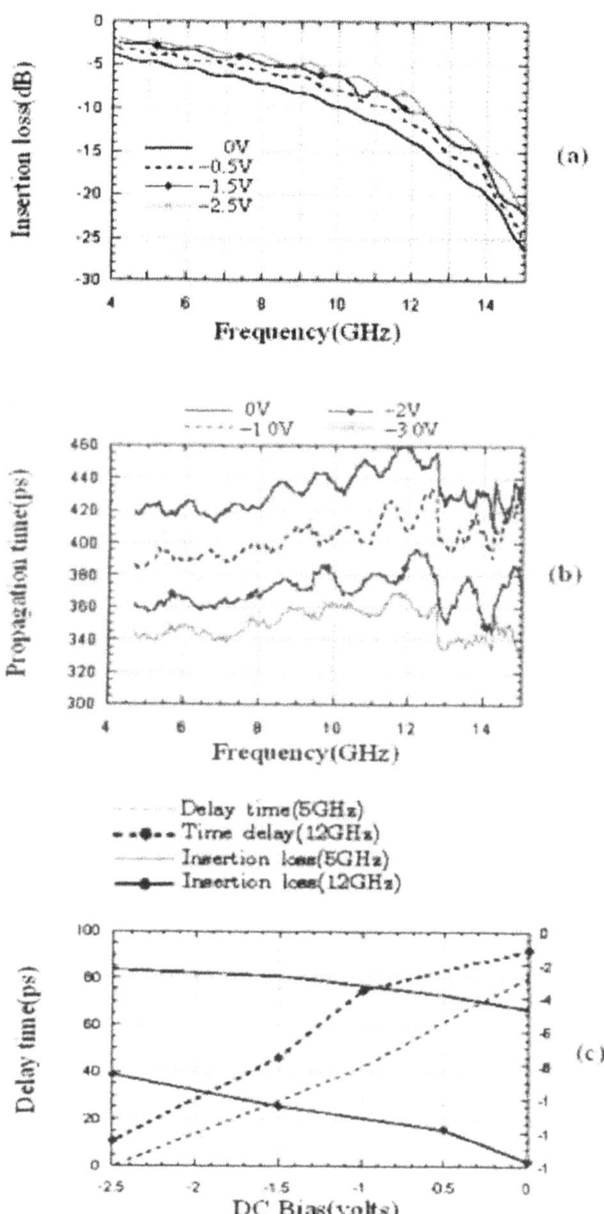

Figure 16.12. (a) Measured insertion loss versus frequency, (b) measured time delay versus frequency, (c) measured insertion loss and time delay versus dc bias.

4. Nonlinear Delay Line-Based Phased Antenna Array

4.1 Hybrid NDL-Based PAAs

NDLs have been employed by UC-Davis in phased antenna arrays to serve as true time delay units. Fig.16.13 shows the first phased antenna array with nonlinear delay line-controlled true time delay, implemented by Richard Hsia in 1997 [7]. Microwave signal is first fed via an eight-way Wilkinson power divider, and then passed through a bias board which provides dc bias for active elements of the array. The CPW-based, nonlinear delay line circuits following the bias board provide true time delay. Finally, the RF signal passes through CPW-to-slotline transitions to directly feed planar tapered slot antennas.

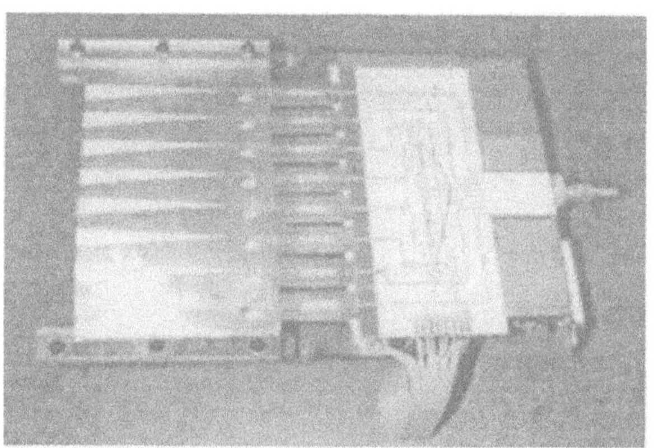

Figure 16.13. One-dimensional hybrid NDL-based phased antenna array.

Linear tapered slot antennas(LTSAs) are chosen as antenna elements to provide symmetric E and H-plane beam patterns. The required delay time is at least 100 ps if steering angle of $\pm 10°$ is to be achieved. Far-field pattern of the antenna element has been simulated by using the three-dimensional, full-wave EM simulator, HFSS, from Agilent. A uniplanar design including a CPW short and 90° radial slotline opens is used for CPW-to-slotline transition. To avoid any loss caused by impedance mismatch, RT/Duroid 6010.5 with 10 mil of thickness and dielectric constant of 10.5 is chosen as substrate, which is the same as used for the hybrid NDLs. A nickel wire is soldered between two ground planes as an air bridge at each CPW-to-slotline junction to ensure broadband operation. The bias board not only provides dc bias voltage to the

NDLs, but also provides a transition from microstrip line to CPW. The circuit is fabricated on a board 25 mil thick with $\epsilon_r = 10.2$. The eight-way Wilkinson power divider is fabricated by Applied Thin Film Products, using 99.6 % alumina as substrate material($\epsilon_r = 9.6$, thickness of 20 mil), PtN/TiW/Au as metallization, and gold of 3.5 μm thick as interconnections. All the components and bias circuitry are integrated onto a Lexan mount with $\epsilon_r = 2.98$. In order to make electrical connections between these components, 1-mil gold wire is bonded at the junction of CPW output on the bias board to the NDL and also between NDL and antenna. This system has demonstrated maximum scanning angles of $\pm 19°$ as shown in the measured patterns of Fig.16.14 [14].

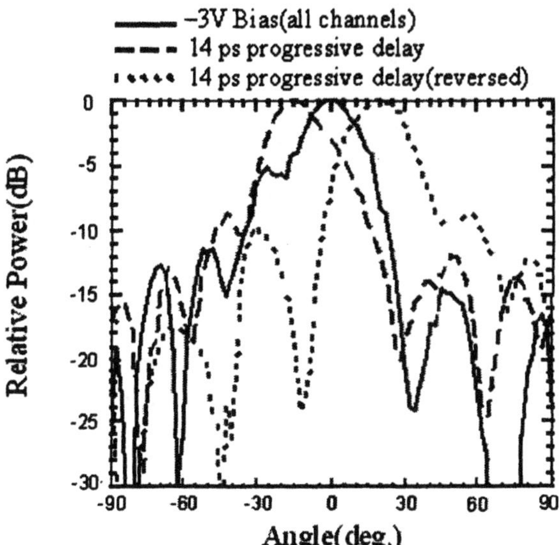

Figure 16.14. Beam steering measurements at 5 GHz of hybrid NDL-based PAA system([7], ©1998 IEEE).

To further confirm the theoretical predictions, a two-dimensional 4×4 PAA based on hybrid NDL system is fabricated as shown in Fig.16.15. Total size of the entire structure is approximately 30 cm $\times$ 18 cm $\times$ 15 cm. The measurement data indicate that it provides approximately $\pm 9°$ beam steering in the E-plane patterns and $\pm 7°$ beam steering in the H-plane patterns [20].

Figure 16.15. Two-dimensional hybrid NDL-based 4×4 phased antenna array.

4.2 Monolithic NDL-Based PAAs

Monolithic NDLs have also been employed in an one-dimensional, 1×8 phased antenna as shown in Fig.16.16 to demonstrate their beam steering capability [18]. The integrated system is similar to the hybrid NDL-based PAAs described in the previous Section. In order to reduce system assembly difficulties, the antenna array, bias network and microstrip-to-CPW transition are designed on a single PC board. Both the monolithic NDL chip and the eight-way Wilkinson power divider are mounted on board by using silver epoxy.

Fig.16.17 shows both the predicted patterns using HFSS and the measured patterns at 9.5 GHz. The beam steering capability is originally designed to be $\pm20°$. Measurement shows that the beams are steered from $-8°$ to $+10°$ because premature breakdown of the varactors reduces the permissible bias range from 8 V to only 4 V, and consequently reduces the controllable delay time. However, the measurement is still in good agreement with the simulated patterns under the same delay time conditions. The asymmetric antenna patterns are due to extra reflection from cable and connector. In future designs, dc bias connector and cable should be placed at the end of array to eliminate these reflections.

5. Conclusions and Future Developments

Devices with nonlinear delay line and related applications are continuing to receive attention and are becoming more mature. Hybrid NDLs

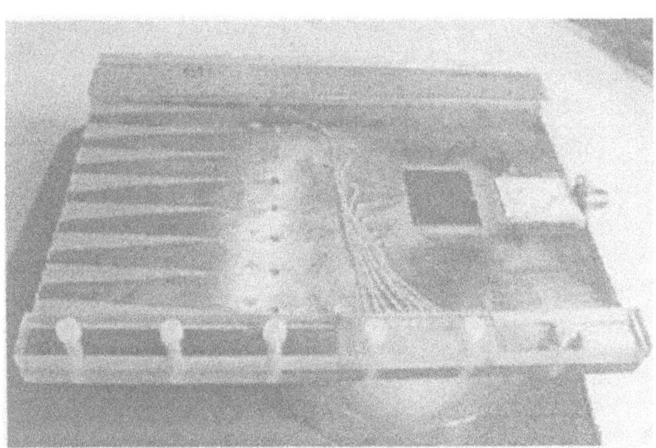

Figure 16.16. **One-dimensional monolithic NDL-based phased antenna array.**

have been successfully demonstrated both in one and two-dimensional PAA systems. To achieve higher operating frequencies, monolithic NDLs have been developed and applied in one-dimensional PAA systems. Recently, a beam shaping PAA based on monolithic NDLs has been proposed as an electronic lens in a plasma reflectometric diagnostic application and is currently under design and fabrication [20]. Comparing both hybrid and monolithic NDL devices, the hybrid NDL is easier to fabricate and can provide longer time delay using fewer sections. However, monolithic NDL can significantly reduce the diode size to largely increase the operating frequency. Yield and nonuniform fabrication may cause early breakdown, which will reduce the reverse bias range, and thus the usable frequency and delay time ranges.

There are still several challenges to be overcome in NDL design. One is the impedance mismatch problem. The impedance of loaded line varies under different dc bias voltages, which results in difficulty of input/output impedance matching. A possible solution is to use cascaded sections of variable impedance to mitigate mismatch. The NDL currently being pursued actually operates at small signals, it is difficult to satisfy the high output power requirement in some applications without adding extra MMIC amplifiers. To overcome this problem, MEMS switches of bridge type instead of varactor diodes are currently under consideration in NDL design based on their high power handling capability [21], [22].

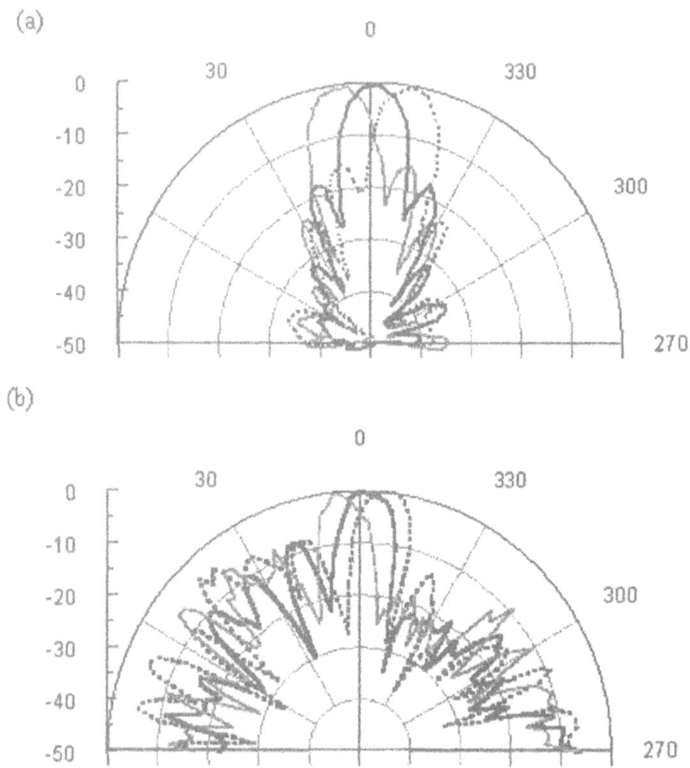

Figure 16.17. **(a) Predicted antenna patterns by using HFSS, (b) beam steering measurements at 9.5 GHz of the monolithic NDL-based PAA system.**

Acknowledgments

The authors acknowledge the support of the U.S. Department of Energy under contract no.DE-FG0-95ER-54295. The authors also like to thank Dr. Weimin Zhang, Dr. Richard. P. Hsia and Dr. Calvin Domier for their contributions in this work.

References

[1] R. J. Mailloux, *Phased Array Antenna Handbook*, Norwood, MA: Artech House, 1994.

[2] Y. Ayasli, "Microwave switching with GaAs FETs," *Microwave J.*, vol.25, no.11, pp.61-74, Nov. 1982.

[3] I. Frigyes and A. J. Seeds, "Optically generated true-time delay in phased-array antennas," *IEEE Trans. Microwave Theory Tech.*,

vol.43, no.9, pp.2378-2386, Sept. 1995.

[4] W. Ng, A. A. Walston, G. L. Tangonan, J. J. Lee, I. L. Newberg, and N. Bernstein, "The first demonstration of an optically steered microwave phased array antenna using true-time-delay," *IEEE J. Lightwave Tech.*, vol.9, no.9, pp.1124-1131, 1991.

[5] D. Dolfi, M. Labeyrie, P. Joffre, and J. P. Huignard, "Liquid crystal microwave phase shifter," *Electron. Lett.*, vol.29, no.10, 1993.

[6] W. M. Zhang, R. P. Hsia, C. Liang, G. Song, C. W. Domier, and N. C. Luhmann, Jr., "Novel low-loss delay line for broadband phased antenna array applications," *IEEE Microwave Guided Wave Lett.*, vol.6, no.11, pp.395-397, 1996.

[7] R. P. Hsia, W. M. Zhang, C. W. Domier, and N. C. Luhmann, Jr., "A hybrid nonlinear delay line-based broad-band phased antenna array system," *IEEE Microwave Guided Wave Lett.*, vol.8, no.5, pp.182-184, 1998.

[8] T. Yun and K. Chang, "A low-cost 8 to 26.5 GHz phased array antenna using a piezoelectric transducer controlled phase shifter," *IEEE Trans. Antennas Propagat.*, vol.49, no.9, pp.1290-98, Sept. 2001.

[9] D. Jeger and J. P. Becker, "Distributed variable-capacitance microstrip lines for microwave applications," *Appl. Phys.*, vol.9, no.9, pp.1124-1131, 1977.

[10] C. J. Madden, M. J. W. Rodwell, R. A. Marsland, Y. C. Pao, and D. M. Bloom, "Generation of 3.5 ps fall time shock-waves on a monolithic GaAs nonlinear transmission line," *IEEE Electron. Dev. Lett.*, vol.9, pp.303-305, June 1988.

[11] C. J. Madden, R. A. Marsland, M. J. W. Rodwell, and D. M. Bloom, "Hyperabrupt-doped GaAs nonlinear transmission line for picosecond shockwave generation," *Appl. Phys. Lett.*, vol.54, no.13, Mar. 1989.

[12] M. J. W. Rodwell, M. Kamegawa, R. Yu, M. Case, E. Carman, and K. S. Giboney, "GaAs nonlinear transmission lines for picosecond pulse generation and millimeter-wave sampling," *IEEE Trans. Microwave Theory Tech.*, vol.39, no.7, 1991.

[13] W. M. Zhang, "Nonlinear transmission lines and applications," Ph.D. thesis, UCLA, 1996.

[14] A. S. Nagra, J. Xu, E. Erker, and R. A. York, "Monolithic GaAs phase shifter circuit with low insertion loss and continuous 0-360° phase shift at 20 GHz," *IEEE Microwave Guided Wave Lett.*, vol.9, no.1, pp.31-33, Jan. 1999.

[15] W. Birk, H. Kibbel, C. Warns, A. Trasser, and H. Schumacher, "Efficient transient compression using an all-Silicon nonlinear transmission line," *IEEE Microwave Guided Wave Lett.*, vol.8, no.5, pp.196-198, May 1998.

[16] I. Bahl and P. Bhartia, *Microwave Solid State Circuit Design*, Wiley-Interscience, 1988.

[17] R. P. Hsia, "Microwave and millimeter wave radar and imaging technology and applications," Ph.D. thesis, UCLA, 1998.

[18] C. Liang, "Novel microwave and millimeterwave technologies for radar applications," Ph.D. thesis, UCD, 2001.

[19] R. Williams, *Modern GaAs Processing Methods*, Artech House, 1990.

[20] C. C. Chang, C. Liang, R. P. Hsia, C. W. Domier, and N.C. Luhmann, Jr., "True time phased antenna array system based on nonlinear delay line technology," *Asia-Pacific Microwave Conf.*, pp.795-800, Taiwan, 2001.

[21] N. S. Barker and G. M. Rebeiz, "Distributed MEMS true-time delay phase shifters and wide-band switches," *IEEE Trans. Microwave Theory Tech.*, vol.46, no.11, pp.1881-1998, Nov. 1998.

[22] A. Borgioli, Y. Liu, A. S. Nagra, and R. A. York, "Low-loss distributed MEMS phase shifter," *IEEE Microwave Guided Wave Lett.*, vol.10, no.1, pp.7-9, Jan. 2000.

Chapter 17

SPATIAL POWER COMBINERS USING ACTIVE PLANAR ARRAYS

Marek E. Bialkowski
School of Information Technology and Electrical Engineering
University of Queensland
St. Lucia, Brisbane, Australia

Abstract A review of spatial power combiners employing tile or tray configurations of active planar arrays is presented. First, reasons for researching space-level combiners of millimeter wave solid-state devices are explained. Then, alternative combining structures are compared as to issues like ease of manufacturing, integration with signal launching/receiving devices, operational bandwidth and heat removal.

Keywords: power combiner, amplifier combiner, tile configuration, tray configuration, hybrid tile-tray configuration, tapered slot antenna, slab-beam combiner, transmit-array, reflect-array.

1. Review of Basic Concepts

Very rapid growth of terrestrial and satellite communication in the last decades of 20th century and the resulting heavy congestion at low microwave bands have been the major driving forces for exploring upper microwave and millimeter wave frequencies. One of the major requirements for a successful shift to the new frequency spectrum is the availability of high-power transmitters. Traditional devices for high-power signal generation and amplification in this frequency range include vacuum tubes such as traveling wave tubes(TWT). These devices offer kW power levels and require high-voltage supplies(often in the 1,000V range) and considerable warm-up time, which are drawbacks in many applications. In contrast to TWT, solid-state devices such as diodes or transistors do not require high-voltage supplies for their operation and their warm-up time is negligible. Current advances in solid-state technology offer two-terminal devices(such as Gunn and IMPATT diodes) operating in excess of 100 GHz and three-terminal devices(such as MESFET, HEMT or HBT) reaching 100 GHz. Due to their working principles, solid-state devices are lumped in appearance, with dimensions being only a fraction of the operational wavelength. As generated power is related to the size of a solid-state device, requirement for small size/wavelength ratio is one of the reasons for its limited power capability at millimeter wave frequencies. It becomes apparent that in order to boost the output power levels at these frequencies, signals generated by individual devices have to be combined.

Combining powers from individual microwave or millimeter wave sources(oscillators or amplifiers) has been the subject of many research works and results in different techniques that serve this purpose [1]-[3]. Depending on the choice of either oscillators or amplifiers, power combiners can be categorized into two configurations as shown in Fig.17.1.

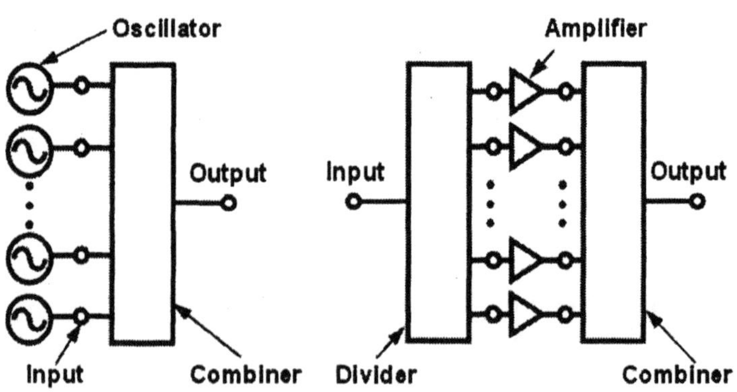

Figure 17.1. General power combining configurations using (a) oscillators and (b) amplifiers.

The first configuration combines a set of oscillators into a more powerful synchronized source. In the second configuration, there is a single low-power source with the output signal split and applied to the inputs of amplifiers whose outputs are combined into a common port to create a more powerful source. The following idealized equations for combined power emphasize the fundamental differences between these two types of power combining configurations. If N identical and isolated oscillators(each generating power P_{indiv}) form the combining structure of Fig.17.1(a), the expected combined power is $P_{\text{comb}} = NP_{\text{indiv}}$. In the combining structure of Fig.17.1(b), with individual amplifiers having the same gain of G_{indiv}, the expected combined output power is $P_{\text{comb}} = G_{\text{indiv}}P_{\text{in}}$, where P_{in} is the power at the divider's input. By comparing these two equations, it is observed that in oscillator combiners, increasing the number of oscillators increases the combiner's output power. In amplifier combiners, the output power is the product of input power and gain of individual amplifiers, and it does not depend on the number of amplifiers. However, increasing the number of amplifiers helps generating high power levels.

From the perspective of wave guiding medium, power combiners can be classified into three major categories: chip-level, circuit-level and space-level combiners as illustrated in Fig.17.2.

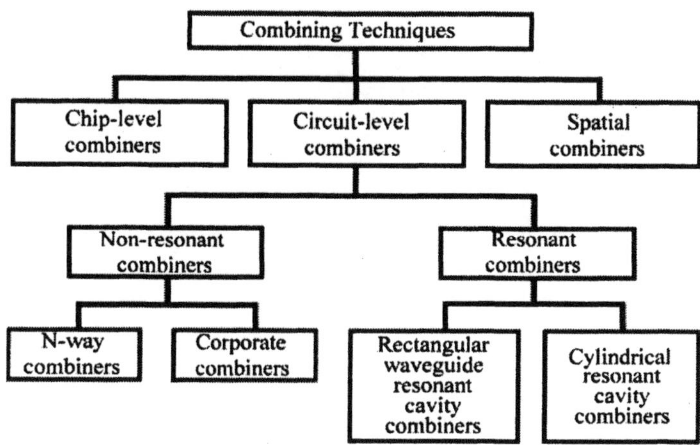

Figure 17.2. Classification of power combining techniques.

The chip-level combiners can be viewed as a subset of circuit-level combiners, in which the separation between active devices is small. The circuit-level combiners exploit a variety of metal/dielectric guides to connect active devices and utilize wider spacing than the chip-level combiners to prevent catastrophic failure.

Success of circuit-level combiners has been noticed at lower microwave frequencies. At upper microwave and millimeter wave frequencies, these types of combiners suffer from high conductor and dielectric losses, especially when a large number of active devices are combined. This is because a long path is often required to connect separate devices. Conductor and dielectric losses, tight manufacturing tolerances and multimoding(excitation of undesired higher-order modes that can adversely affect combiner performance) have been the main drives for exploring virtually conductor and dielectric loss-free space-level combiners. The main difference between space-level combiners and their circuit-level counterparts is that in addition to active devices and their biasing and impedance matching circuits, antennas become their integral part. Integration of active device with an antenna renders spatial power combiners similar to active antenna array [4]. Different types of antennas including dipoles, patches, notch antennas and open-ended waveguides can be used to this purpose, resulting in a variety of space-level combiners. Fig.17.3 shows one of the earliest configurations introduced in microwave literatures that led to developing the concept of spatial power combiners [4]. In this case, the combiner operates as an active array with antenna elements driven by individual amplifiers.

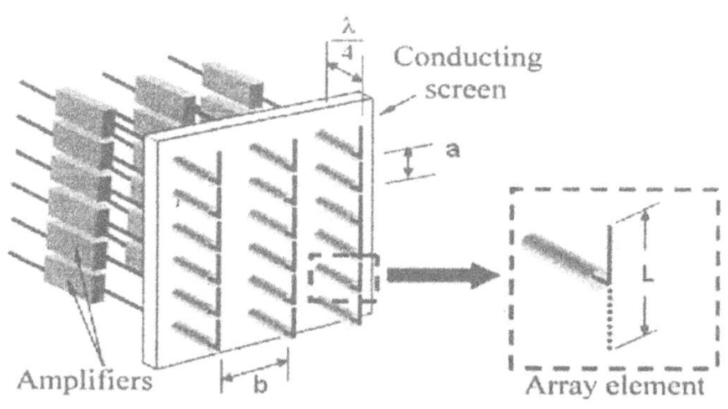

Figure 17.3. Configuration of space-level combined amplifiers [4].

In the following Sections, space-level power combiners that use tile or tray configurations of planar antenna arrays to obtain powerful solid-state microwave and millimeter wave sources will be discussed. The merits and drawbacks of these two alternative configurations as to ease of manufacturing, integration with signal launching/receiving devices, operational bandwidth and heat removal will be thoroughly discussed. Special attention is given to combining amplifiers instead of oscillators, and the reason behind is explained.

2. Oscillators versus Amplifiers

Similar to their circuit-level counterparts, spatial combiners can combine power from individual oscillators or amplifiers. Thus, the two combining configurations introduced in Fig.17.1 also apply to spatial combiners. Notice that most of the early works on spatial combiners was devoted to oscillators [5]-[9]. One possible reason for this progress might be the crude expectation that output power of the oscillator combiner could be almost infinitely increased by merely increasing the number of identical oscillators. The other reason was perhaps due to the concept of quasi-optical oscillator power combiner introduced by Mink [5], which was successfully followed by many other researchers. The term quasi-optical denotes the extension of techniques for very short wavelength(optical) to relatively longer wavelength(microwave or millimeter wave).

The quasi-optical combiner introduced by Mink consists of a Fabry-Perot resonator with two reflectors(one is nontransparent behind the grid-oscillator array, and the other is partially transparent and placed in front of the array) to synchronize individual oscillators and to combine their signals in the broadside direction of array [5]. This power combiner configuration is illustrated in Fig.17.4. It is assumed that the power generated by the array of oscillators is coupled to the lowest order Gaussian mode of resonator. Practical realization of this concept is demonstrated using bar-grid oscillators [6], [7]. One problem noted with the Fabry-Perot resonator is that both frequency and power levels are strongly dependent on the location of reflectors, which is an undesired feature in practical applications. In order to control the frequency and power independently, new approaches are required.

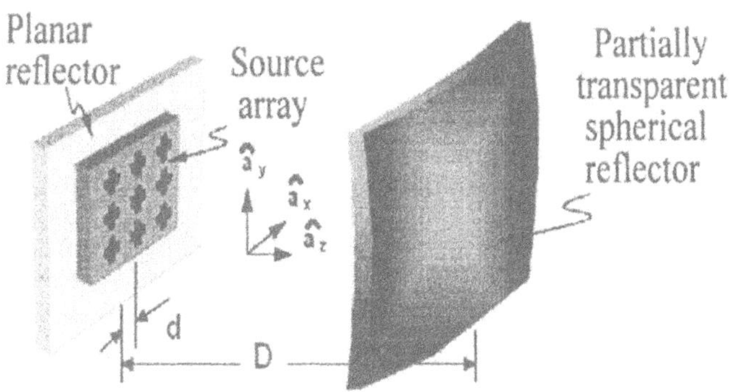

Figure 17.4. **Quasi-optical power combining configuration using open resonator structure [5].**

One alternative suggested in [7] is injection locking technique using external sources. However, practical realization of this technique with weakly or strongly coupled oscillators is challenging [8], [9]. Because active elements operating at the same frequency require identical phasing while maintaining suitable power level of injected signal, which is difficult to accomplish in practice. During these investigations, it is realized that in finite arrays, mutually coupled(via surface or space waves) oscillators exhibit nonidentical surrounding environments. This is particularly true when comparing central and side elements of the array. Consequently, when oscillators are synchronized to the same frequency, their phases are not necessarily identical. This unequal phase distribution across the array is responsible for beam divergence from the broadside direction, which is considered as power leakage from the combiner. This adverse phenomenon, however, is positively utilized in beam steering of coupled oscillator arrays without phase shifter [10].

It is shown that by forcefully introducing phase differences to extreme side elements of the array, its beam can be suitably steered. This concept is explained in [11] by considering linear arrays of evenly spaced inter-injection locked oscillators equipped in the radiating elements. In order to steer the beam, these oscillators require suitable phase progression per oscillator. From the theory of coupled oscillators, it is known that the maximum phase progression is always less than 90°. When it approaches 90°, the coupling between adjacent oscillators sets to zero. In practice, phase progression of up to 60° has been demonstrated. For active elements(an oscillator plus a radiating element) spaced one half-wavelength, phase progression of 60° leads to about ±20° beam steering. Larger beam scan angles can be realized using smaller spacing. This requires employing reduced-size radiating elements such as patch antennas on high-permittivity substrate or patches internally shorted by conducting pins.

The problem that requires further investigations is the rate at which beam steering angle can be switched without excessive phase error or loss of lock. Another issue of concern is the extension of one-dimensional array of inter-injection locked oscillators to two-dimensional arrays. Research works in this area is ongoing due to the anticipated benefits of low-cost beam steering technique [12]. However, as far as power combiners are concerned, practical challenges met in the oscillator arrays(frequency and phase locking, rate of frequency change, and independent control of frequency and power level) have been the decisive factors behind the choice of amplifiers over oscillators in space-level power combiners.

3. Tile Configurations of Amplifier Combiners

The oscillator combiners discussed so far are assumed to operate as active arrays radiating in free space. In many applications, the generated power is delivered to a waveguide port to be used for other purposes than emission to free space. In such cases, the designer needs to choose a suitable method for collecting the generated power. Many researchers on quasi-optical power combiners assume the availability of suitable lenses which will focus the generated beam into a receiving horn antenna. However, size of such a system and free space losses(due to inability to fully intercept the generated power) are not fully addressed.

For quasi-optical combiners of oscillators, accurate evaluation of losses is a cumbersome task. This is due to the difficulty of establishing reference level of input power, as only the output port of system is available for measurements. However, it becomes more feasible for amplifiers, as the input and output ports are accessible for tests [13]. The performance figures of merit of spatial power combiners, including different types of losses, are addressed in [14]. More recent works concerning losses in the spatial power combiners of amplifiers can be found in [15]-[17]. From basic electromagnetic field theories, one can deduce that for efficient power transfer from transmitter to receiver, devices used in the spatial power combining process such as lens, active array and horn need to have their fields matched in terms of polarization, amplitude and phase. This, in general, has to be fulfilled across infinite transverse planes due to the open configuration of combiner, or at least across portions of these planes where the density of transferred power is tangible.

When an active stage is formed by a planar array of identical unit-cell amplifiers(including receiving and transmitting antenna elements), one intends to illuminate the array by a normally incident uniform plane wave so that they can be excited by identical signals. With such an excitation, the signals amplified by unit cells can have equal amplitude and phase, which helps to reach high output power levels. Individual amplifiers can saturate simultaneously, offering the maximum added power and hence maximum dynamic range.

To achieve such conditions of uniform excitation, many researchers illuminate the active stages from their far-field region to emulate normal incidence of uniform plane wave. To minimize free space losses, the use of focusing lenses is suggested. Typical power combining structure involving focusing lenses and horns is shown in Fig.17.5.

As observed in Fig.17.5, in order to maintain an approximately uniform field across the active array, its size has to be comparable with the waist of the focused beam. For Ka-band system(with center frequency of about 30 GHz), the requirement of having the beam waist matching a 2 cm $\times$ 2 cm active stage leads to the use of lenses with 30 cm of diameter and spaced by 1.5 m. The resulting illumination flatness of 1 dB across the active stage is at the expense of 5 dB insertion loss between input and output stages of the system. This example shows the practical difficulties of realizing a compact open-space beam focusing system employing lenses, which can produce uniform illumination on an active stage with minimum free-space insertion loss.

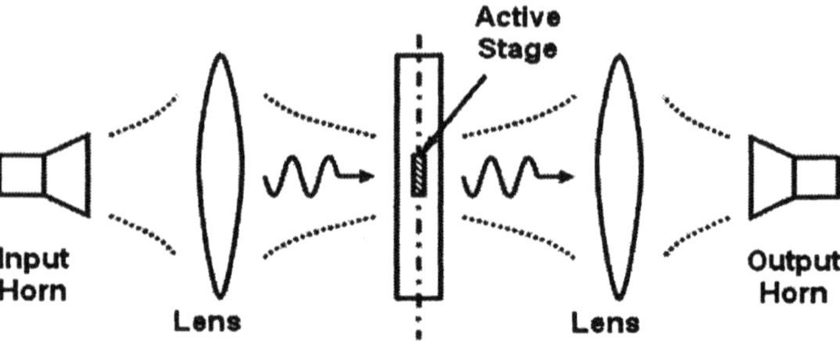

Figure 17.5. Spatial power combiner configuration including focusing lenses and horn antennas.

One may pose a more general question concerning the minimum insertion loss of power combining system consisting of a transmit array and a launching/receiving system with antennas positioned in the far-field region of the array. This problem is investigated in [17], in which the array elements are additionally phased to collimate the incident wave into the receiving antenna. The answer to this question depends on the assumed degree of illumination uniformity and the insertion loss. For example, when one aims for −10 dB illumination(with respect to the array center) of the side elements of a circular transmit array, minimum insertion loss between the ports of launching and receiving horns is approximately 6 dB. A higher insertion loss of about 8 dB is incurred if one aims for −8 dB illumination of the side elements. These results show that when signal launching and receiving devices are placed in the far-field region of active array, the insertion losses can be considerable, and in some instances may exceed the gain of active stage.

Minimizing free-space insertion loss to about 3 dB can be attempted by using launching and receiving antennas positioned in the near-field zone of active array. To achieve this, sectoral or pyramidal horns with or without internal dielectric lenses can be used. This arrangement is more challenging to the designer due to stronger interaction between active array and horn's conducting surface. As a result, more stringent requirements on matching the fields of array and horn for optimal power transfer need to be fulfilled. Assuming that the active elements are identical and uniformly distributed across the array, special type of horn antenna is required, which provides uniform field distribution across the array aperture [18]. For an ordinary pyramidal horn operating in the dominant mode(for example, excited by a rectangular waveguide operating in the dominant TE_{10} mode), the electric field distribution is approximately sinusoidal because the tangential electric field has to vanish on conducting side walls of the horn.

In order to improve the electric field uniformity across the waveguide aperture, regions close to the walls need to be modified. This can be achieved via the use of photonic band gap(PBG) material [19], [20] or dielectric slabs [18]. The PBG surface does not require the tangential electric field to vanish over it. If PBG is used on the horn's side walls, the tangential electric field does not approach zero and hence the field can be more uniform across the horn aperture. In an alternative arrangement, the use of dielectric slabs on the horn's side walls decreases the width of the region across which the tangential electric field has to diminish to zero. This is equivalent to increasing the width of the region over which the electric field distribution is approximately uniform.

In practice, implementation of PBG requires forming a double wall with the first layer of PBG backed by a solid conducting wall [20]. Hence, it is more involved in terms of labor compared with simple inclusion of dielectric slabs. Due to these reasons, the horn with dielectric slabs seems to be a more convenient option to obtain uniform field as illustrated in Fig.17.6. Using dielectric slabs, field nonuniformity of 2 dB in amplitude and 20° in phase is claimed in [21].

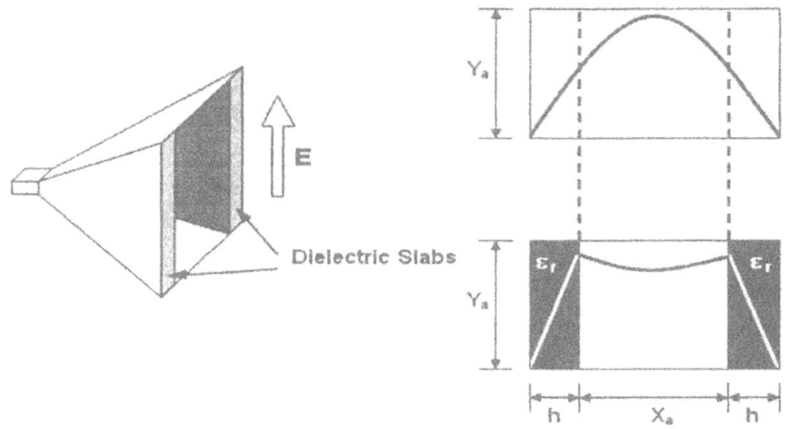

Figure 17.6. Configuration of a hard horn and the electric field distributions across the aperture of conventional horn and hard horn, respectively.

Using a pyramidal horn antenna in the near-field region to produce uniform illumination, special attention has to be paid to avoid adverse interactions between edge elements of the array and the conducting surface of horn [22]. Because dipoles, patches or slots used in the tile configuration leak their fields in transversal direction, exciting higher-order(nonuniform) modes in the horn aperture. By viewing the horn as an over-sized waveguide, similar multimoding can be expected as observed in power combiners of over-sized waveguide type. Excitation of higher-order modes is usually pronounced in the form of dips in the plot of combined amplifier gain. The resulting interactions may reduce the operational bandwidth of a combiner, which is already narrow due to the use of resonant type of radiating elements. The problem of adverse near-field interactions between active array and horn has been briefly addressed in [22]. In order to reduce such undesired effects, their spacing can be increased [22]. However, this is achieved at the expense of higher insertion loss, which leads to overall gain reductoin of the power combiner.

Operational bandwidth and insertion loss of tile amplifier combiner depend on the type and location of radiating elements(such as grids/wires, slots or patches) forming the active array. In contrast to an oscillator array, active stages in the amplifier array need to be equipped with both receiving and transmitting antennas. In addition, both sides of the ground plane may be required to accommodate these antennas. In order to avoid oscillations which may arise due to undesired feedback, suitable isolation between input and output ports of each active stage has to be maintained. Note that to avoid oscillations, other stability criteria applied to standard amplifier design need also be fulfilled. This means that antennas connected to the input and output ports of amplifiers have to achieve a suitable level of isolation, generally greater than the gain of active stage, with some margin to avoid gain ripple. This can be achieved using orthogonal polarizations and by separating receiving and transmitting antennas with a suitable distance. The resulting cell size and spacing are determined by the size of receiving and transmitting antennas, and to a lesser degree, by the size of active stage and its biasing circuits which are usually smaller than the antennas. For bar-grid cells which use dipole antennas, cell dimensions can be as small as 0.2 free-space wavelength [13]. Typical configuration of bar-grid amplifier combiner revealing details of the unit cell is shown in Fig.17.7. Small size of the cell is achieved using printed dipole antennas which are driven by ports 1-2 and 3-4, respectively.

Slot or patch antennas used in other varieties of planar spatial power combiners are bigger in size, and the cell spacing is on the order of 0.7 free-space wavelength. If wider spacing(for example, more than one wavelength) is chosen, capability for the array to receive signal becomes lower. Because larger portion of the ground plane is exposed to the incident signal and more signal is reflected back to the launcher.

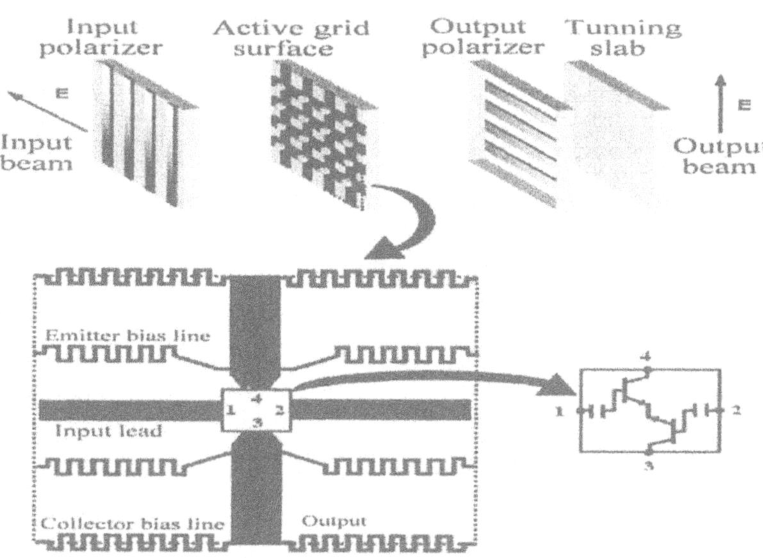

Figure 17.7. **Configuration of grid amplifier combiner showing polarizers, tuning slabs and active stages with orthogonal input/output printed dipole antennas [7].**

Examples of spatial amplifier combiners utilizing patch antennas are shown in Fig.17.8. All of the amplifier arrays shown in Fig.17.8 are of transmission type. One side of the array receives signal and passes it to the active stage. Following amplification, the signal is delivered(using suitable coupling mechanism) to antenna on the other side for radiation. The transmit-array configurations can also employ grids(wires) or slot antennas [24]-[26]. Such radiating elements have the capability of transmitting and receiving signals on both sides of the array, and their function is determined by the sense of polarization which has to match that of launching or collecting device. In contrast, the patch antennas radiate only on one side of the ground plane.

In order to receive and transmit signals, back-to-back patch tile arrays require suitable coupling mechanism between two sides of the ground plane. Fig.17.8 shows small holes/slots in the ground plane to serve this purpose. This arrangement offers broadband, low insertion loss coupling between two microstrip lines located on opposite sides of the ground plane. Due to the isolation offered by the ground plane, cross-polarization of the radiating elements is not strictly required. However, to further increase the isolation, cross-polarized antennas are used for receiving and transmitting, respectively.

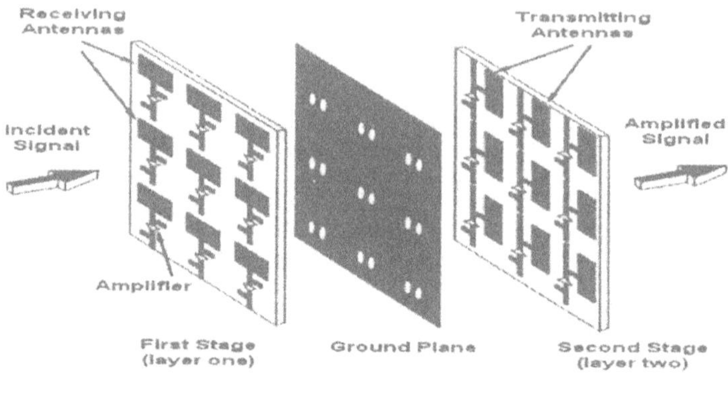

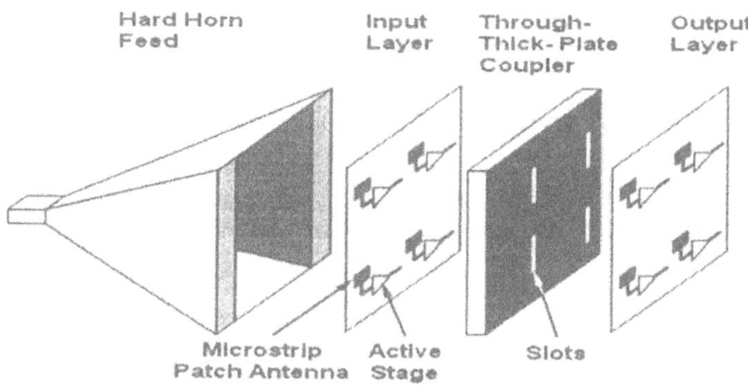

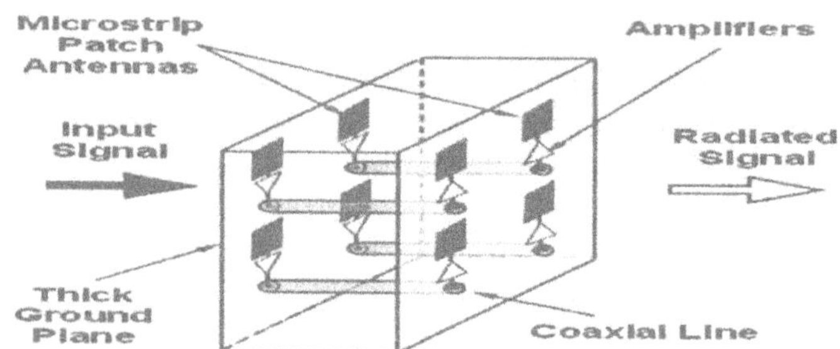

Figure 17.8. Variations of amplifier power combiners using two-stage back-to-back arrays of patches with coupling mechanism of (a) hole [18], (b) slot [30], and (c) coaxial line [27].

In an alternative arrangement shown in Fig.17.8(b), slots in the ground plane are used to transfer signals between two sides of the ground plane. This arrangement allows for a thick conducting ground plane which offers effective removal of heat which is generated by the active stages.

In order to reduce the amount of generated heat, only one side of the planar array is fed by amplifiers, for example, with edge-fed antennas connected to their input ports. The other side is paved with aperture or slot coupled antennas for transmitting the amplified signals [22].

A more efficient heat removal method is offered via coaxial lines connecting the two back-to-back active arrays, as shown in Fig.17.8(c). This configuration allows for introducing cooling liquid between the two ground planes. However, the system is more laborious to develop due to its hybrid structure. Fig.17.9 shows details of the cooling system presented in [27].

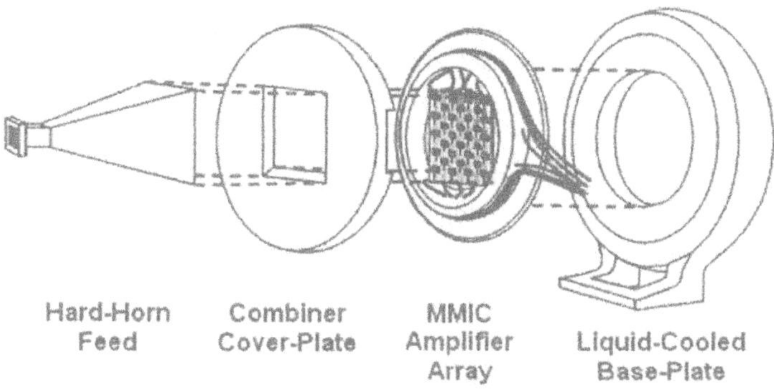

Figure 17.9. **Configuration of 25 W Martin Lockheed/NCSU/Caltech. power combiner including cooling system [27].**

Recognizing the difficulty of removing generated heat in the transmit-arrays, reflect-array configurations for accommodating active stages are proposed and demonstrated in [28], [29]. In contrast to the transmit-array configuration, the reflect-array uses only one side to receive and transmit signals. Hence, the other side can be connected to a cooling system without interrupting the radiation layer. The method of using reflect-array and orthomode horn(either in near or far-field regions) to combine amplified signals is illustrated in Fig.17.10.

The microstrip patches connected to active stages of the reflect-array are dual linearly polarized. One polarization is used for receiving signals while the other is used for transmitting the amplified signals. The purpose of using dual polarized patches instead of separate patches for two orthogonal polarizations is to minimize unit cell spacing for efficient power combining. The orthomode horn, located either in the near or far-field region, launches a linearly polarized spherical wave onto the reflect-array. The patches receive the incident wave with one polarization. The signal is passed to the active stage for amplification. The

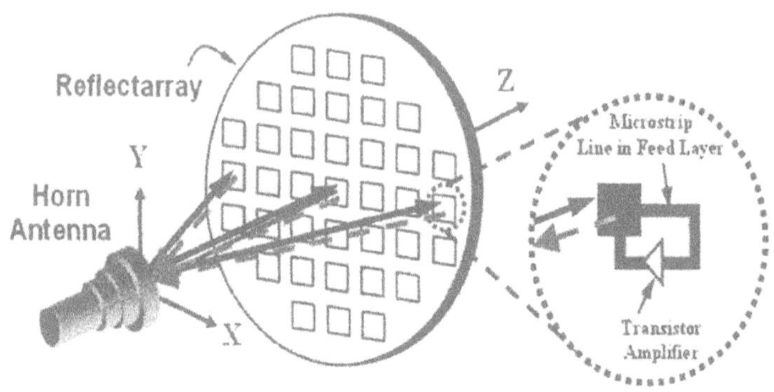

Figure 17.10. **Configuration of spatial power combiner utilizing active reflect-array and orthomode horn for signal launching and collecting.**

amplified signal is then transmitted back to the launcher with an orthogonal polarization. In order to convert the received spherical wave back to a cofocal spherical wave, unit cells of the reflect-array need to be suitably phased. This can be achieved by applying suitable phasing mechanism as generally applied to passive reflect-arrays. In [28], [29], transmission lines with variable length connecting orthogonal ports of the patch are employed to perform this function.

Phasing of unit cell is not necessary if a dielectric lens(positioned internally or externally with respect to the launching/receiving horn) is used to convert an incident spherical wave into a uniform plane wave, which is then intercepted by the reflect-array.

The key to successful realization of power combining using reflect-array is the availability of dual-feed microstrip patch antenna elements featuring good return loss and large isolation between their orthogonal ports. An aperture coupled microstrip patch antenna exhibiting about 10% of return loss bandwidth with an isolation greater than 35 dB is described in [22] and [29]. Using such an antenna element, active stages can be placed on the nonradiating side of reflect-array, where proper heat removal can be arranged.

4. Tray Configurations of Amplifier Combiners

Prior to inclusion in a combiner, amplifiers are often designed and tested using a 50 Ω coaxial line connected to a vector network analyzer. Usually, such tests show that amplifiers feature relatively wide operational bandwidth. However, when narrow band, resonant type of antennas are used to replace the 50 Ω input/output coaxial lines, their operational bandwidth is reduced. The operational bandwidth of amplifier can be maintained if antenna elements with wide impedance bandwidth are used. Traveling wave antennas, for example, feature wide operational bandwidth. The method of employing tapered slot antenna was first demonstrated in [31].

Integrating traveling wave antenna elements into an array requires a tray configuration because these antennas radiate in the end-fire direction. Two antennas are needed for space-level amplifier combining, one connected to the input and the other connected to the output of an amplifier. In contrast to the tile configuration, the spacing between two antennas is not constrained. Therefore, active stages of considerable size can be included in the array. In particular, multistage monolithic amplifiers can be fitted in. As antenna elements are oriented in opposite directions, large isolation can be achieved between the input and output ports of the active stage.

Two antennas connected back-to-back to the input and output ports of an amplifier constitute an active unit cell of the tray array. Repeating cells in a plane forms a tray. Fig.17.11(a) shows a practical implementation of this concept [32]. Following the design presented in [32], a single tray shown in Fig.17.11(a) houses four packaged MMIC amplifiers. Accommodating a larger number of active elements is not feasible due to problems associated with heat removal from central elements of the tray. Fig.17.11(b) shows a power combiner with vertically stacked trays. This vertical arrangement makes tapered slot antennas act as tapered fin lines inside a rectangular waveguide. The input/output ports of such a power combiner are ready for connecting to a rectangular waveguide.

It is worth mentioning that tapered slot antennas are not the only type of antennas that can be utilized in the tray type of power combiners. Fig.17.12 shows other types of antenna elements that can be used to design tray type of power combiners. Fig.17.12(a) shows quasi-Yagi antennas as elements of a tray power combiner [33]. These elements feature approximately one octave of return loss bandwidth, enabling similar bandwidth for the entire combiner. Fig.17.12(b) shows a hybrid tray-tile combiner. In this case, the active stages are assembled using tray configuration, while coupling to pyramidal horns is achieved using passive patch arrays arranged in tile configuration. Fig.17.12(c) shows a hybrid slab-beam combiner where the amplifiers are positioned between focusing lenses [35]. This type of combiner has shown a positive gain. This is in contrast to the space-level power combining systems discussed earlier, where focusing lenses are located in the far-field region of an active stage, and an overall negative gain is usually obtained. The configurations shown in Fig.17.12 offer increased space for accommodat-

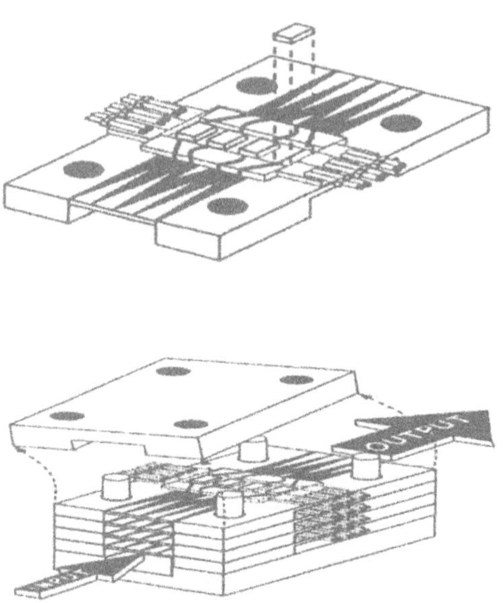

Figure 17.11. **Tray configuration of power combiner using tapered slot antennas, (a) single tray, (b) stack of trays using thick conductor ground planes for heat removal.**

ing active stages, which is the major advantage comparing with the tile configurations discussed earlier.

In order to accommodate a larger number of active stages across the tray, a hybrid solution is demonstrated in Fig.17.13 [36]. In this power combiner, the RF input is supplied to a sectoral horn that distributes the signal to 19 tray assemblies. Each tray consists of a receiving printed dipole antenna. Cascades of Wilkinson power dividers(including phase adjustment circuits) distribute the received signal to 16 multistage amplifiers, each equipped with a printed dipole antenna. Nineteen such trays consisting of drivers and power amplifiers are stacked vertically, and their output signals are space combined to a pyramidal horn antenna using a total of 304(19 $\times$ 16) printed dipoles.

For efficient power combining, a field approximately uniform in phase and amplitude is generated across the input(19 dipoles) and output(304 dipoles) ports of the stacked trays. In order to compensate for the spherical phase front, lenses inside the launching and receiving horns are used to generate planar phase front. Highly uniform field in amplitude and phase is maintained in the E and H-planes, respectively, of the pyramidal horn. Less uniform field is expected in the sectoral horn. The quoted power combining efficiency is approximately 45% for the entire power combiner. The resulting 61 GHz hybrid, circuit/space-level,

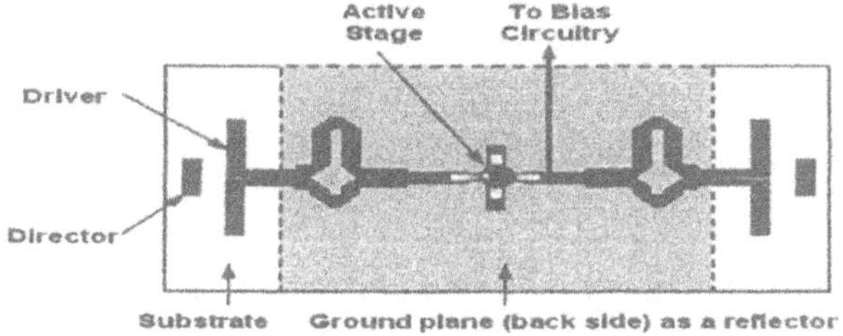

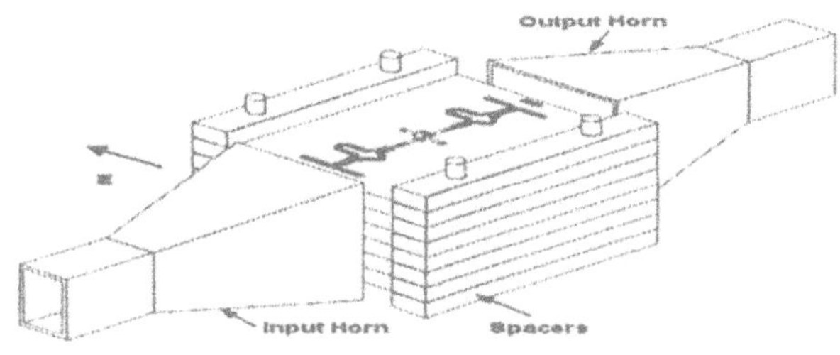

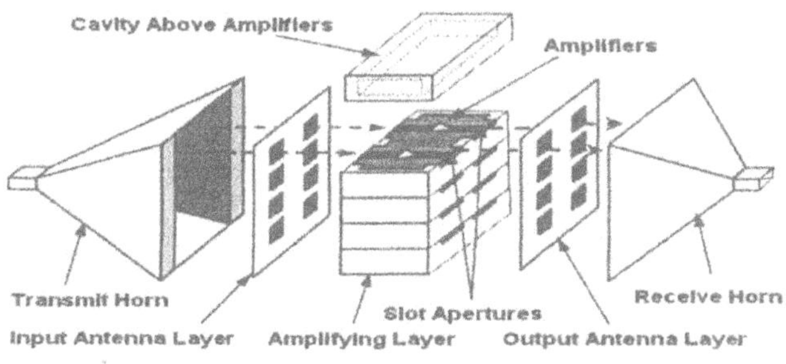

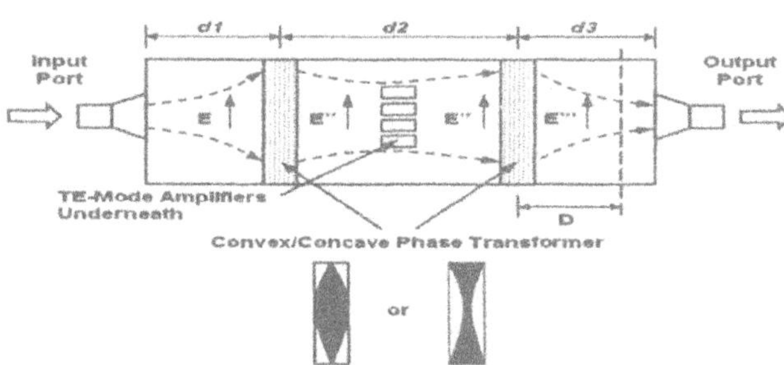

Figure 17.12. (a) Stacked tray combiner housing planar quasi-Yagi antennas [33], (b) waveguide-based aperture coupled patch amplifier array [34], and (c) hybrid slab-beam combiner [35].

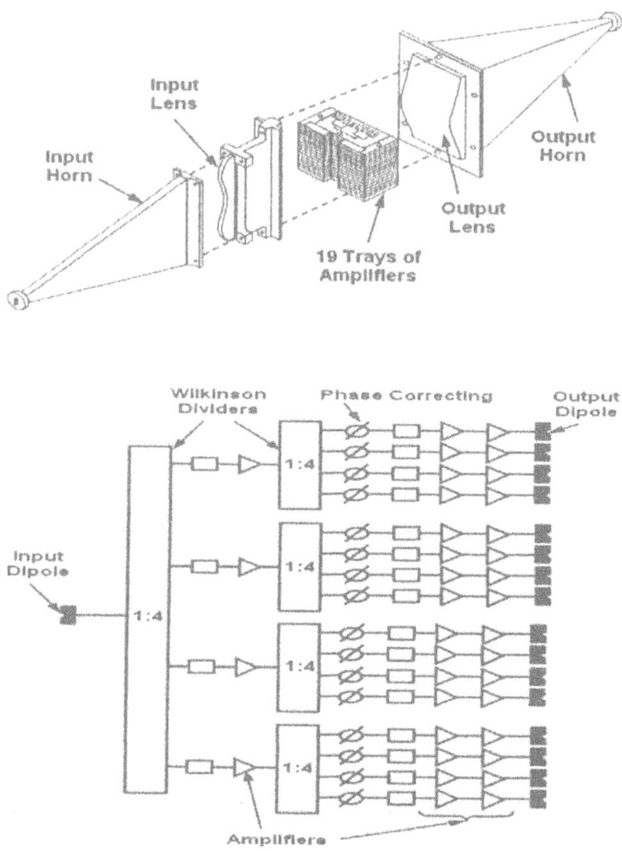

Figure 17.13. (a) Configuration of a tray type of spatial combiner including launching and collecting horns and (b) schematic of a single tray.

power combining system is able to generate 36 W CW power. This is the best result available for solid-state amplifier spatial combiners in this frequency range.

5. Discussions and Conclusions

Most space-level combiner configurations described in microwave literatures evolve from the concept of quasi-optical combiner involving tile configurations of planar arrays. Major advantage of this configuration is the ease of manufacturing. In particular, the planar form is very convenient for integrating a large number of active stages and antennas using monolithic manufacturing techniques. Alternatively, passive array can be built using photolithographic techniques and pretuned monolithic active stages can be added onto this structure to form an active array. A disadvantage of tile configuration is its narrow band operation due to the narrow band resonant type of antenna elements(dipoles, slots or patches) that are used in this configuration. The narrow operational bandwidth is also due to the near-field interactions between active array, signal launching/collecting devices and the over-sized waveguide enclosure accommodating the active stages. Such waveguiding structures are prone to multimoding, which was reported earlier in over-sized waveguide type of power combiners.

Other disadvantages include the difficulties of having multistage amplifiers in a unit cell due to limited space available in the tile arrays. Management of heat generated by active stages is another problem. Surface of the array is already fully packed with antennas and amplifiers, and very little space is left for inclusion of heat sinks. As a result, the generated heat has to be removed through the edges of a thick ground plane or via a dielectric backplane(only for grid type of amplifiers). The situation is more manageable for combiners using microstrip patch array configurations. However, as shown in the examples of tile configuration of power combiners, intricate through-plate transmission lines are needed to connect two back-to-back arrays. One alternative to overcome the difficulties associated with heat removal in tile configurations is via the use of reflect-arrays. This type of arrays can host both receiving and transmitting antennas on one side of the supporting ground plane, while active stages can be accommodated on the other side. Cooling arrangements can then be applied to the active stages without perturbing the radiating layer.

Despite the fact of being underrepresented in the spatial power combining literatures, trays of planar arrays offer better arrangements for housing multistage amplifiers [24]-[26]. Also, integration with horn antennas for distributing and collecting the combined signal results in broader bandwidth than using tile configurations. With regard to heat removal, the tray configurations seem to be more convenient than the tile configurations. Because the distances between the center amplifiers and the array edge in the tray configurations are shorter than those in the tile configurations. The tray configurations can also accommodate circuit-level combiners, which is a very attractive solution. Because when a small number of(medium power level) amplifiers are combined, conductor loss due to the connecting transmission lines is small, thus there is no need for space combining. At the same stage, the circuits offer simple means for adjusting phases of the combined signals to achieve in-phase signal combining. By accommodating resistive elements(such as in Wilkinson divider/combiners), excitation of higher-order modes in the enclosure of trays can be reduced. Space-level combining becomes advantageous in the final stages when signals from the trays are combined [36]. Because high power level signals can be conveniently guided towards the collecting ports without incurring conductor loss.

This Chapter has reviewed research results in the area of power combining, which mostly has been performed within an university environment. It has been shown that the developed concepts are mature enough to attract considerable interest from industries. In fact, some of the presented power combiners have been designed and developed with considerable involvement from these companies. The interest and involvement of such institutions clearly indicate that space-level power combining is not a purely academic research, but has great potential to deliver new products to the communication industries.

Acknowledgments

The author would like to express his gratitude to the Australian Research Council for financing his project on spatial power combiners. Assistance of the author's Ph.D. student, Feng-Chi E. Tsai, in preparation of this Chapter is also acknowledged.

References

[1] K. J. Russell, "Microwave power combining techniques," *IEEE Trans. Microwave Theory Tech.*, vol.27, pp.472-478, May 1979.

[2] K. Chang and C. Sun, "Millimetre-wave power-combining techniques," *IEEE Trans. Microwave Theory Tech.*, vol.31, pp.91-107, Feb. 1983.

[3] M. E. Bialkowski, "Power combiners and dividers," *Wiley Encyclopedia of Electrical and Electronic Engineering*, J. G. Webster, ed., vol.16, pp.585-602, 1999.

[4] D. Staiman, M. E. Breese, and W. T. Patton, "New technique for combining solid-state sources," *IEEE J. Solid-State Circuits*, vol.3, pp.238-243, Sept. 1968.

[5] J. W. Mink, "Quasi-optical power combining of solid state millimeter wave sources," *IEEE Trans. Microwave Theory Tech.*, vol.34, pp.273-279, Feb. 1986.

[6] Z. B. Popovic, R. M. Weikle, M. Kim, et al., "Bar-grid oscillators," *IEEE Trans. Microwave Theory Tech.*, vol.38, pp.225-229, Mar. 1990.

[7] Z. B. Popovic, R. M. Weikle, M. Kim, and D. B. Rutledge, "A 100-MESFET planar grid oscillator," *IEEE Trans. Microwave Theory Tech.*, vol.39, pp.193-200, Feb. 1991.

[8] R. A. York and R. C. Compton, "Quasi-optical power combining using mutually synchronised oscillator arrays," *IEEE Trans. Microwave Theory Tech.*, vol.39, pp.1000-1009, June 1991.

[9] J. Birkeland and T. Itoh, "A 16-element quasi-optical FET oscilator power combining array with external injection locking," *IEEE Trans. Microwave Theory Tech.*, vol.40, no.3, pp.475-481, Mar. 1992.

[10] P. Liao and R. A. York, "A new phase-shifterless beam-scanning technique using arrays of coupled oscillators," *IEEE Trans. Microwave Theory Tech.*, vol.41, no.10, pp.1810-1815, Oct. 1993.

[11] K. D. Stephan, "Inter-injection-locked oscillators for power combining and phased arrays," *IEEE Trans. Microwave Theory Tech.*, vol.34, no.10, pp.1017-1025, Oct. 1986.

[12] R. J. Pogorzelski, R. P. Scaramastra, J. Huang, R. J. Beckon, S. M. Petree, and C. M. Chavez, "A seven-element S-band coupled-oscillator controlled agile-beam phased array," *IEEE Trans. Microwave Theory Tech.*, vol.48, no.8, pp.1375-1384, Aug. 2000.

[13] M. Kim, E. A. Sovero, J. B. Hacker, et al., "A 100-element HBT grid amplifier," *IEEE Trans. Microwave Theory Tech.*, vol.41, no.10, pp.1762-1771, Oct. 1993.

[14] M. Gouker, "Toward standard figures-of-merit for spatial and quasi-optical power-combined arrays," *IEEE Trans. Microwave Theory Tech.*, vol.43, no.7, pp.1614-1616, July 1995.

[15] H. Hwang et al., "A dielectric slab waveguide with four planar power amplifiers," *IEEE MTT-S Int. Microwave Symp. Dig.*, pp.921-924, May 1995.

[16] B. Deckman et al., "A 5-Watt, 37-GHz monolithic grid amplifier," *IEEE MTT-S Int. Microwave Symp. Dig.*, pp.805-808, Boston, MA, June 11-16, 2000.

[17] M. E. Bialkowski, H. J. Song, K. M. Luk, and C. H. Chan, "Theory of an active transmit/reflect array of patch antennas operating as a spatial power combiner," *IEEE AP-S Int. Conf. Dig.*, pp.764-767, Boston, July 8-13, 2001.

[18] T. Ivanov and A. Mortazawi, "A two-stage spatial amplifier with hard horn feeds," *IEEE Microwave Guided Wave Lett.*, vol.6, no.2, pp.88-90, Feb. 1996.

[19] D. Sivenpiper et al., "High impedance electromagnetic surfaces with a forbidden frequency band," *IEEE Trans. Microwave Theory Tech.*, vol.47, no.11, pp.2059-2074, Nov. 1999.

[20] F.-R. Yang et al., "A novel TEM waveguide using uniplanar compact photonic-bandgap(UC-PBG) structure," *IEEE Trans. Microwave Theory Tech.*, vol.47, no.11, pp.2192-2098, Nov. 1999.

[21] M. Ali, S. Ortiz, T. Ivanov, and A. Mortazawi, "Analysis and measurement of hard horn feeds for the excitation of quasi-optical amplifiers," *IEEE MTT-S Int. Microwave Symp. Dig.*, pp.1469-1472, June 1998.

[22] H. J. Song and M. E. Bialkowski, "Transmit array of transistor amplifiers illuminated by a passive patch array in the near field reactive region," *IEEE Trans. Microwave Theory Tech.*, vol.49, no.3, pp.470-476, Mar. 2001.

[23] H. J. Song and M. E. Bialkowski, "Spatial power combiner using an active reflectarray of dual-feed aperture coupled microstrip patch antennas," *IEEE AP-S Int. Conf. Dig.*, Boston, July 8-13, 2001.

[24] J. A. Navarro and K. Chang, *Integrated Active Antennas and Spatial Power Combining*, Wiley Inter-Science, 1996.

[25] R. A. York and Z. B. Popovic, *Active and Quasi-Optical Arrays for Solid-State Power Combining*, Wiley Interscience, 1997.

[26] A. Mortazawi, T. Itoh, and J. Harvey, *Active Antennas and Quasi-Optical Arrays*, ed., Piscataway, NJ: IEEE Press, 1998.

[27] S. Ortiz, J. Hubert, L. Mirth, E. Schlecht, and A. Mortazawi, "A 25-watt and 50-watt Ka-band quasi-optical amplifier," *IEEE MTT-S Int. Microwave Symp. Dig.*, pp.797-800, Boston, MA, June 11-16, 2000.

[28] A. W. Robinson et al., "A 137 element active reflect-array with dual-feed microstrip patch elements," *Microwave Opt. Technol. Lett.*, vol.26, no.3, pp.147-151, Aug. 5, 2000.

[29] M. E. Bialkowski, A. W. Robinson, and H. J. Song, "Design, development and testing of X-band amplifying reflectarrays," *IEEE Trans. Antennas Propagat.*, vol.50, no.8, pp.1065-1076, Aug. 2002.

[30] J. Hubert, L. Merth, S. Ortiz, and A. Mortazawi, "A 4-watt Ka-band quasi-optical amplifier," *IEEE MTT-S Int. Microwave Symp. Dig.*, pp.2386-2389, Anaheim, CA, June 13-19, 1999.

[31] J. A. Navarro et al., "Broadband electronically tunable planar active radiating elements and spatial power combiners using notch antennas," *IEEE Trans. Microwave Theory Tech.*, vol.40, no.2, pp.323-328, Feb. 1992.

[32] N. S. Cheng, T. P. Dao, M. G. Case, D. B. Rensch, and R. A. York, "A 120-W X-band spatially combined solid-state amplifier," *IEEE Trans. Microwave Theory Tech.*, vol.47, no.12, pp.2557-2561, Dec. 1999.

[33] M. E. Bialkowski and H. J. Song, "Investigations into a power combining structure formed by uniplanar quasi-Yagi microstrip antennas," *Microwave Opt. Technol. Lett.*, vol.27, no.1, pp.50-53, Oct. 5, 2000.

[34] A. B. Yakovlev, S. Ortiz, M. Ozkar, A. Mortazawi, and M. B. Steer, "A waveguide-based aperture-coupled patch amplifier array, full-wave system analysis and experimental validation," *IEEE Trans. Microwave Theory Tech.*, vol.48, pp.2692-2699, Dec. 2000.

[35] H. Hwang, G. P. Monahan, M. B. Steer, J. W. Mink, J. Harvey, A. Paollea, and F. K. Schwering, "A dielectric slab waveguide with four planar power amplifiers," *IEEE MTT-S Int. Microwave Symp. Dig.*, pp.921-924, May 1995.

[36] J. J. Sowers, D. Pritchard, A. White, W. Kong, and O. Tang, "A 36-W V-band, solid-state source," *IEEE MTT-S Int. Microwave Symp. Dig.*, pp.235-238, Anaheim, CA, June 13-19, 1999.

Chapter 18

INTRODUCTION TO SMART ANTENNA SYSTEMS

Jean-Fu Kiang and Chun-Wei Wu
Department of Electrical Engineering and
Graduate Institute of Communication Engineering
National Taiwan University
Taipei, Taiwan, ROC

Abstract In this Chapter, smart antenna systems are briefly reviewed. With the fast growth of wireless communication in recent years, smart antennas attract much attention. Smart antennas have the potential to improve system performance in terms of capacity, coverage and quality of service. Accurate and computationally efficient estimations of direction-of-arrival are critical factors in designing a smart antenna system. The weighting vectors obtained by adaptive algorithms can be used to synthesize desired pattern. Applications of smart antenna systems including wideband arrays and position location techniques are also addressed.

Keywords: Smart antenna, adaptive antenna, propagation, direction-of-arrival, beamforming network, wideband array, position location.

1. Introduction

Major impairments on wireless systems include fading, delay spread, co-channel interference, and so on. Several techniques have been developed to alleviate these impairments. Antenna diversity was adopted in the early stage to improve the performance of wireless systems. Smart antenna has the potential to mitigate co-channel interference and increase the capacity of furture wireless systems. Research on antenna design has been focused on the selection of feasible radiating elements and antenna architecture that can satisfy the physical and electrical requirements of system. Signal processing of smart antenna systems has been concentrated on the development of efficient algorithms for direction-of-arrival(DOA) estimation and adaptive beam-forming.

2. Propagation Issues

Physical phenomena associated with electromagnetic wave propagation can be roughly categorized as reflection, diffraction and scattering. In a typical mobile propagation environment, there exist multiple propagation paths between transmitter and receiver. Due to multiple reflections from various objects, waves may travel along different paths with varying length, causing delayed versions of the desired signal to arrive at the receiver and resulting in fading or time dispersion. As a mobile station moves over a short distance, the instantaneous received signal is the sum of many contributions coming from different directions with random phase, hence the signal amplitude varies wildly. Such a phenomenon is called multipath fading.

Large-scale propagation models have been developed to predict the average received signal strength at a given location, which is useful in assessing radio coverage of a transmitter. Most of these models are derived using combination of analytical and empirical methods.

2.1 Outdoor Propagation Models

Propagation properties in a mobile communication system is strongly affected by irregular terrains due to the presence of buildings, hills and other objects. Outdoor propagation models can be roughly categorized into four types: empirical models like Okumura-Hata model [1], structure-based models like Walfisch-Bertoni model [2], semi-empirical models like COST-231-Walfisch-Ikegami model [3], [4], and deterministic models like IHE model [5]. In the empirical models, model parameters are estimated by regression analysis applied to extensive measurement data. In the structure-based models, propagation loss is derived based on simplified terrian profile and theoretical analysis. In the semi-empirical models, analytic parameters are empirically adjusted by measurement data. In the deterministic models, fields are computed by solving an intergral equation or by using ray tracing techniques to exploit terrain information. Ray tracing technique keeps track of propagation rays from the transmitter through reflecting objects into the receiver. Shadows, multiple reflections and obstacle projections need to be considered in the algorithms.

In the Okumura-Hata model, the received field, E_r, is expressed as

$$E_r = E_0 + E_{\text{sys}} + \gamma_{\text{sys}} \log R \qquad \text{dB}_{\mu\text{V/m}}$$

where $E_0 = 35.55$ is an offset parameter, R is the distance between transmitter and receiver, $E_{sys} = P_{BS} - 6.16 \log f + 13.82 \log h_b + a(h_m)$ is the system design parameter which is valid for $f \geq 400$ MHz, h_m is the mobile station height, h_b is the base station height, f is operating frequency in MHz, $a(h_m) = 4.97 - 3.2 \left|\log(11.75 h_m)\right|^2$. The slope parameter is $\gamma_{sys} = -\gamma(44.9 - 6.55 \log h_b)$ where $\gamma = 1$ for $R \leq 20$ km and $\gamma = 1 + (0.14 + 1.87 \times 10^{-4} f + 1.07 \times 10^{-3} h_b)[\log(R/20)]^{0.8}$ for $20 \leq R \leq 100$ km. The Okumura-Hata model is based on extensive measurement data, which is well documented and widely adopted around the world using correction factors to extend its applicability to regions other than Tokyo. However, this model is inaccurate in rural area. Least squares method can be used to fine tune the offset and slope [6]. The system parameter is more difficult to adjust because the dependence of field strength on parameters like antenna height and frequency may change from area to area. In most cases, the parameters converge within ±5% of the adjusted value using 15 to 20 measurement training data.

2.2 Indoor Propagation Models

Coverage of an indoor cellular system is much smaller than that of an outdoor one, and the indoor environment is much more versatile than the outdoor. Propagation properties inside buildings are affected by building floor plan, material, decoration, furnitures, and so on, which should be considered in the deployment of indoor wireless communication systems.

In general, the received power can be expressed as $P_r = P_t + G_b + G_m - PL(\text{dB})$ where P_r is the received power, P_t is the transmitted power, G_b and G_m are antenna gains of the base station and mobile station, respectively, and PL is the path loss. Ray tracing techniques can be used to predict the signal strength. Conventional three-dimensional ray tracing techniques trace all possible rays and are computationally expensive [7]. Reduction to two-dimensional model can save the computational cost at the risk of lower accuracy. However, cell plan can still cause problem in two-dimensional models. Using regular cell size turns out to be inefficient. Hence, irregular cells are proposed to reduce computation cost to 1% while obtaining similar level of accuracy [8].

2.3 Advantages of Smart Antennas

Smart antenna technology bears several advantages, including range extension, null filling, building penetration, robustness, link quality enhancement and system capacity increase. In the early phase of cellular system deployment, sufficient number of base stations must be installed to meet the initial coverage requirement. Using smart antenna technology requires less base stations and hence lower infrastructure cost. Smart antenna system provides higher gain than traditional antennas, hence endows wireless system to tolerate greater path loss or require less transmitting power. Smart antenna can be used to mitigate the impact of multipath effects or even exploit multipath signals to enhance its performance. Multiple access interference(MAI) can also be reduced to increase the number of users supported by each base station. Smart antenna can adapt its radiation pattern to support hot spots, it can also separate signals from different sectors to increase the system capacity.

CDMA systems require power control to ensure that all the signals arriving at a base station have roughly the same power level, which forces all the users to transmit minimum amount of power needed to achieve acceptable signal quality at the base station. In analyzing smart antennas, it is usually assumed that the spacing between array elements is small enough that no amplitude variation exists among the received signals, no mutual coupling exists between elements, all the incident fields can be decomposed into a finite number of plane waves, and the bandwidth of signal incident on the array is small compared with the carrier frequency.

3. Detection Algorithms for Direction-of-Arrival

A smart antenna usually consists of an array of antenna elements which serve as sensors to detect incident signals, and also serve as radiators to transmit signals to specific directions. Accurate and real-time prediction of the direction-of-arrival(DOA) is the key for smart antenna to function effectively. Array-based estimation techniques for direction-of-arrival can be categorized into conventional, subspace-based, maximum likelihood, and their combinations.

3.1 Conventional Techniques

Conventional techniques require a large number of sensors to achieve high resolution. The delay-and-sum method is based on the premise that pointing the strongest beam in a specific direction yields the best estimate of signal arriving in that direction. For a typical narrow band antenna array for DOA detection, the output signal, $y(t)$, is a linear combination of the sensors' output as

$$y(t) = \bar{W}^{\dagger} \cdot \bar{u}(t)$$

where $\bar{W} = [w_0, \cdots, w_{M-1}]^t$ is the weighting vector containing weighting factors on all sensors, $\bar{W}^{\dagger}$ is the Hermitian of $\bar{W}$, $\bar{u} = [u_0, \cdots, u_{M-1}]^t$

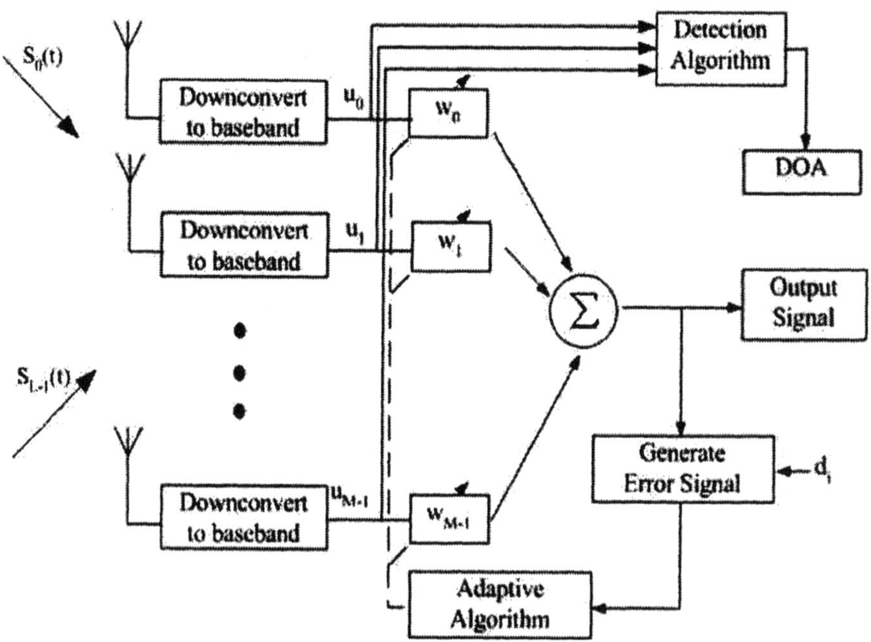

Figure 18.1. Basic architecture of smart antenna system.

is the received signal vector consisting of received signals from all sensors with $\bar{u}(t) = \bar{\bar{A}}(\phi) \cdot \bar{s}(t) + \bar{n}(t)$, $\bar{n}(t) = [n_0(t), n_1(t), \cdots, n_{M-1}(t)]^t$ is the noise vector consisting of received noise at all sensors, $\bar{s}(t) = [s_0(t), s_1(t), \cdots, s_{L-1}(t)]$ is the incident signal vector consisting of all incident signals from L different directions, $\bar{\bar{A}}(\bar{\phi}) = [\bar{a}(\phi_0), \bar{a}(\phi_1), \cdots, \bar{a}(\phi_{L-1})]$ is the steering matrix consisting of all the steering vectors at different incident directions, and $\bar{\phi} = [\phi_0, \phi_1, \cdots, \phi_{L-1}]$ is the incident angle vector consisting of L incident directions. The steering vector, $\bar{a}(\phi_\ell)$, at the incident angle ϕ_ℓ is defined as $[1, \hat{g}_1(\phi_\ell)\gamma_1 e^{j\Psi_1}, \cdots, \hat{g}_{M-1}(\phi_\ell)\gamma_{M-1} e^{j\Psi_{M-1}}]^t$ where $\hat{g}_i(\phi_\ell)$ is the gain of sensor i at direction ϕ_ℓ, and $\gamma_i e^{j\Psi_i}$ is the narrow band channel response of sensor i.

These channel parameters are estimated by using proper propagation models or measurements. The received signals are measured from at least M directions. The input autocorrelation matrix can be estimated as

$$\bar{\bar{R}}_{uu} = E\left[\bar{u}(t)\,\bar{u}^\dagger(t)\right]$$

where $E(\cdot)$ is the ensemble average of a process, which can be approximated by its time average. The total output power can be expressed as

$$P_{DS} = E\left[\left|\bar{W}^{\dagger}\cdot\bar{u}(t)\right|^2\right] = \bar{W}^{\dagger}\cdot E\left[\bar{u}(t)\bar{u}^{\dagger}(t)\right]\cdot\bar{W} = \bar{W}^{\dagger}\cdot\bar{\bar{R}}_{uu}\cdot\bar{W}$$

When a signal is incident upon the array at angle ϕ_0, it is proved that the output power is maximized when $\bar{W} = \bar{a}(\phi_0)$. Having available estimate of the input autocorrelation matrix and steering vectors for all angles of interest, one may estimate the output power as a function of DOA, which is also referred to as angular spectrum. When the angular spectrum is swept over azimuthal angle, ϕ, the weighting vector, $\bar{W}$, is determined by $\bar{a}(\phi)$.

The delay-and-sum method has poor resolution. Finite beamwidth and sidelobe level further restrict its effectiveness when signals arrive from multiple directions. This can be improved to some extent by adding more sensors.

In the Capon's minimum variance method [9], weighting vector is properly chosen to improve over the delay-and-sum method in dealing with multipath signals. It applys certain degrees of freedom to enhance signals in the desired directions, and uses the remaining degrees of freedom to reduce signals in other directions. This technique minimizes the contribution of undesired interferences by minimizing the output power while maintaining the gain in the desired directions to be unity, namely

$$\min E[|y(t)|^2] = \min\left(\bar{W}^{\dagger}\cdot\bar{\bar{R}}_{uu}\cdot\bar{W}\right), \quad \text{subject to } \bar{W}^{\dagger}\cdot\bar{a}(\phi_0) = 1$$

The weighting vector thus obtained is called the minimum variance distortionless response(MVDR) beam-forming weights.

$$\bar{W} = \frac{\bar{\bar{R}}_{uu}^{-1}\cdot\bar{a}(\phi)}{\bar{a}^{\dagger}(\phi)\cdot\bar{\bar{R}}_{uu}^{-1}\cdot\bar{a}(\phi)}$$

Output power of the array as a function of DOA is given by Capon's angular spectrum as

$$P_{\text{Capon}}(\phi) = \frac{1}{\bar{a}^{\dagger}(\phi)\cdot\bar{\bar{R}}_{uu}^{-1}\cdot\bar{a}(\phi)}$$

The DOAs can be estimated by locating peaks in the Capon's angular spectrum over the whole range of ϕ.

The Capon's minimum variance method fails if there exist other signals that are correlated with the signal of interest. This method inadvertently uses that correlation to reduce the processor output power without spatially nulling it.

3.2 Subspace-Based Techniques

Subspace-based techniques are suboptimal techniques which exploit eigenstructure of the input signal matrix to acquire high resolution. The MUSIC(multiple signal classification) and ESPRIT(estimation of signal parameter via rotation invariance technique) algorithms will be briefly reviewed.

In the MUSIC algorithm [10], an eigen decomposition on $\bar{\bar{R}}_{uu}$ is first performed to have

$$\bar{\bar{R}}_{uu} \cdot \bar{\bar{V}} = \bar{\bar{V}} \cdot \bar{\bar{\Lambda}}$$

where $\bar{\bar{\Lambda}} = \text{diag}\{\lambda_0, \lambda_1, \cdots, \lambda_{M-1}\}$ with eigenvalues of $\bar{\bar{R}}_{uu}$, $\lambda_0 \geq \lambda_1 \geq \cdots \geq \lambda_{M-1}$ and associated eigenvectors $\bar{\bar{V}} = [\bar{q}_0, \bar{q}_1, \cdots, \bar{q}_{M-1}]$. Usually, there exists a magnitude drop between $\{\lambda_0, \cdots, \lambda_{L-1}\}$ and $\{\lambda_L, \cdots, \lambda_{M-1}\}$ if there are L incident signals. The second set of eigenvalues are close to one another, and are named noise eigenvalues. This algorithm estimates the number of signal, $\hat{L}$, from multiplicity, K, of the clustered small eigenvalues as $\hat{L} = M - K$. The MUSIC angular spectrum at azimuthal angle, ϕ, is defined as

$$\hat{P}_{\text{MUSIC}}(\phi) = \frac{\bar{a}^\dagger(\phi) \cdot \bar{a}(\phi)}{\bar{a}^\dagger(\phi) \cdot \bar{\bar{V}}_n \cdot \bar{\bar{V}}_n^\dagger \cdot \bar{a}(\phi)}$$

where $\bar{\bar{V}}_n = [\bar{q}_L, \bar{q}_{L+1}, \cdots, \bar{q}_{M-1}]$ is the matrix containing all eigenvectors associated with noise eigenvalues. Inherent orthogonality between $\bar{a}(\phi)$ and $\bar{V}_n$ tends to minimize the denominator and hence gives rise to peaks in the MUSIC angular spectrum. The $\hat{L}$ largest peaks in the MUSIC angular spectrum correspond to signals incident upon the array. Once the DOAs are determined from the MUSIC angular spectrum, the signal autocorrelation matrix, $\bar{\bar{R}}_{ss}$, can be estimated as

$$\bar{\bar{R}}_{ss} = \left[\bar{\bar{A}}^\dagger(\bar{\phi}) \cdot \bar{\bar{A}}(\bar{\phi})\right]^{-1} \cdot \bar{\bar{A}}^\dagger(\bar{\phi}) \cdot \left[\bar{\bar{R}}_{uu} - \lambda_{\min}\bar{\bar{I}}\right] \cdot \bar{\bar{A}}(\bar{\phi}) \cdot \left[\bar{\bar{A}}^\dagger(\bar{\phi}) \cdot \bar{\bar{A}}(\bar{\phi})\right]^{-1}$$

The MUSIC algorithm exploits eigenstructure of the input autocorrelation matrix. Major advantages of the MUSIC are good resolution and less computational complexity. It does not need to estimate the signal power associated with each arrival angle. However, the MUSIC algorithm requires precise and accurate array calibration. When components of $\bar{s}(t)$ are highly correlated, $\bar{\bar{R}}_{ss}$ becomes nearly singular, and the rank of $\bar{\bar{R}}_{uu}$ is reduced. Various extensions have been developed, including root MUSIC [11] and cyclic MUSIC [12].

The ESPRIT algorithm significantly reduces the computational and storage requirements of the MUSIC algorithm [13]. In the ESPRIT, it is presumed that the array can be decomposed into two equal sized, identical subarrays with two associated elements, one in each subarray, displaced from each other by a fixed translational distance. The basic idea behind ESPRIT is to exploit rotation invariance of the underlying signal subspace introduced by translational invariance of the sensor array.

Consider a linear array of M sensors which are decomposed into subarrays $S_0 = \{e_0, e_1, \cdots, e_{M/2-1}\}$ and $S_1 = \{e_{M/2}, e_{M/2+1}, \cdots, e_{M-1}\}$, where sensor e_i is paired with sensor $e_{M/2+i}$ with a separation of Δx. Received signal vectors at the two subarrays can be expressed as

$$\begin{aligned} \tilde{u}_0(t) &= \tilde{\bar{A}}(\bar{\phi}) \cdot \bar{s}(t) + \tilde{n}_0(t) \\ \tilde{u}_1(t) &= \tilde{\bar{A}}(\bar{\phi}) \cdot \bar{\bar{\Phi}} \cdot \bar{s}(t) + \tilde{n}_1(t) \end{aligned} \tag{18.1}$$

where $\tilde{u}_0(t) = [u_0(t), u_1(t), \cdots, u_{M/2-1}(t)]^t$, $\tilde{u}_1(t) = [u_{M/2}(t), u_{M/2+1}(t), \cdots, u_{M-1}(t)]^t$, $\tilde{n}_0(t) = [n_0(t), \cdots, n_{M/2-1}(t)]^t$, $\tilde{n}_1(t) = [n_{M/2}(t), \cdots, n_{M-1}(t)]^t$, and $\bar{\bar{\Phi}} = \text{diag}\{\exp(j\gamma_0), \exp(j\gamma_1), \cdots, \exp(j\gamma_{L-1})\}$ with $\gamma_k = k_0 \Delta x \cos\phi_k$, k_0 is the wave number in free space. Eq.(18.1) can be further summarized as $\bar{u}(t) = \bar{\bar{A}}(\bar{\phi}) \cdot \bar{s}(t) + \bar{n}(t)$ where

$$\bar{u}(t) = \begin{bmatrix} \tilde{u}_0(t) \\ \\ \tilde{u}_1(t) \end{bmatrix}, \quad \bar{\bar{A}}(\bar{\phi}) = \begin{bmatrix} \tilde{\bar{A}}(\bar{\phi}) \\ \\ \tilde{\bar{A}}(\bar{\phi}) \cdot \bar{\bar{\Phi}}(\bar{\phi}) \end{bmatrix}, \quad \bar{n}(t) = \begin{bmatrix} \tilde{n}_0(t) \\ \\ \tilde{n}_1(t) \end{bmatrix}$$

The overall signal space is the union of subspaces expanded by the received signal vectors $\tilde{u}_0(t)$ and $\tilde{u}_1(t)$, respectively. Let $\bar{\bar{V}}_0$ and $\bar{\bar{V}}_1$ be the signal subspaces expanded by subarrays S_0 and S_1, respectively. The signal subspace can be derived from the input autocorrelation matrix, $\bar{\bar{R}}_{uu} = \bar{\bar{A}}(\bar{\phi}) \cdot \bar{\bar{R}}_{ss} \cdot \bar{\bar{A}}(\bar{\phi}) + \sigma^2 \bar{\bar{I}}$. If $L \leq M$, the $M - L$ smallest eigenvalues of $\bar{\bar{R}}_{uu}$ are roughly equal to σ_n^2. Eigenvectors in $\bar{\bar{V}}_s$ associated with the L largest eigenvalues have the same range as that of $\bar{\bar{A}}(\bar{\phi})$, namely

$$\text{Range}\{\bar{\bar{V}}_s\} = \text{Range}\{\bar{\bar{A}}(\bar{\phi})\}$$

Hence, there exists a unique nonsingular matrix, $\bar{\bar{T}}$, such that $\bar{\bar{V}}_s = \bar{\bar{A}}(\bar{\phi}) \cdot \bar{\bar{T}}$. Decompose $\bar{\bar{V}}_s$ into $\bar{\bar{V}}_0$ and $\bar{\bar{V}}_1$ such that $\bar{\bar{V}}_0 = \tilde{\bar{A}}(\bar{\phi}) \cdot \bar{\bar{T}}$ and $\bar{\bar{V}}_1 = \tilde{\bar{A}}(\bar{\phi}) \cdot \bar{\bar{\Phi}} \cdot \bar{\bar{T}}$, and it is proved that $\bar{\bar{V}}_0$ and $\bar{\bar{V}}_1$ share a common column space. Define $\bar{\bar{V}}_{01} = [\bar{\bar{V}}_0 \mid \bar{\bar{V}}_1]$, it is proved that a matrix $\bar{\bar{F}} = [\bar{\bar{F}}_0 \mid \bar{\bar{F}}_1]^t$ exists to have

$$\bar{\bar{V}}_0 \cdot \bar{\bar{F}}_0 + \bar{\bar{V}}_1 \cdot \bar{\bar{F}}_1 = 0 \tag{18.2}$$

where $\bar{\bar{F}}$ spans the null space of $\bar{\bar{V}}_{01}$. By defining $\bar{\bar{\Psi}} = -\bar{\bar{F}}_0 \cdot \bar{\bar{F}}_1^{-1}$, (18.2) is reduced to

$$\tilde{\bar{A}}(\bar{\phi}) \cdot \bar{\bar{T}} \cdot \bar{\bar{\Psi}} \cdot \bar{\bar{T}}^{-1} = \tilde{\bar{A}}(\bar{\phi}) \cdot \bar{\bar{\Phi}}(\bar{\phi}) \tag{18.3}$$

or

$$\bar{\bar{V}}_0 \cdot \bar{\bar{\Psi}} = \bar{\bar{V}}_1$$

Assume that $\tilde{\bar{A}}(\bar{\phi})$ is of full rank, which is true if the DOAs of all signals are distinct from one another, (18.3) is thus reduced to

$$\bar{\bar{T}} \cdot \bar{\bar{\Psi}} = \bar{\bar{\Phi}} \cdot \bar{\bar{T}}$$

This implies that the eigenvalues of $\bar{\bar{\Psi}}$ are equal to the diagonal elements of $\bar{\bar{\Phi}}$, and the columns of $\bar{\bar{T}}$ are the eigenvectors of $\bar{\bar{\Psi}}$. Due to noises and uncertainties, approximate solution to $\bar{\bar{V}}_0 \cdot \bar{\bar{\Psi}} = \bar{\bar{V}}_1$ can be obtained by using the least squares method to have

$$\bar{\bar{\Psi}} = (\bar{\bar{V}}_0^\dagger \cdot \bar{\bar{V}}_0)^{-1} \cdot (\bar{\bar{V}}_0^\dagger \cdot \bar{\bar{V}}_1)$$

Assume that the eigenvalues of $\bar{\bar{\Psi}}$ are $\{\psi_k\}$ with $0 \le k < L$, which are the same as the diagonal elements of $\bar{\bar{\Phi}}$. The DOAs can be estimated as

$$\phi_k = \cos^{-1}\left[\frac{\angle \Phi_k}{k_0 \Delta x}\right]$$

Since least squares method is used in the ESPRIT, stringent array calibration requirements can be relieved.

3.3 Maximum Likelihood Techniques

With low signal-to-noise ratio or when the number of signal samples is small, performance of the maximum likelihood(ML) technique is superior to the subspace-based techniques [14]. The ML estimator can be derived from N received data vectors as

$$\bar{\bar{U}} = \bar{\bar{A}}(\bar{\phi}) \cdot \bar{\bar{S}} + \bar{\bar{N}}$$

where $\bar{\bar{U}} = [\bar{u}(t_0), \cdots, \bar{u}(t_{N-1})]$ is the collection of received data vectors at N instants, $t_0, t_1, \cdots, t_{N-1}$, respectively, $\bar{\bar{S}} = [\bar{s}(t_0), \cdots, \bar{s}(t_{N-1})]$

is the collection of incident signal vectors at N instants, and $\bar{\bar{N}} = [\bar{n}(t_0), \cdots, \bar{n}(t_{N-1})]$ is the collection of noise vectors at N instants.

It is assumed that the number of incident signals is smaller than the number of sensors. The set of steering vectors are assumed to be linearly independent. The noise samples are assumed to follow statistically independent ergodic complex-valued Gaussian process with zero mean and covariance $\sigma^2 \bar{\bar{I}}$, where σ^2 is an unknown scalar and $\bar{\bar{I}}$ is the identity matrix. The signals are viewed as samples of an unknown deterministic rather than random process. Joint probability density function of the samples can be expressed as

$$f(\bar{\bar{U}}) = \prod_{k=0}^{N-1} \frac{1}{\pi \det[\sigma^2 \bar{\bar{I}}]} \exp\left(-\frac{1}{\sigma^2} \left|\bar{u}(t_k) - \bar{\bar{A}}(\bar{\phi}) \cdot \bar{s}(t_k)\right|^2\right) \tag{18.4}$$

where $\det[\bar{\bar{Z}}]$ is the determinant of matrix $\bar{\bar{Z}}$. Take the logarithm of (18.4), and ignore the constant terms to have the log likelihood function as

$$-NL \log \sigma^2 - \frac{1}{\sigma^2} \sum_{k=0}^{N-1} \left|\bar{u}(t_k) - \bar{\bar{A}}(\bar{\phi}) \cdot \bar{s}(t_k)\right|^2$$

The maximum likelihood estimator is obtained by minimizing the following function with respect to the unknown parameters.

$$\min_{(\bar{\phi}, \bar{\bar{S}})} \sum_{k=0}^{N-1} \left|\bar{u}(t_k) - \bar{\bar{A}}(\bar{\phi}) \cdot \bar{s}(t_k)\right|^2 \tag{18.5}$$

The least squares solution of $\bar{s}(t_k)$ at a given angular vector, $\bar{\phi}$, is

$$\hat{\bar{s}}(t_k) = \bar{\bar{A}}_{\mathrm{LS}}(\bar{\phi}) \cdot \bar{u}(t_k)$$

where $\bar{\bar{A}}_{\mathrm{LS}}(\bar{\phi}) = \bar{\bar{A}}(\bar{\phi}) \cdot [\bar{\bar{A}}^\dagger(\bar{\phi}) \cdot \bar{\bar{A}}(\bar{\phi})]^{-1} \cdot \bar{\bar{A}}^\dagger(\bar{\phi})$. Thus, (18.5) is reduced to a minimization problem over $\bar{\phi}$ as

$$\min_{\bar{\phi}} \sum_{k=0}^{N-1} \left|\bar{u}(t_k) - \bar{\bar{A}}_{\mathrm{LS}}(\bar{\phi}) \cdot \bar{u}(t_k)\right|^2$$

Hence, the ML estimator of DOA, $\hat{\bar{\phi}} = \{\hat{\phi}_1, \cdots, \hat{\phi}_{L-1}\}$, is obtained by minimizing the following function with respect to $\bar{\phi}$

$$C(\bar{\phi}) = \sum_{k=0}^{N-1} \left|\bar{\bar{A}}_{\mathrm{LS}}(\bar{\phi}) \cdot \bar{u}(t_k)\right|^2 = \mathrm{trace}\left[\bar{\bar{A}}_{\mathrm{LS}}(\bar{\phi}) \cdot \bar{\bar{R}}_{uu}\right]$$

Maximization of log likelihood function is a nonlinear, multidimensional maximization problem which is computationally intensive, thus limiting the applications of ML techniques. Many efficient algorithms have been proposed to simplify the solution procedure. For example, property restoral-based techniques separate multiple signals and estimate their spatial signatures from which their DOAs can be determined using subspace techniques. Many approaches have been proposed to enhance DOA estimation. In [15], a linear prediction method in conjunction with Capon's method to accurately estimate DOA gives about 80% reduction in computational cost as compared with the MUSIC.

3.4 Other Issues

In all the aforementioned methods, mutual coupling is neglected, which may cause amplitude and phase deviation among array elements. In [16], mutual coupling is considered to offer more accurate estimation of DOA at the cost of longer computation time. In spite of the potential advantages of high resolution techniques like MUSIC and ESPRIT, application to real systems has been rather limited due to their relatively high sensitivity to various errors.

In [17], downlink power distribution techniques based on DOA estimation can be categorized as dominant DOA(DomDOA), pseudo-inverse DOA(PseDOA), spatial signature(SS) and pseudo-inverse SS(PseSS) methods. The DomDOA method transmits downlink power in the direction with the strongest uplink power. The PseDOA method transmits downlink power in the direction with the strongest uplink signal as the DomDOA method. Meanwhile, nulls are placed in nondominant DOAs in the radiation pattern. The SS method distributes the signal power in all directions according to the spatial signature of the desired mobile user instead of focusing signal power in the dominant DOA.

Compared with the SS method which maximizes the signal-to-noise ratio(SNR) without inserting nulls, the PseSS method maximizes the SNR while inserting nulls. Logically, nulls are inserted at the DOAs of undesired mobile users. Notice that SS and PseSS methods do not use steering vectors at all when forming downlink beams. Performance of direction finding can be improved by reducing sidelobe levels, which allows for easier distinction of signal sources. Better DOA estimation improves the downlink signal-to-interference ratio(SIR), thus creating a DOA-based downlink beam-forming method(like PseDOA method) to work comparably to SS-based methods.

4. Switched Beam Systems

Current mobile communication systems utilize sectorization to reduce interference and increase capacity. Each cell is broken into three or six sectors, with dedicated antennas and radio coverage in each sector. Using more sectors reduces the number of interference sources in each sector. However, system efficiency decreases due to overlap of antenna patterns, and more in-cell handoffs are experienced by mobile stations.

Spatial selection diversity provides an effective and economical means of reducing narrow band fading. Beam-forming is performed independently by each array, and the selection diversity scheme chooses the array output with the highest signal quality. Both adaptive antenna and switched beam antennas can be viewed as simplfied versions of smart antenna systems. Switched beam antenna provides a number of fixed beams [18]. In general, the receiver selects the beam that enhances signal and reduces interference to the maximum extent.

A beam-forming network(BFN) capable of generating S beams with M radiators can be characterized by a weighting matrix, $\bar{\bar{W}}$, which relates the signal vector at network output to the received signal vector as

$$\bar{y}(t) = \bar{\bar{W}}^{\dagger} \cdot \bar{u}(t)$$

where $\bar{y}(t) = [y_0(t), \cdots, y_{S-1}(t)]^t$ is the signal vector, assuming there are S incident signals, $\bar{\bar{W}} = [\bar{W}_0, \bar{W}_1, \cdots, \bar{W}_{S-1}]$ with $\bar{W}_k = [w_{k0}, w_{k1}, \cdots, w_{k(M-1)}]^t$. The nth output is the inner product of received signal vector with $\bar{W}_n$. BFN can be exploited in the switched beam system and is usually followed by a bank of adaptive array processors.

5. Adaptive Antenna Systems

Adaptive antenna system is capable of changing its radiation pattern dynamically to avoid receiving interference, to track mobile users, to enhance received signals, or to transmit signals in specifc directions. Its performance can be further enhanced by increasing the complexity of array signal processor. To optimize the beam pattern, the weighting vector can be determined by minimizing the cost function using minimum mean square error(MMSE) [19] or least squares(LS) techniques.

In deriving the MMSE solution, it is assumed that the environment is stationary, and the cost function is defined as

$$C_{\text{MMSE}}(\bar{W}_k) = E\left[|z_k(iT_s) - d_k(iT_s)|^2\right]$$

where $z_k(t) = \bar{W}_k^{\dagger} \cdot \bar{u}(t)$ is the array output associated with the kth signal, $d_k(t)$ is local estimate of the kth signal, and T_s is sampling period. Define the ℓth component of $\nabla C(\bar{W}_k)$ as

$$[\nabla C(\bar{W}_k)]_\ell = \frac{\partial C(\bar{W}_k)}{\partial w'_{k\ell}} + i\frac{\partial C(\bar{W}_k)}{\partial w''_{k\ell}}, \quad 0 \le k < S, 0 \le \ell < M$$

where $w'_{k\ell}$ and $w''_{k\ell}$ are real and imaginary part, respectively, of $w_{k\ell}$. Gradient of the cost function thus becomes

$$\nabla C_{\text{MMSE}}(\bar{W}_k) = 2\bar{\bar{R}}_{uu} \cdot \bar{W}_k - 2E\left[\bar{u}(iT_s)d_k^*(iT_s)\right]$$

Notice that $\nabla\left[\bar{W}_k^\dagger \cdot \bar{\bar{B}}(\bar{\phi}) \cdot \bar{W}_k\right] = 2\bar{\bar{B}}(\bar{\phi}) \cdot \bar{W}_k$, $\nabla(\bar{W}_k^\dagger \cdot \bar{c}) = 2\bar{c}$, and $\nabla(\bar{c}^\dagger \cdot \bar{W}_k) = 0$, where $\bar{\bar{B}}(\bar{\phi})$ is an arbitrary matrix independent of $\bar{W}_k$, and $\bar{c}$ is an arbitrary constant vector. The weighting vector, $\bar{W}_k$, is obtained by setting the gradient of cost function to zero. Thus, we have $\bar{W}_k = \bar{\bar{R}}_{uu}^{-1} \cdot E[\bar{u}(iT_s)d_k^*(iT_s)]$.

In deriving the least squares solution, the cost function is defined as

$$C_{\text{LS}}(\bar{W}_k) = \left|\sum_{m=0}^{P-1} \bar{W}_k^\dagger \cdot \bar{u}^m(iT_s) - d_k^m(iT_s)\right|^2$$

where P samples of signal vectors, $\bar{u}^0(iT_s), \bar{u}^1(iT_s), \cdots, \bar{u}^{P-1}(iT_s)$, are collected. The gradient of cost function becomes

$$\begin{aligned}\nabla C_{\text{LS}}(\bar{W}_k) = 2\sum_{m=0}^{P-1}\sum_{n=0}^{P-1} \bar{u}^m(iT_s) \cdot \bar{u}^n(iT_s)^\dagger \cdot \bar{W}_k \\ -2\sum_{m=0}^{P-1}\sum_{n=0}^{P-1} \bar{u}^m(iT_s)d_k^{m*}(iT_s)\end{aligned}$$

Define the received signal matrix as $\bar{\bar{U}}(iT_s) = [\bar{u}^0(iT_s), \bar{u}^1(iT_s), \cdots, \bar{u}^{P-1}(iT_s)]$, and the desired signal vector as $\bar{d}_k = [d_k^0(iT_s), d_k^1(iT_s), \cdots, d_k^{P-1}(iT_s)]^t$, the solution is obtained by setting the gradient of cost function to zero as

$$\bar{W}_k = \left[\bar{\bar{U}}^\dagger(\bar{\phi}) \cdot \bar{\bar{U}}(\bar{\phi})\right]^{-1} \cdot \bar{\bar{U}}^\dagger(\bar{\phi}) \cdot \bar{d}_k$$

Rather than solving the weighting vector directly, iterative approach is often used to update the weighting vector, $\bar{W}_k(iT_s)$, at each sampling instant. Typically, these algorithms have lower complexity than the direct solution, and can be used to track non-stationary solution. The modified cost function associated with these algoritms is

$$\bar{W}_k[(i+1)T_s] = \bar{W}_k(iT_s) - \frac{1}{2}\mu\nabla C[\bar{W}_k(iT_s)]$$

where μ is the convergence factor which controls the rate of adaptation.

Both LS and MMSE techniques require the estimate of desired angular filter ouput, which is often conducted by periodically sending a training sequence. Because sending training sequences takes bandwidth, an alternative is to use decision-directed adaptation, in which estimate of the desired signal samples, $d_k(iT_s)$, is reconstructed based on array output and signal demodulator. However, when error occurs at the demodulator, reconstructed estimate of the desired signal becomes poor, and inappropriate weighting vector may lead to degradation of transmitting or receiving quality. Combining these two techniques can improve performance. Training sequence can be used for initial adaptation, decision-directed adaptation can then be used to track time variation in the radio channel.

6. Wideband Array

Wideband array is an adaptive array which makes use of temporal filtering to suppress interference in the frequency domain as well as in the angular domain. Wideband array captures multipath components arriving at significantly different delays by using tapped delay line as adaptive interference rejecting filter [20].

Tapped delay line allows each element to have a phase response that varies with frequency. Hence, it can be viewed as an equalizer which equalizes array response at different frequencies. It compensates for the difference that high frequency compoments incur larger phase shifts than low frequency compoments as they travel the same distance. Although it is not possible to achieve the same radiation pattern across all frequencies, it is possible to use a tapped delay line to reduce response deviation at different frequencies.

CDMA RAKE receiver is an example of wideband spatial processing array, with complexity level similar to that of narrow band arrays [21]. In outdoor environments, delay between multipath components is usually long. However, if chip rate is properly selected, low autocorrelation property of the CDMA spreading codes can assure that the multipath components will appear nearly uncorrelated with one another. Assume that Q correlators are used in a CDMA receiver to capture the Q strongest multipath components. Weighting matrix is used to calculate a linear combination of correlator outputs for bit detection. The first correlator is synchronized to the strongest multipath component, m_1, the second correlator is synchronized to m_2, which has low correlation with m_1. The multipath component m_2 arrives later than component m_1 by τ_1. In RAKE receiver, if the output from one correlator is corrupted by fading, the signal can still be recovered through weighting process. Combination of Q separate decisions offered by the RAKE receiver provides a multipath diversity which can overcome fading and improve CDMA reception. The Q decision statistics are weighted to form an overall decision statistic with the weighting coefficients derived from power or SNR of each correlator output. As in adaptive equalizers and diversity combining, there are many ways to assign the weighting coefficients. Due to multiple access interference, RAKE fingers with strong signal may not necessarily provide strong output after correlation. Choosing weighting coefficients based on actual outputs of the correlators usually yields better performance.

7. Position Location Techniques

Position location techniques using radio waves can be categorized as direction finding and range-based, which may be used separately or in combination. In the direction finding technique, mobile station is located by measuring the angle-of-arrival(AOA) of mobile signal. Triangulation method can be used to locate the position of mobile station at the intersection of line-of-bearings(LOBs). Two receiving base stations are required, but more than two are needed in practice due to finite angular resolution, multiplath effects and noise.

Range-based techniques include ranging, range-sum, range-difference, and so on. Several base stations are used to record the propagation delay or range between mobile and base stations. System using ranging technique locates a mobile station by measuring the distance between mobile and base stations. The distance traveled by wave is given by $d = c\tau$ where τ is the propagation delay and c is the speed of light. Hence, distance between the mobile station and each base station can be determined by measuring the time-of-arrival(TOA) of each signal [22]. The TOA estimate can be used to define a sphere of constant range around each base station. Then, the intersection of spheres defined from multiple base stations determines the position of mobile station. Sometimes, practical ranging systems are unable to measure the exact range between mobile and base stations due to bias, which can be corrected by using range measurement from another base station.

The range-sum technique measures the sum of ranges between a mobile station and two base stations by measuring the time-sum-of-arrival (TSOA) of the associated signals. An ellipsoid is thus defined with the two base stations as foci, and the intersection of ellipsoids from multiple range-sum measurements determines the location of mobile station.

The range-difference technique measures the difference in ranges between a mobile station and two base stations by measuring the time-difference-of-arrival(TDOA) of the associated signals [23]. A hyperboloid is thus defined with the two base stations as foci, and the intersection of hyperboloids from multiple range-difference measurements determines the location of mobile station. Major sources of error in the TDOA technique are measurement noise and non-line-of-sight(NLOS) range error. Measurement noise is usually modeled as a zero-mean Gaussian random process, while the NLOS range error usually has an unknown distribution with positive mean [24]. Field test shows that average NLOS range error can be as large as 589 m in IS-95 CDMA system [25].

Major challenge for a system to provide location service is to achieve high location accuracy at low implementation cost, taking into account hostile propagation environments. Direction finding and range-based techniques have their own advantages and limitations [26]. For example, the TDOA technique requires at least three properly located base stations, and generally provides better accuracy than the AOA technique. On the other hand, the AOA technique requires only two base stations, but it is highly range sensitive. Small error in angle measurement may result in large position error when the mobile station is far away from the base station.

Position location systems which are capable of measuring frequency shift in the received signal due to mobile station movement are called Doppler systems. Doppler shift is proportional to the relative velocity of mobile movement. Doppler systems function more effectively when the velocity of mobile station is high enough to create reslovable Doppler shift. Examples are satellite-based position location systems or high-speed vehicles moving in high-tier cellular systems.

References

[1] M. Hata, "Empirical formula for propagation loss in land mobile radio sercvices," *IEEE Trans. Veh. Technol.*, vol.29, no.3, pp.317-325, Aug. 1980.

[2] J. Walfisch and H. L. Bertoni, "A theoretical model for UHF propagation in urban environments," *IEEE Trans. Antennas Propagat.*, vol.36, pp.1788-1796, Dec. 1988.

[3] E. Damosso and L. Correia,"Digital mobile radio towards future generation systems," *COST231* Final Rep., European Commission, pp.474, 1999.

[4] F. Ikegami, S. Yoshida, T. Takeuchi, and M. Umehira, "Propagation factors controlling mean field strength on urban streets," *IEEE*

Trans. Antennas Propagat., vol.32, pp.822-829, Aug. 1984.

[5] T. C. Becker, M. Dottling, and W. Wiesbeck, "Fast determination of channel impulse responses by 3D wave propagation modeling," *IEEE Int. Conf. Converg. Technol. Appl.*, vol.1, pp.287-291, 1996.

[6] A. Medeisis and A. Kajackas, "On the use of the universal Okumura-Hata propagation prediction model in rural areas," *Proc. IEEE Veh. Technol. Conf.*, vol.3, pp.1815-1818, Tokyo, 2000.

[7] J. W. McJown and R. L. Hamilton, Jr., "Ray-tracing as a design tool for radio networks," *IEEE Network Mag.*, vol.5, no.6, pp.27-30, Nov. 1991.

[8] Z. Ji, B. H. Li, H. X. Wang, H. Y. Chen, and T. K. Sarkar, "Efficient ray-tracing methods for propagation prediction for indoor wireless communications," *IEEE Antennas Propagat. Mag.*, vol.43, no.2, pp.41-49, Apr. 2001.

[9] J. Capon, "High resolution frequency wave number spectrum analysis," *Proc.IEEE*, vol.57, pp.1408-1418, Aug. 1969.

[10] R. O. Schmidt, "Multiple emitter location and signal parameter estimation," *IEEE Trans. Antennas Propagat.*, vol.34, no.3, pp.281-290, Mar. 1986.

[11] A. J. Barabell, "Improving the resolution performance of eigenstructure-based direction finding algorithms," *Proc. IEEE Int. Conf. Acoust., Speech, Signal Process.*, pp.336-339, 1983.

[12] W. A. Gardner, "Simplification of MUSIC and ESPRIT by exploitation of cyclostationary," *Proc.IEEE*, vol.76, pp.845-847, July 1988.

[13] R. Roy and T. Kailath, "ESPRIT-estimation of signal parameters via rotational invariance techniques," *IEEE Trans. Acoust., Speech, Signal Process.*, vol.37, no.7, pp.984-995, July 1989.

[14] I. Ziskind and M. Wax, "Maximum likelihood localization of multiple sources by alternating projection," *IEEE Trans. Acoust., Speech, Signal Process.*, vol.36, no.10, pp.1553-1560, Oct. 1988.

[15] M. Hirakawa, H. Tsuji, and A. Sano, "Computationally efficient DOA estimation based on linear prediction with Capon method," *Proc. IEEE Int. Conf. Acoust., Speech, Signal Process.*, vol.5, pp.3009-3012, 2001.

[16] L. R. Dandekar, L. Hao, and G. Xu, "Experimental study of mutual coupling compensation in smart antenna applications," *IEEE Trans. Wireless Commun.*, vol.1, no.3, pp.480-487, July 2002.

[17] D. Segovia-Vargas, R. Martin-Cuerdo, and M. Sierra-Perez, "Mutual coupling effects correction in microstrip arrays for direction-

of-arrival(DOA) estimation," *Proc. Inst. Elect. Eng.*, pt.H, vol.149, no.2, pp.113-118, Apr. 2002.

[18] G. V. Tsoulos, "Smart antennas for mobile communication systems: benefits and challenges," *J. Electron. Commun. Eng.*, vol.11, pp.84-94, Apr. 1999.

[19] B. Yang, C. Zhigang, and K. B. Letaief, "Analysis of low-complexity windowed DFT-based MMSE channel estimator for OFDM systems," *IEEE Trans. Commun.*, vol.49, pp.1977-1987, Nov. 2001.

[20] D. Cassioli, M. Z. Win, and A. R. Molisch, "The ultra-wide bandwidth indoor channel: from statistical model to simulations," *IEEE J. Select. Areas Commun.*, vol.20, pp.1247-1257 Aug. 2002.

[21] R. Price and P. E. Green, Jr., "A communication technique for multipath channels," *Proc.IRE*, vol.46, pp.555-570, Mar. 1958.

[22] P. S. Ray, "A novel pulse TOA analysis technique for radar identification," *IEEE Trans. Aerospace Electron. Syst.*, vol.34, pp.716-721, July 1998.

[23] D. H. Shin and T. K. Sung, "Comparisons of error characteristics between TOA and TDOA positioning," *IEEE Trans. Aerospace Electron. Syst.*, vol.38, pp.307-311, Jan. 2002.

[24] J. Caffery, Jr. and G. L. Stuber, "Subscriber location in CDMA cellular networks," *IEEE Trans. Veh. Technol.*, vol.47, no.2, pp.406-416, May 1998.

[25] S. S. Woo, H. R. You, and J. S. Koh, "The NLOS mitigation technique for position location using IS-95 CDMA networks," *Proc. IEEE Veh. Technol. Conf*, vol.4, pp.2556-2560, Sept. 2000.

[26] C. Li and W. Huang, "Hybrid TDOA/AOA mobile user location for wideband CDMA cellular systems," *IEEE Trans. Wireless Commun.*, vol.1, no.3, pp.439-447, July 2002.

Chapter 19

GENERATION OF WIDEBAND ELECTROMAGNETIC RESPONSES USING EARLY-TIME AND LOW-FREQUENCY DATA

Tapan K. Sarkar,[1] Jinhwan Koh,[1] and Magdalena Salazar Palma[2]

[1] *Department of Electrical Engineering and Computer Science*
Syracuse University
Syracuse, New York, USA

[2] *Grupo de Microondas y Radar*
Departamento de Senales, Sistemas y Radiocomunicaciones
Escuela Tecnica Superior de Ingenieros de Telecomunicacion
Universidad Politecnica de Madrid
Madrid, Spain

Abstract This Chapter introduces a new approach to generate wideband and temporal response for three-dimensional conducting structures. It is accomplished through the use of a hybrid method that involves generation of early-time and low-frequency information for the electromagnetic structure of interest. The early-time and low-frequency information are complementary and contain all the necessary information for the ultra-wideband electromagnetic response of the system at a specific spatial location. The time domain response is modeled by a Laguerre series expansion. The frequency domain response is also expressed in an analytic form using the same expansion coefficients used in modeling the time domain response. Data in both domains are used to solve for the polynomial coefficients by data fitting. Once the unknown polynomial coefficients are obtained, data of interest are concurrently extrapolated in both domains. This approach is attractive because expansions with a few terms give good extrapolation in both time and frequency domains, with minimum computation time.

Keywords: wideband response, Laguerre expansion, EFIE, MoM, MOT, causality, extrapolation.

1. Introduction

Electromagnetic analysis in the frequency domain usually assumes that the field quantities are time-harmonic. Method of moments(MoM) technique, which is based on an integral equation formulation, can be used to perform the frequency domain analysis [1]. However, this approach becomes very computationally intensive for broadband analysis because the MoM code needs to be executed at each frequency of interest, and the size of matrix can be very large at high frequencies.

Time domain approach is preferred for broadband analysis. Other advantages of time domain formulation include easier modeling of nonlinear and time-varying media and use of gating to eliminate unwanted reflections. For time domain integral equation formulation, the method of marching-on-in-time(MOT) is usually employed [2]-[4]. A serious drawback of this algorithm is the occurrence of late-time instabilities in the form of high-frequency oscillations.

In this Chapter, a technique is proposed to overcome late-time oscillations. Using early-time and low-frequency data, stable late-time and broadband information can be obtained. While the frequency domain MoM approach can generate low-frequency data efficiently, the MOT algorithm can be used to obtain stable early-time data quickly. The overall analysis is thus very computationally efficient.

Time and frequency domain responses of three-dimensional conducting objects are considered in this Chapter. It is assumed that the conducting objects are excited by bandlimited waveforms, such that both the time and frequency domain responses are approximately of finite extent. From a strictly mathematical point of view, a causal time domain response cannot be strictly bandlimited, and vice versa. However, a response strictly limited in time can be assumed to be approximately bandlimited if the amplitude of frequency response outside the band of interest is too small to be of any practical consequence [5].

In computational electromagnetics, one may want to obtain the electromagnetic fingerprint of an object. This is equivalent to obtaining the entire impulse response in the time domain or obtaining the transfer function over the entire frequency band. Either information requires tremendous computational resources. Here, we propose a hybrid approach which will minimize the computational efforts. The goal is attained by generating early-time and low-frequency information which is not computationally demanding. Then, a Laguerre series is used to fit data in the time domain and its transform, which are a set of polynomials in the frequency domain, to extrapolate the response in both time and frequency domains concurrently. In this approach, new information is not created, but existing information available in complementary domains is used to carry out the extrapolations.

An optimal choice of basis functions to characterize the time and frequency domain responses are those which have compact extents in both domains. The Laguerre series is well suited for practical signals with compact extent in the time domain. The fact that the Fourier transform of a Laguerre function is an analytic function allows us to manipulate time and frequency domain data concurrently.

It is better to use the Laguerre polynomials instead of the associate Hermite functions [6], even though the latter are eigenfunctions of the Fourier transform operator. The Hermite expansion is two sided over $(-\infty, \infty)$, hence the choice of origin of Hermite expansion for a causal function is very critical. In contrast, the Laguerre series is defined over the interval $[0, \infty)$, hence it is more suitable for modeling causal responses.

In the next Section, we will introduce the Laguerre functions and derive relevant equations for the problem. Some numerical results will be demonstrated in Section 3, followed by the conclusions.

2. Formulation

The Laguerre functions of order n is defined as

$$L_n(t) = \frac{1}{n!} e^t \frac{d^n(t^n e^{-t})}{dt^n}, \qquad n \geq 0, t \geq 0$$

Since they are defined over $t \geq 0$, they are causal. They can also be computed recursively as [7]

$$L_0(t) = 1, \qquad L_1(t) = 1 - t$$
$$L_n(t) = \frac{2n-1-t}{n} L_{n-1}(t) - \frac{n-1}{n} L_{n-2}(t), \qquad n \geq 2$$

Fig.19.1 shows the Laguerre functions of orders 0 to 3, respectively.

The Laguerre functions satisfy the orthogonality conditions that

$$\int_0^\infty e^{-t} L_n(t) L_m(t) dt = \delta_{mn}$$

where δ_{mn} is the Kronecker delta function which equals 1 when $m = n$ and equals 0 otherwise. A set of orthonormal basis functions can thus be derived from the Laguerre functions as

$$\phi_n(t, \ell_1) = e^{-t/2} L_n(t/\ell_1)$$

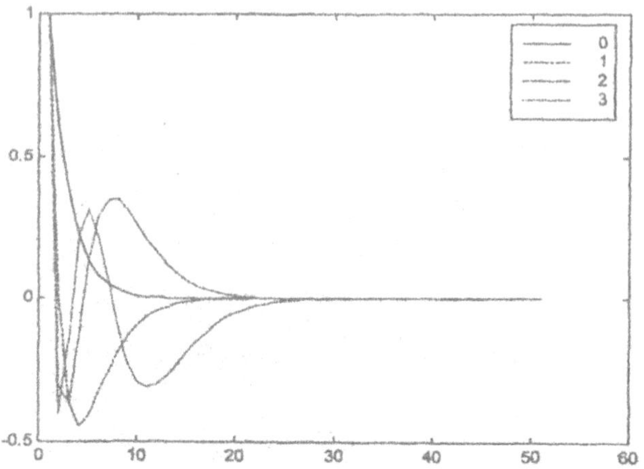

Figure 19.1. Laguerre functions of orders 0 to 3.

where ℓ_1 is a scaling factor. A causal response function, $x(t)$, with $t \geq 0$ at a specific location can be expanded in a Laguerre series as

$$x(t) = \sum_{n=0}^{N-1} a_n \phi_n(t, \ell_1)$$

which can approximate causal responses quite well. Extent of the expansion can be adjusted by varying the scaling factor, ℓ_1.

Fourier transform of the above expression becomes

$$X(f) = \sum_{n=0}^{N-1} \frac{a_n}{2\pi\ell_2} \frac{(-1/2 + jf/\ell_2)^n}{(1/2 + jf/\ell_2)^{n+1}}$$

where $j = \sqrt{-1}$ and $\ell_2 = (2\pi\ell_1)^{-1}$. The choice of scaling factor, ℓ_1, is crucial because it also affects ℓ_2. These two scaling factors decide the extent of Laguerre functions in the time and frequency domain responses, respectively. Given about 50-60% of early-time data and an equal amount of low-frequency data, it is possible to concurrently extrapolate the responses in both domains with a proper choice of the order of expansion, N, and the scaling factor, ℓ_1. The value of N can be decided a priori by considering the time-bandwidth product of waveforms. Choosing an unnecessarily large N will introduce oscillations in the extrapolation region. The coefficients of Laguerre expansion are obtained by solving a least squares problem through the use of singular value decomposition(SVD) [8].

Let M_1 and M_2 be the number of time and frequency domain samples in functions $x(t)$ and $X(f)$, respectively. Then, matrix representation of the time domain data becomes

$$\bar{\bar{\Phi}} \cdot \bar{a} = \bar{x}$$

where $\bar{a} = [a_0, a_1, \cdots, a_{N-1}]^t$, $\bar{x} = [x(t_1), x(t_2), \cdots, x(t_{M_1})]^t$, and

$$\bar{\bar{\Phi}} = \begin{bmatrix} \phi_0(t_1,\ell_1) & \phi_1(t_1,\ell_1) & \cdots & \phi_{N-1}(t_1,\ell_1) \\ \phi_0(t_2,\ell_1) & \phi_1(t_2,\ell_1) & \cdots & \phi_{N-1}(t_2,\ell_1) \\ & & \cdots & \\ \phi_0(t_{M_1},\ell_1) & \phi_1(t_{M_1},\ell_1) & \cdots & \phi_{N-1}(t_{M_1},\ell_1) \end{bmatrix}$$

Similarly, in the frequency domain, we have

$$\bar{\bar{\Psi}} \cdot \bar{a} = \bar{X}$$

where $\bar{X} = 2\pi\ell_2[X(f_1), X(f_2), \cdots, X(f_{M_2})]^t$, and

$$\bar{\bar{\Psi}} = \begin{bmatrix} \frac{1}{1/2+jf_1/\ell_2} & \frac{(-1/2+jf_1/\ell_2)}{(1/2+jf_1/\ell_2)^2} & \cdots & \frac{(-1/2+jf_1/\ell_2)^{N-1}}{(1/2+jf_1/\ell_2)^N} \\ \frac{1}{1/2+jf_2/\ell_2} & \frac{(-1/2+jf_2/\ell_2)}{(1/2+jf_2/\ell_2)^2} & \cdots & \frac{(-1/2+jf_2/\ell_2)^{N-1}}{(1/2+jf_2/\ell_2)^N} \\ & & \cdots & \\ \frac{1}{1/2+jf_{M_2}/\ell_2} & \frac{(-1/2+jf_{M_2}/\ell_2)}{(1/2+jf_{M_2}/\ell_2)^2} & \cdots & \frac{(-1/2+jf_{M_2}/\ell_2)^{N-1}}{(1/2+jf_{M_2}/\ell_2)^N} \end{bmatrix}$$

By concatenating these two equations, we have

$$\begin{bmatrix} \bar{\bar{\Phi}} \\ \bar{\bar{\Psi}} \end{bmatrix} \cdot \bar{a} = \begin{bmatrix} \bar{x} \\ \bar{X} \end{bmatrix} \tag{19.1}$$

where the matrix size is $(M_1 + M_2) \times N$. The N unknown coefficients of expansion are obtained by solving this matrix equation.

3. Numerical Examples

In this Section, four examples are presented to demonstrate this technique. Codes have been developed to evaluate the currents on an arbitrarily shaped, closed or open body using electric field integral equations(EFIE) both in time [3] and frequency domains [1], and a triangular patching scheme is used to discretize the surface [3]. We apply the same surface patching scheme in both domains to eliminate some of the discretization effects. In the triangular patching scheme, surface of a scatterer is approximated with a set of triangles. The current perpendicular to each nonboundary edge is assigned an unknown. The frequency domain data are generated using the program in [1].

A linearly polarized plane wave with Gaussian profile in time is chosen as the excitation, namely,

$$\bar{E}^{\mathrm{inc}} = \hat{u}_i \frac{1}{\sigma\sqrt{\pi}} E_0 e^{-\gamma^2}$$

where $\gamma = (t - t_0 - \bar{k} \cdot \bar{r})/\sigma$, $\hat{u}_i$ is the unit polarization vector of the incident plane wave, E_0 is the amplitude of incident wave, t_0 is a delay to shift the time origin of the waveform to $t = 0$, $\bar{r}$ is the observation position, $\bar{k}$ is the wave number vector defining the direction of incident plane wave, and σ^2 is the spread factor of Gaussian pulse. The frequency response to this Gaussian plane wave is the frequency response of system multiplied by the spectrum of Gaussian plane wave which is

$$F(f) = \frac{E_0}{c} e^{-j\omega t_0 - (\omega\sigma/c)^2/4}$$

where $\omega = 2\pi f$.

Electromagnetic responses are demonstrated by extrapolating from the currents induced on parallel-plate resonator, plate-sphere combination, resonating cavity and a cone-hemisphere combination. All the objects are assumed to be perfectly conducting. In all the following computations, E_0 is chosen to be 377 V/m. The time step, Δt, is dictated by discretization used in modeling the object in each example, the frequency step, Δf, is 2 MHz.

In all the examples, the extrapolated time domain response is compared to that obtained using the marching-on-in-time(MOT) code [2]. The extrapolated frequency domain response is compared to that obtained using the method of moments(MoM) code [1]. In all the plots, dashed curves refer to the extrapolated response using Laguerre expansions, and solid curves refer to the data obtained using the MOT or MoM codes.

Example 1: Plates with Different Sizes. Fig.19.2 shows two square plates of zero thickness and widths of 1 m and 2 m, respectively, placed parallel to the x-y plane with a separation of 2 m. A plane wave is incident along the $-\hat{z}$ direction, and $\hat{u}_i$ is along the x axis. The time step used in the MOT code is 196.8 ps, the other parameters are $\sigma = 0.79$ ns and $t_0 = 9.2$ ns. The edge under consideration is located at the center of the smaller plate, and is oriented in the $\hat{y}$ direction.

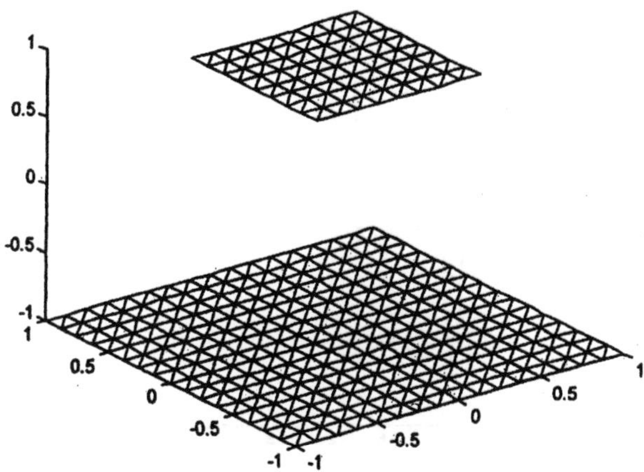

Figure 19.2. Triangulation scheme of two plates with different sizes.

The time domain data are calculated from $t = 0$ to $t = 50$ ns, with 250 data points, using the MOT code. The frequency domain data are calculated from dc to $f = 998$ MHz with 500 data points. The order of expansion is chosen to be 60, considering the time-bandwidth product of waveform. Assume that only the first 90 data points in the time domain (up to $t = 18$ ns) and the first 120 data points in the frequency domain (up to $f = 238$ MHz) are available. The polynomial coefficients are determined by solving the matrix equation (19.1) using these available data. The coefficients are then used to extrapolate the time domain response up to 250 points(50 ns) and the frequency domain response up to 500 points(998 MHz).

From Fig.19.3, it is observed that the time domain reconstruction is almost indistinguishable from the reference MOT data. Reconstruction in the frequency domain is also very good, as shown in Figs.19.4 and 19.5.

Example 2: Plate-Sphere Combination. Fig.19.6 shows a plate-sphere combination. The sphere is centered at the origin with radius of 1 m, the plate is displaced from the origin by 5 m. The incident wave arrives along the $-\hat{x}$ direction, $\sigma = 2.359$ ns, $t_0 = 9.2$ ns, and the time step used in the MOT code is 0.484 ns. The edge under consideration is close to the center of plate and is oriented along the $\hat{y}$ direction.

The time domain data are calculated using the MOT code from $t = 0$ to $t = 145$ ns(300 data points), and the frequency domain data are calculated using the MoM code from dc to $f = 298$ MHz(150 data points).

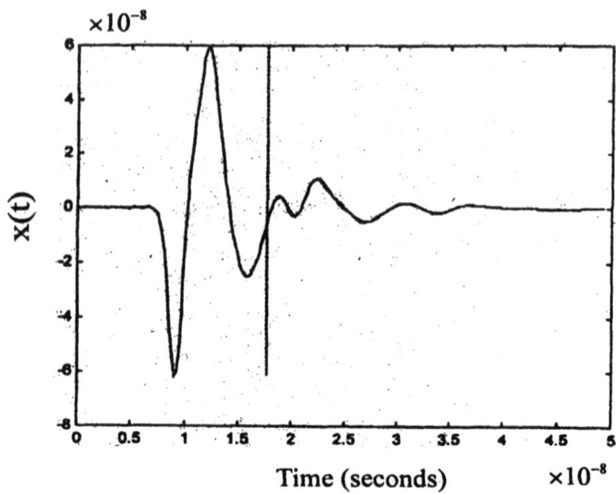

Figure 19.3. Time domain response at the center of the smaller plate.

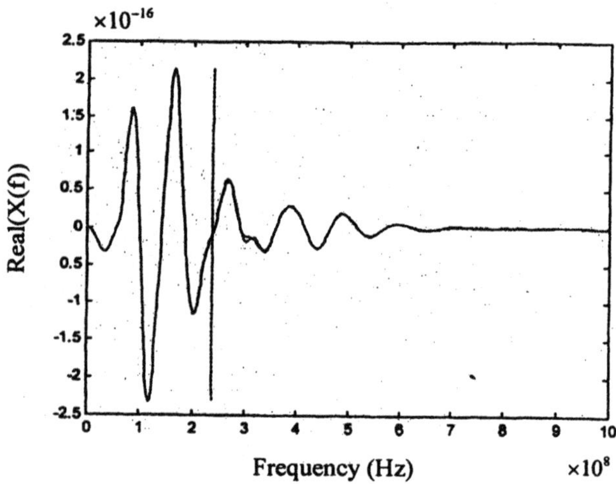

Figure 19.4. Real part of frequency response at the center of the smaller plate.

The order of expansion is chosen to be 50. Using the first 80 data points in the time domain(up to 38.67 ns) and the first 50 data points in the frequency domain(up to 98 MHz), the time domain response is extrapolated to 300 points and the frequency domain response is extrapolated to 150 points.

Fig.19.7 shows that the time domain reconstruction match reasonably well to the reference MOT data. Fig.19.8 shows that the real part of

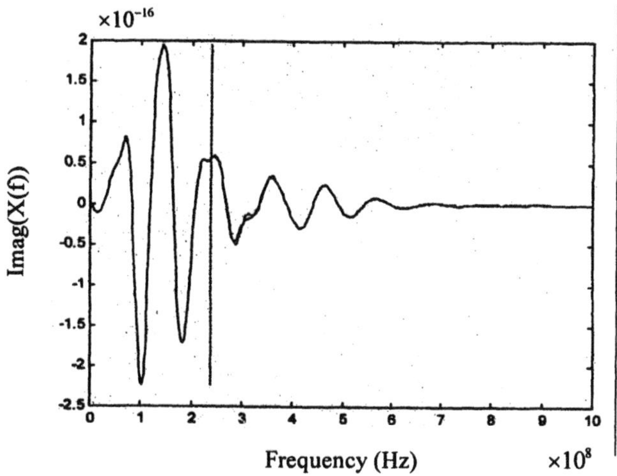

Figure 19.5. Imaginary part of frequency response at the center of the smaller plate.

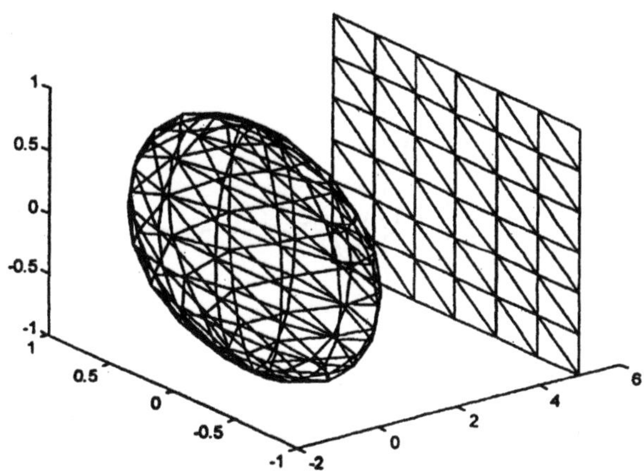

Figure 19.6. Triangulation scheme of a plate-sphere combination.

frequency response has a reasonably good reconstruction. The imaginary part shows a similar match, and is not presented.

Example 3: Rectangular Cavity. Consider a rectangular cavity of dimensions 1 m × 1 m × 4 m, centered at the origin, with its edges lined up with the three coordinate axes, and its longest side is parallel to the x axis. The face at $x = 2$ m is open. The incident wave arrives from the direction of $\theta = 5\pi/6$, $\phi = \pi/6$, $\hat{u}_i$ is parallel to the direction of

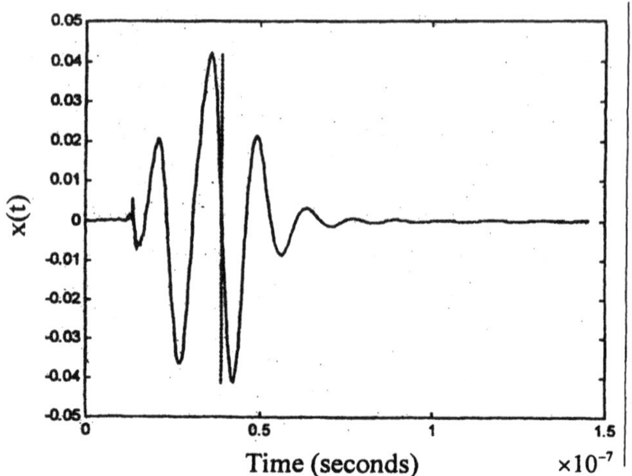

Figure 19.7. Time domain response at the center of the plate.

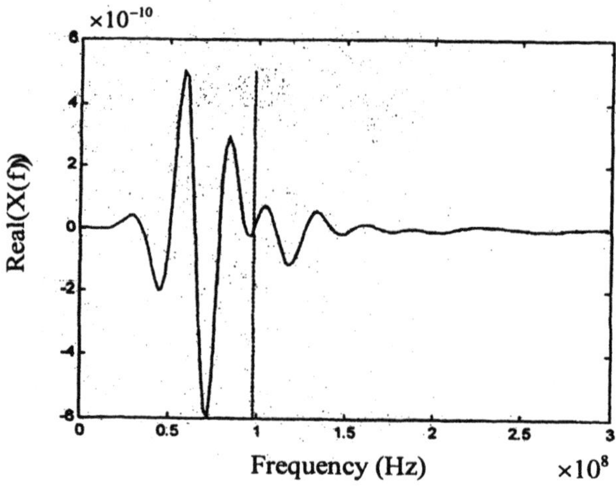

Figure 19.8. Real part of frequency response at the center of the plate.

$\theta = \pi/6$, $\phi = 5\pi/6$, σ= 1.18 ns and t_0= 4.56 ns. The time step used in the MOT code is 0.267 ns. The time domain data are calculated using the MOT code from $t = 0$ to $t = 133.33$ ns(500 data points), and the frequency domain data are calculated using the MoM code from dc to $f = 498$ MHz(250 data points). The order of expansion is chosen to be 40. Assuming that the first 50 data points in the time domain(up to $t = 13.33$ ns) and the first 60 data points in the frequency domain(up to

$f = 118$ MHz) are available, the time domain response is extrapolated to 500 points and the frequency domain response is extrapolated to 250 points. Fig.19.9 shows that the time domain response closely matches with the reference MOT data. The frequency domain reconstruction is also good, with the real part shown in Fig.19.10.

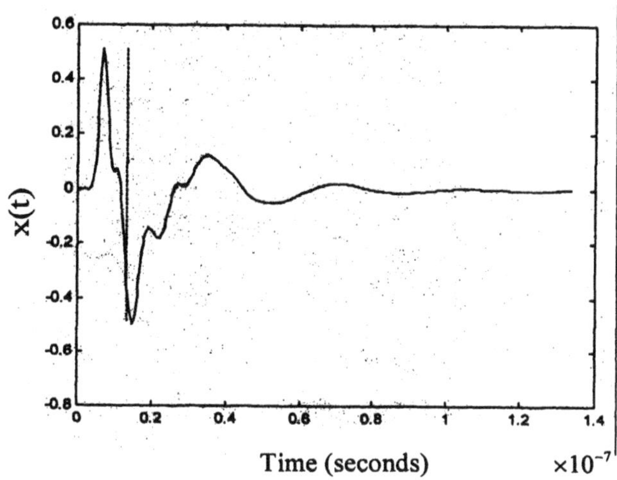

Figure 19.9. Time domain response at one point on the cavity wall.

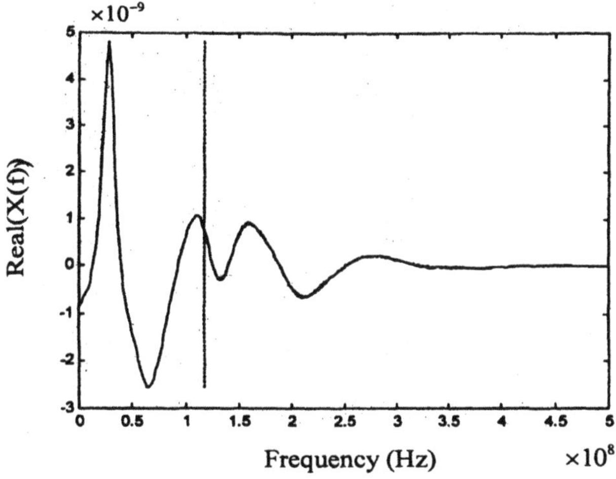

Figure 19.10. Real part of frequency domain response at the same point as in Fig.19.9.

Example 4: Cone-Hemisphere Combination. Consider a combination of a cone and a hemisphere, where the hemisphere is attached to the base of cone with its axis oriented along the $\hat{x}$ direction. The bases of cone and hemisphere have radii of 1 m, and the height of cone is 4 m. The incident wave arrives along the $-\hat{z}$ direction, $\hat{u}_i$ points in the $\hat{y}$ direction, σ= 2.948 ns and t_0= 11.495 ns. The time step used is 206.67 ps and the frequency interval is 2 MHz.

The time domain response is calculated using the MOT code from $t = 0$ to $t = 93$ ns(450 data points), and the frequency domain response is calculated with the MoM code from dc to $f = 448$ MHz(225 data points). The order of extrapolation is chosen to be 40. Assuming that the first 130 data points in the time domain(up to $t = 26.66$ ns) and the first 40 data points in the frequency domain(up to $f = 78$ MHz) are available, the time domain response is extrapolated to 450 points and the frequency domain response is extrapolated to 225 points. Fig.19.11 shows that the time domain response closely matches with the reference MOT data. The frequency domain reconstruction is also good, with the real part shown in Fig.19.12.

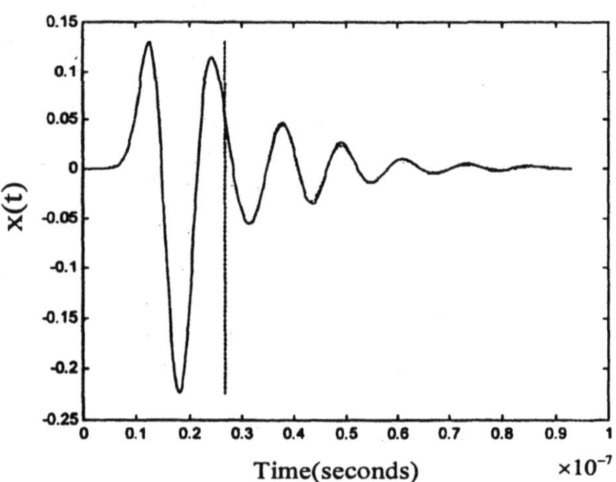

Figure 19.11. Time domain response at one point on a cone-hemisphere combination.

4. Conclusions

This Chapter introduces concurrent extrapolation in both time and frequency domains using only early-time and low-frequency data. It is accomplished with the Laguerre expansion which is inherently causal and thus fit the time domain data better than the associate Hermite func-

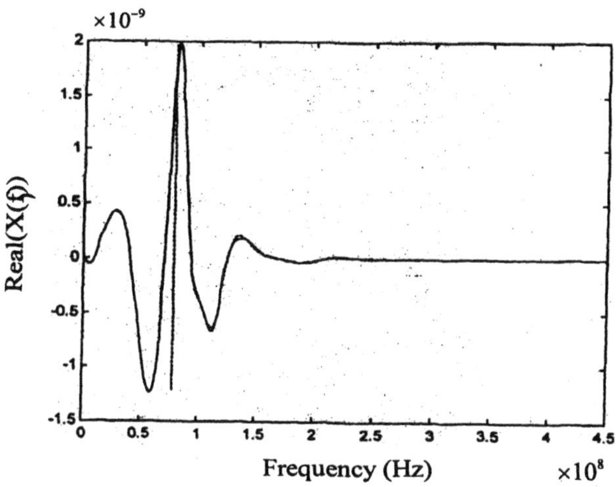

Figure 19.12. Real part of frequency domain response at the same point as in Fig.19.11.

tions. This method is computationally efficient because only early-time and low-frequency information are required. The size of matrix required to solve for the expansion coefficients are small and expansions of order around 50 give good representation of the signals in both domains.

This new technique has been applied to extrapolate the current on scatterers excited by uniform plane wave. Four examples have been presented to demonstrate the extrapolation of responses in both the time and frequency domains.

References

[1] B. M. Kolundzjia, J. S. Ognjanovic, and T. K. Sarkar, *WIPL-D, Software for Electromagnetic Modeling of Composite Metallic and Dielectric Structures*, Norwood, MA: Artech House, 2000.

[2] D. A. Vechinski, S. M. Rao, and T. K. Sarkar, "Transient scattering from three dimensional arbitrarily shaped dielectric bodies," *J. Opt. Soc. Am. A*, vol.1, no.4, pp.1458-1470, 1994.

[3] T. K. Sarkar, W. Lee, and S. M. Rao, "Analysis of transient scattering from composite arbitrarily shaped complex structures," *IEEE Trans. Antennas. Propagat.*, vol.48, no.10, pp.1625-1634, Oct. 2000.

[4] R. S. Adve, T. K. Sarkar, O. M. Pereira-Filho, and S. M. Rao, "Extrapolation of time domain responses from three dimensional con-

ducing objects utilizing the matrix pencil technique," *IEEE Trans. Antennas Propagat.*, vol.45, no.1, pp.147-156, Jan. 1997.

[5] S. Narayana, T. K. Sarkar, and R. S. Adve, "A comparison of two techniques for the interpolation / extrapolation of frequency domain responses," *Digital Signal Process.*, vol.6, no.1, pp.51-67, Jan. 1996.

[6] M. M. Rao, T. K. Sarkar, T. Anjali, and R. S. Adve, "Simultaneous extrapolation in time and frequency domains using Hermite expansion," *IEEE Trans. Antennas Propagat.*, vol.47, pp.1108-1115, June 1999.

[7] A. D. Poularikas, *The Transforms and Applications Handbook*, IEEE Press, 1996.

[8] G. H. Golub and C. F. Van Loan, *Matrix Computations*, Johns Hopkins University Press, 1991.

Chapter 20

ASYMPTOTIC HIGH FREQUENCY METHODS

Hsi-Tseng Chou[1] and Teh-Hong Lee[2]

[1] *Department of Communications Engineering*
Yuan-Ze University
Chung-Li, Taiwan, ROC

[2] *ElectroScience Laboratory*
The Ohio State University
Columbus, Ohio, USA

Abstract Fundamental concepts of asymptotic high frequency(HF) techniques including geometrical optics(GO), physical optics(PO), geometrical theory of diffraction(GTD) and its uniform version(UTD), and physical theory of diffractions(PTD) are reviewed in this Chapter. Instead of going through mathematical derivations, important concepts will be summarized only to demonstrate the physical natures of asymptotic HF techniques. Examples of practical application will be presented to appreciate their usefulness.

Keywords: Asymptotic high frequency techniques, geometrical optics, physical optics, geometrical theory of diffraction, physical theory of diffraction, shooting and bouncing rays, generalized ray expansion.

1. Introduction

Fast analysis of electromagnetic(EM) radiation, propagation and scattering problems associated with electrically large objects has been of great interest. This Chapter presents a tutorial review of asymptotic high frequency(HF) techniques, including geometrical optics(GO) [1]-[3], physical optics(PO), geometrical theory of diffraction(GTD) [4], [5] and its uniform version(UTD) [6]-[12], and physical theory of diffraction(PTD) [13], [14], that are commonly used in practical applications.

Major focuses are placed on their conceptual descriptions and practical applications. Most low frequency numerical methods, such as method of moments(MoM), finite difference time domain(FDTD) and finite element method(FEM), become inefficient as dimension of the objects under analysis is electrically large. In contrast, HF techniques have the advantages of maintaining their efficiency and accuracy by describing their solutions in terms of ray fields, and are particularly suitable for the analysis of large objects because dominating rays are generally few in number and remain unchanged despite of the increase of operating frequency. Furthermore, they provide clear physical pictures to interpret the mechanisms of EM wave propagation in practical applications. The HF techniques can be further integrated with other numerical methods to extend the applicability of both high and low-frequency techniques.

Developments of HF techniques originate from the asymptotic evaluations of existing Green's functions on specific geometries and problems, which interpret the EM field in terms of the mechanisms of reflection, diffraction, and so on [15], [16]. After the mechanisms are identified, they are then generalized to treat similar geometries with more arbitrary shapes based on the local phenomenon of ray optics.

Instead of going through cumbersome derivations, this Chapter will first describe the fundamental concepts of GO and its direct applications including shooting and bouncing rays(SBR)[17], generalized ray expansion(GRE) methods [18]-[21] in Section 2. A direct extension of GO based on equivalent current concepts, which is referred to as physical optics(PO), is described in Section 3. Physical optics can potentially extend the applicability of GO via numerical integration of radiation integral over a set of equivalent currents defined by GO approximations. It can provide good approximations of field behaviors in the specular regions of scattering problems. On the other hand, the GTD/UTD concepts are described in Section 4, which are theoretic extension of GO and provide more complete description of high frequency wave propagation mechanism. In general, GO only describes the field behavior of reflections, and results in discontinuous field behaviors. The GTD/UTD tends to compensate GO by including diffraction mechanism. Application examples using these high-frequency solutions are presented in Section 5. Potential extension of GTD/UTD to deal with more complicated problems is their hybrid combination with conventional numerical methods. Section 6 presents two typical approaches based on the integration of GTD/UTD with conventional MoM. Application examples are also presented. Section 7 presents fundamental PTD concepts. With similar motivation as GTD/UTD correct the GO, PTD intends to correct the PO and provides accurate field behavior outside the specular region.

It is assumed throughout this Chapter that the scatters and surrounding media are isotropic and homogeneous. A time harmonic of $e^{j\omega t}$ is assumed for the fields and is omitted.

2. Fundamental Concepts of Geometrical Optics

2.1 Geometrical Ray Optical Field

Geometrical optics(GO) is an asymptotic solution to Maxwell's equations [2], [3], [22], which employs ray concepts to describe EM wave propagation mechanisms. The GO field propagates along the ray paths by following the Fermat's principle with power and polarization conservations. The Fermat's principle requires that the length of propagation ray path is stationary and minimized, and can be used to determine the ray paths for various scattering mechanisms including propagation in free space, reflection and diffraction. In particular, the ray path in a homogeneous medium is a straight line as illustrated in Fig.20.1 where a GO ray propagating from the reference point, O, to a field point, P, along a straight path can be expressed as

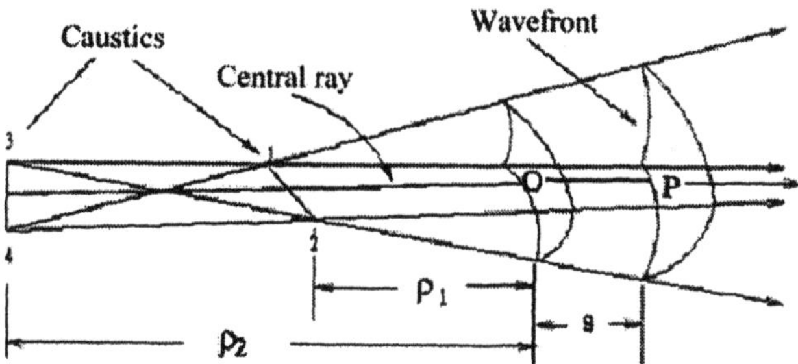

Figure 20.1. An astigmatic GO ray tube used to illustrate GO propagation and its wavefront variations.

$$\bar{E}(P) \simeq \bar{E}(O)\sqrt{\frac{\rho_1\rho_2}{(\rho_1+s)(\rho_2+s)}}\, e^{-jks} \tag{20.1}$$

where s is the length of $|\overline{OP}|$, ρ_1 and ρ_2 are the radii of principal and orthogonal wavefront curvatures, respectively, and can be either positive or negative according to the location of O. Note that s is positive in the direction of propagation and negative otherwise. If $\hat{s}$ denotes the propagation direction, then the EM field will satisfy the following relations

$$\hat{s}\cdot\bar{E}(P) = 0, \qquad \bar{H}(P) \simeq \hat{s}\times Y\bar{E}(P)$$

where Y is the intrinsic admittance of the medium. Note that $\hat{s}$, $\hat{E}$ and $\hat{H}$ are perpendicular to one another. The locations at $s = -\rho_1$ and $s = -\rho_2$ are referred to as ray caustics, as indicated by lines 1-2 and 3-4, respectively, in Fig.20.1, where field singularities occur and cause invalid GO ray field representations. Three special cases can be derived from (20.1) in terms of ρ_1 and ρ_2. If $\rho_1 = \rho_2$, (20.1) becomes a spherical wave. If either ρ_1 or ρ_2 is infinite, (20.1) becomes a cylindrical wave. If both ρ_1 and ρ_2 are infinite, (20.1) becomes a plane wave. The computation of (20.1) involves branch selection of square roots that can be defined as

$$\sqrt{\frac{\rho_i}{\rho_i + s}} = \begin{cases} \left|\sqrt{\rho_i/(\rho_i + s)}\right|, & \rho_i/(\rho_i + s) > 0 \\ j\left|\sqrt{\rho_i/(\rho_i + s)}\right|, & \rho_i/(\rho_i + s) < 0 \end{cases}$$

with $i = 1, 2$. Note that a phase jump of $\pi/2$ occurs while the propagation path crosses a ray caustic.

The GO expression in (20.1) can be simplified as

$$\bar{E}(P) \sim \bar{E}(O)\sqrt{\frac{\left|\bar{\bar{Q}}(s)\right|}{\left|\bar{\bar{Q}}(0)\right|}}e^{-jks}$$

where

$$\bar{\bar{Q}}(s) = \begin{bmatrix} 1/(\rho_1 + s) & 0 \\ 0 & 1/(\rho_2 + s) \end{bmatrix} \tag{20.2}$$

which satisfies

$$\bar{\bar{Q}}^{-1}(s) = \bar{\bar{Q}}^{-1}(0) + s\bar{\bar{I}} \tag{20.3}$$

Note that (20.3) can be employed in arbitrary ray coordinate systems, defined by any two orthogonal directions transversal to $\hat{s}$, and will reduce to (20.2) if the two transversal directions are in the planes of principal wavefront curvature. In general, (20.3) is not diagonal unless planes of principal wavefront curvature are selected.

2.2 Transformation of Source Radiation into GO Field

In most practical applications, EM fields radiated from current(either electric or magnetic) sources need to be transformed into GO ray field before (20.1) is applicable. Note that GO requires the EM field to satisfy the characteristics of ray optics in which a point source is generally assumed. With this concept, far field of a radiation source can generally exhibit the nature of ray optics, which can be expressed in a general form of

$$\bar{E}(\bar{r}) = \left[F_\theta(\theta,\phi)\hat{\theta} + F_\phi(\theta,\phi)\hat{\phi}\right]\frac{e^{-jkr}}{r} \tag{20.4}$$

where the spherical coordinate system, (r, θ, ϕ), satisfies the requirement of ray coordinate system with $\hat{r}$ being the direction of ray propagation. Comparing (20.4) with (20.1) by letting $\rho_1 = \rho_2 = \rho_0$ and $r = \rho_0 + s$ for a spherical wave, we obtain

$$\bar{E}(O)\rho_0 e^{jk\rho_0} = \left[F_\theta(\theta,\phi)\hat{\theta} + F_\phi(\theta,\phi)\hat{\phi}\right] \tag{20.5}$$

It is observed that (20.5) is dependent on the direction of propagation, (θ, ϕ). Note that ρ_0 can be chosen arbitrarily without affecting the subsequent derivations. Thus, one may simply set $\rho_0 = 1$, and $\bar{\bar{Q}}(s)$ in (20.2) is reduced to

$$\bar{\bar{Q}}(s) = \begin{bmatrix} 1/r & 0 \\ 0 & 1/r \end{bmatrix} \tag{20.6}$$

2.3 Ray Tracing Techniques and GO Reflections

Reflection is the most important propagation mechanism in GO. After the EM waves radiated from the sources are transformed into GO ray fields, a ray tracing procedure based on the reflection mechanism needs to be followed in order to find the scattering field. Note that reflection occurs only in the portions of obstacles that are directly illuminated by the incident field. In other words, GO gives null field in the shadow region, and causes discontinuities at the incident and reflection shadow boundaries(ISB/RSB) as illustrated in Fig.20.2.

The ray tracing procedure follows the Fermat's principle or Snell's law to determine the propagation ray paths and the reflection point hit by the incident ray. The Snell's law requires that the incident and reflected angles formed by the incident and reflected rays, respectively, with the surface normal vector are the same at the reflection point as illustrated in Fig.20.3, namely $\theta^r = \theta^i$.

Note that the incident and reflected rays as well as the unit normal vector, $\hat{z}_0$, at the reflection point lie on the same plane. The associated transformation between the incident and reflected ray wavefronts are shown in Fig.20.4. Once the reflection point, P_R, is found, the surface

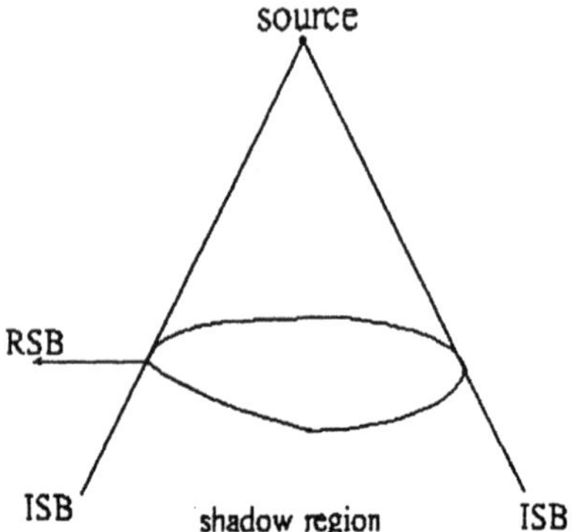

Figure 20.2. GO causes discontinuities at ISB and RSB since null field exists in the shadow region and no reflection in the regions beyond RSB.

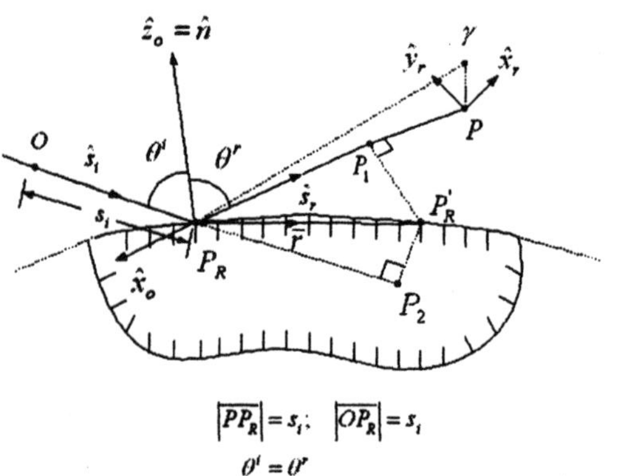

Figure 20.3. GO field reflected from a smooth curved boundary.

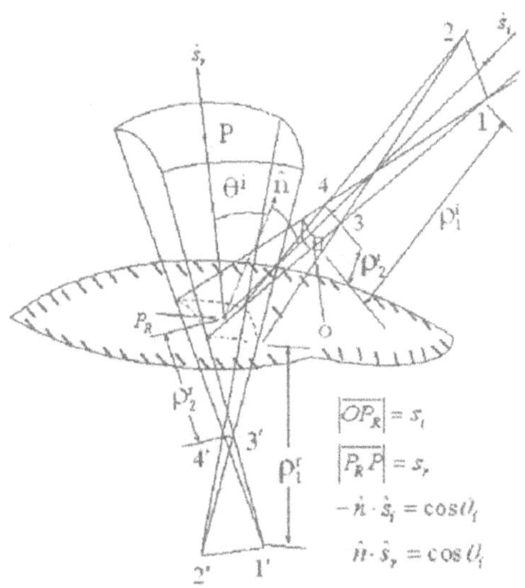

Figure 20.4. Transformation of incident and reflected ray wavefronts.

can then be locally modeled as

$$z_0 = -\frac{1}{2}\left[x_0, y_0\right] \cdot \bar{\bar{Q}}'(P_R) \cdot \left[x_0, y_0\right]^t \tag{20.7}$$

where $(x_0\ , y_0, z_0)$ are the local coordinates with the origin located at P_R, and $\bar{\bar{Q}}'(P_R)$ is a matrix describing the surface curvatures. The reflected field is assumed to be also a GO ray field, and its associated parameters can be obtained by matching boundary conditions on the surface in the vicinity of P_R. In particular, the incident and reflected GO rays can be denoted as

$$\bar{E}^i(P_R) \simeq \bar{E}^i(O)\sqrt{\frac{\left|\bar{\bar{Q}}^i(P_R)\right|}{\left|\bar{\bar{Q}}^i(O)\right|}}e^{-jks_i} \tag{20.8}$$

$$\bar{E}^r(P) \simeq \bar{E}^r(P_R)\sqrt{\frac{\left|\bar{\bar{Q}}^r(P)\right|}{\left|\bar{\bar{Q}}^r(P_R)\right|}}e^{-jks_r} \tag{20.9}$$

where $s_i = \left|\overline{OP_R}\right|$, $s_r = \left|\overline{P_RP}\right|$, the matrices $\bar{\bar{Q}}^i$ and $\bar{\bar{Q}}^r$ are derived based on the same rule as reaching (20.3). The parameters in (20.9) can be derived from (20.7) and (20.8) as [22], [23]

$$\hat{s}_r = \hat{s}_i - 2(\hat{s}_i \cdot \hat{z}_0)\hat{z}_0$$

$$2\cos\theta^i \bar{\bar{Q}}^s(P_R) + \bar{\bar{\Theta}}_i^t \cdot \bar{\bar{Q}}^i(P_R) \cdot \bar{\bar{\Theta}}_i = \bar{\bar{\Theta}}_r^t \cdot \bar{\bar{Q}}^r(P_R) \cdot \bar{\bar{\Theta}}_r$$

where $\bar{\bar{\Theta}}_i(\bar{\bar{\Theta}}_r)$ is the transformation matrix between the coordinate system of incident(reflected) GO ray and that of the local surface. In particular, they can be expressed as

$$\bar{\bar{\Theta}}_{i,r} = \begin{bmatrix} \hat{x}_{i,r} \cdot \hat{x}_0 & \hat{x}_{i,r} \cdot \hat{y}_0 \\ \hat{y}_{i,r} \cdot \hat{x}_0 & \hat{y}_{i,r} \cdot \hat{y}_0 \end{bmatrix}$$

with $(x_{i,r}, y_{i,r}, s_{i,r})$ being the coordinates for the incident and reflected rays, respectively.

The initial amplitude of reflected GO ray field, $\bar{E}_r(P_R)$, in (20.9) can be found by relating the incident ray with the surface reflection coefficient, $\bar{\bar{R}}(P_R)$, at P_R as

$$\bar{E}^r(P_R) = \bar{E}^i(P_R) \cdot \bar{\bar{R}}(P_R)$$

The dyadic reflection coefficient, $\bar{\bar{R}}(P_R)$, is in general a 2×2 matrix. However, if the coordinate systems of incident and reflected rays are selected properly as illustrated in Fig.20.5, it can be simplified and diagonalized as

$$\bar{\bar{R}}(P_R) = \hat{e}_{||}^i \hat{e}_{||}^r R_{||}(P_R) + \hat{e}_{\perp} \hat{e}_{\perp} R_{\perp}(P_R) \tag{20.10}$$

where the unit vectors are indicated in Fig.20.5, $R_{||}(P_R)$ and $R_{\perp}(P_R)$ are the Fresnel reflection coefficients of the TE and TM modes, respectively. The unit vectors in (20.10) may not coincide with the directions of principal wavefront curvatures for the incident and reflected rays. Namely, $\bar{\bar{Q}}^r(P_R)$ in (20.9) may not be diagonal as in (20.2). Matrix diagonalization should be performed if the radii of principal wavefront curvatures are of interest [7]. Finally, the total GO field is given by

$$\bar{E}^{GO}(P) \simeq \bar{E}^i(P)U_i + \bar{E}^r(P)U_r \tag{20.11}$$

where U_i and U_r are the Heaviside step functions. Note that U_i is determined by the incident shadow boundary, $U_i = 1$ if the field point is visible to the incident ray and $U_i = 0$ otherwise. Similarly, U_r is determined by the reflection shadow boundary, $U_r = 1$ if reflected rays exist and $U_r = 0$ otherwise. Obviously, (20.11) shows discontinuities across the incident and reflection shadow boundaries.

2.4 Shooting and Bouncing Rays Method

Ray tracing procedure is in general cumbersome and complicated. Shooting and bouncing rays(SBR) method simplifies the ray tracing pro-

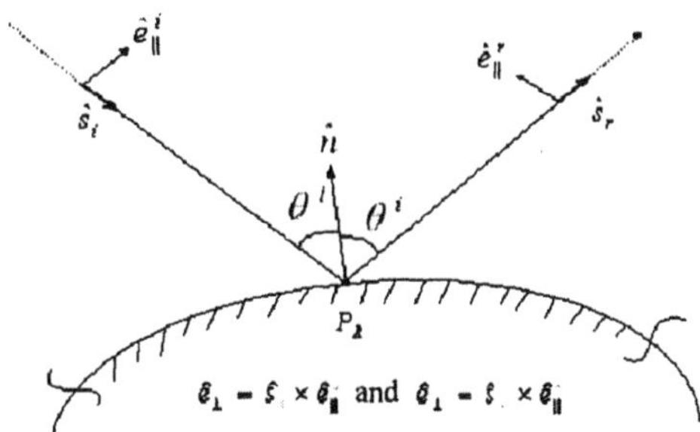

Figure 20.5. Unit vectors associated with incident and reflected fields.

cedure by representing the source radiation in terms of discrete ray tubes, and then tracing each ray tube independently through ray path of the central ray [17]. The direction of each ray tube is fixed and known, and each ray tube represents a pulse type of basis function that has uniform amplitude and quadratic phase variation over its cross section. The GO ray tube can thus be described as [22]

$$\bar{E}(P) \sim \bar{E}(O)\sqrt{\frac{\left|\bar{\bar{Q}}(s)\right|}{\left|\bar{\bar{Q}}(0)\right|}}e^{-jk\left(s+(1/2)[x,y]\cdot\bar{\bar{Q}}(s)\cdot[x,y]^t\right)} \tag{20.12}$$

where the coordinate system, (x, y, s), is defined along the central ray. In general, the area of each ray tube should be sufficiently small so that the approximation of (20.12) remains valid. One may either assume arbitrary cross sectional shapes or a polygonal shape for each ray [19], [23], [24]. Size and shape of the cross section will change continuously while the ray tubes are propagating in order to observe power conservation.

The SBR method is outlined as follows. First, represent the source radiation in terms of ray tubes through the use of (20.5) and (20.6). In general, a set of discrete angular grids are assumed to determine the ray propagating directions. The second step is to determine the shape of ray cross section. Circular, triangular or quadratic shapes are popularly employed. The third step is to trace each ray propagating path and its associated shape of cross section. Two different ways have been proposed. The first one traces the central ray only and assumes a cross sectional shape that observe power conservation regardless the shape change during propagation. Subdivision of ray tubes might be required if the cross section area becomes too large [17]. The fundamental concepts are illustrated in Fig.20.6(a). The second approach is illustrated in Fig.20.6(b) where polygonal shapes of ray tubes are assumed. Corner rays as well as the central ray are traced in order to determine the shape and parameters simultaneously. Since the shapes of ray tubes are determined by corner rays, power conservation is assumed to be satisfied except for some numerical errors.

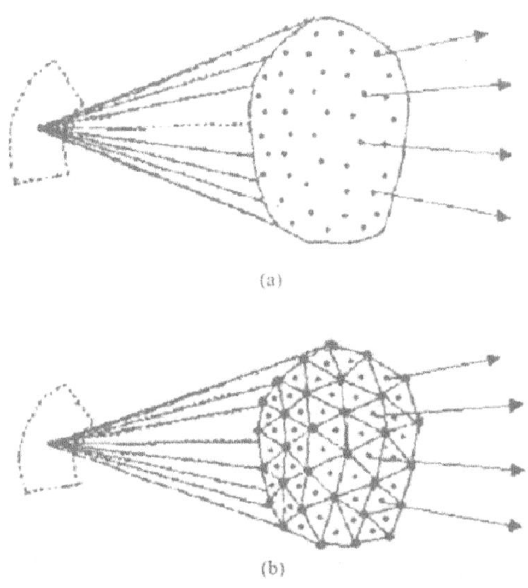

Figure 20.6. **Field radiation is represented in terms of GO ray tubes, (a) only central rays are traced with circular ray tube shape to conserve power, (b) ray tube shape is triangular in which both central and corner rays are traced.**

Fig.20.7 demonstrates the scattering problem of a cavity with an open end, which is illuminated by an incident plane wave [17], [25]. The plane wave at the aperture of the open end is discretized in terms of a few ray tubes propagating in the direction of incident field. In this example, only central rays are traced, and the shapes of ray tubes are assumed to be circular with their size determined by power conservation. Each ray is traced inside the cavity until it hits the open-end aperture again. The aperture fields are determined by superposition of contributions from all the ray tubes. A set of equivalent currents are defined on the aperture by $\bar{J}_{\rm ap} = \hat{n} \times \bar{H}_{\rm ray}$ and $\bar{M}_{\rm ap} = \bar{E}_{\rm ray} \times \hat{n}$, where $\bar{E}_{\rm ray}$ and $\bar{H}_{\rm ray}$ are the total fields calculated by the SBR method, and $\hat{n}$ is an unit vector normal to the aperture. These equivalent currents radiate the scattering field via

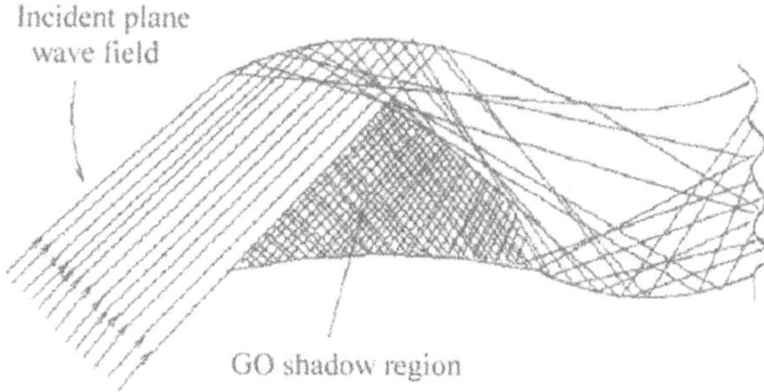

Figure 20.7. **Implementation of SBR in the analysis of a cavity scattering problem, where the GO shadow region has null field.**

regular radiation integral as [18], [20], [21], [26], [27]

$$\bar{E}^s(\bar{r}) \simeq \frac{jkZ_0}{4\pi} \iint\limits_S \left[\hat{R} \times \hat{R} \times \bar{J}_{\rm ap}(\bar{r}') + Y_0 \hat{R} \times \bar{M}_{\rm ap}(\bar{r}')\right] \frac{e^{-jkR}}{R} d\bar{r}' \quad (20.13)$$

where $\bar{R} = R\hat{R}$ is measured from the open-end aperture to the field point. This expression remains valid as long as the field point is not located extremely close to the aperture. This expression can also be used to compute the near fields.

Fig.20.8 shows the RCS of a circular cavity with radius of 3λ and length of 6λ. Surface curvatures of the cavity walls will change the wavefront curvatures of reflected rays. Ray caustics of the reflected ray will also occur inside the cavity, and as a result, errors begin to increase with wider incident angles. As shown in Fig.20.8, even though errors are observed, the resulting curves are similar to those obtained by using hybrid modal expansion. Due to the influence of ray caustics, errors increase in longer cavity in which more bounces of rays occur.

2.5 Generalized Ray Expansion Method

Generalized ray expansion(GRE) method is essentially an extension of SBR through a two-step ray expansion [18]-[21]. A set of equivalent currents, $(\bar{J}_{eq}, \bar{M}_{eq})$, are defined by the EM fields radiated from original sources. Ray expansion is performed on the field radiated from $(\bar{J}_{eq}, \bar{M}_{eq})$. An advantage of this two-step expansion is that the ray paths launched from the locations of $(\bar{J}_{eq}, \bar{M}_{eq})$ remain the same regardless of the variation of original sources. GRE is particularly useful in situations where the ray tracing procedure needs to be repeated while the GRE method needs to trace the ray paths only once. The concept is demon-

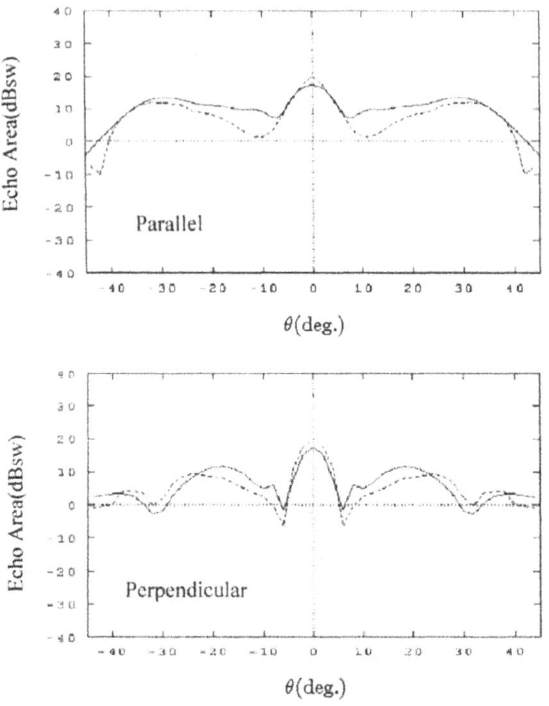

Figure 20.8. **RCS of a circular PEC cavity with radius of 3λ and length of 6λ calculated by using the SBR method, —: hybrid model, – –: SBR method.**

strated in Fig.20.9 where a spherical surface is defined from the phase center of the equivalent currents, $(\bar{J}_{eq}, \bar{M}_{eq})$. The spherical surface is divided into subapertures from where ray expansion is performed, and one ray is launched from the phase center of each subaperture. In practical applications, ray paths are traced and stored, and can be reused with each ray weighted according to the new sources. The final results can thus be acquired efficiently. However, one disadvantage of the GRE method is that the number of rays is much greater than that required in the SBR since all possible ray directions need to be considered and traced. As a result, GRE method will be more efficient only if the number of repeated computation is sufficiently large. However, through utilization of the equivalent currents, part of the diffraction effects can be included in the ray tracing if those effects have been considered in the equivalent currents.

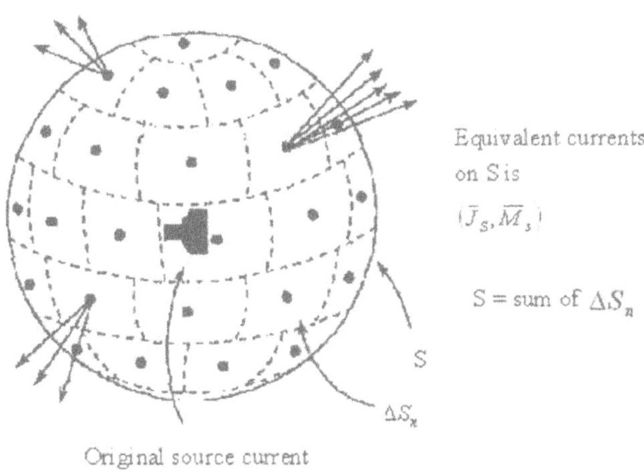

Figure 20.9. Illustration of field expansion in GRE method where the source radiation defines a set of equivalent currents whose radiations are subsequently represented by GO rays. The GO rays are launched from the phase center of each subaperture.

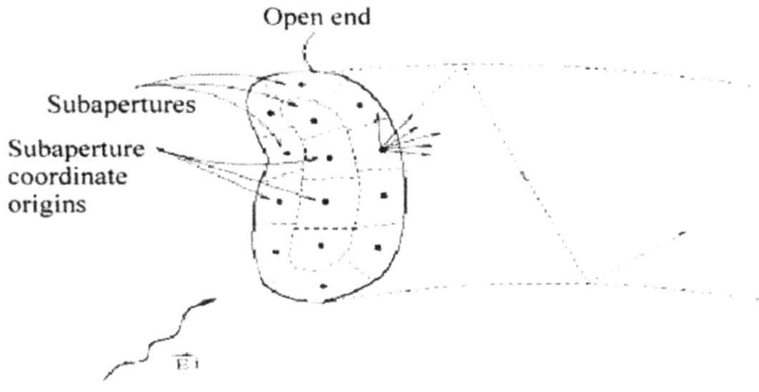

Figure 20.10. Implementation of GRE method in modeling EM scattering from a cavity with an open-end aperture.

Fig.20.10 shows the implementation of GRE method to the same cavity scattering problem as was analyzed by using SBR. Aperture of the open end is subdivided into subapertures and rays are launched radially from the phase center of each subaperture. The same set of rays are emloyed in the expansion of radiation from the aperture regardless of the directions of incident waves. Simulation results are shown in Fig.20.11.

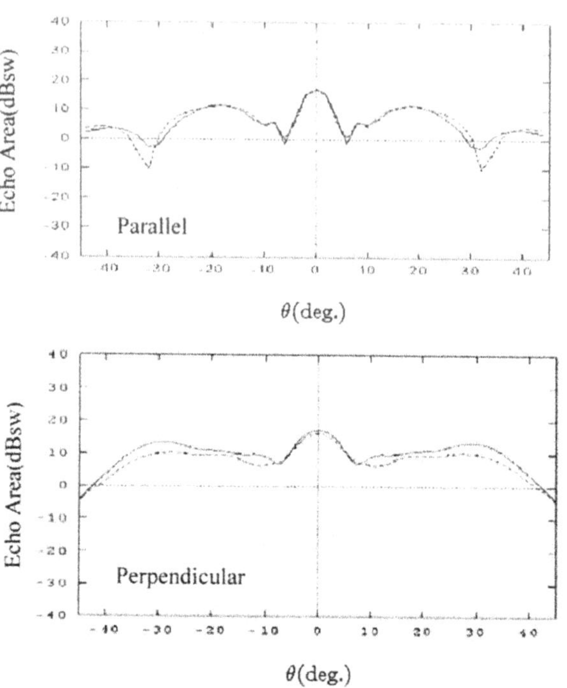

Figure 20.11. **RCS of a circular PEC cavity with radius of 3λ and length of 6λ calculated by using the GRE, —: hybrid model, – –: GRE-GO.**

3. Physical Optics

Physical optics(PO) is an equivalent current method based on GO approximation and extends its applications [28]. Equivalent electric and magnetic currents are defined from the total magnetic and electrical fields obtained by GO. PO uses the radiation integral in (20.13) to calculate the scattering fields that, in summation with the incident fields, become the approximated total fields. The equivalent currents radiate scattering field in the absence of objects. The total EM fields on the object surface obtained by GO approximation can be expressed as

$$\bar{E}(\bar{r}') \simeq \bar{E}^i(\bar{r}') + \bar{E}^r(\bar{r}') = \bar{E}^i(\bar{r}') + \bar{E}^i(\bar{r}') \cdot \bar{\bar{R}}_E(\bar{r}')$$
$$\bar{H}(\bar{r}') \simeq \bar{H}^i(\bar{r}') + \bar{H}^r(\bar{r}') = \bar{H}^i(\bar{r}') + \bar{H}^i(\bar{r}') \cdot \bar{\bar{R}}_H(\bar{r}')$$

where $\bar{\bar{R}}_E(\bar{r}')$ is given in (20.10) and $\bar{\bar{R}}_H(\bar{r}')$ is the dyadic reflection coefficient for the magnetic field, which can be expressed as

$$\bar{\bar{R}}_H(\bar{r}') = \hat{e}^i_{||}\hat{e}^r_{||}R_\perp(\bar{r}') + \hat{e}_\perp\hat{e}_\perp R_{||}(\bar{r}') \tag{20.14}$$

where the parameters in (20.14) are identical to those in (20.10). Note that currents only exist in the lit regions. For a perfect electric conductor(PEC) that results in null tangential electric field on its surface, magnetic current vanishes and the electrical current can be approximated as

$$\bar{J}_s(\bar{r}') \simeq \hat{n}' \times \bar{H}(\bar{r}')$$

where the GO scattering field can be predicted by applying image theory on a local planar surface tangent at P_R, which is valid at high frequencies because of large surface radii of curvature compared with wavelength. As a result, the scattering field outside the object can be expressed as

$$\bar{E}^s_{\mathrm{PO}}(\bar{r}) \simeq \frac{jkZ_0}{4\pi}\iint_S \left[\hat{R} \times \hat{R} \times (2\hat{n}' \times \bar{H}(\bar{r}'))\right] \frac{e^{-jkR}}{R} d\bar{r}' \tag{20.15}$$

where numerical integration is generally required. Several distinct characteristics of PO should be discussed. Firstly, because GO only predicts the scattering fields in lit regions of the incident field, the surface of object considered in the integral of (20.15) is limited to the lit regions. Secondly, the interfaces between lit and shadow regions form edge type of boundaries which result in artificial edge diffraction contributions to the scattering fields. Thirdly, the current distributions in PO approximation do not include any effects caused by edges or shadow region surfaces if finite objects are considered. However, if the scattering field integral in (20.15) is evaluated asymptotically by stationary phase method, the saddle point contribution will give rise to the reflected field as predicted by GO [29]-[32]. This indicates that PO can only predict accurate fields in the specular regions where reflection contributions dominate. However, PO provides uniform field distributions and transitions at any place, which potentially extends the applicability of GO, if numerical integration is performed. Besides, it provides reasonably accurate results.

4. Geometrical Theory of Diffraction and Its Uniform Version

GO results in discontinuous fields across the ISB and RSB, and null fields in the shadow regions. GTD, first introduced by Keller and later

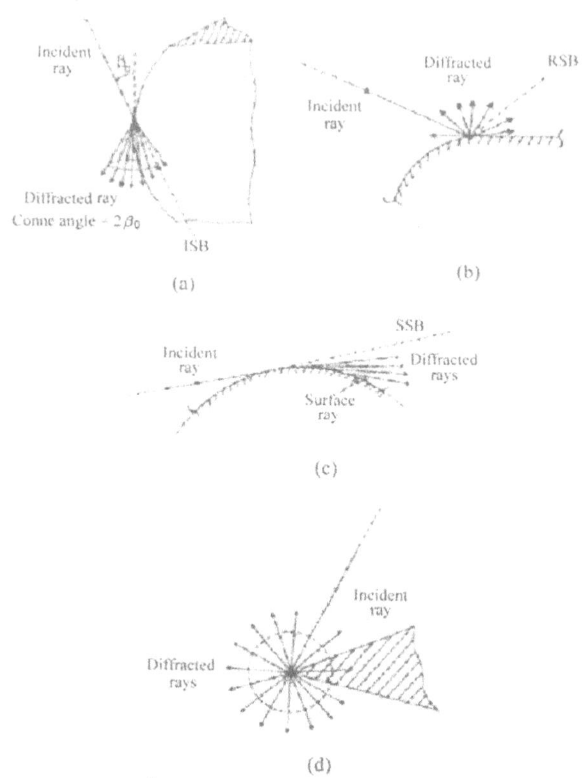

Figure 20.12. **Typical diffraction mechanisms: (a) curved edge diffraction, (b) edge diffraction caused by two surfaces with discontinuous radii of surface curvatures, (c) surface diffraction, (d) corner diffraction.**

modified by Pathak and Kouyoumjian to provide uniform transitions across the shadow boundaries, intends to correct the GO based on interpretation of diffraction mechanisms. The GTD/UTD solutions are generally obtained based on asymptotic solutions of the Green's functions available for some canonical geometries, and then generalized to treat objects with relatively arbitrary shapes. The generalization is based on the natures of canonical problems where diffraction is a local phenomenon depending on the scatterer surfaces and incident field in the neighborhood of diffraction point.

Diffraction mechanisms in scattering and radiation problems can be categorized into several types according to the locations of diffraction point as illustrated in Fig.20.12. The total field is the superposition of GO field and diffracted field, namely, $\bar{E}(P) = \bar{E}^{\mathrm{GO}}(P) + \bar{E}^{d}(P)$. The diffracted field not only provides contribution in the shadow region, but also provides uniform field distribution across ISB/RSB. The overall propagation mechanism is illustrated in Fig.20.13. In particular, the diffracted fields for the scattering problem can be expressed in a general form as

$$\bar{E}^d(P) = \bar{E}^i(P_d) \cdot \bar{\bar{D}} A(s) e^{-jks}$$

where $s = \left|\overline{PP_d}\right|$ with P_d being the diffraction point, $\bar{\bar{D}}$ is the diffraction coefficient that depends on the local nature of surface in the neighborhood of P_d, and $A(s)$ is a divergence factor due to ray propagation. The form of $\bar{\bar{D}}$ and $A(s)$ varies with different diffraction mechanisms.

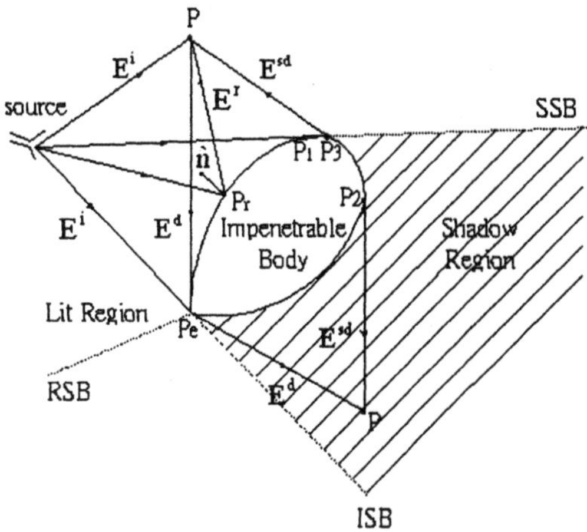

Figure 20.13. Various reflection and diffraction mechanisms and associated rays including incident, reflection, edge diffraction and surface diffraction. The diffraction field compensates the incident and reflected fields at ISB and RSB to obtain continuous field distributions.

One of the most popular diffraction mechanisms is edge diffraction as illustrated in Fig.20.12(a) with its associated wavefront transformation around a wedge with PEC walls illustrated in Fig.20.14. In this case, the diffraction point, P_e, is located on the edge, and the divergent factor becomes

$$A(s_d) = \sqrt{\frac{\rho_e}{s_d(s_d + \rho_e)}}$$

where ρ_e is the radius of wavefront curvature. Note that one of the GO caustics is located at P_e. The diffraction ray path is determined based on the Fermat's principle. Given an incident ray, a diffraction point and a wedge, the diffracted rays will emanate into a cone shape, referred

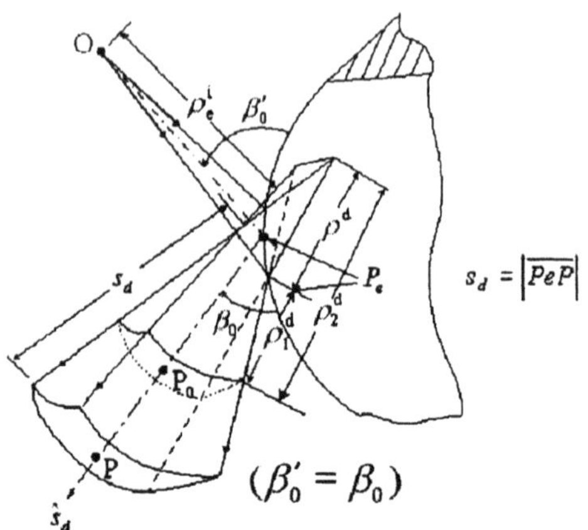

Figure 20.14. Transformation of wavefronts between incident and edge diffracted rays.

to as Keller's cone, by observing $\hat{s}_i \cdot \hat{e} = \hat{s}_d \cdot \hat{e}$ where $\hat{e}$ is the tangent unit vector along the edge at P_e. The diffraction coefficient, $\bar{\bar{D}}_e$, is a 2×2 matrix of nonzero entries if arbitrary ray coordinate systems are selected for the incident and reflected rays. Let (s_i, β_i, ϕ_i) and (s_d, β_d, ϕ_d) be the coordinate systems for the incident and diffracted rays, respectively, with $\hat{\beta}_i = \hat{s}_i \times \hat{\phi}_i$ and $\hat{\beta}_d = \hat{s}_d \times \hat{\phi}_d$. If $\hat{\phi}_i$ and $\hat{\phi}_d$ are chosen on a plane orthogonal to $\hat{e}$, then $\bar{\bar{D}}_e$ can be diagonalized and expressed as $\bar{\bar{D}}_e = -D_{es}\hat{\beta}_i\hat{\beta}_d - D_{eh}\hat{\phi}_i\hat{\phi}_d$. For a PEC wedge, we have [7], [33]

$$
\begin{aligned}
D_{\substack{es\\eh}}(\phi,\phi',\beta_0) = &-\frac{e^{-j\pi/4}}{2n\sqrt{2\pi k}\sin\beta_0} \\
&\left\{\cot\left[\frac{\pi+(\phi-\phi')}{2n}\right] F\left[kL^i a^+(\phi-\phi')\right]\right. \\
&+\cot\left[\frac{\pi-(\phi-\phi')}{2n}\right] F\left[kL^i a^-(\phi-\phi')\right] \\
&\mp\cot\left[\frac{\pi+(\phi+\phi')}{2n}\right] F\left[kL^{rn} a^+(\phi+\phi')\right] \\
&\left.\mp\cot\left[\frac{\pi-(\phi+\phi')}{2n}\right] F\left[kL^{ro} a^-(\phi+\phi')\right]\right\}
\end{aligned}
\tag{20.16}
$$

where the parameters ϕ_d, ϕ_i, β_0 and n are defined in Fig.20.15, and F is a transition function that renders uniform total field distribution while the field point moves across ISB/RSB. Explicitly, the transition function can be expressed as

$$F(x) = 2j\sqrt{x}e^{jx} \int_{\sqrt{x}}^{\infty} d\tau e^{-j\tau^2}, \qquad -\frac{\pi}{2} < \mathrm{Im}(x) < \frac{3\pi}{2}$$

In (20.16), $a^{\pm}(\rho) = 2\cos^2(n\pi N^{\pm} - \beta/2)$, where $N^{\pm}$ is an integer that

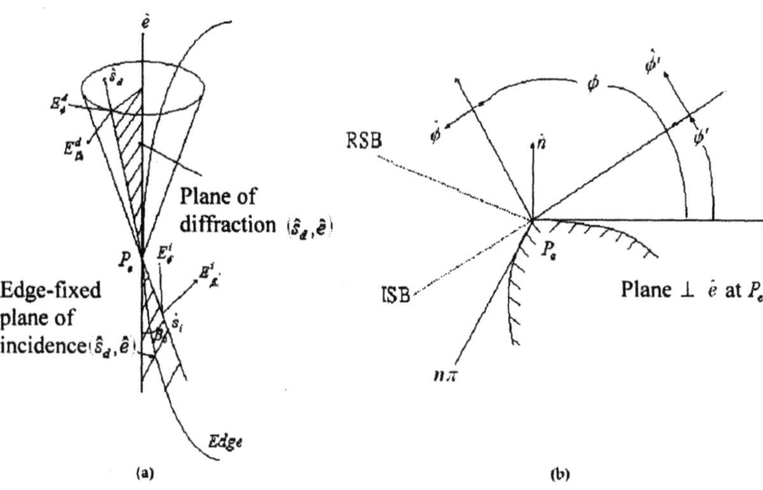

Figure 20.15. Edge-fixed planes of incidence and diffraction, and associated parameters, (a) edge diffracted rays, (b) diffracted ray projected onto a plane perpendicular to the edge direction.

most closely satisfies the relation $2\pi n N^{\pm} - \beta = \pm\pi$ with $\beta = \phi_d \pm \phi_i$. The distance parameters, L^i, L^{ro} and L^{rn}, are used to determine the size of transition region which is associated with the incident and reflection shadow boundaries of the two wedge faces. Their functional forms are in general complicated. However, if the incident field is a spherical wave, L^i can be simplified as

$$L^i = \frac{s^i s^d}{s^i + s^d} \sin^2 \beta_0$$

and L^{ro} and L^{ri} are equal to L^i if the wedge has a straight edge. While the field point, P, is far away from the ISB/RSB, $F(x)$ approaches unity and $\bar{\bar{D}}_e$ reduces to the nonuniform Keller's GTD diffraction coefficients. In the vicinity of ISB/RSB at $\phi_d \pm \phi_i = 0$, small-argument expression of $F(x)$ can be used to compensate for the singularity in the diffraction coefficients. In particular, diffracted field in the vicinity of ISB/RSB is reduced to [33]

$$\bar{E}^d\Big|_{\text{ISB,RSB}} = \mp\frac{1}{2}\bar{E}^{i,r} + \text{(continuous higher order terms)}$$

where the upper sign applies on the lit side of ISB/RSB, and the lower sign applies on the shadow side of ISB/RSB. The diffracted field is added to the incident or reflected field to result in $(1/2)\bar{E}^{i,r}$, and ensures the continuity of fields at ISB/RSB.

5. Practical Applications of High-Frequency Techniques

The asymptotic high frequency techniques discussed in this Chapter can be applied to solve many practical electromagnetic problems. Many of them involve antennas or structures that are electrically large and are impractical or computationally inefficient to be solved by numerical methods like method of moments or finite element method. Some of these problems include pattern analysis of large aperture antennas like pyramidal horn and reflector antennas, as well as antennas near complicated structures like ships, satellites or buildings. In this Section, several examples will be demonstrated to show how these high frequency methods are applied.

5.1 Far-Field Pattern of Pyramidal Horns

Geometry of a pyramidal horn antenna is shown in Fig.20.16. This type of antenna has been widely used as standard-gain horn for antenna pattern and gain measurements as well as the feed of reflector antennas. A pyramidal horn is typically characterized as an aperture antenna and aperture field integration method is normally used to determine its far-field radiation pattern. Based on equivalence principle, one has to obtain both the electric and magnetic currents(or fields) over the entire antenna surface to calculate the radiation pattern of antennas. The challenge involved in this approach is the difficulty to actually determine the currents(or fields) over the entire antenna surface. Consequently, approximation has to be made to obtain the radiation patterns. One approach is to integrate the electric and magnetic currents(or fields) of a rectangular waveguide TE_{10} mode with a quadratic phase distribution present at the horn aperture to account for the horn flare angle. However, this approach can only be used to obtain the antenna patterns in the forward hemispherical region, the main beam in particular. Al-

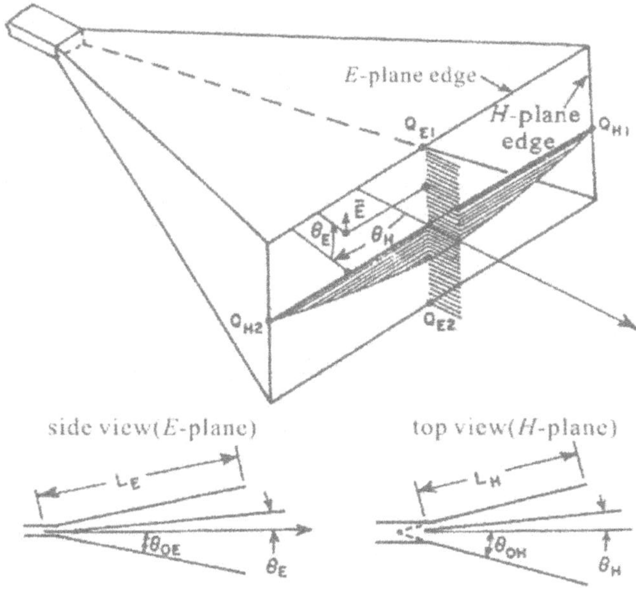

Figure 20.16. Geometry of a pyramidal horn antenna.

ternatively, one can use uniform geometrical theory of diffraction(UTD) method to calculate the full E and H-plane patterns, including front and back regions of the antenna.

***E*-Plane Pattern Calculation.** Using edge diffraction theory to calculate the E-plane pattern of pyramidal horn antenna was first presented by Yu, Rudduck and Peters [34]. Similar to the aperture field method, the dominant propagation mode within the horn is approximated as a spherical wave with TE_{10} mode distribution given by

$$E^{\mathrm{GO}} = \frac{e^{-jkR_0}}{R_0} \cos\left(\frac{\pi}{2} \frac{\tan\theta_H}{\tan\theta_{0H}}\right)$$

where R_0 is the distance from the apex of horn to the observation point, and θ_H is the angle measured in the H-plane, as shown in Fig.20.16. The amplitude distribution is uniform in the E-plane and is a cosine function in the H-plane. In the E-plane, $\theta_H = 0$, the geometrical optics field reduces to

$$E^{\mathrm{GO}} = \frac{e^{-jkR_0}}{R_0}$$

E-plane radiation pattern of the pyramidal horn can then be determined by the direct radiated field from the source located at the apex and two edge diffracted fields located at intersections of the E-plane with the horn, as shown in Fig.20.17. The total field in the E-plane is in general given by

$$\bar{E}^{\text{total}} = \bar{E}^{\text{GO}} + \bar{E}_1^d + \bar{E}_2^d$$

where $\bar{E}_1^d$ and $\bar{E}_2^d$ are edge diffracted fields from the diffraction points,

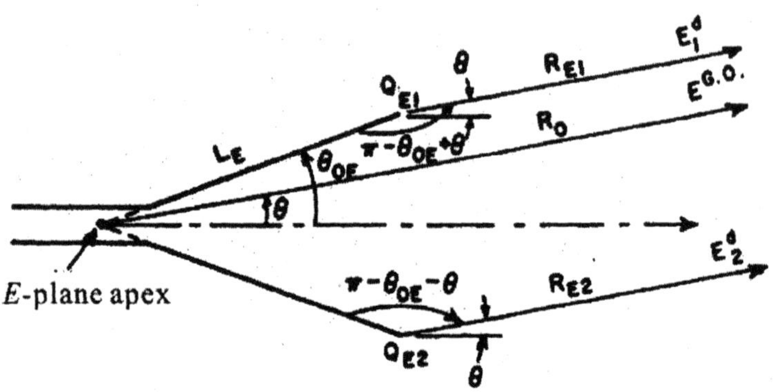

Figure 20.17. E-plane cross section of a pyramidal horn for far-field pattern calculation.

Q_{E1} and Q_{E2}, respectively. Depending on the location of observation points, not all of the above field terms contribute to the radiation pattern in a specific direction. For example, if the observation point is located in region I as shown in Fig.20.18, the total E-plane pattern includes geometrical optics term as well as two diffracted field terms. On the other hand, if the observation point is located in regions II, IV and VI, only two diffracted field terms contribute to the antenna pattern. In regions III and V, only one of the edge diffracted field terms contributes to the pattern.

***H*-Plane Pattern Calculation.** Although one can use similar approach as in the E-plane to calculate the H-plane pattern, it is found that incident fields at the edge diffraction points, Q_{H1} and Q_{H2}, as shown in Fig.20.19, are zero because the incident fields are parallel to the diffracting edges which are perpendicular to H-plane of the horn. This leaves geometrical optics field the only term contributing to the H-plane pattern. However, it is not accurate since the slopes of incident fields at the diffracting edges create slope diffracted fields that contribute to the antenna pattern [3]. Hence, the total field in the H-plane is modified as

$$\bar{E}^{\text{total}} = \bar{E}^{\text{GO}} + \bar{E}_1^{sd} + \bar{E}_2^{sd}$$

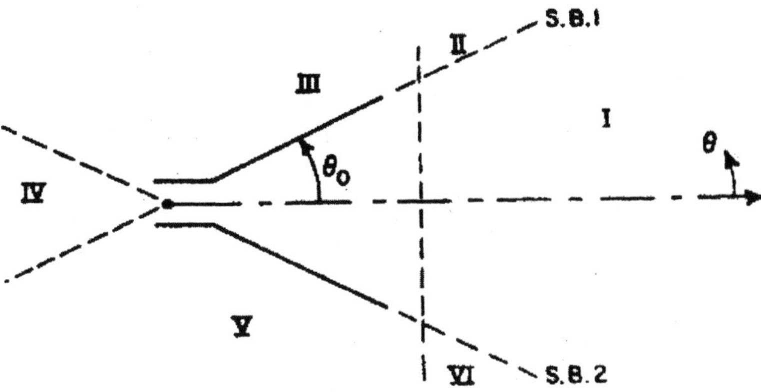

Figure 20.18. Regions with different combinations of geometrical optics and diffracted fields.

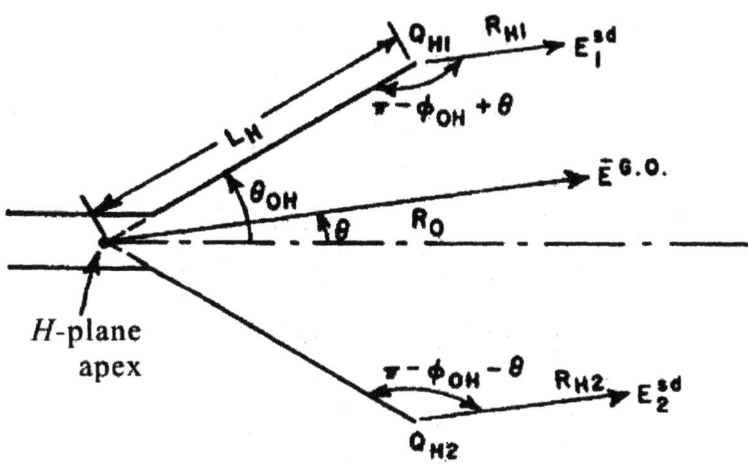

Figure 20.19. H-plane cross section of the horn antenna for far-field pattern calculation.

where $\bar{E}_1^{sd}$ and $\bar{E}_2^{sd}$ are slope diffracted fields from the diffraction points, Q_{H1} and Q_{H2}, respectively. Again, depending on the location of observation point, not all of the three terms contribute to the pattern in a specific direction.

By comparing measured and calculated patterns, it has been found that the UTD approach is very accurate in analyzing pyramidal horn antennas [3], [34].

5.2 Pattern Analysis of Reflector Antennas

Reflector antennas are electrically large and can be analyzed by using asymptotic high frequency method. Reflector antennas have been widely used in microwave communication in the past few decades, which requires antennas with high gain and high pointing accuracy to maintain the quality of communication link. In global satellite communication, reflector antennas on board of a satellite are designed to cover a given geographical area on earth. Conventional reflector antennas include direct-fed reflector and dual Cassegrain/Gregorian reflectors. Typically, paraboloids are used in direct-fed systems and as the main reflector in dual-reflector systems. Hyperboloids are used as subreflector in the Cassegrain systems, and ellipsoids are used as subreflector in the Gregorian systems. Most of the classical and important works related to reflector antennas can be found in [24].

Size of a reflector can be easily over 40 wavelengths or even more than a few hundred wavelengths. Using integral equation-based approach to calculate the antenna patterns becomes impractical because it will take huge computer memory and long computation time. Consequently, asymptotic high frequency methods are more appropriate to analyze such reflector antennas. Since reflector antenna is also characterized as aperture antenna, one can use the aperture field/surface current method to calculate its radiation patterns. Physical optics(PO) and aperture integration(AI) techniques have been widely used to calculate the radiation pattern of parabolic reflector antennas [35]. These methods can successfully predict the main beam and the first few sidelobes. Note that the UTD fails to predict the main beam and the first few sidelobes since these regions are either within or close to the reflection caustics of reflector surface, and geometrical optics fails to provide correct antenna patterns in these regions. On the other hand, similar to the horn antenna problem, both PO and AI cannot predict accurate results for wide-angle sidelobes and back lobes of reflector antennas. Consequently, combination of PO/AI and UTD is feasible to calculate the full pattern of reflector antennas [1].

5.3 Physical Optics Analysis

PO is a high frequency method that can be used to determine the scattered fields from an electrically large object. Reflector antennas are electrically large and the radiation directions of interest are near and within the specular reflection region of reflector. With PO approach, incident magnetic fields on the reflector surface can be found by ray tracing technique starting from the feed antenna. The induced surface currents can then be calculated and integrated to obtain the radiation pattern. This approach provides high accuracy in the main beam and near sidelobes, but poor accuracy at wide angles. Since the reflector surface is electrically large, the number of patches for PO integration, as indicated in Fig.20.20, can become huge and consequently, requires significant amount of computation time even with today's fast computers.

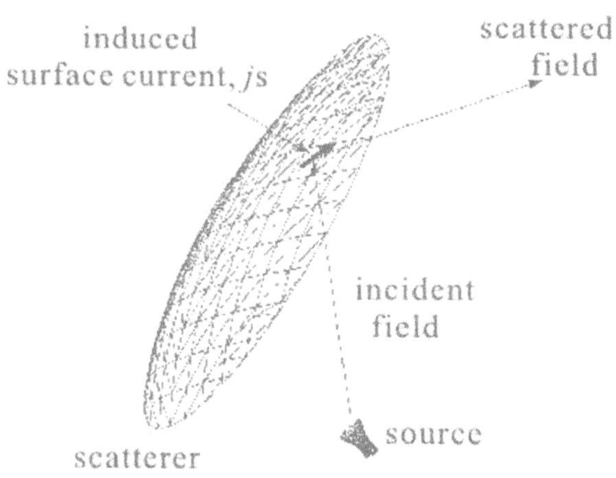

Figure 20.20. Physical optics method for reflector pattern analysis.

5.4 UTD Analysis

PO and AI do not predict correct sidelobes at wide angles and back lobes of parabolic reflector antennas. To complement these methods, uniform geometrical theory of diffraction has been used to include diffracted fields from the reflector rim to predict accurate sidelobes at wide angles and back lobes [1], [7], [20], [36], [37]. UTD has also been used to calculate the full pattern of hyperbolic and elliptic subreflectors in the Cassegrain and Gregorian dual-reflector systems, respectively [16], [27]. If the reflector rim is not smooth, UTD corner diffraction is also needed to predict the sidelobe levels [26]. The scattered field, $\bar{E}_s$, from a given reflector at a given observation point, $\bar{r}$, as shown in Fig.20.21, can be determined by UTD as

$$\bar{E}_s(\bar{r}) = \bar{E}_r(\bar{r}) + \bar{E}_d(\bar{r})$$

where $\bar{E}_r(\bar{r})$ and $\bar{E}_d(\bar{r})$ are the reflected and diffracted fields, respectively, from the reflector. Note that for parabolic reflector, the main beam can be calculated by PO and the sidelobes at wide angles are contributed by the edge diffracted fields only. Total field of the whole reflector can be obtained by adding the direct radiated field from the feed to the scattered field, $\bar{E}_s(\bar{r})$.

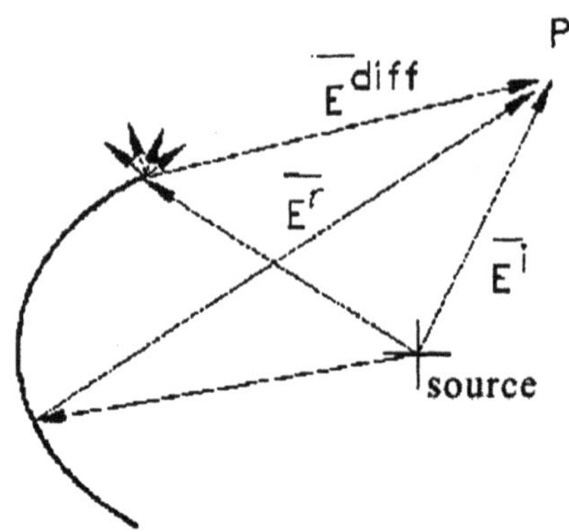

Figure 20.21. UTD analysis of a reflector.

Fig.20.22 shows the calculated patterns at 11 GHz of a circularly symmetric parabolic reflector with diameter of 24" and focal length of 8". Only the patterns within $|\theta| = 12.2°$ are obtained by PO. Rest of the patterns are calculated by using UTD. The combination of PO/AI and UTD has been used to design and analyze various antenna systems. For example, Lee, Rudduck and Lambert [8], [38] have shown that this approach can accurately predict the antenna patterns and gains by validating the computed results with measurement data.

5.5 Scattering Analysis in a Complicated Structure

Another problem that can be analyzed by the asymptotic high frequency method is to analyze the performance of antennas in the presence of complicated structures. With recent demand on global satellite communication, many antennas have been implemented on satellite, and there is growing concern regarding the interaction between antennas and their surrounding structures [39]. In addition, with the growth of wireless communication services, there is also a great concern about the performance of communication links in urban areas, such as the one modeled in Fig.20.23 where a transmitting antenna and a receiving antenna are located in a complicated environment. The only feasible approach to solve these types of problems is to use UTD ray tracing techniques to include all possible ray paths between transmitting and receiving anten-

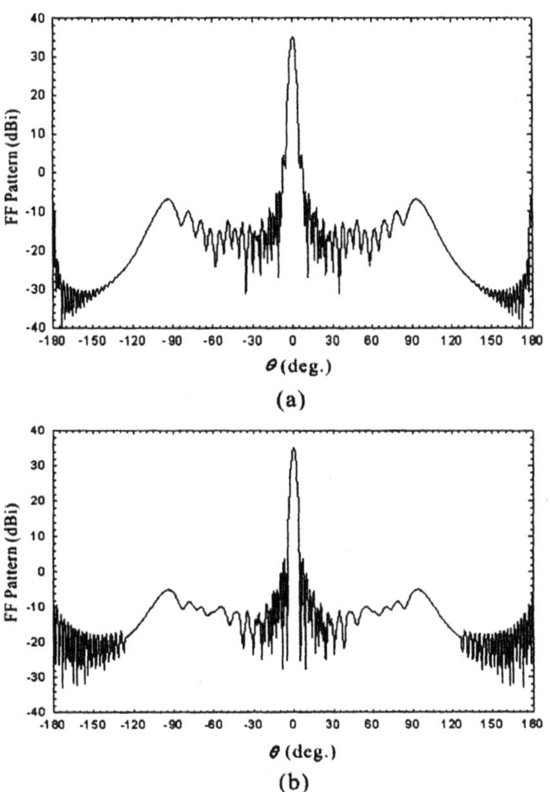

Figure 20.22. Calculated patterns at 11 GHz of a parabolic reflector with diameter of 24" and focal length of 8", (a) H-plane pattern, (b) E-plane pattern.

nas. These rays include direct line-of-sight signal as well as reflection, diffraction and multiple bounces of these rays. Fig.20.24 shows the coupling between transmitting and receiving antennas without and with the presence of environmental structures. The impact of structures on communication link performance is obvious.

6. Hybridization of UTD with Numerical Methods

UTD is applicable only when the objects under consideration are relatively simple in shape and electrically large. Moreover, diffraction coefficients in the UTD formulations need to be available and valid, which generally depend on features of the objects under analysis and are usually difficult to obtain in many practical applications. These limitations can be overcome to some extent via hybrid integration with conven-

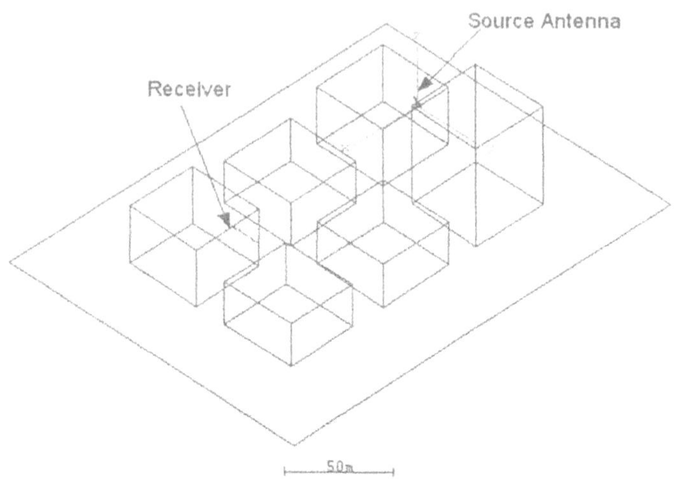

Figure 20.23. **Configuration of transmitting and receiving antennas in a complicated environment.**

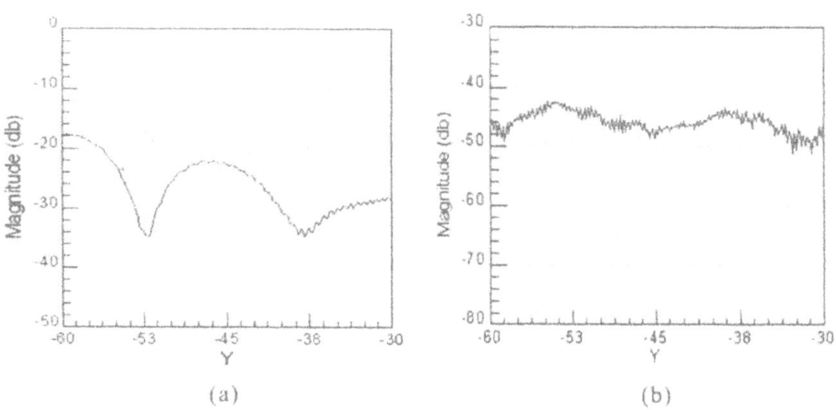

Figure 20.24. **Coupling between transmitting and receiving antennas, (a) without structure, (b) with structure.**

tional numerical methods such as MoM, FDTD and FEM. Among those, the hybrid combination with MoM is most popular even though hybrid combinations with FDTD and FEM begin to attract more attentions and developments. Two hybridizations have been employed with MoM. One involves the utilization of UTD-type of basis functions to reduce the number of MoM unknowns, the other utilizes the UTD solutions as Green's functions in the MoM computation [40].

Hybrid UTD-MoM is an effort to reduce the number of unknowns required in the MoM modeling of large objects by using UTD-type of solutions as basis functions to replace most of regular MoM basis functions. Unknowns of the UTD-type basis functions are related to the diffraction coefficients. Regular MoM basis functions such as pulse functions can be employed in regions where the UTD-type of basis functions are not valid, usually near the edges. Thus, the number of unknowns can be significantly reduced.

Consider the geometry illustrated in Fig.20.25 where a two-dimensional finite PEC strip is illuminated by a TM incident plane wave. Conventional MoM may employ pulse functions to represent the induced current as

$$J(x') = \sum_{n=1}^{N} A_n P_n(x')$$

where the pulse function, $P_n(x)$, is defined as

$$P_n(x') = \begin{cases} 1, & x' \in n\text{th pulse} \\ 0, & \text{elsewhere} \end{cases}$$

By solving the electric field integral equation(EFIE), one obtain a matrix equation as

$$\bar{\bar{Z}} \cdot \bar{I} = \bar{V} \tag{20.17}$$

where $\bar{I} = [A_1, \cdots, A_N]^t$ contains the unknown coefficients, $\bar{V} = [V_1, \cdots, V_N]^t$ is related to the excitation or incident field on each basis, and $\bar{\bar{Z}}$ is an $N \times N$ matrix consisting of mutual impedances from MoM modeling.

The basis functions are divided into two groups as shown in Fig.20.25. In the interior region where HF solutions are valid, currents can be represented by the UTD-type of basis functions, and regular MoM basis functions are employed in the exterior regions. Thus, A_n in the UTD region is approximated by

$$A_n = C_1 \frac{e^{-jk\ell_n}}{\sqrt{\ell_n}} + C_2 \frac{e^{-jk\ell_n}}{(\sqrt{\ell_n})^3} + C_3 \frac{e^{-jkr_n}}{\sqrt{r_n}} + C_4 \frac{e^{-jkr_n}}{(\sqrt{r_n})^3} \tag{20.18}$$

where ℓ_n and r_n are the distances measured from the nth pulse basis to the left and right edges, respectively. C_1 to C_4 are the unknown coefficients that will be solved by MoM, which are related to the UTD

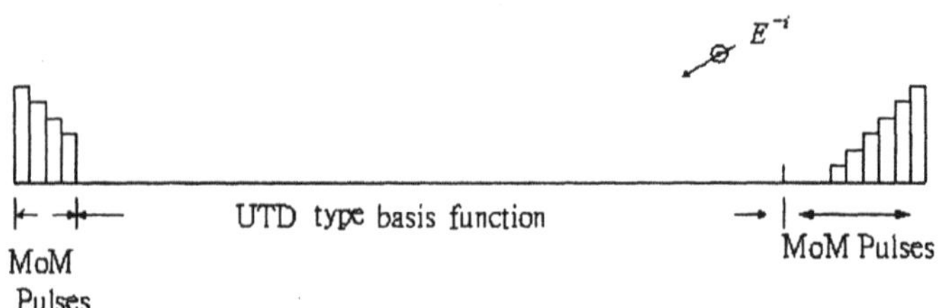

Figure 20.25. Hybrid UTD-MoM analysis on a two-dimensional strip illuminated by an incident plane wave.

diffraction coefficients. Alternatively, (20.18) can be expressed in a matrix form as

$$\bar{I} = \bar{\bar{B}} \cdot \bar{L} \tag{20.19}$$

where $\bar{L} = [A_1, \cdots, C_1, C_2, C_3, C_4, \cdots, A_N]^t$, and the elements in $\bar{\bar{B}}$ can be deduced from (20.18). Note that the number of elements in $\bar{L}$(denoted by M hereafter) is much smaller than N. Substituting (20.19) into (20.17) gives

$$\left[\bar{\bar{Z}}\Big|_{N\times N} \cdot \bar{\bar{B}}\Big|_{N\times M} \right] \cdot \bar{L}|_{M\times 1} = \bar{V}|_{N\times 1}$$

where there are N linear equations to be solved for M unknowns. In practice, one may either randomly select M equations or apply least squares method to solve for the unknowns. Once C_1 to C_4 are obtained, the associated coefficients of A_n can be deduced using (20.18).

Fig.20.26 shows the induced current on the strip surface obtained by this hybrid method and by regular MoM. The results match very well. Width of the strip is 25λ, and the induced current is represented by 100 pulse functions. The hybrid method uses only 14 basis functions(five pulses in each edge region and four GTD basis functions) to expand the current.

7. Physical Theory of Diffraction

As mentioned in Section 3, PO provides accurate prediction of the scattering field only in the specular reflection region. The prediction becomes erroneous away from the specular reflection region since diffraction effects are not included. Physical theory of diffraction(PTD), first introduced by Ufimtsev for analyzing bodies with edges, intends to correct PO with diffraction contribution in a similar fashion as GTD/UTD

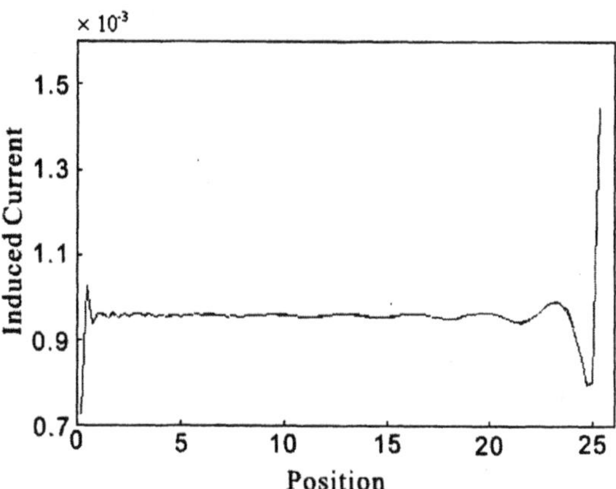

Figure 20.26. Induced current on the strip in Fig.20.25, two curves match closely.

correct the GO. Similar to the distinction between PO and GO, PTD also requires a numerical integration over the induced current on the scattering bodies to distinguish it from the ray techniques of GTD/UTD.

The correction that PTD provides to PO can be described in two aspects. In the induced current aspect, PTD provides nonuniform current correction caused by the edge effects to the PO current as

$$\bar{J}_s(\bar{r}') = \bar{J}_{\text{PO}}(\bar{r}') + \bar{J}_{\text{PTD,correction}}(\bar{r}')$$

In the field aspect, PTD provides correction on the PO edge diffractions by expressing the total field as

$$\bar{E}(\bar{r}) \simeq \bar{E}^i(\bar{r}) + \bar{E}^s_{\text{PO}}(\bar{r}) + \bar{E}^d_{\text{PTD,correction}}(\bar{r}) \tag{20.20}$$

Note that $\bar{E}^s_{\text{PO}}(\bar{r})$ consists of reflections and incomplete edge diffractions asymptotically, and $\bar{E}^d_{\text{PTD,correction}}(\bar{r})$ is related to the radiation from $\bar{J}_{\text{PTD,correction}}(\bar{r}')$, which is not the same as the edge diffraction in GTD/UTD.

Similar to PO, the PTD involves a numerical integration over surface currents, and remains valid even in the ray caustic regions of GTD/UTD if the currents are predicted correctly. Rigorous derivation of current corrections can be found in [13], [14], [41]-[44]. However, one simple approach to obtain the current of PTD correction is to relate PTD with GTD/UTD by field distribution where correct and accurate fields are assumed in both approaches. This approach expresses the PO scattering field in terms of reflections and diffractions asymptotically as

$$\bar{E}^s_{\text{PO}}(\bar{r}) = \bar{E}^r_{\text{PO}}(\bar{r}) + \bar{E}^d_{\text{PO}}(\bar{r}) \tag{20.21}$$

where $\bar{E}^r_{\rm PO}(\bar{r})$ and $\bar{E}^d_{\rm PO}(\bar{r})$ are the reflected and diffracted fields, respectively. From (20.20) and (20.21), one expresses the PTD field correction as

$$\bar{E}^d_{\rm PTD,correction}(\bar{r}) = \bar{E}^d_{\rm GTD}(\bar{r}) - \bar{E}^d_{\rm PO}(\bar{r}) \tag{20.22}$$

where $\bar{E}^d_{\rm GTD}(\bar{r})$ is the edge diffraction predicted by GTD/UTD. Note that both $\bar{E}^d_{\rm GTD}(\bar{r})$ and $\bar{E}^d_{\rm PO}(\bar{r})$ are known. Hence, an equivalent current can be assumed to render (20.22). In general, a simple line current source is assumed, even though $\bar{J}_{\rm PTD,correction}(\bar{r}')$ is a surface current, if one is only interested in its radiating fields away from surfaces. This approach is also referred to as equivalent current method(ECM).

8. Future Developments

The HF techniques described in the previous Sections are particularly useful in the fast analysis of scattering problems associated with electrically large objects. The techniques are mostly based on GTD/UTD solutions. Other GTD/UTD type of solutions that have not been described in this Chapter include typical diffraction mechanisms in radiation(source is on a curved surface and field point is off the surface) [33], mutual coupling(both source and field points are on the surface) [33] problems as well as their time domain versions [45]-[48]. Those solutions can be generally expressed in the form of

$$\bar{E}(P) = \left\{ \begin{array}{c} d\bar{p}_e \\ \\ d\bar{p}_m \end{array} \right\} \cdot \bar{\bar{D}} A(s) e^{-jks}$$

In general, $\bar{\bar{D}}$ is a 2×2 matrix for radiation problems, and a 3×3 matrix for mutual coupling problems. Due to the requirements to compute near coupling effects in practical applications, more higher-order terms are included to obtain more accurate results.

Currently, existing GTD/UTD type of solutions can be employed to solve many practical problems. Solutions that are under development include edge excited surface waves, more complete solutions to compensate for the discontinuities caused by current GTD/UTD solutions such as multiple shadow boundary overlaps, and solutions with the field points located in the paraxial regions of GTD/UTD solutions. Moreover, extension of GTD/UTD approach to analyze scattering/radiation from nonmetallic penetrable/impenetrable objects are of interest in the development of HF solutions, as well as their hybrid integration with numerical methods such as MoM and FEM. Other solutions include uniform asymptotic theory(UAT) [49], [50], spectral theory of diffraction(STD) and incremental theory of diffraction(ITD) [51].

References

[1] J. Keller, "Geometrical theory of diffraction," *J. Opt. Soc. Am.*, vol.52, pp.116-130, Feb. 1962.

[2] M. Kline, "An asymptotic solution of Maxwell's equation," *Pure Appl. Math.*, Commun., vol.4, pp.225-262, 1951.

[3] R. Luneberg, *Mathematical Theory of Optics*, Providence, Rl: Brown Univ. Press, 1964.

[4] J. Keller, "Geometrical theory of diffraction," in *Calculus of Variations and Applications*, L. M. Graves, ed., New York: McGraw-Hill, pp.27-52, 1958.

[5] B. Levy and J. B. Keller, "Diffraction by a smooth object," *Pure Appl. Math.*, Commun., vol.12, pp.159-209, 1959.

[6] G. L. James, *Geometrical Theory of Diffraction for Electromagnetic Waves*, Inst. Elect. Eng. EM Waves Series 1, Stevenage, UK: Peter Peregrinus, 1976.

[7] R. Kouyoumjian and P. Pathak, "A uniform geometrical theory of diffraction for an edge of a perfectly conducting surface," *Proc. IEEE*, vol.62, pp.1448-1461, Nov. 1974.

[8] R. Kouyoumjian, P. Pathak, and W. Burnside, "A uniform GTD for the diffraction by edges, vertices and convex surfaces," in *Theoretical Methods for Determining the Interaction of Electromagnetic Waves with Structures*, J. K. Skwirzynski, ed., Netherlands: Sijthoff and Noordhoff, 1981.

[9] D. McNamara, C. Pistorius, and J. Malherbe, *Introduction to the Uniform Geometrical Theory of Diffraction*, Norwood, MA: Artech House, 1990.

[10] P. H. Pathak, "An asymptotic analysis of the scattering of plane waves by a smooth convex cylinder," *Radio Sci.*, vol.14, pp.419-435, 1979.

[11] P. H. Pathak and M. C. Liang, "On a uniform asymptotic solution valid across smooth caustics of rays reflected by smoothly indented boundaries," *IEEE Trans. Antennas Propagat.*, vol.38, pp.1192-1203, 1990.

[12] P. H. Pathak, W. D. Burnside, and R. J. Marhefka, "A uniform UTD analysis of the diffraction of electromagnetic waves by a smooth convex surface," *IEEE Trans. Antennas Propagat.*, vol.28, pp.609-622, 1980.

[13] P. Ufimtsev, "Method of edge waves in the physical theory of diffraction," translation from Russian "Medthod Krayevykh voin v

fizicheskoy teorii difraktsii," 1962, prepared by U.S. Air Force Foreign Technol. Div., Wright-Patterson AFB, Ohio, Sept. 1971.

[14] P. Ufimtsev, "Elementary edge waves and the physical theory of diffraction," *Electromagnetics*, vol.11, pp.125-160, Apr.-June 1991.

[15] L. Felsen and N. Marcuvitz, *Radiation and Scattering of Waves*, Englewood Cliffs, NJ: Prentice Hall, 1973.

[16] R. G. Kouyoumjian, "Asymptotic high-frequency methods," *Proc. IEEE*, vol.53, pp.864-876, 1965.

[17] H. Ling, R. C. Chou, and S. W. Lee, "Shooting and bouncing rays: Calculating the RCS of an arbitrarily shaped cavity," *IEEE Trans. Antennas Propagat.*, vol.37, pp.194-205, 1989.

[18] R. J. Burkholder, "High-frequency asymptotic methods for analyzing the EM scattering by open-ended waveguide cavities," Ph.D. thesis, Ohio State Univ., June 1989.

[19] H. T. Chou, R. J. Burkholder, P. H. Pathak, and D. Andersh, "Generalized ray expansion with elliptic ray tubes in hybrid analysis of EM scattering by open cavities," *IEICE Trans. Electron.*, Japan, vol.E78-C, no.10, Oct. 1995.

[20] P. H. Pathak and R. J. Burkholder, "High-frequency electromagnetic scattering by open-ended waveguide cavities," *Radio Sci.*, vol.26, no.1, pp.211-218, Jan.-Feb. 1991.

[21] P. H. Pathak, R. J. Burkholder, and R.-C. Chou, "Some extensions to the GRE analysis of EM scattering by non-uniform open waveguide cavities," Rept. 719631-5, ElectroScience Lab., Ohio State Univ., Dec. 1991.

[22] G. A. Deschamps, "Ray techniques in electromagnetics," *Proc. IEEE*, vol.60, pp.1022-1035, 1972.

[23] H. T. Chou and P. H. Pathak, "Use of Gaussian ray basic functions in the ray tracing methods for the application to high frequency wave propagation problems," *Proc. Inst. Elect. Eng.*, pt.H, vol.147, no.2, pp.77-81, Apr. 2000.

[24] S. W. Lee, H. Ling, and R. C. Chou, "Ray tube integration in shooting and bouncing ray method," *Microwave Opt. Technol. Lett.*, vol.1, pp.286-289, 1988.

[25] J. Baldauf, S. W. Lee, L. Lin, S. K. Jeng, S. M. Scarborough, and C. L. Yu, "High frequency scattering from trihedral corner reflectors and other benchmark targets: SBR versus experiment," *IEEE Trans. Antennas Propagat.*, vol.39, pp.1345-1351, 1991.

[26] R. J. Burkholder, R.-C. Chou, and P. H. Pathak, "Two ray shooting methods for computing the EM scattering by large open-ended cavities," *Computer Phys. Commun.*, vol.68, pp.353-365, 1991.

[27] P. H. Pathak and R. J. Burkholder, "Modal, ray and beam techniques for analyzing the EM scattering by open-ended waveguide cavities," *IEEE Trans. Antennas Propagat.*, vol.37, pp.635-647, May 1989.

[28] R. F. Harrington, *Time-Harmonic Electromagnetic Fields*, New York: McGraw-Hill, 1961.

[29] J. J. Bowman, T. B. A. Senior, and P. L. E. Usienghi, *Electromagnetic and Acoustic Scattering by Simple Shapes*, ed., Ch.18, Amsterdam, North-Holland, 1969.

[30] L. Felsen, "Asymptotic expansion of the diffracted wave for a semi-infinite cone," *IRE Trans. Antennas Propagat.*, vol.5, pp.402-404, 1957.

[31] L. Felsen, "Plane wave scattering by small angles cones," *IRE Trans. Antennas Propagat.*, vol.5, pp.121-129, 1957.

[32] L. B. Felsen, "Alternative representations in regions bounded by spheres, cones, and planes," *IRE Trans. Antennas Propagat.*, vol.5, pp.109-121, 1957.

[33] P. Pathak, "Techniques for high-frequency problems," Ch.4 in *Antenna Handbook*, Y. T. Lo and S. W. Lee, ed., New York: Van Nostrand Reinhold, 1988.

[34] A. Sommerfeld, "Mathematische theorie der diffraction," *Math. Ann.*, vol.47, pp.314-374, 1986.

[35] R. Rojas and Z. Al-hekail, "Generalized impedance/resistive boundary conditions for electromagnetic scattering problems," *Radio Sci.*, vol.24, pp.1-12, Jan.-Feb. 1989.

[36] A. Ludwig, "The definition of cross polarization," *IEEE Trans. Antennas Propagat.*, vol.21, pp.116-119, Jan. 1973.

[37] M. Otero and R. Rojas, "EM scattering from a complex material cylinder in the presence of an impedance wedge," *J. Electromagn. Waves Appl.*, vol.10, pp.1563-1581, 1996.

[38] J. Kauffman, W. Crosswell, and L. Powers, "Analysis of radiation patterns of reflector antennas," *IEEE Trans. Antennas Propagat.*, vol.24, pp.53-65, Jan. 1976.

[39] K. Lambert, R. Rudduck, and T. Lee, "A new method for obtaining antenna gain from backscatter measurements," *IEEE Trans. Antennas Propagat.*, vol.38, 896-902, June 1990.

[40] H. T. Chou, M. Hsu, P. H. Pathak, and R. J. Burkholder, "Hybrid combination of numerical and aysmptotic techniques for analysis of high frequency radiation / scattering by generic aircraft / missile shapes," *URSI Natl. Radio Sci. Meeting*, Boulder, Jan. 9-13, 1996.

[41] D. 1. Butorin. and P. Ufimtsev, *Sov. Phys. Acoust.*, vol.32, 1986.

[42] E. Knott, "The relationship between Mitzner's ILDC and Michaeli's equivalent currents," *IEEE Trans. Antennas Propagat.*, vol.33, pp.112-114, 1985.

[43] A. Michaeli, "Equivalent edge currents for arbitrary aspects of observation," *IEEE Trans. Antennas Propagat.*, vol.32, pp.252-258, Mar. 1984.

[44] A. Michaeli, "Elimination of infinities in equivalent edge currents, Part 1: fringe current components," *IEEE Trans. Antennas Propagat.*, vol.34, pp.912-918, July 1986.

[45] H. T. Chou, P. H. Pathak, and P. R. Rousseau, "A TD-UTD for transient mutual coupling by pulsed antennas placed directly on a smooth convex surface," *URSI Natl. Radio Sci. Meeting*, Atlanta, Georgia, 1998.

[46] H. T. Chou, P. H. Pathak, and P. R. Rousseau, "On the development of TD-UTD ray solutions for fields of pulsed antennas in the presence of smooth convex surfaces," *URSI Radio Sci. Gen. Assembly Meeting*, Toronto, CA, Aug. 1999.

[47] H. T. Chou, P. H. Pathak, and P. R. Rousseau, "A time domain version of some asymptotic high frequency methods," *URSI Radio Sci. Gen. Assembly Meeting*, Toronto, CA, Aug. 1999.

[48] P. R. Rousseau and P. H. Pathak, "Time-domain uniform geometrical theory of diffractior curved wedge," *IEEE Trans. Antennas Propagat.*, pp.1375-1382, Dec. 1995.

[49] D. S. Ahuwalia, R. M. Lewis, and J. Boersma, "Uniform asymptotic theory of diffraction by a plane screen," *J. Appl. Math.*, vol.16, pp.783-807, 1968.

[50] S. Lee and G. Deschamps, "A uniform asymptotic theory of EM diffraction by a curved wedge," *IEEE Trans. Antennas Propagat.*, vol.24, pp.25-34, Jan. 1976.

[51] R. Tiberio, S. Maci, and A. Toccafondi, "An incremental theory of diffraction: Electromagnetic formulation," *IEEE Trans. Antennas Propagat.*, vol.43, pp.87-96, 1995.

Chapter 21

KA-BAND CHANNEL MODELS FOR LOW-EARTH-ORBIT SATELLITES

Jean-Fu Kiang,[1] Yung-Chang Chen,[2] Le-Gen Shi,[2] and Ying-Jie Su[2]

[1] *Department of Electrical Engineering and*
Graduate Institute of Communication Engineering
National Taiwan University
Taipei, Taiwan, ROC

[2] *Department of Electrical Engineering*
National Tsing Hua University
Hsin-Chu, Taiwan, ROC

Abstract ROCSAT-1 is the first low-earth-orbit(LEO) satellite planned by the National Space Program Office(NSPO), Taiwan, ROC. Ka-band communication experiment has been conducted by using ROCSAT-1 to explore high-frequency technology and wideband applications. A channel model has been developed to study propagation characteristics of the link between ROCSAT-1 and ground terminal. Attenuation models including rain, gaseous and scintillation losses are considered. Orbital parameters, power level, radiation patterns of the satellite antennas, transmission data rate, and internal noise sources are incorporated. The proposed model is useful in assessing the communication performance of ROCSAT-1.

Keywords: Ka-band, LEO, orbit dynamics, rain attenuation, link budget, channel characteristics, simulation.

1. Introduction

Using low-earth-orbit(LEO) satellites to provide communication services becomes popular in recent years. Major advantage of using LEO satellites as compared with geosynchronous-earth-orbit(GEO) satellites is the proximity of LEO satellites to earth, hence requiring less power [1],

[2]. However, LEO satellites have certain disadvantages. Firstly, their orbital period around earth is much shorter than one day, hence the link time with any fixed ground terminal is only on the order of minutes in each pass. Secondly, quick-response tracking antenna is required at the ground terminal to track the satellite. To provide a continuous coverage as required in personal communication services, satellite constellation like IRIDIUM system is required [3].

Ka-band has been explored for satellite communication to provide wideband applications(e.g., video broadcast) and to add new frequency resources to relieve the crowded usage in lower bands. However, rain attenuation in the Ka-band is stronger than in the lower bands, and may significantly degrade communication quality [4]-[14].

ROCSAT-1 is the first LEO satellite planned by the National Space Program Office(NSPO) of Taiwan, the Republic of China, to conduct scientific experiments. In the Ka-band communication experiment, a video clip is converted into bit stream, transmitted to the satellite at the rate from 3.3 Mbps to 6.6 Mbps, then echoed back to the ground terminal by the satellite transponder. Carrier frequency of the uplink is 28.25 GHz, while that of the downlink is 18.45 GHz. Fig.21.1 shows the scenario of experiment where various types of noise sources are indicated.

We have developed a channel model for LEO satellites, in which characteristics of rain attenuation in the Ka-band are incorporated. Realistic orbital parameters, power level, antenna radiation patterns, transmission data rate and internal noise sources have been considered. Simulation results are used to assess the performance of LEO satellite communication links, and compared with measurements after ROCSAT-1 is launched.

In the next Section, orbit of the satellite will be introduced. Position and velocity of the satellite must be known a priori to estimate the attenuation loss. Attenuation mechanisms in the Ka-band include rain attenuation, gaseous attenuation, scintillation and internal device noises. Then, we will discuss link budget calculation to assess the carrier-to-noise ratio(C/N) at the receiver. Simulation procedure and results will be presented in the last Section, followed by Conclusions.

2. Orbit Models for Satellites

LEO satellite moves fast with respect to its ground terminal. Hence, precise description of the relative motion between satellite and ground terminal is necessary to conduct simulation. The orbit dynamics used to predict satellite motion will be reviewed. We will first review some

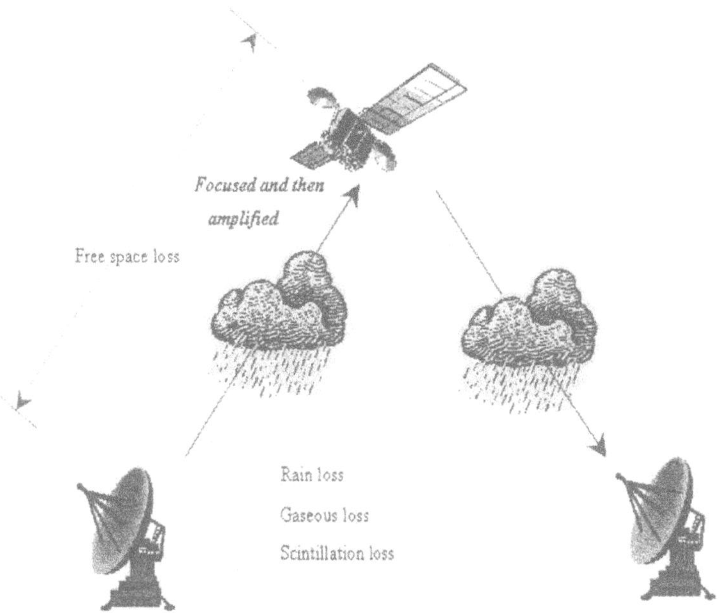

Figure 21.1. Scenario of LEO satellite communication experiment.

frequently used coordinate systems and their parameters like elevation angle, azimuthal angle, latitude, longitude, and so on. Next, classical orbital elements used to describe satellite orbit will be reviewed. Then, satellite motion in a given coordinate system will be briefly described.

2.1 Orbit Dynamics

Fig.21.2 shows the geometric relation between earth with mass M and a satellite with mass m. The law of gravity states that any two bodies attract each other with a force proportional to the product of their masses and inversely proportional to the square of distance between them, namely

$$\bar{F}_g = -\frac{GMm}{r^2}\frac{\bar{r}}{r} \tag{21.1}$$

where the mass of satellite is much smaller than that of earth. Newton's second law of motion states that

$$\bar{F}_g = m\ddot{\bar{r}} \tag{21.2}$$

where $\bar{F}_g$ is the force on mass m due to mass M, and $\bar{r}$ is the vector pointing from M to m. For simplicity, neglect the forces from sun, moon

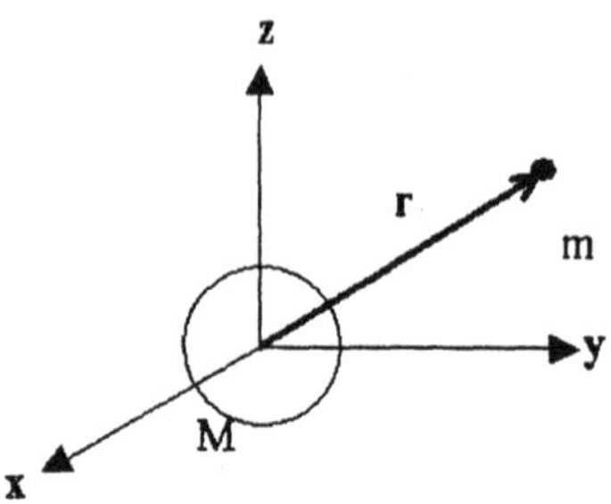

Figure 21.2. Relative position of earth(M) and satellite(m).

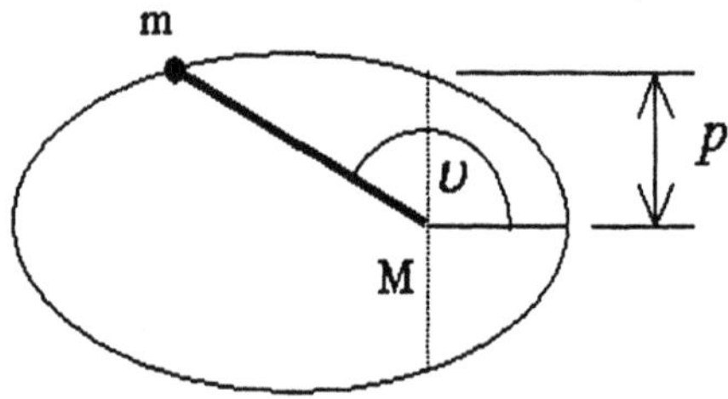

Figure 21.3. Elliptic orbit of m with $0 < e < 1$.

and other planets. Relative motion of earth and satellite can thus be viewed as a two-body problem [15]. Equating (21.1) and (21.2), we have

$$\ddot{\bar{r}} + \frac{K}{r^3}\bar{r} = 0 \tag{21.3}$$

where $K = GM$. By using (21.3) and the law of conservation of angular momentum, $\bar{L} = \bar{r} \times \bar{v}$, we obtain the orbit equation

$$r = \frac{p}{1 + e\cos\nu} \tag{21.4}$$

where $p = L^2/K$ is a geometrical constant of the conic section, called semi-latus rectum, $e = B/K(B$ is an integration constant) is the eccentricity which determines the type of conic section represented by (21.4), and ν is the true anomaly. Fig.21.3 shows the elliptic orbit described by (21.4) with $0 < e < 1$.

2.2 Coordinate Systems

A number of coordinate systems are often used to describe the satellite orbit.

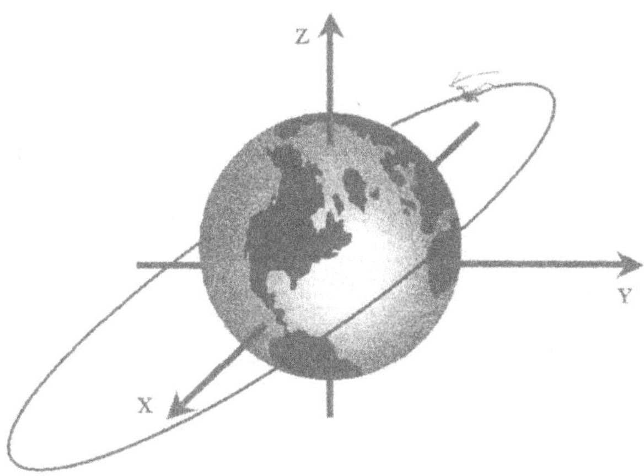

Figure 21.4. Celestial coordinate system.

Celestial Coordinate System. This coordinate system is fixed with respect to stars. As shown in Fig.21.4, the x-y plane is coincident with the earth equator. The orbit focus is at the center of earth, the x axis points towards the first star in the Aries, the z axis points toward the north pole, and the y axis is defined by rotating the x axis around z axis by 90°.

Geocentric Coordinate System. This coordinate system is similar to the celestial one, except it is fixed to earth. The x axis points to the intersection of equator and the Greenwich meridian(0° longitude), the y axis points to the intersection of equator and the 90°E meridian, and the z axis points towards the north pole. Location on earth surface is represented by longitude and latitude.

Perifocal Coordinate System. The orbital plane of a satellite is defined as the PQ plane as shown in Fig.21.5. Origin of this coordinate system is at the center of earth, which is also one of the foci of the elliptic orbit. The P axis points towards the perigee, the Q axis points at 90° from the perigee along the orbit in the sense of satellite motion, and the L axis is perpendicular to the orbital plane.

Topocentric-Horizon Coordinate. As shown in Fig.21.6, origin of this coordinate system is the ground terminal on earth surface. The horizontal plane contains the S axis which points south and the E axis which points east. The U axis is perpendicular to the horizontal plane and points upward. The elevation angle, θ, and azimuthal angle, ϕ, are

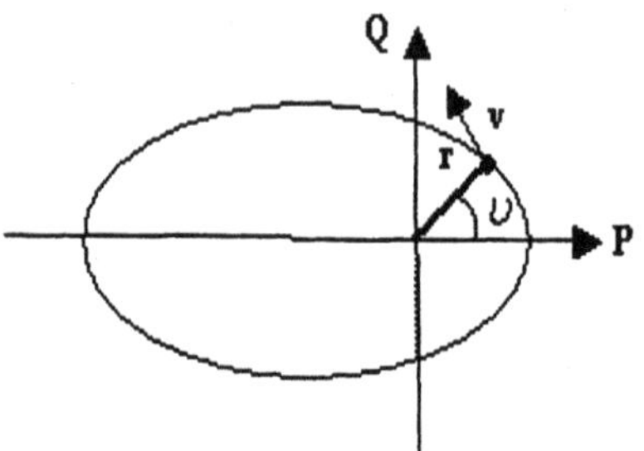

Figure 21.5. Perifocal coordinate system.

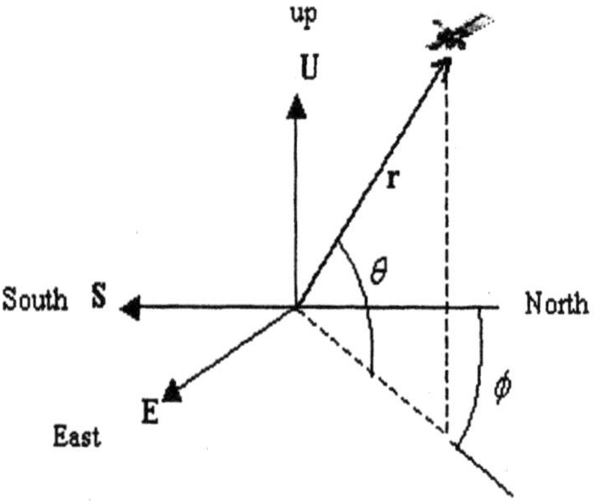

Figure 21.6. Topocentric-horizon coordinate system.

defined as

$$\theta = \sin^{-1} \frac{\bar{r} \cdot \hat{U}}{|\bar{r}|}$$

$$\phi = \cos^{-1} \frac{-\bar{r} \cdot \hat{S}}{|\bar{r}| \cos\theta} = \sin^{-1} \frac{\bar{r} \cdot \hat{E}}{|\bar{r}| \cos\theta}$$

2.3 Classical Orbital Parameters

Fig.21.7 shows a typical elliptic orbit. The semi-major axis(a) defines orbit size, the eccentricity(e) defines orbit shape, the orbital inclination(i) is the angle between unit vector $\hat{z}$ and the angular momentum vector, $\bar{L}$. The right ascension of ascending node(RAAN)(Ω) is the angle on the x-y plane, between unit vector $\hat{x}$ and the point where satellite crosses the

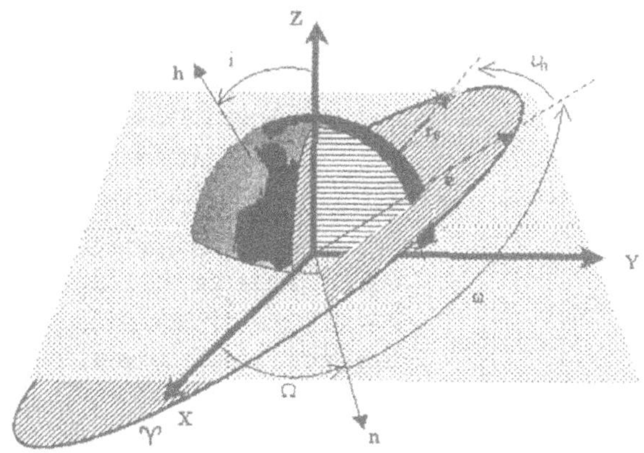

Figure 21.7. Typical elliptic orbit and its associated parameters.

equatorial plane in a north bound sense, measured counterclockwise from the north side of equatorial plane. The argument of periapsis(ARGP)(ω) is the angle in the orbital plane, between the ascending node and the periapsis point, measured in the sense of satellite motion. The time of periapsis passage(t) is the time when the satellite passes by a given point along its orbit. The first five parameters completely describe the size, shape and orientation of satellite orbit. The time of periapsis passage is required to locate the satellite along its orbit.

These parameters are applicable to describe satellite orbit in the geocentric-equatorial or heliocentric-ecliptic coordinate system. Note that $p = a(1 - e^2)$ in (21.4). Satellite orbit in the perifocal coordinate system can be expressed as

$$\bar{r} = r\cos\nu\hat{P} + r\sin\nu\hat{Q} \tag{21.5}$$

from which we obtain

$$\bar{v} = (\dot{r}\cos\nu - r\dot{\nu}\sin\nu)\hat{P} + (\dot{r}\sin\nu + r\dot{\nu}\cos\nu)\hat{Q} \tag{21.6}$$

Taking the time derivative of (21.4) using $L = r^2\dot{\nu}$ and $p = L^2/K$, we obtain

$$\dot{r} = e\sqrt{\frac{K}{p}}\sin\nu, \qquad r\dot{\nu} = \sqrt{\frac{K}{p}}(1 + e\cos\nu)$$

Substituting them into (21.6) yields

$$\bar{v} = \sqrt{\frac{K}{p}}\left[-\sin\nu\hat{P} + (e + \cos\nu)\hat{Q}\right]$$

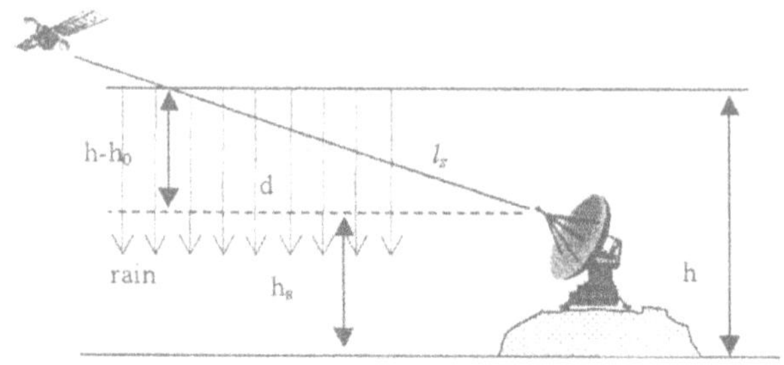

Figure 21.8. Schematic of a path in rain.

3. Characteristics of Ka-Band Channel

Quality of communication link is mainly determined by the carrier-to-noise ratio(C/N) at receiver. Received power level is determined by the transmitted power, free space loss, transponder gain, rain attenuation, gaseous attenuation, internal loss and other loss mechanisms.

3.1 Rain Attenuation

Extensive observations have been conducted and global-scale prediction models have been constructed on rain characteristics [4]-[14]. Most of the prediction models are of statistical nature, based on the observation data collected at sporadic measurement sites. Three rain models are frequently used, which are Crane model [10], ITU-R model [11] and DAH model [13]. The Crane and ITU-R models have been widely used for global prediction. However, the newly developed DAH model claims better prediction than the other two models.

Crane Model. Referring to Fig.21.8, the Crane model is an empirical model based on rain distribution(point rain rate), vertical rain extent(height of 0°C isotherm), path length immersed in the rain, and frequency dependent coefficients derived from raindrop characteristics [10]. Point rain rate can be looked up from tables categorized according to rain climate region [2].

ITU-R Model. The ITU-R model predicts rain attenuation based on the effective rain height(function of ground terminal latitude), path length through rain cell, rain intensity at 0.01 percentile, wave polarization, and frequency dependent coefficients which can be interpolated from tables [11].

DAH model. The DAH model is a modification of ITU-R model, where two adjustment factors are proposed to improve its prediction accuracy [13].

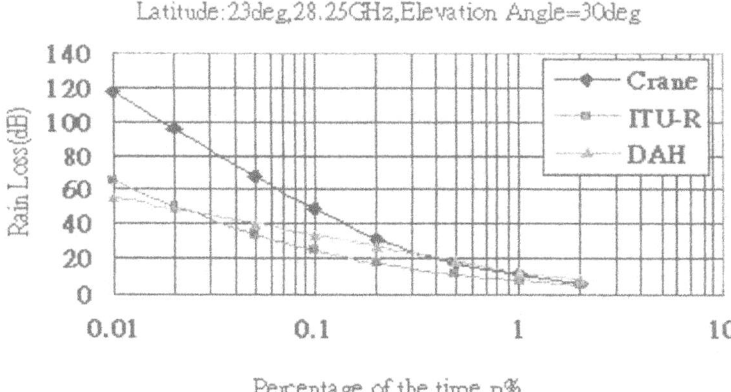

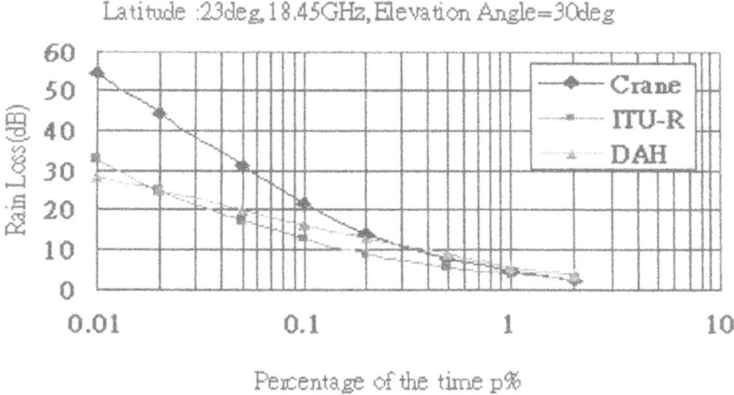

Figure 21.9. **Rain attenuation predicted by three rain models.**

Among these three rain models, the ITU-R model and DAH model use the rain rate, $R_{0.01}$, of an average year to estimate the attenuation at other rain rates, while the Crane model uses the expected rain rate to estimate rain attenuation. The attenuations predicted by these three models are shown in Fig.21.9, where the climate rain rate distribution provided by the Crane model is used.

Summarized from many researches conducted around the world, the DAH model with the climate rain rate distribution from Crane model seems to have the best fit with the rain attenuation data over most regions of the world [12]. However, fit between the rain models and attenuation data is not guaranteed at the ground terminal at Tainan. Hence, all the three models are considered in our simulation, and the local rain rate distribution at Tainan is used instead of the rain rate distribution in the Crane model.

3.2 Gaseous Attenuation

In a clear sky, gaseous absorption, L_{gas}, is the major source of link noise. It consists of oxygen absorption, L_{oxy}, and water vapor absorption, L_{wat}, which can be estimated empirically as follows [13], [16].

Oxygen Loss. Specific attenuation due to oxygen is estimated as

$$\gamma_{\text{oxy}} = \left[7.19 \times 10^{-3} + \frac{6.09}{f^2 + 0.227} + \frac{4.81}{(f-57)^2 + 1.5}\right] \times f^2 \times 10^{-3}$$

in dB/km, where f is the operating frequency in GHz, H_{oxy} is the equivalent height of the oxygen layer, which is about 6 km in dry days. The total loss due to oxygen becomes

$$L_{\text{xoy}} = \frac{H_{\text{oxy}}\gamma_{\text{oxy}}}{\sin\theta}$$

where θ is the elevation angle pointing at the satellite.

Water Vapor Loss. Specific attenuation due to water vapor is estimated as

$$\gamma_{\text{wat}} = \left[0.05 + 0.0021\mu + \frac{3.6}{(f-22.2)^2 + 8.5} + \frac{10.6}{(f-183.3)^2 + 9.0} + \frac{8.9}{(f-325.4)^2 + 26.3}\right] \times f^2 \times 10^{-4}$$

in dB/km, where f is the operating frequency in GHz, μ is water vapor density in g/m^3. The equivalent height of water vapor can be estimated by an empirical formula as [13]

$$H_{\text{wat}} = H_{\text{w0}}\left[1 + \frac{3}{(f-22.2)^2 + 5} + \frac{5}{(f-183.3)^2 + 6} + \frac{2.5}{(f-325.4)^2 + 4}\right]$$

in km, where H_{w0} is set to 1.6 km in clear sky. The total loss due to water vapor becomes

$$L_{\text{wat}} = \frac{H_{\text{wat}}\gamma_{\text{wat}}}{\sin\theta}$$

At $f = 30$ GHz and at an elevation angle of 20°, the water vapor absorption is less than 1 dB even with a high water vapor density of 10 g/m^3.

The total gaseous attenuation is expressed as the sum of oxygen loss and water vapor loss as $L_{\text{gas}} = L_{\text{oxy}} + L_{\text{wat}}$. Fig.21.10 shows the variation of gaseous attenuation with water vapor density.

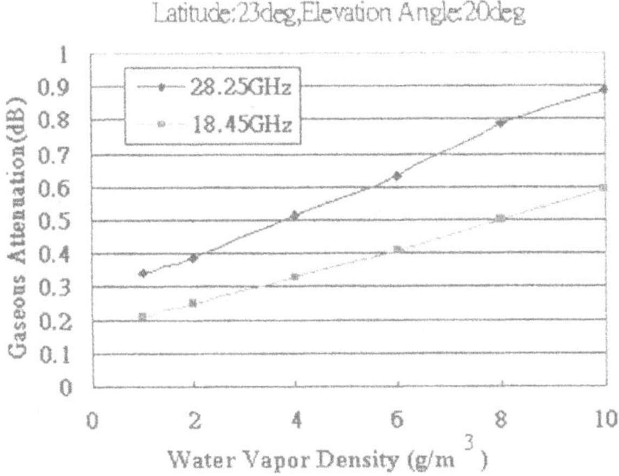

Figure 21.10. Variation of gaseous attenuation with water vapor density.

3.3 Scintillation Effects

Rapid fluctuation in amplitude of a transmitted signal is caused by time-dependent irregularities in the transmission path. When a wave passes through rain drops, its phase will be different from that through free space. Such phase difference causes fluctuations in signal amptitude. Although scintillation is contributed by both ionosphere and troposphere, mechanisms and characteristics in both regions are different and generally impact different frequency bands. Above 2 GHz, tropospheric scintillation is often stronger than ionospheric scintillation [17], [18]. Ionospheric scintillation is significant only in the high-latitude and equatorial areas [19]. Hence, only tropospheric scintillation is consideed in this Section.

All the scintillation models are established by applying statistical methods on measurement data. The Karasawa model [20] and ITU-R model [21] are most frequently used, but they can only describe the variations of scintillation intensity over seasons. TKS model is modified from the Karasawa model, which takes cumulus cloud information into consideration, and can describe the diurnal variation of scintillation [22]. Scintillation intensity becomes higher in rain [18]. However, effects of rain attenuation is much stronger so that the scintillation effects can be ignored in the rain.

Scintillation is generally related to clouds, especially the cumulus type. Fig.21.11 shows the cumulative distributions of scintillation intensity.

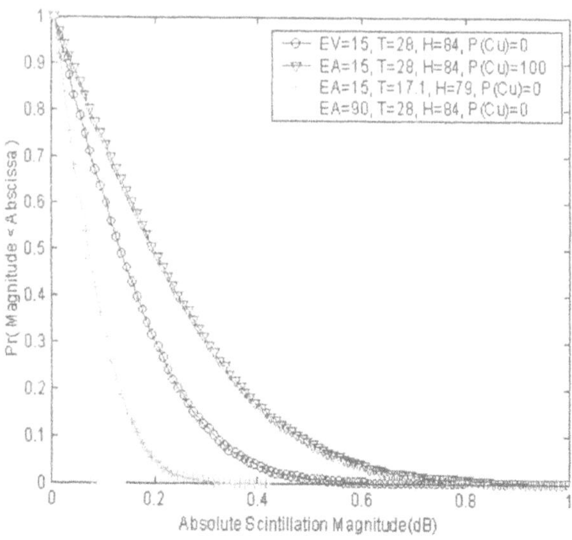

Figure 21.11. Distribution of scintillation intensity.

The scintillation intensity becomes higher if there are clouds in the signal transmission path. The Karasawa model and ITU-R model incorporate only monthly average weather parameters. In short terms, the TKS model is expected to provide better prediction accuracy than the other two models.

3.4 Noise Characteristics

In spite of various noise sources, thermal noise temperature, T, is always used to represent the noise power received by antenna. Noise power at the receiver can be expressed as $N = \kappa T_{\text{sys}} B$, where κ is the Boltzmann's constant, T_{sys} is system noise temperature in °K, and B is noise bandwidth. The system noise temperature is contributed by the environment(T_{ant}, measured at the antenna aperture) and electronics(T_e, internal circuits behind the antenna). As shown in Fig.21.12, the antenna noises may include cosmic background noise, galactic noise, solar noise, earth noise, rain noise, and so on [1]. Among them, rain noise is the major noise source in Ka-band.

For rain cell or raining cloud, noise temperature(T_{rain} in °K) and ambient temperature(T_m in °K) are related to each other by rain attenuation (L_{rain} in dB) as [1]

$$T_{\text{rain}} = T_m \left(1 - 10^{-L_{\text{rain}}/10}\right)$$

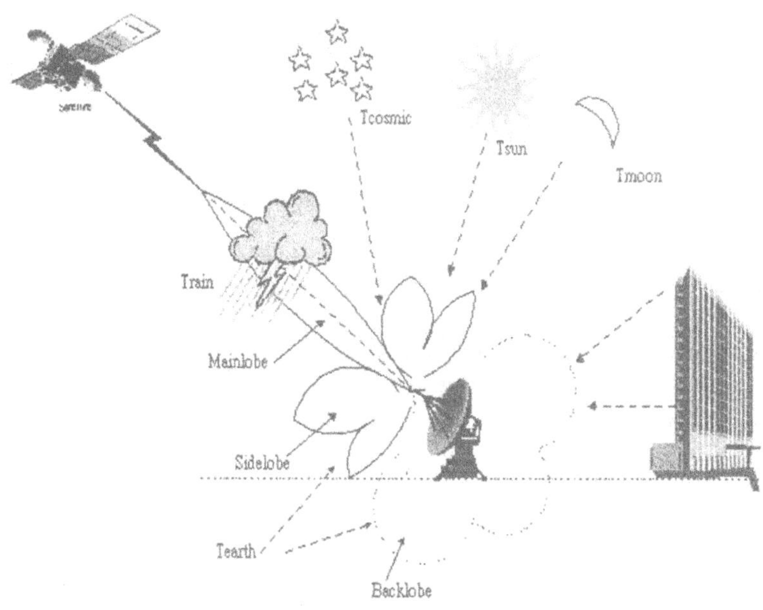

Figure 21.12. Contributing factors of antenna noise.

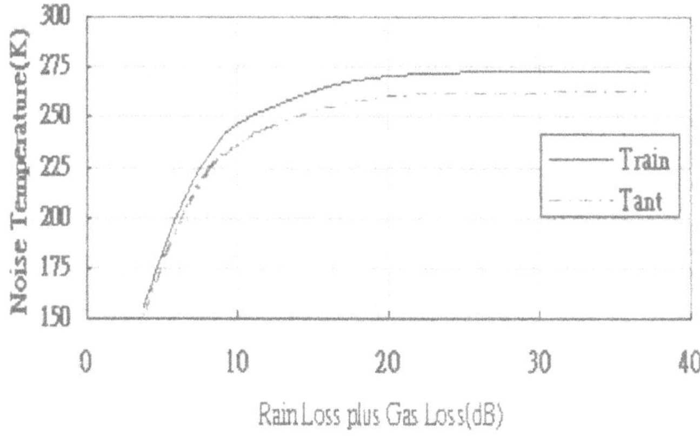

Figure 21.13. Relation between antenna temperature and rain attenuation, $\beta = 0.95$, $T_{\text{other}} = 70°\text{K}$.

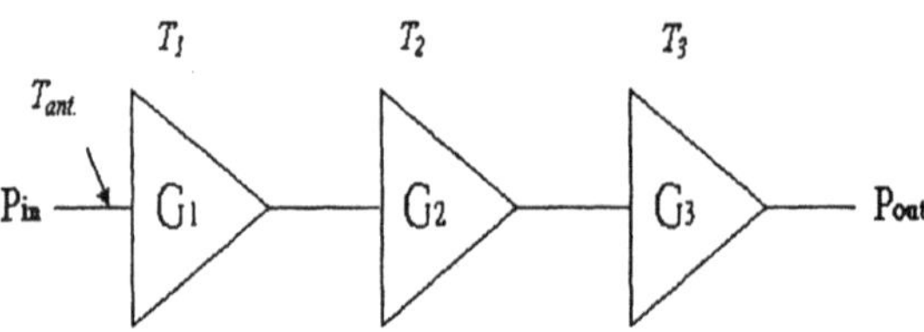

Figure 21.14. Three stages of amplifiers.

With higher rain attenuation, the noise temperature will be closer to the ambient temperature. Gaseous absorption accompanying the rainfall is usually included in the rain attenuation. Assuming that the main lobe of an antenna points at a raining cloud with ambient temperature of 273 °K, then the antenna temperature, T_{ant}, can be expressed as a linear combination of T_{rain} and other noise source, T_{other}, as

$$T_{\text{ant}} = \beta T_{\text{rain}} + (1-\beta)T_{\text{other}}$$

The weighting factor, β, is determined by shapes of the main lobe, side-lobes and back lobes of the antenna. Fig.21.13 demonstrates the relation between antenna temperature and rain(plus gaseous) attenuation, where $\beta = 0.95$ and $T_{\text{other}} = 70°\text{K}$.

Low-noise amplifiers(LNA) and other active or passive components are generally connected between the receiving antenna and the receiver. The amplifier enhances not only the input signal but also noise. Fig.21.14 shows three stages of amplifiers. The input noise temperature is T_{ant}, the gain of stage i is G_i, and the noise temperature at stage i is T_i. The output noise power can be expressed as

$$\begin{aligned}
N_{\text{out}} &= G_1G_2G_3\kappa T_{\text{ant}}B + G_1G_2G_3\kappa T_1B + G_2G_3\kappa T_2B + G_3\kappa T_3B \\
&= G_1G_2G_3\kappa\left(T_{\text{ant}} + T_1 + \frac{T_2}{G_1} + \frac{T_3}{G_1G_2}\right)B \\
&= G_1G_2G_3\kappa\left(T_{\text{ant}} + T_e\right)B
\end{aligned}$$

where B is noise bandwidth and T_e is the effective noise temperature at the input of the first stage. Explicitly,

$$T_e = T_1 + \frac{T_2}{G_1} + \frac{T_3}{G_1G_2}$$

Hence, the receiver noise temperature, T_{sys}, can be expressed as $T_{\text{sys}} = T_{\text{ant}} + T_e$ [23].

4. Link Budget Estimation

The function of a satellite transponder is to receive signal from the ground terminal and then retransmit the signal back to the ground terminal. A satellite communication system has limited transmitting power and bandwidth. It must be designed to meet certain minimum performance requirements such as bit error rate(BER) which is related to S/N, C/N, C/N_0 or E_b/N_0. To estimate these ratios in a given system, one usually needs to know the transmitting antenna gain(G_t), receiving antenna gain(G_r), satellite antenna gain(G_s), equivalent isotropic radiation power(EIRP), free space loss(L_f), transmission loss in the atmosphere(L_{atm}) and system noise(N_{sys}).

4.1 Definition of Parameters

The power transmitted from an isotropic antenna through free space to another antenna can be expressed as

$$P_r = \frac{P_t}{4\pi r^2} A_e \text{ (W)} \tag{21.7}$$

where P_r is the received power, P_t is the transmitted power, r is the distance between transmitting and receiving antennas, and A_e is the effective aperture area of receiving antenna. For paraboloid antenna of diameter D, we have

$$A_e = \eta \left(\frac{\pi D^2}{4} \right) \text{ (m}^2\text{)} \tag{21.8}$$

where η is the antenna efficiency which depends on manufacture imperfection and operation conditions. Substituting (21.8) into (21.7), we obtain

$$P_r = P_t \eta \left(\frac{\pi D}{\lambda} \right)^2 \left(\frac{\lambda}{4\pi r} \right)^2 \text{ (W)} \tag{21.9}$$

Antenna Gain. For an isotropic antenna, the transmitted power density is constant in every direction. Focusing capability of an antenna can be described by the antenna gain defined as

$$G_t = \frac{P_{\text{ant}}}{P_{\text{iso}}}$$

where P_{ant} is the power density in the main beam direction, and P_{iso} is the power density radiated by an isotropic antenna. Thus, (21.9) can be rewritten as

$$P_r = G_t P_t \eta \left(\frac{\pi D}{\lambda} \right)^2 \left(\frac{\lambda}{4\pi r} \right)^2 \text{ (W)} \tag{21.10}$$

Note that the gain of any receiving antenna is related to its effective aperture area as

$$G_r = \frac{4\pi}{\lambda^2} A_e = \eta \left(\frac{\pi D}{\lambda}\right)^2$$

where λ is wavelength.

Unlike ground terminal antennas, the satellite antennas have a very wide beamwidth to cover wide area. To compensate for the variation of free space loss as a LEO satellite passes by a ground terminal, satellite antennas are required to have secant-like radiation patterns. However, due to blockage and other imperfections on antenna reflector and feed horn, ripples of radiation pattern exist in the E-plane, H-plane, and other cross sections. In our simulations, measured radiation patterns of the satellite antennas are recorded and interpolated to obtain the antenna gain in any specific direction. Antenna gains of both the transmitting and receiving antennas on the satellite are combined as $G_s = G_{sr} G_{st}$.

Equivalent Isotropic Radiation Power(EIRP). The EIRP is defined as EIRP $= G_t P_t$ (W). If an isotropic antenna radiates with the same power density toward the receiving antenna as the transmitting antenna does, then the amount of power required will be EIPR.

Free Space Loss. Free space loss refers to the decrease of power density due to the distance between transmitting and receiving antennas. It is defined as

$$L_f = \left(\frac{\lambda}{4\pi r}\right)^2$$

Using these definitions, (21.10) can be reduced to

$$P_r = P_t G_t G_r L_f = (\text{EIRP}) G_r L_f \text{ (W)} \tag{21.11}$$

Transmission Loss in Atmosphere. Loss in the atmosphere includes rain attenuation, gaseous attenuation and scintillation effects, which can be expressed in natural unit as

$$A_{\text{atm}} = A_{\text{rain}} A_{\text{gas}} A_{\text{scin}}$$

or in log scale as

$$L_{\text{atm}} = 10 \log A_{\text{atm}} = L_{\text{rain}} + L_{\text{gas}} + L_{\text{scin}} \text{ (dB)}$$

Thus, (21.11) can be modified as

$$P_r = (\text{EIRP}) G_r L_f A_{\text{atm}} \text{ (W)}$$

or expressed in log scale as

$$P_r = (\text{EIRP}) + G_r + L_f + L_{\text{atm}}\ (\text{dB}_\text{W})$$

System Noise Temperature. Total noise includes the noise from environment to antenna and the noise generated in components of the system, namely, $T_{\text{sys}} = T_{\text{ant}} + T_e$.

4.2 Signal-to-Noise Ratio

In satellite communication, link quality is determined by the ratio between signal power and noise power. Alternative parameters like C/N, C/N_0 or E_b/N_0 are also used. While C/N and C/N_0 are preferred to analyze system performance, E_b/N_0 is more frequently used to assess the quality of a transmission sequence. These parameters are defined as

$$\frac{C}{N} = \frac{C}{\kappa T_{\text{sys}} B}, \qquad \frac{C}{N_0} = \frac{C}{\kappa T_{\text{sys}}}, \qquad \frac{E_b}{N_0} = \frac{(C/R)}{\kappa T_{\text{sys}}}$$

where C is carrier power, N is noise power, N_0 is noise power density, and R is transmission data rate. C/N of the complete link(from ground terminal to satellite, then back to the ground terminal) determines the quality of communication, and can be decomposed into C/N of the up-link and downlink, respectively, as [1]

$$\left(\frac{C_c}{N_c}\right)^{-1} = \left(\frac{C_u}{N_u}\right)^{-1} + \left(\frac{C_d}{N_d}\right)^{-1}$$

where the subscripts c, u and d stand for complete link, uplink and downlink, respectively.

Modulation scheme of BPSK or QPSK is used in the experiments of ROCSAT-1, which implies a bit error rate of $10^{-3}(10^{-5})$ when E_b/N_0 = 6.7 dB(9.7 dB). When E_b/N_0 = 6.7 dB and the bit rate is 6.6 Mbps, the associated C/N_0 is 74.9 dB_Hz. Suppose the data rate is decreased to 3.3 Mbps with the same carrier power, then E_b/N_0 is increased to 9.7 dB and the bit error rate is reduced to 10^{-5}. In case an error correction scheme is implemented into the bit stream, the probability of bit error in the content bits can be further reduced. With the error correction capability implemented in ROCSAT-1, the required E_b/N_0 is reduced to 4.7 dB.

5. Simulation Procedure and Results

Unit normal vector at the ground terminal on earth surface can be expressed in the geocentric coordinate system as

$$\hat{n}_R = \cos\theta_e \cos\Omega_e \hat{x} + \cos\theta_e \sin\Omega_e \hat{y} + \sin\theta_e \hat{z}$$

where Ω_e is the angular speed of earth spinning, θ_e is the latitude of ground terminal. If the satellite orbit is h above the mean sea level, then the radius of satellite orbit is $R_s = R_e + h$, where R_e is the earth radius. From the balance of centrifugal force and gravitational force, the satellite speed, v_s, can be determined from

$$\frac{v_s^2}{R_s} = g_0 \frac{R_e^2}{R_s^2}$$

where g_0 is the gravitational acceleration at mean sea level. With $h =$ 650/600/550 km, the satellite moves at about 7.546/7.573/7.600 km/s and takes about 97.84/96.80/95.77 minutes to move one round along its orbit. The ARGP of ROCSAT-1 orbit is about 270°.

Let $\hat{h}$ be the unit normal vector of the satellite orbital plane. The unit vector $\hat{m}$ points at where the orbit intercepts the x-z plane. Thus, $\hat{h}$ and $\hat{m}$ can be expressed as

$$\hat{h} = -\sin\alpha\hat{x} + \cos\alpha\hat{z}$$
$$\hat{m} = \cos\alpha\hat{x} + \sin\alpha\hat{z}$$

where α is the inclination angle of the orbital plane. These two vectors are orthogonal to each other, and the other orthogonal unit vector on the orbital plane can be determined as

$$\hat{\ell} = \hat{m} \times \hat{h} = -\hat{y}$$

Assume that the satellite moves at an angular speed of Ω_s radians/s, then the position of satellite can be expressed explicitly as

$$\bar{r}_s = R_s \cos\Omega_s t\hat{\ell} + R_s \sin\Omega_s t\hat{m}$$

Note that the satellite moves in the same sense as the earth spinning. Since orbital period of the LEO satellite is much shorter than the spinning period of earth, positions of the ground terminal and LEO satellite are updated at any time step of the simulation procedure.

Only along portion of the orbit do the satellite and ground terminal establish a line-of-sight(LOS) link. This condition can be checked as

$$\cos^{-1}\left[\hat{n}_R \cdot \frac{(\bar{r}_s - \bar{r}_g)}{|\bar{r}_s - \bar{r}_g|}\right] = \begin{cases} > 90^\circ, & \text{beyond the horizon} \\ 75^\circ \sim 90^\circ, & \text{LOS link, nontrackable} \\ 70^\circ \sim 75^\circ, & \text{LOS link, trackable} \\ < 70^\circ, & \text{LOS link, communication} \end{cases}$$

where $\bar{r}_g$ is the position of ground terminal. Tracking antenna on the ground terminal tries to track the satellite when it shows up on the horizon(antenna elevation angle between 0° and 15°). Communication link is secured when the elevation angle is higher than 20°.

The velocity of satellite relative to the ground terminal can be calculated as

$$\bar{v}_{sg} = \frac{d}{dt}(\bar{r}_s - \bar{r}_g) = -R_s\Omega_s \cos\Omega_s t\hat{m} + R_s\Omega_s \sin\Omega_s t\hat{n}$$
$$+R_e\Omega_e \cos\theta_e \sin\Omega_e t\hat{x} - R_e\Omega_e \cos\theta_e \cos\Omega_e t\hat{y}$$

It is obvious that elevation angle is a crucial parameter that affects the communication experiment of ROCSAT-1. Low elevation angle implies high free space loss. With the link budget limitation imposed upon ROCSAT-1, 15° is the minimum elevation angle that the ground terminal can lock onto the satellite, then communication experiments can be conducted when the elevation angle is greater than 20°. Some passes may have elevation angles lower than 20° all over the pass, hence no communication link can be established. In our experiments, 50% of the passes have their maximum elevation angle lie between 15° to 40°. The link time may vary abruptly between two consecutive passes, and greater maximum elevation angle implies longer link time.

Fig.21.15 shows the simulation procedure in which a bit stream is transmitted from the ground terminal to ROCSAT-1, amplified by its transponder, then sent back to the ground terminal. The bit error rate and associated quality factors can thus be assessed by running this simulation procedure.

Consider a specific pass in which the altitude of satellite is 600 km and the orbit inclination angle is 35°. In-sight time(elevation angle greater than 0°) of this pass is 474 seconds, link time(elevation angle greater than 20°) is 396 seconds, and the maximum elevation angle is 86°. Fig.21.16 shows the free space loss over this pass at both uplink(28.25 GHz) and downlink(18.45 GHz) frequencies. The abscissa indicates the time elapse as satellite flies by, and is marked by the elevation angle observed from the ground terminal.

The uplink and downlink rain loss over the same pass are shown in Fig.21.17 at the rain rate of 2 mm/hr., and in Fig.21.18 at the rain rate of 5 mm/hr. It is observed that when the rain rate is 2 mm/hr., the DAH prediction is the highest among all three models, and the ITU-R prediction has the least variation over all elevation angles. For the uplink(downlink), the ITU-R prediction is lower than the Crane prediction when the elevation is lower than 35°(25°), and vice versa when the elevation is higher than 35°(25°). When the rain rate is increased to 5

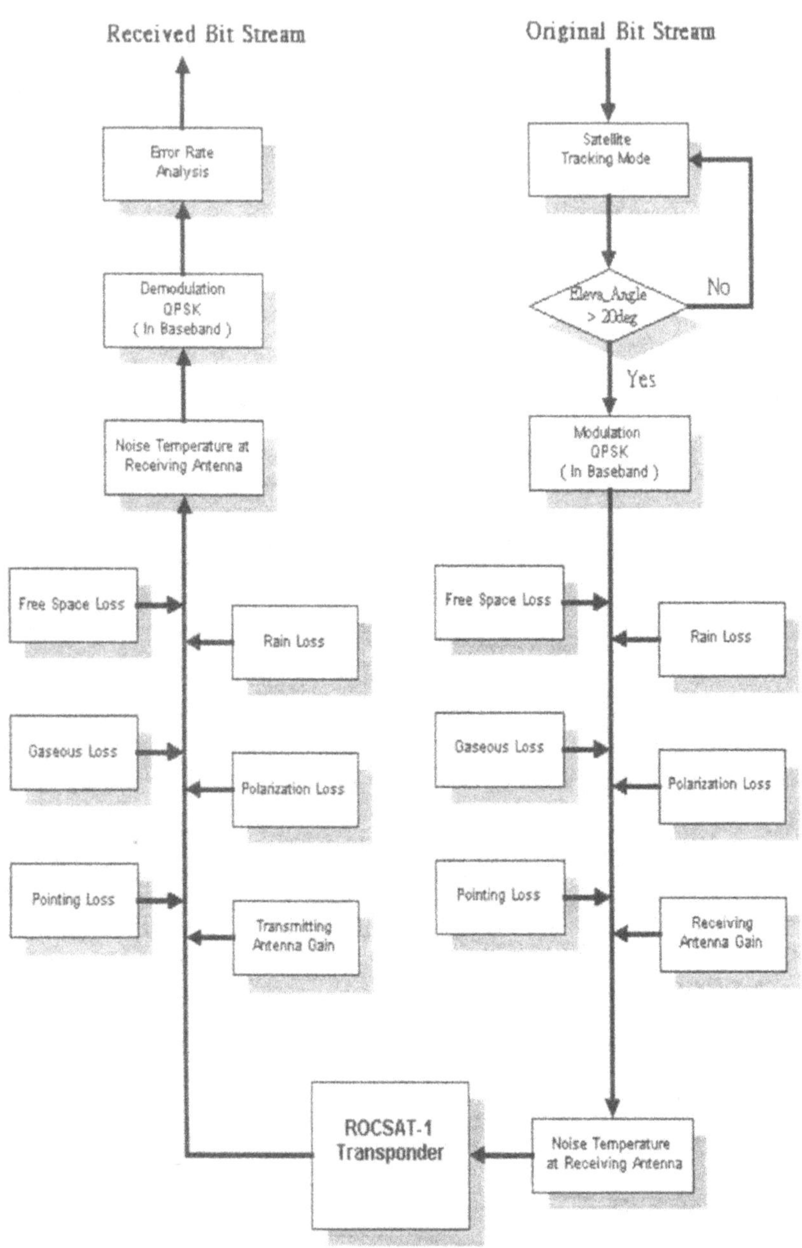

Figure 21.15. Simulation procedure of ROCSAT-1 communication experiment.

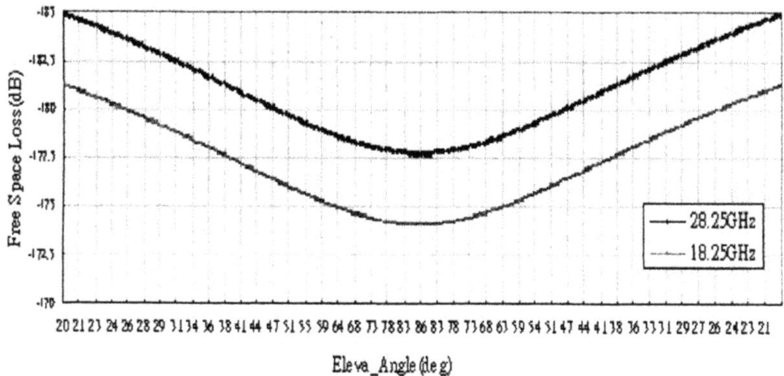

Figure 21.16. Free space loss over a specific pass, $h = 600$ km, $\alpha = 35°$.

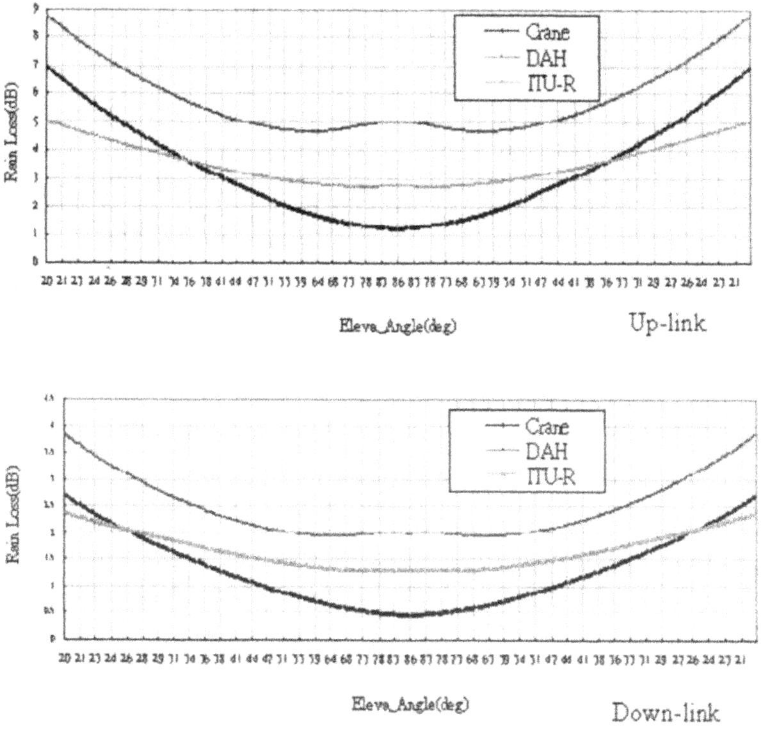

Figure 21.17. Rain loss at rain rate of 2 mm/hr. over the same pass as in Fig.21.16.

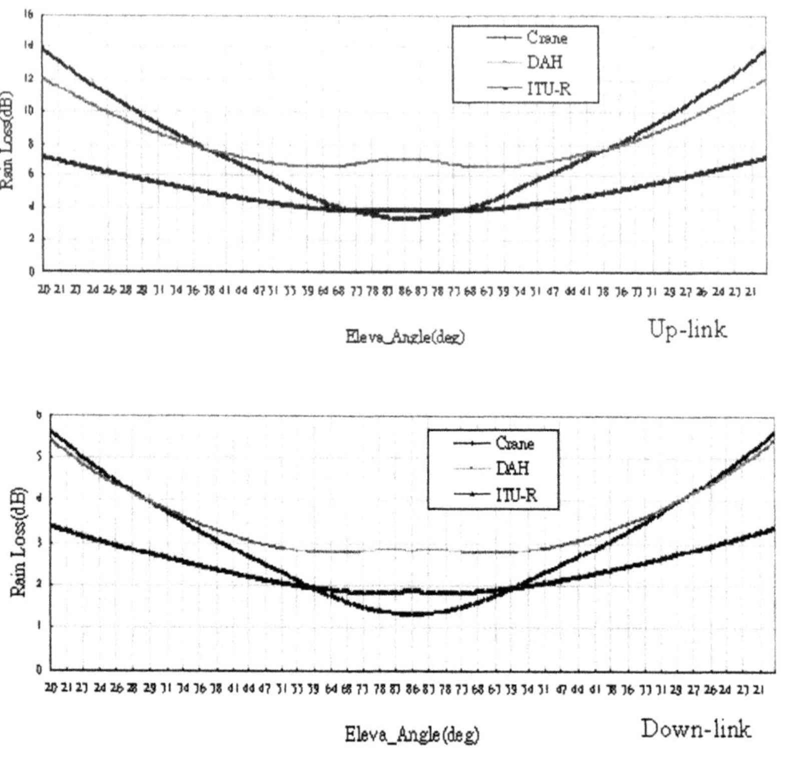

Figure 21.18. Rain loss at rain rate of 5 mm/hr. over the same pass as in Fig.21.16.

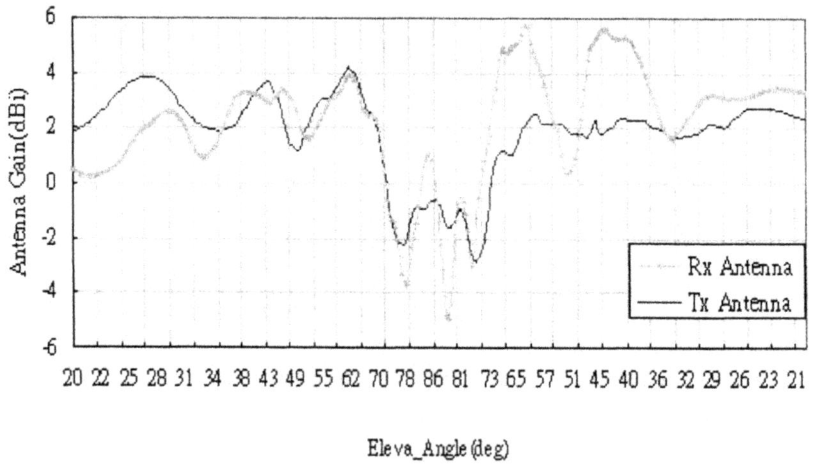

Figure 21.19. Gains of ROCSAT-1 antennas over the same pass as in Fig.21.16.

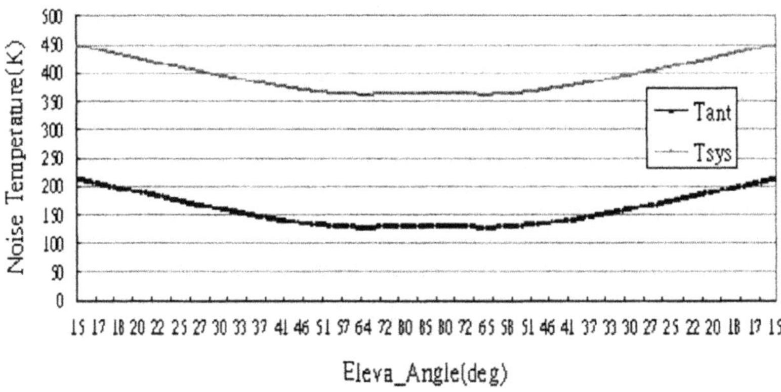

Figure 21.20. Noise temperature over the same pass as in Fig.21.16.

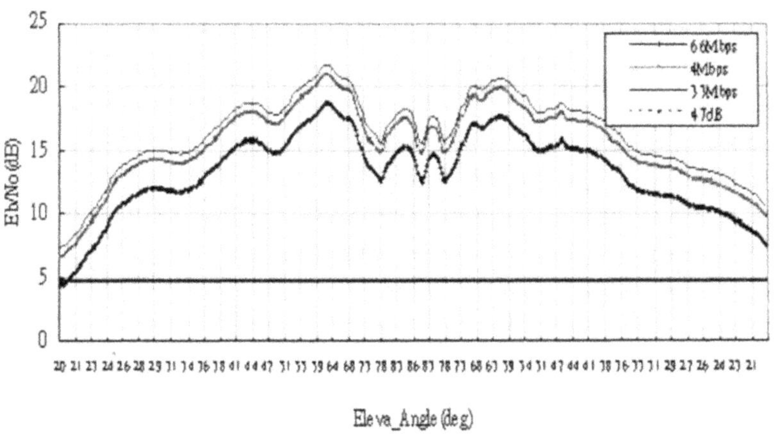

Figure 21.21. E_b/N_0 over the same pass as in Fig.21.16.

mm/hr., the DAH prediction and Crane prediction become closer at low elevation angles. Upper and lower bounds of prediction can be deduced from these three models.

The gaseous loss accompanying rainfall is relatively negligible. Simulation shows that the gaseous loss at an elevation angle of 20° is 0.3 dB at the rain rate of 2 mm/hr, and is 0.64 dB at the rain rate of 53 mm/hr.

As the satellite flies by the ground terminal, cone angle and clock angle of the satellite antenna patterns change quickly, especially at high elevation angles. Radiation patterns of the satellite antennas measured

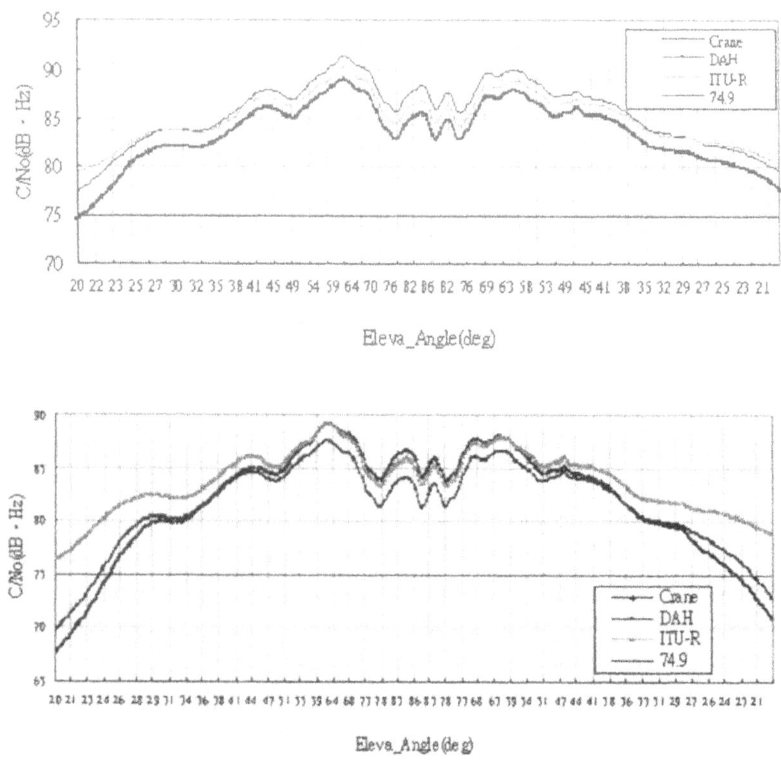

Figure 21.22. C/N_0 over the same pass as in Fig.21.16.

at predetermined directions are recorded, and interpolation is used to calculate the antenna gain at any direction. Fig.21.19 shows the antenna gain pointing toward the ground terminal as the satellite flies by. Ripples are observed because of blockage and imperfections of the antennas as mentioned in the last Section.

Fig.21.20 shows the noise temperature over the same pass. Assume that the amplifier noise temperature is 235.39 °K, and reference temperature of the cold rain cloud is 273 °K.

To achieve the transmission data rate of 6.6 Mbps with an average bit error rate of $10^{-3}(E_b/N_0 = 6.7 \text{ dB})$, C/N_0 is required to be at least 74.9 $\text{dB}_{\text{H}_\text{z}}$. Fig.21.21 shows the variation of E_b/N_0 over the same pass. It is observed that the margin above threshold is wider at higher elevation angles. Lowering the transmission data rate increases bit energy, hence increases the margin above threshold.

In raining weather, the C/N_0 can be lower than threshold, especially at low elevation angles. Fig.21.22 shows the predictions by using the

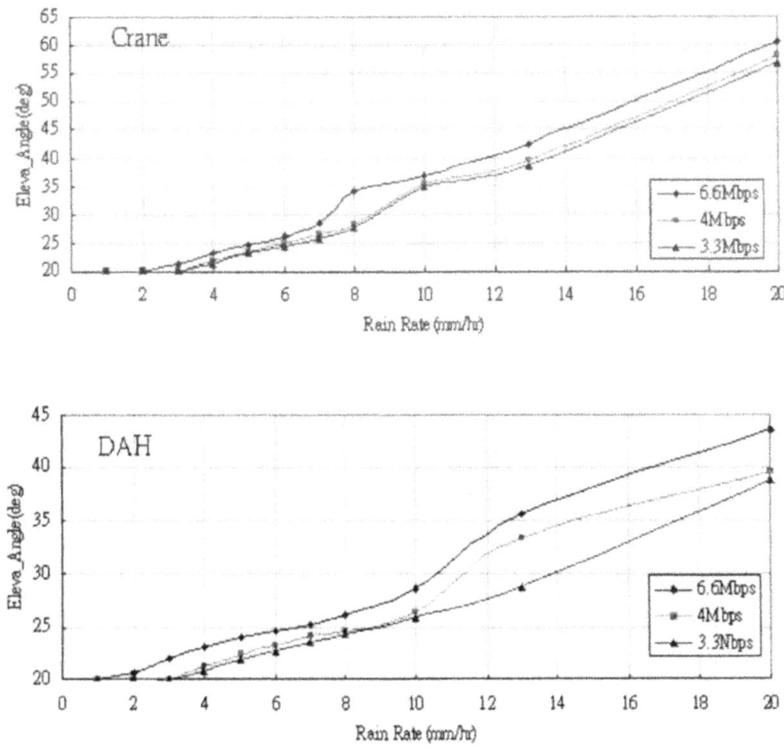

Figure 21.23. Minimum starting elevation angle as a function of rain rate.

three rain models. Since the transmitting power of transponder is constrained, one can either reduce the transmission data rate or use only portion of the link with greater elevation angles at which C/N_0 is higher, to keep the corresponding E_b/N_0 above threshold.

Fig.21.23 shows the minimum elevation angles as a function of rain rate at which C/N_0 is above threshold. It is observed that the Crane prediction is more conservative than the DAH prediction.

6. Conclusions

A channel simulation program for Ka-band has been developed to study characteristics of the communication link between ground terminal and the ROCSAT-1. The effects of orbital parameters, power level, antenna radiation patterns, rain attenuation, transmission data rate and internal noise source on the carrier-to-noise ratio are studied. The simulation results can be used to assess the communication performance of ROCSAT-1 and other LEO satellites in flight.

References

[1] G. D. Gordon and W. L. Morgan, *Principles of Communication Satellite*, New York: John Wiley, 1993.

[2] T. T. Ha, *Digital Satellite Communications*, New York: McGraw-Hill, 1990.

[3] J. V. Evans, "Satellite system for personal communications," *Proc. IEEE*, vol.86, pp.1325-1341, July 1998.

[4] D. C. Hogg and T.-S. Chu, "The role of rain in satellite communications," *Proc. IEEE*, vol.63, pp.1308-1331, Sept. 1975.

[5] T. Oguchi, "Electromagnetic wave propagation and scattering in rain and other hydrometeors," *Proc. IEEE*, vol.71, pp.1029-1078, Sept. 1983.

[6] T. Manabe, T. Ihara, J. Awaka, and Y. Furuhama, "The relationship of raindrop-size distribution to attenuations experienced at 50, 80, 140, and 240 GHz," *IEEE Trans. Antennas Propagat.*, vol.35, pp.1326-1330, Nov. 1987.

[7] M. Sekine, C.-D. Chen, and T. Musha, "Rain attenuation from log-normal and Weibull raindrop-size distribution," *IEEE Trans. Antennas Propagat.*, vol.35, pp.358-359, Mar. 1987.

[8] J. A. G. Lopez, J. M. Hernando, and J. M. Selga, "Simple rain attenuation prediction method for satellite radio links," *IEEE Trans. Antennas Propagat.*, vol.36, pp.444-448, Mar. 1988.

[9] R. L. Olsen, D. V. Rogers, and D. B. Hodge, "The a relation in the calculation of rain attenuation," *IEEE Trans. Antennas Propagat.*, vol.26, pp.318-328, Mar. 1987.

[10] R. K. Crane, "Prediction of attenuation by rain," *IEEE Trans. Commun.*, vol.28, pp.1717-1733, Sept. 1980.

[11] ITU Recom. 838, ITU, Geneva, 1992.

[12] R. K. Crene and A. W. Dissanayake, "ACTS propagation experiment: Attenuation distribution observations and prediction model comparisons," *Proc. IEEE*, vol.85, pp.879-892, June 1997.

[13] A. Dissanayake, J. Allnutt, and F. Haidara, "A prediction model that combines rain attenuation and other propagation impairment along earth-satellite paths," *IEEE Trans. Antennas Propagat.*, vol.45, pp.1564-1558, Oct. 1997.

[14] D. G. Sweeney and C. W. Bostian, "The dynamics of rain-induced fades," *IEEE Trans. Antennas Propagat.*, vol.40, pp.275-278, Mar. 1992.

[15] R. R. Bate, D. D. Mueller, and J. E. White, *Fundamentals of Astrodynamics*, New York: McGraw-Hill, 1971.

[16] E. R. Westwater, J. B. Snider, and M. J. Falls, "Ground-based radinmetric observations of atmospheric emission and attenuation at 20.6, 31.65, and 90.0 GHz: A comparison of measurements and theory," *IEEE Trans. Antennas Propagat.*, vol.38, pp.1569-1580, Oct. 1990.

[17] T. J. Moulsley and E. Vilar, "Experimental and theoretical statistics of microwave amplitude scintillations on satellite down-links," *IEEE Trans. Antennas Propagat.*, vol.30, no.6, pp.1099-1106, Nov. 1982.

[18] D. C. Cox, H. W. Arnold, and H. H. Hoffman, "Observations of cloud-produced amplitude scintillation on 19- and 28-GHz earth-space paths," *Radio Sci.*, vol.16, no.5, pp.885-907, Sept.-Oct. 1981.

[19] W. Choosaeng, S. Kitsadawanich, and N. Hemmakorn, "C-band scintillation satellite signal effected by electron density irregularities in ionospheric atomosphere," *Asia-Pacific Microwave Conf.*, pp.949-952, 1998.

[20] Y. Karasawa, M. Yamada, and J. E. Allnutt, "A new prediction method for tropospheric scintillation on earth-space paths," *IEEE Trans. Antennas Propagat.*, vol.36, no.11, pp.1608-1614, Nov. 1988.

[21] M. M. B. M. Yusoff, N. Sengupta, C. Alder, I. A. Glover, P. A. Watson, R. G. Howell, and D. L. Bryant, "Evidence for the presence of turbulent attenuation on low-elevation angle earth-space paths - part I : Comparison of CCIR recommendation and scintillation observations on a 3.3° path," *IEEE Trans. Antennas Propagat.*, vol.45, no.1, pp.73-84, Jan. 1997.

[22] J. K. Tervoneon, M. M. J. L. van de Kamp, and E. T. Salonen, "Prediction model for the diurnal behavior of the tropospheric scintillation variance," *IEEE Trans. Antennas Propagat.*, vol.46, no.9, pp.1372-1378, Sept. 1998.

[23] R. E. Ziemer and W. H. Tranter, *Principles of Communications: Systems, Modulations, and Noise*, Boston: Houghton Mifflin, 1988.

[24] CCIR Rep. 564-4, ITU, Geneva, 1992.

[25] M.-S. Alouini, S. A. Borgsmiller, and P. G. Steffes, "Channel characterization and modeling for Ka-band very small aperture terminals," *Proc. IEEE*, vol.85, pp.981-997, June 1997.

Chapter 22

CHANNEL MODELING FOR LAND-MOBILE COMMUNICATION SYSTEMS

Jenn-Hwan Tarng
Department of Communication Engineering
National Chiao-Tung University
Hsin-Chu, Taiwan, ROC

Abstract In this Chapter, models characterizing various propagation properties such as path loss, signal amplitude fluctuation(shadow and small-scale fadings), time and angular dispersions are reviewed. These models include empirical models, statistical models, deterministic physical models and hybrid physical-statistical models. Each model is suitable for describing certain propagation or radio channel for land-mobile communication systems.

Keywords: radio channel model, path loss model, fading statistics, channel impulse response, spatio-temporal channel model, ray tracing technique.

1. Introduction

One of the major goals to develop propagation and radio channel models is to support the development of mobile communication systems. Propagation models for the first generation analog systems mainly describe power and Doppler characteristics of the radio signal. For the second generation digital systems, both multipath channel impulse response(CIR) and Doppler shift are considered. To improve frequency efficiency, development of spatial channel model is needed. Currently, there is demand for models that can provide spatial and temporal information to analyze and design relevant systems such as smart antenna systems. Also, ultrawideband(UWB) channel models are desired since UWB radios are gaining more attention recently.

2. Types of Propagation Models

Radio propagation prediction for land-mobile communication is a highly complicated electromagnetic problem because there are many objects like buildings, trees and vehicles in the environment. Reflection, transmission, diffraction and scattering are four major propagation mechanisms. The first two mechanisms occur when radio wave is incident upon a smooth dielectric surface(transmission does not occur on metallic surface) which is electrically large. Diffraction may occur when wave is incident on an edge or a vertex, or is grazing on to a curved surface. Rayleigh scattering occurs when wave is incident on particles with size much smaller than the incident wavelength. Wave may also be scattered by rough surfaces, small objects or other irregularities. With these propagation mechanisms, duplicated versions of the signal most often arrive via more than one path. Interference among these multipath waves leads to rapid fluctuation of the received signal amplitude, which is known as multipath fading.

Complicated propagation envircnment(random in some aspects) makes it very difficult to predict the received signal strength with accuracy. Different models have been developed to increase the prediction accuracy, including physical models, empirical models, statistical models and physical-statistical models [1].

Physical models are deterministic methods to predict path loss, time and angle dispersions, and so on. Knowledge in electromagnetic theory and physical mechanism are required to build a physical model. Empirical models are constructed by fitting appropriate functional form to measurement data through regression. Statistical models are most popular, in which theoretical or empirical approaches can be incorporated.

3. Empirical Models for Median Path Loss

Although an empirical model is relatively easy to use, it may not be applicable when conditions are different from those on which it was established. Usually, median path loss is predicted, which is the average of many samples taken over small area(20-40 wavelengths) to filter out small-scale fading(fast fading) effects.

3.1 Overview of Empirical Models

Okumura's model is widely used for median path loss prediction in macrocellular systems [2]. The model is applicable in the frequency range of 150 to 1,920 MHz and over the propagation range of 1 to 100 km. The base station antenna height ranges from 30 to 1,000 m. This model is based on measurement data collected in Tokyo, Japan and its surrounding suburban areas. Okumura et al. developed a set of curves for median path loss in excess of free-space loss. These curves have been approximated by a set of formulas proposed by Hata [1]. Okumura's predictions have been found useful in many cases, particularly in suburban areas [3]. However, other measurements have shown disagreement with these predictions. Fig.22.1 shows that the Okumura-Hata model overestimates the path loss in Taipei. Major cause of error is the environment difference between Taipei and Tokyo.

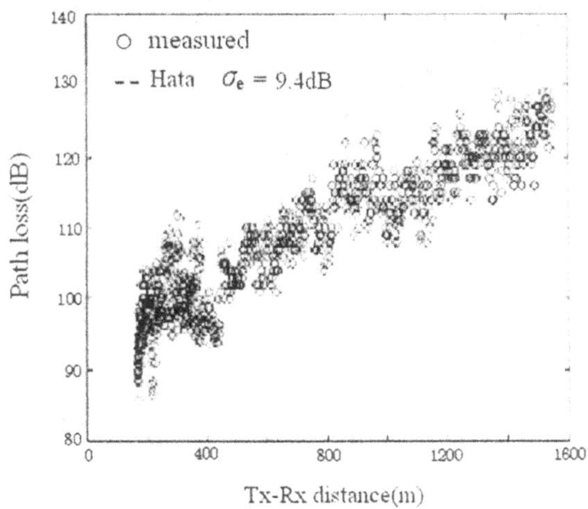

Figure 22.1. Measured and predicted path losses at 2.44 GHz as a function of distance. Measurement is along Chinan Road, standard deviation of prediction error is 9.4 dB.

Lee's model is a power law model with parameters estimated from measurements in a number of locations, together with a procedure for calculating effective antenna height of base station, where terrain variation is considered [4].

3.2 Environmental Classification

In developing empirical models, it is important to properly categorize the environment where the systems are operating. For example, the Okumura-Hata model applies four categories: large cities, medium and small cities, suburban areas and open areas. The difference among them may be ambiguous. The Ibrahim and Parsons model applies a numerical approach to categorize terrains [5]. Nevertheless, there is no guarantee that there is a one-to-one correspondence between specific propagation characteristics and category. In order to find more appropriate parameters for the model to yield better prediction, radio propagation models based on physics are preferred for mobile systems.

4. Ray Tracing-Based Site-Specific Models

In the past few years, site-specific propagation models have been proposed and widely used in studying radio propagation in indoor and outdoor environments [6]. Radio channel is modeled in a deterministic way by using site-specific information such as geographical/architecture database. Geometrical and/or physical optics are applied in modeling reflection, diffraction and scattering effects at high frequencies because they are easy to implement with ray tracing techniques and are more numerically efficient than low frequency techniques.

4.1 Site-Specific Propagation Models

The received field can be determined by using ray tracing techniques, and the contribution by the ith ray can be expressed as

$$\bar{E}_i = \bar{E}_o \cdot G_{ti} G_{ri} L_i(d) L_D(\phi_i) \prod_j \bar{\bar{D}}_{ji} \cdot \prod_k \bar{\bar{\Gamma}}_{ki}$$

where $\bar{E}_o$ is the electric field at 1 m away from the transmitting antenna, G_{ti} and G_{ri} are the radiation patterns of transmitting and receiving antennas, respectively, $L_i(d)$ and $L_D(\phi_i)$ are path losses of the ith ray with path length d and incident angle ϕ_i, respectively, $\bar{\bar{D}}_{ji}$ and $\bar{\bar{\Gamma}}_{ki}$ are the jth diffraction coefficient and kth reflection coefficient, respectively, of the ith ray.

4.2 Multiple Diffractions over Obstacles

For macrocellular systems operating in an urban area, multiple diffractions over rooftops may be the dominant propagation mechanism. Saunders proposed a flat edge model to simplify the situation by assuming that all the buildings are of equal height and spacing [7]. This formulation is applicable with any elevation angle and many buildings. Walfisch-Bertoni model is the first one to actually demonstrate that multiple diffractions by buildings can account for various path lengths which are observed in measurements [8]. In this model, numerical computation of Kirchhoff-Huygens integral is used and a power-law formula is fitted to the field over rooftop. In the COST231 project [9], Walfisch-Bertoni model is combined with Ikegami model [10] to predict the fields down to the street level, and some empirical correction factors are used to match with measurements. In [11], ray transmission matrix(RTM) method is proposed to effectively incorporate multiple diffractions due to major and secondary obstacles in the vertical transmitter-receiver plane. Given the transmission loss of each obstacle, total field at the receiver can be derived by application of RTM method to calculate the path losses of all relevant rays.

4.3 Ray Tracing Techniques

Ray tracing techniques can be categorized into two groups: techniques based on shooting and bouncing rays(SBR) method [12] and techniques based on image theory. The core of ray tracing technique is the algorithm to determine intersections of rays with environmental objects, which will be repeatedly used in simulation.

In a typical cellular environment, hundreds of objects such as buildings can be modeled by facets and edges. In applying the intersection algorithm, a facet of arbitrary shape is first decomposed into planar polygons, then ray-plane intersection is calculated to check whether the intersection point is inside or outside a polygon. In real environment, the number of ray traces is very large. Some techniques are generally used to enhance the tracing efficiency, including binary space partitioning(BSP), space volumetric partitioning(SVP) and angular Z-buffer(AZB) [13]. The former one is suitable for propagation model based on image theory and the latter two are appropriate for both SBR method and image theory.

However, even with ray tracing techniques, site-specific models are not easy to implement due to the lack of detailed terrain and building database and intensive computational load for three-dimensional problems. Some approaches may be applied to overcome these difficulties. For example, one may develop an effective and efficient procedure for generating terrain and building data, incorporate a statistical approach to describe the detailed effects of local scattering, and develop ray tracing accelerating algorithms.

4.4 Model Validation

Fig.22.2 shows the ray trajectories of signal paths using ray tracing technique. Direct propagation, single reflection, corner diffraction and multiple roof diffractions are considered in delineating these ray trajectories. Fig.22.3 shows the comparison between path losses predicted by site-specific propagation model and measured data along Linsen-Nan Road, Taipei, at the frequency of 2.44 GHz. Standard deviation between measurements and predicted results is equal to 5.39 dB. Measured and predicted delay-azimuth power spectra are shown in Figs.22.4 and 22.5, respectively. Most of the signal paths are clustered and the site-specific model does not make accurate prediction for clusters C, E and F.

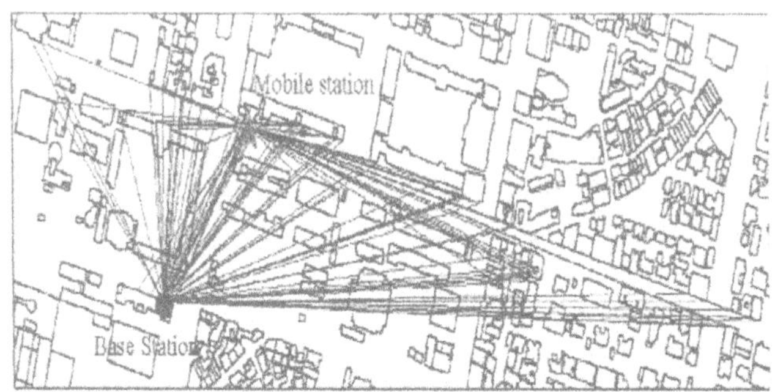

Figure 22.2. Ray trajectories of signal paths in an urban area(Chinan Road, Taipei) predicted by using ray tracing technique. The circle and star represent base and mobile stations, respectively.

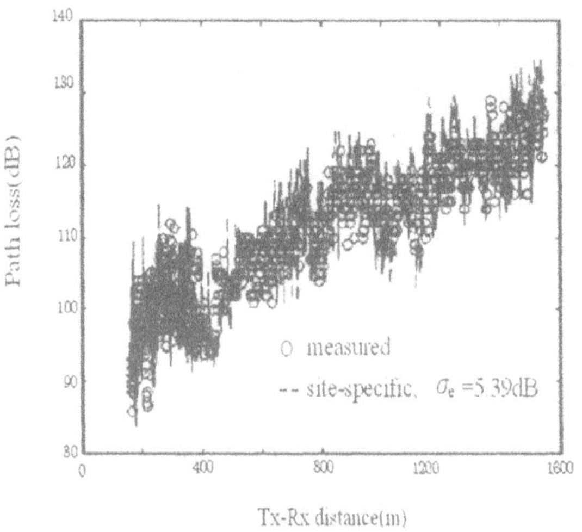

Figure 22.3. Measured and predicted path losses at 2.44 GHz as a function of distance. The path loss is measured along Chinan Road, Taipei.

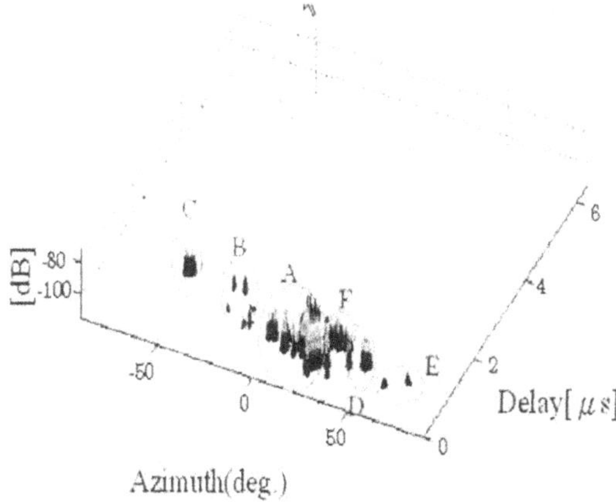

Figure 22.4. Measured delay-azimuth power spectrum along Chinan Road, Taipei.

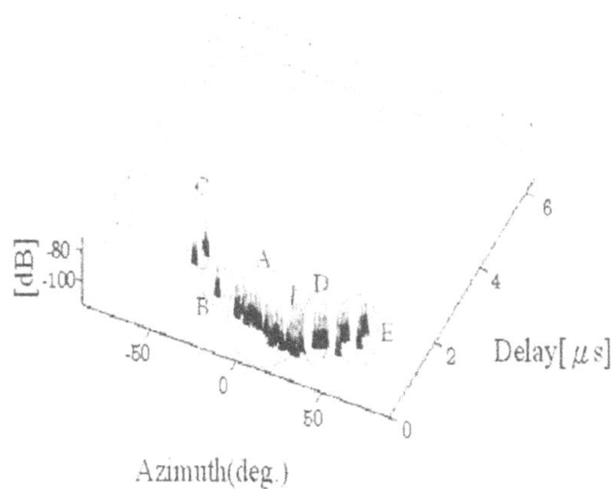

Figure 22.5. Delay-azimuth power spectrum predicted by site-specific propagation model. The signal paths that have similar propagation characteristics arrive in groups, which are labeled from A to F.

5. Statistical Channel Impulse Response Models

Turin models a multipath channel as linear filter with complex-valued impulse response which can be expressed as [14]

$$h(t,\tau) = \sum_{i=0}^{N(t)-1} a_i(t,\tau)e^{j\theta_i(t,\tau)}\delta(t-\tau_i(t)) \tag{22.1}$$

where $a_i(t,\tau)$ and $\tau_i(t)$ are amplitude and excess delay, respectively, of the ith multipath component at time t. The phase term, $\theta_i(t,\tau)$, represents the phase shift due to direct propagation, reflection, diffraction or scattering of the ith multipath component. Usually, $\theta_i(t,\tau)$ is uniformly distributed in $[0, 2\pi]$, N is the total number of multipath components, and $\delta(t - \tau_i(t))$ is the unit impulse function that represents a specific multipath component with an excess delay of $\tau_i(t)$.

These parameters can be expressed in analytical forms using statistical theory. Measurements can be carried out to determine these parameters.

5.1 Distribution of Arrival Time Sequence

A Poisson process is usually used to describe the arrival time, $\tau_i + \tau_o$, with τ_o is the delay time of the first arriving signal. It is assumed that the objects causing multipath propagation delay are located randomly. However, hypothesis made in the Poisson model does not comply with the observation by Turin. Hence, an alternative second-order model, the Δ-K model, is proposed which considers the fact that echoes may arrive in clusters from closely spaced reflecting or scattering objects [14].

The Δ-K model defines two states, S_1 and S_2. The mean arrival rates of echoes are $\lambda_n(t)$ and $K\lambda_n(t)$ in states S_1 and S_2, respectively. The value of K determines whether the echoes cluster together or spread out, and Δ is called delay bin. The model is illustrated in Fig.22.6. The probability that an echo will occur in the next Δ seconds increases if $K > 1$, and vice versa if $K < 1$. The probability that an echo exists in bin 1 is λ_1, hence the probability of no echo is $(1 - \lambda_1)$. If an echo exists in the $(j-1)$st bin, then the probability of an echo in the jth bin is $K\lambda_j$. For example, the probability of having echoes in bins 1, 2 and 4 is $\lambda_1 K\lambda_2(1 - K\lambda_3)\lambda_4$. The underlying probabilities need to be determined from the measurement data.

Fig.22.7 shows values of λ_j over a metropolitan area(Taipei), a medium city(Chung-Li, Taiwan) and suburban of Chung-Li. It is found that the optimal value of λ_j is distributed within a similar range as that extracted by Turin, and K increases as the building density increases.

Hashemi refined the Δ-K model by modifying the branching process of Fig.22.6 to take into account spatial correlation of the arrival times [15]. It is suggested that when there is an echo in the ith bin of a measured sequence, the probability that an echo occurs in the ith bin of the next measured sequence is increased, and vice versa.

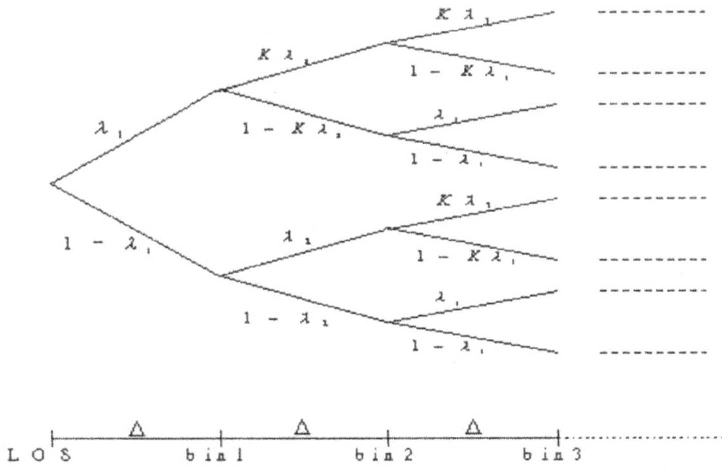

Figure 22.6. Δ-K model.

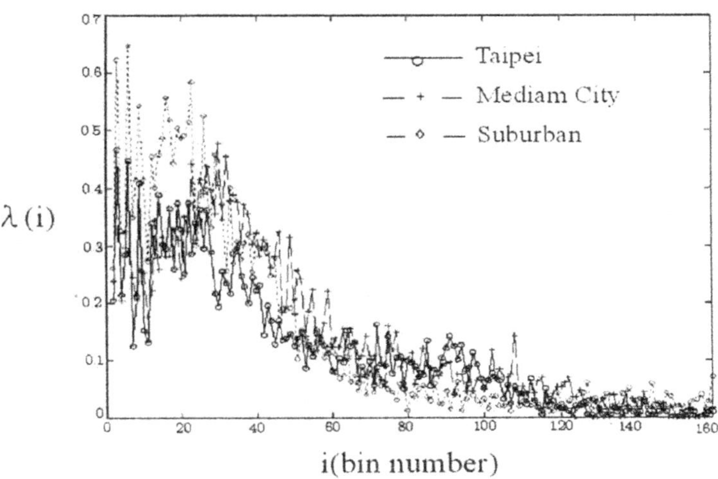

Figure 22.7. $\lambda(i)$ for Taipei($K = 3.2$), a medium city($K = 2.5$) and suburban($K = 1.8$).

5.2 Path Amplitude Distribution

Turin et al. proposed that the signal amplitude over large areas follows a lognormal distribution given by [14]

$$p(r) = \frac{1}{\sqrt{2\pi}\sigma r} e^{-(\ln r - \mu)^2 / 2\sigma^2}, \qquad r > 0$$

where μ and σ are local mean and standard deviation, respectively. This is a generally accepted model for describing shadowing effects in many locations.

Suzuki proposed an amplitude distribution of near line-of-sight(LOS) paths over local areas, which follows Nakagami distribution as [16]

$$p(r) = \frac{2}{\Gamma(m)} \left(\frac{m}{\Omega}\right)^m r^{2m-1} e^{-mr^2/\Omega}, \qquad m \geq 1/2, r \geq 0$$

where $\Gamma(m)$ is the Gamma function, $\Omega = E(r^2)$, and $m = \Omega^2/\sigma^2$ is the shape factor. This formula is reasonably accurate to describe envelope fading in multipath environment. It can closely approximate Rician distribution and reduces to Rayleigh distribution when $m = 1$.

6. Spatio-Temporal Channel Models

To increase system capacity of 3G and B3G communication systems, smart antenna systems and space division multiple access(SDMA) technique have been used to increase signal gain and reduce interference. Hence, spatio-temporal channel models are needed to study space-time channel characteristics such as temporal and azimuth dispersions, which is helpful in developing associated adaptive signal processing techniques. Many spatio-temporal models have been developed based on either geometrical theories or measurements. Overview of these models is given in [17], and Table 22.1 summarizes some typical models.

6.1 Urban Environment

Next, a hybrid(physical-statistical) spatio-temporal channel model for macrocellular urban environments is presented and the concept behind Rayleigh's model is applied. It combines a site-specific(deterministic) propagation model with a stochastic model. The former model is used to calculate the direct wave, reflected waves and diffracted waves incurred by dominant reflectors and objects like buildings. The statistical model incorporates scattering effect caused by randomly located scatterers such as pedestrians, vehicles, trees, street lamps and effective surface roughness of dominant buildings.

Note that physical-statistical model is a hybrid model which takes advantages of both statistical and physical models. Its predictions can be conducted point by point, yet not being associated with specific locations. A physical-statistical model was first proposed by Saunders et

Table 22.1. Summary of typical spatio-temporal models.

Spatio-temporal models	Scatterer distribution model	Shape of scatterer distribution region	Output parameters
Lee's model	using effective scatterers to describe local scatterers	a circular ring	discrete AOA, Doppler shift
Geometrically based circular model	uniformly distributed local scatterers	a circle	AOA, TOA, joint TOA and AOA, Doppler shift and signal amplitude
Gaussian wide sense stationary uncorrelated scattering (GWSSUS)	scatters are grouped into clusters	number and location of scattering clusters are not indicated	a general result of covariance matrix
Time-varying vector channel model (Rayleigh model)	local scatterers plus dominant reflections	positions of dominant reflectors may be site-specific	AOA, TOA, joint TOA and AOA, small-scale fading
Measurement-based model			AOA, TOA
Ray tracing techniques	deterministic model based on electromagnetic and geometrical theories		AOA, TOA, and signal amplitude

al. [18] to estimate the shadowing factor of Lutz model which is used for narrow band satellite-mobile communication [19].

Fig.22.8 shows a typical scatterer distribution. Based on (22.1), the spatio-temporal vector channel impulse response(VCIR) can be expressed as [20]

$$\bar{h}(t,\tau) = \sum_{i=0}^{N(t)-1} \bar{v}(\phi_i)a_i(t,\tau)e^{j\theta_i(t,\tau)}\delta(t-\tau_i(t))$$

where $\bar{v}(\phi_i)$ is the steering vector which is a function of array geometry and angle-of-arrival of the ith signal. Site-specific model includes the mechanisms of LOS propagation, reflections and/or diffractions by dominant reflectors and/or scatterers.

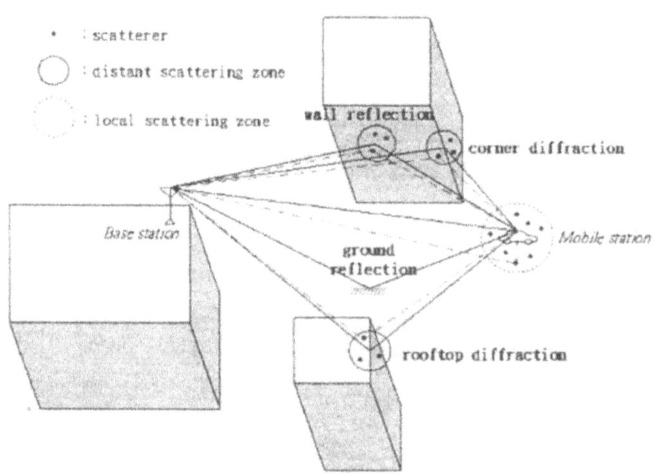

Figure 22.8. Typical scatterer distribution. Distant scattering zones are effective scattering rough surface of dominant buildings.

Local scatterers may be used to model the clustering effect of dominant reflectors. Both distant and local scatterers are described by using geometrically based circular model, namely, they are uniformly distributed within a circle with its center on the surface of dominant buildings or at the mobile station. The radius of circle is called the effective scattering radius. Note that scattering effect due to scatterers around the base station is neglected because the base station in macrocellular environment is usually located on roofs with very few large objects near by. Size of the effective scattering zone and the number of effective scatterers are usually determined by empirical methods.

6.2 Validation of Hybrid Spatio-Temporal Models

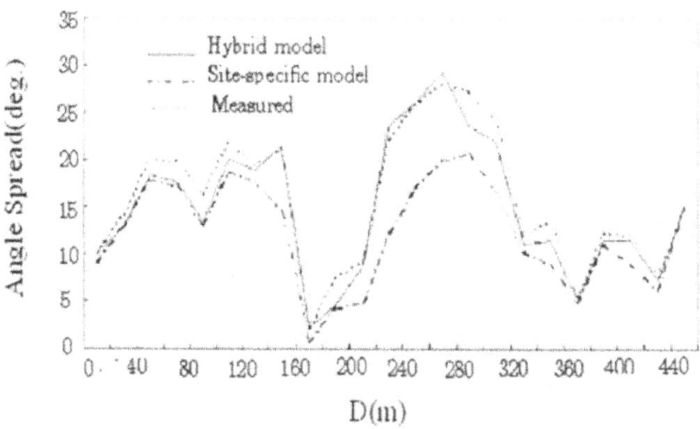

Figure 22.9. Measured and predicted angle spreads along Chinan Road in Taipei, —: hybrid model, – · –: site-specific model, · · ·: measurement.

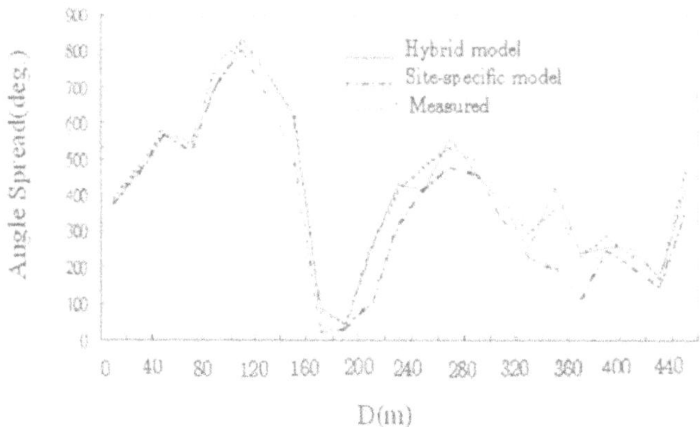

Figure 22.10. Measured and predicted delay spreads along Chinan Road in Taipei, —: hybrid model, – · –: site-specific model, · · ·: measurement.

RUSK vector channel sounder is used to measure the TOAs and AOAs of received signals in multipath environments [21]. The base station is equipped with an eight-element antenna array and the mobile station transmits a 2.44 GHz analog signal with bandwidth of 120

MHz. The measurement is carried out in Taipei along several streets. Fig.22.9 shows the comparison between measured and predicted root-mean-square(rms) angle spread along one of the streets. It is found that the hybrid model yields higher prediction accuracy than site-specific model because the former includes important contribution of the randomly distributed scatterers. Comparison between measured and predicted rms delay spread along the same path is shown in Fig.22.10.

7. Future Developments

Although radio wave propagation is governed by Maxwell's equations, its applications to practical mobile or wireless systems are still emerging. Challenges exist for modeling radio propagation. The aforementioned Δ-K model and ray tracing techniques can be applied to indoor channels. Propagation coefficients such as scattering factor, reflection and transmission coefficients need further studies to be extended to UWB radios. Amplitude fading statistics need to be examined because the number of overlapping multipaths in a time bin decreases in UWB systems. The size, shape and spacing of scatterers are of interest for developing spatio-temporal channel models which incorporate scattering clusters. Channel parameters can be updated by measurements to cover a wider range of operation conditions and frequency bands. Time-varying channel modeling and characterization require further studies.

References

[1] J. Lavergnat and M. Sylvain, *Radio Wave Propagation, Principles and Techniques*, John Wiley, 2000.

[2] M. Hata, "Empirical formula for propagation loss in land mobile radio services," *IEEE Trans. Veh. Technol.*, vol.29, pp.317-325, 1980.

[3] "Digital land mobile radio communications," Final Rep., COST 207 management committee, Commission of European Communities, L-2920, Luxembourg, 1989.

[4] W. C. Lee, *Mobile Design Fundamentals*, John Wiley, 1993.

[5] M. F. Ibrahim and J. D. Parsons, "Signal strength prediction in built-up areas," *Proc. Inst. Elect. Eng.*, vol.130F, no.5, pp.377-384, 1983.

[6] T. S. Rappaport, *Wireless Communications Principles and Practice*, New Jersey: Prentice Hall, 1996.

[7] S. R. Saunders and F. R. Bonar, "Explicit multiple building diffraction attenuation function for mobile radio wave propagation," *Electron. Lett.*, vol.27, no.14, pp.1276-1277, 1991.

[8] J. Walfisch and H. L. Bertoni, "A theoretical model of UHF propagation in urban environments," *IEEE Trans. Antennas Propagat.*, vol.36, no.12, pp.1788-1796, 1988.

[9] "Digital mobile radio: COST 231 view on the evolution towards 3rd generation systems," Final Rep., COST 231, Commission of European Communities and COST Telecommun., Brussels, 1999.

[10] F. Ikegami, T. Takeuchi, and S. Yoshida, "Theoretical prediction of mean field strength for urban mobile radio," *IEEE Trans. Antennas Propagat.*, vol.39, no.3, pp.299-302.

[11] M. Lebherz, W. Wiesbeck, and W. Krank, "A versatile wave propagation model for the VHF/UHF range considering three-dimensional terrain," *IEEE Trans. Antennas Propagat.*, vol.40, pp.1121-1131, Oct. 1992.

[12] A. S. Glassner, *An Introduction to Ray Tracing*, ed., San Diego, CA: Academic Press, 1989.

[13] M. F. Catedra and J. Perez-Arriaga, *Cell Planning for Wireless Communications*, Boston: Artech House, 1999.

[14] G. L. Turin, F. D. Clapp, T. L. Johnston, S. B. Fine and D. Lavry, "A statistical model of urban multi-path propagation," *IEEE Trans. Veh. Technol.*, vol.21, pp.1-9, Feb. 1972.

[15] H. Hashemi, "Simulation of the urban radio propagation channel," *IEEE Trans. Veh. Technol.*, vol.28, pp.213-225, Aug. 1979.

[16] H. Suzuki, "A statistical model for urban radio propagation," *IEEE Trans. Commun.*, vol.25, pp.673-680, July 1977.

[17] R. B. Ertel, P. Dieri, K. W. Sowerby, T. S. Rappaport, and J. H. Reed, "Overview of spatial channel models for antenna array communication systems," *IEEE Pers. Commun.*, pp.10-22, 1998.

[18] S. R. Saunders and B. G. Evans, "A physical model of shadowing probability for land-mobile satellite systems," *Electron. Lett.*, vol.33, no.15, pp.1548-1549, 1996.

[19] E. Lutz, "A Markov model for correlated land mobile satellite channels," *Int. J. Satel. Commun.*, vol.14, pp.333-339, 1996.

[20] J. C. Liberti, Jr. and T. S. Rappaport, *Smart Antennas for Wireless Communications: IS-95 and Third Generation CDMA Applications*, New Jersey: Prentice Hall, 1999.

[21] R. S. Thoma, D. Hampicke, A. Richter, G. Sommerkorn, A. Schneider, U. Trautwein, and W. Wirnitzer, "Identification of time-variant directional mobile radio channels," *IEEE Trans. Instru. Measure.*, vol.49, no.2, pp.357 -364, 2000.

[22] Y. Okumura. E. Ohmori, T. Kawano, and K. Fukuda, "Field strength and its variability in VHF and UHF land mobile radio service," *Rev. Electr. Commun. Lab.*, vol.16, pp.825-873, 1968.

[23] T. S. Rappaport, S. Y. Seidel, and K. R. Schaubach, "Site-specific propagation for PCS system design," in *Wireless Personal Communications*, M. J. Feuerstein and T. S. Rappaport, ed., Boston: Kluwer Academic Publishers, pp.281-315, 1993.

[24] S. O. Rice, "Statistical properties of sine wave plus random noise," *Bell Syst. Tech. J.*, vol.27, pp.109-157, 1948.

Chapter 23

ELECTROMAGNETIC COMPATIBILITY IN MICROWAVE ENGINEERING

Jean-Fu Kiang, Tze-Hsuan Chang, and Monsen Leu
Department of Electrical Engineering and
Graduate Institute of Communication Engineering
National Taiwan University
Taipei, Taiwan, ROC

Abstract This Chapter provides an overview of electromagnetic compatibility issues in microwave engineering, especially in printed circuit boards, electromagnetic shielding, transient and its suppression, reverberation chamber and frequency management. Fundamental concepts and interference mechanisms are reviewd with practical examples wherever possible.

Keywords: electromagnetic compatibility(EMC), electromagnetic interference(EMI), signal integrity(SI), radiated emission(RE), conducted emission(CE), radiated susceptibility(RS), conducted susceptibility(CS), printed circuit board(PCB), crosstalk, simultaneous switching noise, ground bounce, delta-I noise, ground plane, differential mode(DM), common mode(CM), decoupling, parasitic, resonance, filter, isolation, interconnection, layout, loop radiation, current boundary, transfer impedance, shielding effectiveness(SE), transient, surge, spike, burst, electrostatic discharge(ESD), electrical fast transient(EFT), electromagnetic pulse(EMP), lightning, transient suppression device(TSD), reverberation chamber, spectrum management.

1. Introduction

A system is claimed to fulfill electromagnetic compatibility(EMC) requirement when it is in good function order and does not create electromagnetic interference. Rapid advancement in microwave and electronic technologies demands higher operating frequency, clock speed and component density in a system. Hence, electromagnetic interference(EMI) is more likely to occur in systems like printed circuit board(PCB), module or chip. Signal integrity(SI) has to be considered in every system design. If EMC is not well considered at early design phase, more endeavors and cost will be drawn to modify the products to comply with EMC regulations [1].

EMC problems can be categorized into interference(EMI) and susceptibility(EMS) problems, and further categorized into radiated emission(RE), conducted emission(CE), radiated susceptibility(RS) and conducted susceptibility(CS) problems. Limits on emission and susceptibility are different. The difference between actual and required levels is called margin. An emission margin can be established more easily, but a susceptibility margin is more difficult to define because different types of interference sources may cause EMS problem at different interference levels.

Interference occurs when there exists an interfering path and the magnitude of culprit noise exceeds the immunity margin of the victim susceptor. The fundamental factors in any EMI problem are frequency, amplitude, separation and timing, abbreviated as FAST. To cause an interference, frequency band of the culprit must overlap with the operating frequencies of the victim, the culprit and victim must be operating at the same time, the separation between culprit and victim is close enough, and the amplitude of the culprit noise is large enough to affect the vicitm.

Different EMI mechanisms dominate in different frequency bands. Conducted emission like crosstalk or common-impedance coupling dominates when frequency is lower than 30 MHz. Radiated emission from large objects like cables dominates when frequency is between 30 MHz and 300 MHz. Radiated emission from small objects like circuit boards or slots dominates when frequency is higher than 300 MHz. At higher clock speed, circuit layouts become more critical for EMI and EMS considerations. Reflections, crosstalk or radiation may cause interference, especially when the component density is high [1].

2. Printed Circuit Board Design

Higher-order harmonics of the clock signal usually exist in printed circuit board. Commonly used PCBs have four layers where ICs, modules and discrete components are mounted on the top and bottom surfaces, power and ground planes are embedded in the inner two layers, and vias are used as interconnections through layers [2].

General waveform in digital circuits can be approximated as a trapezoidal pulse as shown in Fig.23.1, which can be expressed in a Fourier series as

$$T(t) = \frac{U_g t_h}{2T} + \sum_{n=1}^{\infty} \left(\frac{U_g}{n\pi}\right) \Big[\mathrm{sinc}\,(n\pi\tau_r/T)\, e^{jn\pi\tau_h/T}$$
$$-\mathrm{sinc}\,(n\pi\tau_f/T)\, e^{-jn\pi\tau_h/T}\Big] \cos(n\omega t)$$

where U_g is the pulse amplitude, $\omega = 2\pi/T$ is the clock speed in radians, T is the clock period, τ_h is the half-amplitude width of the pulse, τ_r is the rise time, τ_f is the fall time, and $\mathrm{sinc}(\alpha) = \sin\alpha/\alpha$. If $\tau_r = \tau_f$, the envelope of Fourier coefficients decreases at a rate of 20 dB per decade above $f_1 = 1/\pi\tau_h$, and at 40 dB per decade above $f_2 = 1/\pi\tau_r$. The maximum frequency of practical interest is defined as $f_{\mathrm{knee}} = 1/2\tau_r$. Half of a wavelength at f_{knee} is called the length of rising edge. The low-to-high signal transition just extends over such length. Transmission line effects can be ignored when the trace length is shorter than $\lambda/10$ at the highest frequency or 1/6 of the length of rising edge. If the trace is shorter than the length of rising edge, rising or falling edge of a clock pulse may be reflected by the load back to the source and perturb the original clock pulse. With increasing clock speed, the wavelength of higher-order harmonics become comparable to the trace length, and the traces may act like antennas at those harmonics.

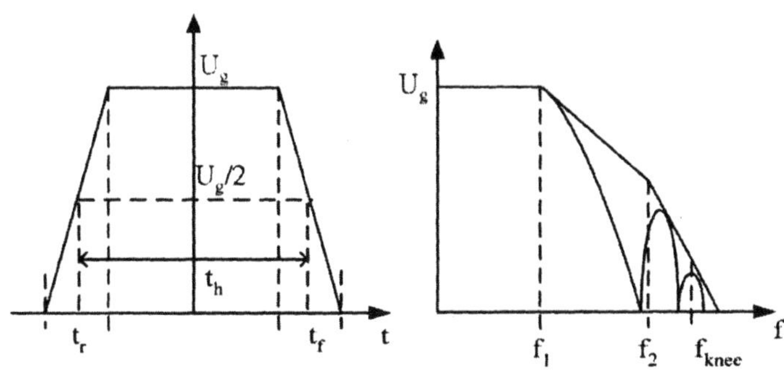

Figure 23.1. Temporal waveform and spectral coefficients of periodic trapezoidal signal.

By the year 2005, 0.1 μm CMOS process will be in manufacture, in which the maximum IC die size is about 2.8 cm $\times$ 2.8 cm with operating frequency up to 20 GHz and higher. Direct radiation from ICs is negligible below 1 GHz due to their small size compared with wavelength. At 15 GHz, the wavelength is about 2 cm in free space and about 6 mm in silicon, which is smaller than the IC package size [3], and direct radiation may no longer be ignored.

Internal EMI in a chip may come from clock pulses, fast transition current, delta-I noise, and so on. External EMI may come from cables, PCB traces, package, bonding wires, lead frame, heat sink, and so on [4].

2.1 Ground Plane

Proper grounding should be considered in the design of digital and analog circuits, which can reduce undesired effects such as ground noise, crosstalk and radiated emission. Better performance is achievable by using intact ground plane, which may be impractical due to the required layout area and the difficulty of interconnection between layers. A meshed ground plane can overcome these drawbacks, which also provides multiple return paths to reduce voltage drop across the ground plane [5].

The Ampere's law implies that wider conductor induces weaker magnetic fields. Thus, flat foil is better than round conductor for fabricating ground planes. The skin depth is inversely proportional to the square root of frequency. Hence, a thin metal sheet is enough to shield high frequency fields. However, ac resistance of conductor increases due to the reduced skin depth at higher frequencies.

Current signals on a signal-return pair can be decomposed into differential mode(DM) and common mode(CM). Signal current and its return are in opposite sense in the differential mode which is usully used in signal transmission. On the other hand, the signal current and its return are in the same sense in the common mode.

Possible radiation mechanisms include dipole-like radiation from cables attached to PCB, multiboard configurations driven by an effective noise source at the interconnection, direct radiation from components, heat sinks, traces or power bus, and so on. Two adjacent power or ground planes in a multilayered structure form a patch-like antenna. Utilizing resistive termination along board edges is expected to reduce the fringing fields and mitigate radiation from the edges of PCB. An approach is reported in [6] to mitigate EMI by connecting multiple ground planes together along the periphery of PCB with closely spaced vias.

Current driven interference is more significant than voltage driven one in PCB connectors [7]. Magnetic flux that encircles the signal-return connector can be modeled by a partial inductance. By constructing a connector that minimizes such partial inductance, current driven EMI can be reduced. Some signal traces are routed to another board, but only a limited number of pins are available for grounding. Signals unbalanced with their returns tend to cause stronger interference.

2.2 Simultaneous Switching Noise

Simultaneous switching noise is also known as ground bounce or delta-I noise. It is an undesired fluctuation of power potential(V_{CC} bounce) or ground potential(ground bounce) induced by switching current spikes drawn to transistors or capacitive loads through parasitic inductances of power or ground planes, respectively. If the number of parallel CMOS gates is increased, the rise time and fall time of signal transit will increase, and the signal transit time between chips will be prolonged and may cause logic errors [8]. Simultaneous switching noise can be reduced by isolating noisy power plane region using decoupling capacitors, V_{CC} regulator, solid power and ground planes, reducing internal power supply impedance, using multiple power and ground pins per chip, optimizing signal trace layout, using smaller devices, damping resistor, separating power and ground planes for I/Os and logic circuits, using parallel ground pins or shorter bonding wires to reduce inductance, and so on. An output buffer can also be used, which includes slew rate control, current control, voltage swing, open drain and time-offset signaling. If power pins are alloted near the end of an IC package, a large internal loop may be formed to create emission. The loop area can be reduced by alloting power pins near the center of package. Simple π-network filter having two capacitors and one ferrite bead or power trace can be used as a decoupling circuit to mitigate dc voltage fluctuations [9]. One must ensure that the decoupling measure is effective for frequencies 10 to 100 times higher than the clock speed of digital circuits [10].

In [9], it is pointed out that at low frequencies, location of decoupling capacitor on board is unimportant for PCB with intact power and ground planes. Because all the decoupling capacitors are shared by the whole plane. However, at high frequencies, locations of decoupling capacitors are critical for power bus design [11]. Typically, the decoupling capacitors are effective below their self-resonant frequencies determined by their capacitance and parasitic inductance. Surface mounted decoupling capacitors have much smaller interconnection inductance to PCB power planes as compared to leaded components, and effectiveness of the former is limited by the inductance associated with traces or vias for frequencies higher than a few MHz. Practical capacitors are accompanied by series parasitic inductance due to interconnections like vias, soldering leads, traces, which cause self-resonance to deteriorate their decoupling effectiveness. Parasitics of buried capacitors mainly come from vias. Typical inductance associated with vias is on the order of a few nH for 10-mil layer spacing [12], and the effectiveness of dc power bus decoupling can be deteriorated by such inductance. Compared to off-chip decoupling capacitors, the on-chip decoupling ones are more effective at higher frequencies due to the lack of bonding wire inductance. To further reduce parasitic effects while having layout flexibility, embedded passive components with multilayered PCB structures can be used [10].

A large capacitor is usually placed close to switching circuit pins as current sink, which may cause parasitic effects at higher frequencies [13]. Several capacitors can be placed around a sensitive IC to form a decoupling capacitor wall at the cost of reduced routing flexibility [14]. Another disadvantage is that new resonance may arise inside the wall at higher frequencies.

Due to high speed of switching currents, very low power supply impedance is required to avoid large supply voltage swing which may deteriorate circuit functions. Intact planes or large-size planes have low impedance [14]. Multilayered PCB can support low-impedance power bus structure by using multiple planes for both power and ground. At low frequencies, ground bounce is mainly determined by parasitic inductances [13]. Two power planes may constitute a parallel-plate waveguide when frequency is higher than 1 GHz, voltage fluctuation due to circuit switchings can be guided over the power planes to perturb V_{CC} [13]. The vias running through power or ground planes may excite parallel-plate waveguide modes which can propagate to PCB edges, couple to apertures, slots, cables to cause interference.

Strong radiated emission from ground and power planes at resonance should be avoided in the design of multilayered PCBs. Measures to suppress resonance include placing decoupling capacitors to IC power pins or along PCB edges, increasing the intrinsic capacitance of power buses in PCB, or using absorbing materials to cover power buses.

Returning current through ground plane with parasitic inductance will develop a voltage fluctuation accross the return path. Such potential difference can drive objects attached to the ground such as cables or conducting chassis to radiate. A few return paths with nonnegligible inductance in a multisignal flexible cable connecting two multilayered PCBs will develop a potential difference between the two connected PCBs. Any cable attached to the boards can be driven to radiate [15].

There have been attempts to shape the switching waveform to reduce delta-I noise. Snubber circuits can be used to reduce dv/dt or di/dt, resonant circuits can be used to generate quasi-sinusoidal waveform instead of rectangular waveform, random pulse width modulation(PWM) can be used to broaden and reduce the strength of EMI spectrum, filters can be used to reduce the noise. Electric field coupling or crosstalk between PCB traces is an important source of EMI in switching power supply. In [16], a genetic algorithm is used to optimize trace layout to reduce EMI.

Digital signal waveforms draw a short pulse of current from power plane. During transition of a CMOS gate in the output stage, one transistor begins to conduct while the other may not be fully turned off yet. A large current spike occurs when many such outputs switch simultaneously. To keep the voltage level of power supply within its normal operating range, a capacitor can be shunt between power and ground planes as shown in Fig.23.2. The minimum capacitance required is estimated as $C_{\min} = \Delta i \Delta t / \Delta V_{\max}$, where Δi is the magnitude of spike current, Δt is spike duration, and $\Delta V_{\max}$ is the maximum allowable voltage fluctuation. Length of interconnection between IC and capacitor should be kept as short as possible to reduce its parasitic inductance.

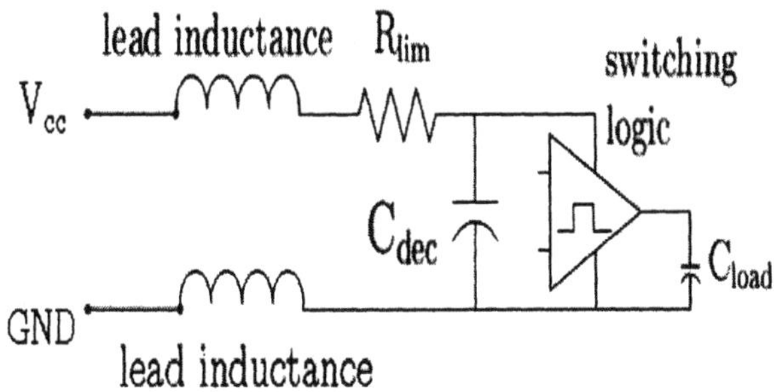

Figure 23.2. On-chip decoupling capacitor.

Voltage fluctuation on the ground plane of PCBs can be represented by an equivalent noise source, U_g. Resistance effects dominate at low frequencies. At higher frequencies, U_g is determined by inductive effects, and increases in proportional to frequency. When frequency goes even higher, U_g approaches a constant, and transmission line effects dominate. The ground plane has an impedance, Z_{noise}, which is in series with the antenna impedance, Z_{ant}, associated with the board or its attached cables.

When vias of IC or leads of decoupling capacitors are placed more tightly, the mutual inductance tends to cancel the self inductance. Thus, self-resonant frequency of the decoupling capacitors is increased, and the decoupling mechanism becomes more effective [14]. Impedance elements like inductors or lines loaded with ferrite beads may be inserted in the power bus to impede current spikes. At low frequencies, any gap on the loop will stop the current flow and consequently eliminate interference. However, paths of high frequency currents are more difficult to interrupt because displacement currents may cross over the gap. Slots or moats may be etched all around sensitive modules to impede or isolate noise. Drawback of this method is that parasitic modes may be excited to deteriorate isolation [13].

Parasitic modes exist in PCBs, which are spurious electromagnetic modes supported by the PCB structure [17]. Conversion of digital or analog signals into parasitic modes is called mode conversion or mode coupling. Examples of parasitic modes include slotline modes excited on splitted reference plane, parallel-plate modes excited by vias or flip-chip leads, and surface wave modes excited at discontinuities of interconnections.

2.3 Interconnections

Signal integrity is maintained when possible distortion to a guided signal does not affect its designated function. Interconnections in a PCB layout include pins, vias, transmission lines, and so on. Discontinuities of interconnection in chip level appear at leads, pads, bonding wires, and so on [18]. Interconnections and discontinuities bear parasitic resistance, inductance and capacitance which may degrade signal integrity [10]. Signal integrity on interconnections are determined by trace length, trace spacing, number of long parallel traces, delayed switching of signals, and so on. Power traces on PCB may be longer than 10 cm, and the signal line length ranges from 2 to 10 cm. Crosstalk on interconnections and its side effects should be considered in PCB layout. For example, crosstalk induced delay may appear as shown in Fig.23.3 if the culprit line and the victim line are switched simultaneously with opposite transitions [19].

When the operating wavelength is ten times or longer than the component size, the component can be treated as a discrete component. Typical examples are resistor, capacitor and inductor, characterized by R, C and L, respectively. At high clock speed, a transmission line can no longer be considered as a zero-length interconnection. Transmission line

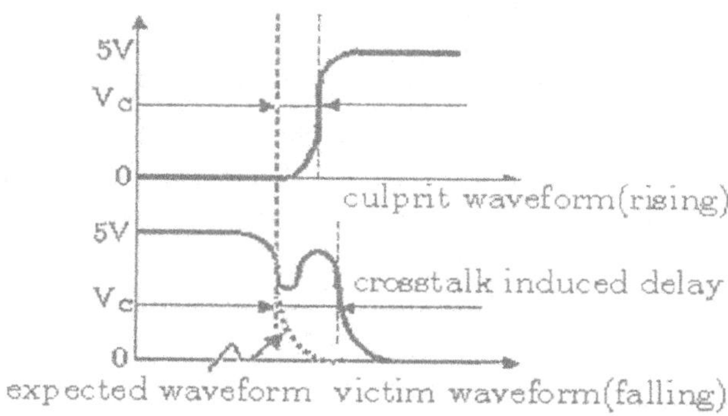

Figure 23.3. Crosstalk induced delay.

attributes like characteristic impedance, phase velocity and delay time have to be considered. When an interconnection has its length comparable to wavelength, its properties can be described by transmission line model with distributed resistance, capacitance and inductance. The capacitance is used to account for the charge stored between two conductors, the inductance is used to account for the magnetic flux encircling two conductors, and the resistance accounts for the loss on conductors. A transmission line can also be viewed as a cascade of short sections, with the length of each section much shorter than the shortest wavelength of interest.

Reflections occur at discontinuities like vias, terminations, bonding wires and connectors as shown in Fig.23.4 where a guided signal is splitted into transmitted and reflected signals. The amplitude of reflected signal is determined by the contrast of characteristic impedances at the discontinuity.

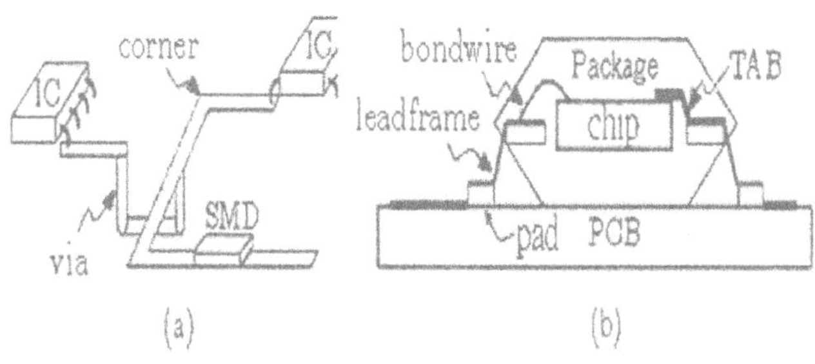

Figure 23.4. Discontinuities in (a) transmission lines, (b) IC package.

A transmission line can be viewed as short or long line as frequency changes. For a short line, the reflected signal mixes with the rising edge of the original signal if the rise time is longer than twice the delay time along the line, and perturbs the original signal waveform. For a long line, the round-trip delay time along transmission line is longer than the rise time, and the reflection along the long line, if not in proper timing, may mistakenly trigger connected ICs due to asynchronization. The transmission line should be matched at both the source and load ends to avoid reflections. The matching network can be a series resistor at the output of a sending stage, parallel or dynamic terminations at the input of a receiving stage.

Typical IC power supply has an impedance of about 5 Ω and can respond quickly to switching signals. The pad drivers have an impedance of higher than 100 Ω, hence their response is relatively slow. In typical IC, lead frame has an inductance of about 10 nH, and the package has an inductance of about 5 nH. Local resonances may occur in a module due to parasitics in chips, power supply, surface mounted devices or other components [8].

Voltage at the load end of a transmission line is usually lower than the source voltage because part of the signal is dissipated by the series resistance of transmission line. A step voltage waveform becomes smoother when it is guided along the line because the distributed capacitances along the transmission line serve as low-pass filters to bypass high frequency components of the original signal. A step voltage waveform at the source side overshoots at the rising edge of waveform because high frequency components tend to be impeded by the distributed inductances along the transmission line. If the distributed inductance is increased, spike at the source end increases and voltage at the load end rises more slowly. Distributed inductance can be reduced by keeping the return line as close as possible to the signal line.

Interconnections and discontinuities in a module introduce high frequency noise on power leads and ground paths, which may couple to other parts of the PCB and cause more interference problems. Hence, high-speed digital circuits require considerable decoupling capacitors to bypass the high frequency switching currents and reduce power supply fluctuations. The decoupling capacitor must be effective for frequencies that are 10 to 100 times higher than the clock speed.

Discrete capacitors adjacent to nearly every IC can be replaced by embedded capacitors which take less space, have smaller parasitic inductance associated with vias or leads [10]. Internal power and ground planes in PCB can also be used as capacitor.

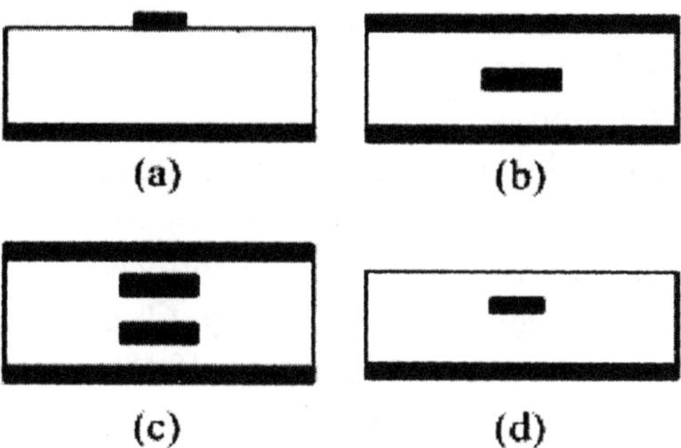

Figure 23.5. Cross section of (a) microstrip line, (b) strip line, (c) dual strip line and (d) buried microstrip line.

Fig.23.5 shows several transmission line structures used in multilayered PCBs. For traces embedded in a layered medium, Green's function approach can be applied to calculate the trace current and associated crosstalk [20]. Use of microstrip lines is rather limited for high-density boards since no trace crossing is allowed. Strip line is a structure with traces embedded between two ground planes. Thus, signal guidance along strip line is less easily affected by other components. Asymmetric dual strip line can also reduce crosstalk while requiring one less ground plane. Two signal layers can be laid out perpendicularly to each other in strip line structure. Buried microstrip line is immersed in the substrate, and has lower line impedance than conventional microstrip lines. Vias are used to connect signal lines between different layers in a PCB. When there is no return path for a via, the return current tends to find alternative path and hence induces reflection and interference. One solution is to use grounding vias to connect different ground planes to form a continuous return path.

Connectors are used to route signals between different PCBs in a multiboard configuration. The signal-return impedance of a connector is typically low but nonnegligible, hence a potential drop is developed across the connector. The connector can thus drive attached ground or power planes to radiate at the clock speed or its harmonics.

2.4 Crosstalk

Crosstalk occurs when a guided signal on a transmission line is coupled to an adjacent line. The coupled signal becomes stronger if transmission lines are moved closer, signal slew rate increases or coupling length increases. Crosstalk mechanisms can be conductive, capacitive, inductive or radiative, and can be assessed by the direction of signal flow. For example, the voltage polarities at both ends of the victim line are the same when capacitive coupling dominates, and are opposite when inductive coupling dominates.

As shown in Fig.23.6, the total current at Z_F is the sum of capacitive and inductive currents, while the total current at Z_N is the difference between them. In capacitive coupling, the induced current, I_{ind}, is related to the exciting voltage, V_{exc}, by $I_{\text{ind}} = C_m dV_{\text{exc}}/dt$ where C_m is the mutual capacitance bewteen two lines. If the load impedances at the near end and far end are Z_N and Z_F, respectively, the induced voltages at the near and far ends become $V_N = (Z_N \parallel Z_F)I_{\text{ind}} = V_F$. In inductive coupling, the induced voltage, V_{ind}, is related to the exciting current, I_{exc}, by $V_{\text{ind}} = L_m dI_{\text{exc}}/dt$ where L_m is the mutual inductance between two lines. The induced voltages at the near and far ends are $V_N = V_{\text{ind}} Z_N/(Z_N + Z_F) = -V_F(Z_N/Z_F)$. To reduce crosstalk, one may shorten the coupling length, widen line separation, add a guard line between signal lines, match the transmission line impedances, design signals with longer rise and fall times.

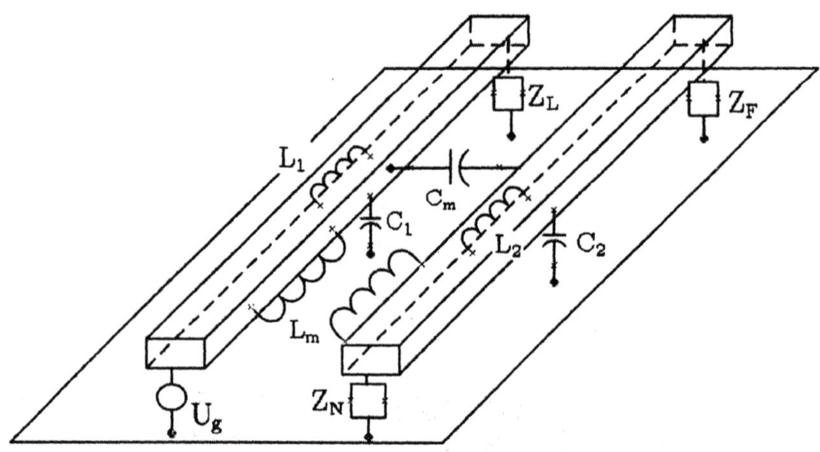

Figure 23.6. Capacitive and inductive crosstalk.

Local properties of transmission line like impedance and dielectric constant determine the amount of energy transferred to the victim line. In common impedance coupling, currents from different circuits flow through a common path such as power plane or ground plane to develop potential drop. Common impedance coupling can be reduced by removing the common path or reducing its impedance by using separate or larger ground planes.

Current path analysis can be used to prevent common impedance coupling. All closed current loops are first categorized as dc or ac loops where all interconnection parasitics are neglected and analog transistors as modeled as ideal current sources. Next, draw current path in all the loops and find branches that are shared by multiple current loops. Finally, move the nodes associated with a common impedance branch as close as possible without changing circuit topology.

Fig.23.7(a) shows a signal trace above its return trace. The inductances per unit length are

$$L_s = \frac{\mu}{2\pi} \ln\left(\frac{\pi d}{W_s + t} + 1\right) \quad \text{(H/m)}$$
$$L_r = \frac{\mu}{2\pi} \ln\left(\frac{\pi d}{W_r + t} + 1\right) \quad \text{(H/m)}$$

where t is trace thickness, d is substrate thickness, W_s and W_r are the width of signal and return trace, respectively. The total inductance per unit length is obtained by adding these two components.

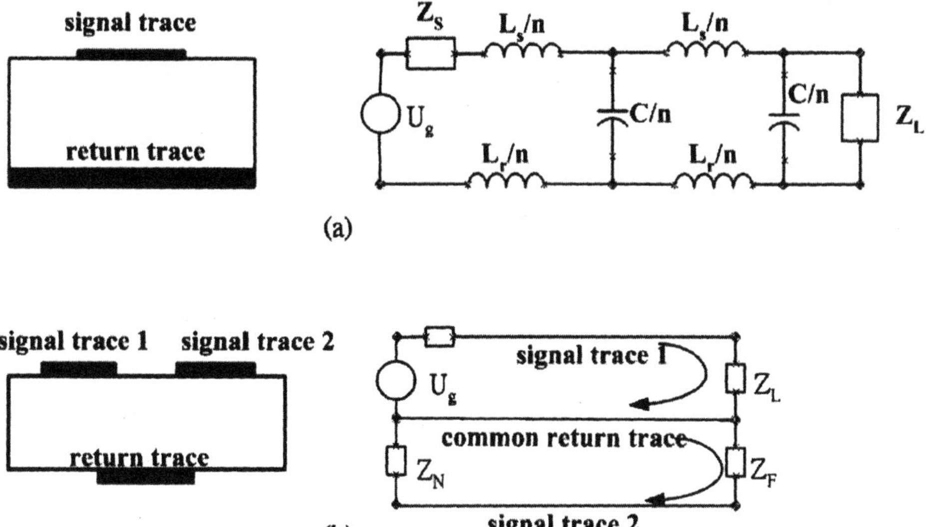

Figure 23.7. (a) Signal trace with wide return trace, (b) two signal traces with common return trace.

Fig.23.7(b) shows two signal traces sharing a common return trace. A wider return trace has smaller impedance than a narrower one, and smaller impedance incurs smaller coupling voltage between two adjacent loops.

To reduce noise and crosstalk in PCB, one may route traces on different sides of a ground plane which serves as shield. A guard trace can be placed between two signal traces. The guard trace is usually connected to ground at both ends to avoid resonances. A better practice is to estimate the highest frequency present on the board, and connect the guard trace to ground at least every quarter wavelength of the highest frequency. Adding an extra return ground on top of the trace to form a strip line can further reduce noise and crosstalk.

Capacitance exists between a signal trace and its return path(C_r) or another trace(C_{12}), where C_r determines the characteristic impedance and signal speed along the trace, C_{12} accounts for the coupling of electric energy between two signal traces. High frequency components of the signal are easier to couple to other traces than the low frequency ones.

Time variation of magnetic flux is responsible for the interaction between different current loops. The Faraday's law indicates that the voltage induced on a loop will drive a current which excites magnetic field to counteract the existing one. Both signal and return currents contribute to the inductance which determines the characteristic impedance and signal speed along the transmission line. The magnetic flux extending into another signal-return loops gives rise to inductive crosstalk. Reducing loop area or moving the loops apart can reduce the mutual inductance.

Return current of each signal tends to flow on the ground plane underneath the signal trace with the extent of current distribution determined by the height of trace above ground plane. The return current distribution also determines the amount of magnetic flux linked to another trace, which is parameterized by the mutual inductance between them. Waveforms of coupled signal via capacitive and inductive coupling look like derivative of the original waveform. For inductive coupling, spike occurs on the rising edge of a pulse signal. In low-impedance digital circuits, inductive mechanism generally dominates.

When a signal propagates from source to load, the spot of energy coupling moves with the signal waveform along the line. The forward-propagating coupled waveform thus moves along the victim line at the same speed with the signal waveform. The backward-propagating coupled waveform arrives at the near end of the victim line at the moment of signal launching. The last piece of waveform arrives at the near end in about twice the propagation time for the signal to propagate over the line. Thus, the duration of crosstalk at the near end is equal to the duration of signal waveform plus twice the propagation time over the line.

Capacitive crosstalk can be reduced by moving signal lines apart, by placing signal line close to the ground plane, or by placing a guard line or capacitive shield to pick up and bypass the crosstalk current. Inductive crosstalk can be reduced by moving loops apart, by reducing loop areas, by placing the ground plane closely under the traces, or by placing a guard line or inductive shield to pick up the magnetic flux. Good shielding can effectively reduce inductive crosstalk in a cable.

2.5 Loop Radiation

High frequency signals are more easily radiated by transmission lines. Although metal boxes can be used to reduce such radiation, however, some energy may still leak through apertures on their surface. The magnitude of leakage depends on frequency, geometry of transmission lines and apertures. The leaky power becomes significant if dimension of the transmission lines or apertures is comparable to signal wavelength.

The magnitude of radiation by closed loops is determined by the loop area. In relative sense, traces in chips form the smallest loops, traces on PCB form medium loops, and cables form the largest loops. Electric field in the far zone radiated by a small loop can be expressed as

$$E = 1.316 \frac{AIf^2}{dS}$$

where $E(\text{mV/m})$ is the electric field strength, $A(\text{cm}^2)$ is the loop area, $I(\text{A})$ is the drive current, $f(\text{MHz})$ is frequency, $d(\text{m})$ is separation, and S is the shielding ratio between source and observation points. Reducing loop area, driving current or frequency will reduce such radiation.

At high frequencies, the return current path is not clearly defined. Besides, signal and return currents are not always well balanced. Unbalanced return current causes reflections, distorts singal waveform, and incurs crosstalk to adjacent traces. If the return path is close to other traces, part of the return current may detour through these traces. If a slot or moat blocks the return path, the return current will be forced to detour and create interference [4], signal reflection and higher-order modes may also occur. Higher frequency components are less affected because the continuity of return current across slot can be maintained by the displacement current.

Unbalanced currents on a signal-return pair incur common mode(CM) current which tends to radiate strong fields that can be several orders of magnitude higher than the DM emission with the same magnitude of current [21]. To reduce CM current, the ground plane can be enlarged to reduce its inductance [22], a ferrite sleeve can be inserted in the CM path to raise the path impedance, or a capacitor can be shunted to bypass the high frequency components to ground.

By reciprocity, current loops can also pick up electromagnetic energy. The voltage induced in a small loop is related to the incident electric field by

$$V_i = 2\pi \frac{AEfB}{300S}$$

where V_i(V) is the voltage induced in the loop, $A(\text{cm}^2)$ is the loop area, E(V/m) is the incident electric field strength, f(MHz) is frequency, B is attenuation factor of the loop and B =1 if the frequency is within the pass-band, S is the shielding effectiveness of the shield. Measures to reduce loop radiation can also be applied to reduce the energy received by the loop.

Emission on chip level can be reduced by decreasing the number of simultaneously switching nodes, prolonging clock rise/fall time, separating power supply and circuits, decoupling noise from power supply pins, placing mixed-signal modules properly, optimizing pad drivers, reducing die area, and so on. Radiation on interconnection level can be reduced by shortening interconnection length, reducing signal-return separation, decreasing current magnitude, increasing signal transition time, modifying signal waveform, shielding interconnection, filtering I/O ports, and so on. Radiation on system level can be reduced by observing the following guidelines: identifying emission sources and coupling paths, simulating emission related parts, assessing emission reduction measures, implementing reduction measures, setting up measurement, evaluating emission reduction measures, and summarizing design guidelines.

To measure radiation from a PCB or any electronic device, radiating sources need to be identified first. Far field can be transformed from near-field measurements in semianechoic chamber based on an equivalent set of electric and magnetic dipoles which radiate the same near field as the original device under test [23].

2.6 Filters

Filters are often uesd at I/O ports of a PCB to suppress currents induced by external interference [24]. Filters are often composed of capacitors and inductors, where shunt capacitors are used to divert high frequency noise to the ground, and series inductors are used to impede high frequency noise from entering the circuits. Low-impedance shunt element is usually adopted to shield high-impedance susceptor, and high-impedance series element is often adopted to protect low-impedance susceptor.

Filter performance at high frequencies is often degraded by parasitics. Parasitic inductance and resistance of a capacitor change its capacitive attribute above its resonant frequency which is determined by the capacitance and parasitic inductance. Inductance of an inductor is mainly determined by the permeability of core material, but parasitic capacitance will change its inductive attribute above its resonant frequency [25]. Magnetic saturation should be avoided in choosing magnetic core material for an inductor. Capacitors in a filter can be broken down by over-voltage, hence filters are preferably protected by nonlinear devices like avalanche diodes.

Current passing through an equipment may cause interference to interior circuitries, which can be prevented by using current boundary. All connectors or I/O cables which may carry CM currents should be bundled together and shielded. The shield tends to carry noise current to another equipment, hence a filter is usually used to bypass all the undesired CM currents to ground instead of penetrating into cabinet.

Current boundaries are implemented by placing additional loops to divide a large module into small submodules. External CM currents are detoured by those current boundaries without creating mutual interference, and signals flowing between separate modules should be well controlled. Each loop should be kept small or a ground plane should be placed near the loops. Each current boundary may function as a common impedance between adjacent modules, hence wide traces should be used. Current boundaries can be placed at physical interfaces to protect internal circuitries from ESD, for example, at the interface between a PCB and interconnecting cables or backplanes. Internal PCB traces carrying high-frequency signals may induce CM current to external cables, and proper shield is required to reduce interference.

Electrical safety regulations prohibit direct connection of mains wires to the cabinet wall. A mains filter is needed to insulate low-frequency power current away from the cabinet wall while providing high-frequency current a low-impedance path to this wall.

2.7 Trace Terminations

Reflection occurs if impedances of the line and the load mismatch. For signal integrity consideration, long interconnection should be terminated with a matched load, especially when the line is longer than 1/6 of the length of rising edge. One disadvantage of using resistive matching load is power dissipation at the load and reduction of load voltage relative to the source voltage. A series resistor may be concatenated between the load resistor and dc voltage source to reduce dc power dissipation. To ac signals transmitted along the line, these two resistors appear to be in parallel with an equivalent resistance matched to the line.

Ac components of a digital signal appear only at the rising and falling edges of waveform. A capacitor in series with loading resistor can be used as termination. This capacitor behaves as a short circuit at the rising and falling edges of waveform, hence this series RC circuit becomes a matched load. The RC time constant should be kept as small as possible, but large enough to remove high frequency components. Plateau portion of the signal waveform quickly charges the capacitor, hence the resistor dissipates almost no dc power. However, bipolar pulses must be choosen to avoid overcharging the capacitor.

Matching termination at the source end will absorb possible reflection from the load, which is useful for clock distribution circuits. A small resistor placed in series with the output of line driver is used to absorb residual reflection from the load, which is only suitable for point-to-point interconnections. Hence, it should be used with care especially when the original signal is small. Voltage divider formed by the series resistor and line impedance will reduce the output voltage.

When a CMOS device is connected to a PCB trace as load in digital circuits, its impedance is very close to that of a small capacitor. Capacitive loading is prominent in bus lines when many CMOS devices are shunted at regular intervals. All these capacitors increase per-unit-length capacitance of the line, hence the characteristic impedance is decreased and the propagation delay is increased. It also requires more energy to switch logic states. The stub connecting each input logic to the bus line should be kept much shorter than the length of rising edge. Daisy chain interconnection can be used to avoid long stubs and discontinuities which give rise to uncontrolled reflections, but capacitive loading effect remains. Capacitive loading of backplane to plug-in boards also exists. An effective measure is to design these boards to have higher line characteristic impedance than specified. The line impedance will decrease to the specified level after being inserted into the backplane.

2.8 Layout

Proper PCB layout can reduce interference. Relevant layout parameters include signal type(clock, data, address, control), component type(trace termination, decoupling capacitor, and so on), pin type, interconnection topology, timing, characteristic impedance of transmission line, and so on. Requirements on geometry, transmission line parameters and signal distortion will be disussed in this Subsection.

In the geometry requirement, components should be separated into groups according to signal form(analog or digital) and frequency. Suitable interconnection topology as exemplified in Fig.23.8 should be chosen. Parameters like trace length, pin-to-pin separation and length of interconnection to power plane contribute to reflections and radiation. I/O components should be placed close to connectors. Hot components should be placed close to aerator at the air inlet or outlet.

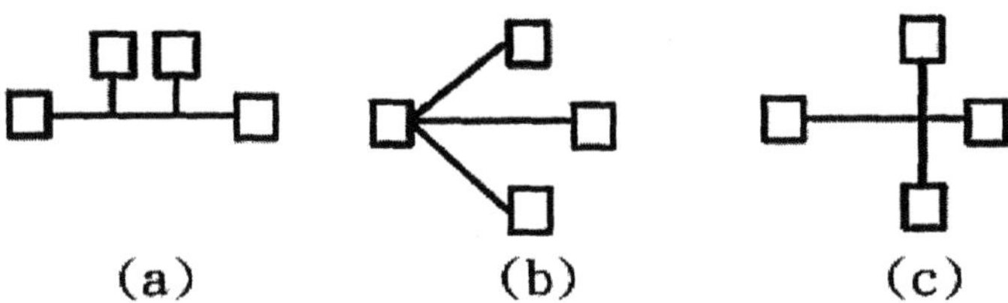

Figure 23.8. Interconnection topologies: (a) serial net, (b) near-end cluster, (c) far-end cluster.

Maximum number of layers that vias are allowed to run through should be limited. Parameters of layered media should be adjusted to have consistent transmission line characteristics in different layers. Signal layers should be close to ground or power planes to reduce crosstalk or radiation. Decoupling capacitors connecting power and ground planes should be properly used to bypass high-frequency noise. Minimum separation should be maintained between noisy components and potential victims. Maximum allowable coupling area or length should not be exceeded. Loop area formed by a signal line and its return path should be minimized.

Transmission lines are required not to exceed the allowable characteristic impedance range or maximum number of discontinuities. Driver impedance should be matched to that of interconnection.

Current levels in analog circuits may vary over orders of magnitude. Existing loops should be sorted in order of decreasing di/dt. The most threatening loops are laid out first, in which components are placed close together and interconnections are kept as short as possible. Then, the next threatening loop is laid out in the same manner.

Rise time and amplitude of signal waveforms in digital circuits are determined by the logic technology chosen. Timing and waveform of signals can be distorted by reflections. Small edge slope of the signal waveform is suggested, and monotonic signal edges should be preserved to avoid erroneous switching. Signal distortion may cause erroneous switching, which can be assessed by analysis or simulation. The onset timing between sensitive and disturbing signals should be offset, voltage fluctuation should be lower than the allowable threshold, and far-field radiation should be assessed for each layout.

Critical signals have to be immune to crosstalk and other signals which are more tolerable to crosstalk. For example, waveforms on all data and address lines in synchronous digital circuits change at the same time, triggered by clock or strobe signals. Crosstalk occurs only at waveform transitions, while the signal state is sampled and decoded one half clock period later. Hence, crosstalk will not affect circuit function. On the other hand, clock and strobe signals should be clear of noise and away from data or address lines.

To reduce high-frequency EMI, one may combine circuits that communicate among themselves at high speed into a module, and try to reduce the speed of communication between different modules. Signals that are used only within a module should be confined to that module, and current boundaries can be used to prevent leakage of signals into wider areas than necessary. Notice that in synchronous logic designs, high-speed system clock signal is distributed over the whole system.

2.9 Cables

A cable may carry CM currents if part of the return current flows through other paths instead of its outer conductor. Susceptible cables may pick up these CM currents and convert them into spurious DM signals inside the cable. Transfer impedance, Z_T, of a cable is defined as the induced DM voltage divided by the CM current flowing over to this cable, to describe the conversion of CM signal into spurious DM signal which is superimposed upon the designed DM signal. Thus, higher Z_T implies stronger coupling between inside and outside of the cable. By reciprocity, the same Z_T also describes possible CM noise outside the cable induced by the DM signal on the cable.

The transfer impedance is frequency dependent in which resistive effect usually dominates at low frequencies, while inductive effect appears at high frequencies. To minimize undesired coupling, one may try to shift the inductive effects to occur above the highest frequency of interest. When the inner conductor of a cable is completely shielded by the outer conductor, skin effect helps to reduce leakage and thus Z_T.

Twisted pairs are interference proof at low frequencies because the induced voltages on adjacent segments are out of phase [26]. Noise on a cable may be coupled into adjacent cables. A longer cable tends to carry higher CM currents, parasitic inductance and capacitance from cable to ground. When cables are connected to PCB or shielding enclosure, CM currents are generally induced to cause interference. When a coaxial cable is connected to a wire pair, wire connected to the outer conductor of cable forms a pigtail which incurs parasitic inductance and disturbs the balance of currents between signal and return lines. Cables thus connected generally have an average radiation resistance of 100 to 150 Ω over the frequency range of 150 kHz to 1 GHz. Ferrite chokes are usually used to limit the CM current on cables [27]. At high frequencies, CM paths become effective antennas to radiate interference or convert external field into CM current.

CM radiation from cables becomes one of the primary EMI sources, and becomes more serious as the clock speed and signal slew rate increase. Radiation from cables attached to PCBs or shielding enclosures becomes significant near the resonant frquency of cable segments [28]. Hence, resonant frequencies of cables should be shifted away from harmonics of clock. Cable guide is usually made of wide metal sheet to provide a close-by return path for the CM current and shield the coaxial cables. It needs to be electrically connected to the cables at all junctions. When the return conductor is widened, transfer impedance of the cable and guide combination drops significantly. Separation of three to five times the cable diameter is a good rule of thumb.

3. Electromagnetic Shielding

Shielding is a straightforward measure to reduce radiated emission and increase radiated susceptibility simultaneously. Shielding effectiveness(SE) is defined as the ratio of the observed field in the absence of shield, E_0, to the observed field in the presence of shield, E_s, measured at the same spot, namely, SE $= 20\log_{10}(E_0/E_s)$ dB [29]. Shielding effectiveness of typical metal enclosure is determined by frequency, dimensions, internal material, openings, measurement spot and antenna configuration. For example, copper foil(1 oz) used on PCB is approximately 1.2 mils thick, which is equal to six skin depths at 100 MHz, and provides about 52 dB of shielding effectiveness.

Holes can significantly deteriorate a shield by detouring the induced current on shield to flow around holes, thus deflecting magnetic field into the holes. On enclosures of high-speed digital circuits, aperture arrays or perforated screens are preferred in reducing interference over large apertures for ventilation and heat removal.

Electronic equipment is often housed in an encolsure with openings like seams, slots, apertures, slits, cracks and joints through which fields may penetrate and deteriorate the shielding effectiveness of enclosure. When frequency is lower than 100 kHz, the shielding effectiveness is mainly determined by the shielding material. When frequency is between

100 kHz and 100 MHz, the shielding effectiveness is determined by both material and leakage. When frequency is higher than 100 MHz, the shielding effectiveness is deteriorated by leakage which is determined by number, size and shape of openings.

Proper design of gaskets, slits, cracks and joints will reduce undesired electromagnetic leakage [30]. However, it is more difficult to mitigate radiation from designed apertures such as those for heat removal, ventilation or I/O cable connections. Circuitries inside a shield are connected to the outside by power lines or I/O cables. Currents may be induced in loops or cables outside the shield, and external field may also penetrate through holes to interfere with internal circuitries. These power lines or I/O cables should be directly connected to the PCBs while connecting the ground wire or outer conductor of the cable to the shield. Filter or choke is suggested along the I/O cables to impede CM current. The filter should be grounded on the enclosure, inside and outside wirings should be completely separated electromagnetically. Ground planes in circuitries should be electrically attached to the shielding enclosure with wide metal strips.

In the low frequency limit, small apertures can be approximated by equivalent electric and magnetic dipoles [31]. EMI from closely spaced electrically short slots is proportional to the number of slots [32]. Radiation from a perforated screen with a large number of apertures may result in EMI problems at frequencies above the fundamental resonant frequency of cavity. Honeycomb configuration can be used instead to reduce such interference. EMI reduction of more than 20 dB has been achieved by using dual perforated screens spaced 1 cm apart when compared to a single perforated screen [33]. However, the space between screens forms a thin cavity, which can lead to significant radiation in specific directions. Damping the resonances by loading the space between screens with lossy material alleviates this problem and achieves more than 20 dB of reduction over a single screen.

Operating frequencies in modern high-speed digital circuits and wireless devices are higher than 1 GHz. The enclosure behaves like a cavity resonator, thus increases emission level, decreases immunity, and causes significant mutual coupling between different modules inside the enclosure. Absorbing materials can be used to reduce the Q factor of enclosure [34], or shift the enclosure resonant frequency to avoid harmful interference. Radiation through slotted screens becomes maximum at resonant frequencies of the slot [35], and radiation at the resonant frequencies of cavity modes through slots and apertures of nonresonant dimensions can be as strong as the former. At frequencies above the fundamental cavity-mode resonance, radiation from enclosures can be stronger than the radiation from I/O cables [32].

For portable electronic devices, light-weight enclosure can be designed by fabricating conductive wire meshes into plastic cases, coating conductive materials on plastics cases, or mixing conductive fillers or powders into plastics matrix. Frequently used conductive materials include magnesium, titanium, alumium, nickel, tin, and so on [36]. Fabricating conductive meshes into plastics may not be desirable because size, weight and cost are increased. The shielding effectiveness of plastics filled with conductive powders such as carbon, silver or copper turns out to be unsatisfactory due to difficulties in bonding and grounding. Moreover, the cost of conductive plastics is higher than that of sheer metal.

Solid metal has the best shielding effectiveness among all materials. Conductive polymer composites are frequently considered to replace metal shield due to their light weight, low cost, design flexibility, versatile electrical properties, superior strength-to-weight and modulus-to-weight ratios. Shielding effectiveness of composites is determined by geometry, distribution, conductivity, absorption coefficients of fillers, also by fiber orientation and laminate thichness in the composite [37], [38].

Placement of openings and sensitive devices is also critical for shielding design. Analog circuits are far more sensitive to electromagnetic interference than digital circuits. Hence, the former need more stringent protection from external interference. Magnetic fields are difficult to shield at low frequencies because only small current is induced on the metal shield to generate counteracting magnetic fields. Materials with high permeability are needed to shield magnetic field.

4. Transients

Transient is a phenomenon which occurs during a short time interval between two consecutive steady states. There are different transient phenomena. Voltage surge is a transient voltage signal propagating along a line, which is characterized by rapid increase followed by slower decrease of voltage level. Surges are high-energy, short-duration pulses caused by lightning or power load switching, which can reach 6 kV indoors, 10 kV and higher outdoors. Spike is a unidirectional pulse with short duration. Burst is a sequence of distinct pulses or an oscillation with limited duration. Electrostatic discharge(ESD) is an abrupt release of charges that have accumulated on an object on which the potential is raised up to 15 kV(20-25 kV in the worst case). Electrical fast transient(EFT) is a fast, impulsive signal induced by external sources like nuclear electromagnetic pulse(N-EMP) or internal sources like surge EMP(SG-EMP). Electrical fast transient/bursts(EFT/B) is a burst of very fast noisy pulses up to several kV or higher, often due to arcing when switches are thrown up .

Energetic events such as lightning, N-EMP or electrical load switching may cause interference, damage of electronic circuits, burning of interface components, breakdown of *pn*-junctions, burning of fuses and protection devices, and so on. Transient may cause glitches in digital signals or distortion in analog signals. Switching an inductive load generates transient phenomenon which involves alternation between capacitive and inductive energies. Such a transient phenomenon may be caused by current flowing through solenoids, motors or relay coils, current induced by capacitive switching, fuse burning or ESD ignition on the line.

When switch of a relay coil is opened, energy stored in the coil is transferred to an effective capacitor and induces high voltage which will cause a flash-over in the switch gap. Then, current is restored and voltage drops to extinguish the sparks. High-voltage switching creates strong radiation to interfere with sensitive electronic equipments. Discharge/flash-over in nanosecond scale gives rise to interference current in a very broad band. Typical waveform of such interference is a damped sinusoidal with its frequency determined by the loop inductance and capacitance.

4.1 Electrostatic Discharge

Electrostatic discharge(ESD) or spark occurs when conductive object like people, furniture, device or equipment is moved close to a charged insulator to cause discharge which may damage nearby electronic circuits. A predischarge corona may occur to delay the discharge process. Internal ESD may occur if the induced voltage across an insulator exceeds its dielectric breakdown voltage. For example, ESD may occur if the voltage difference between a PCB and its enclosure is higher than the air breakdown voltage [20]. The amount of energy dissipated during a discharge process may only be several joules. However, the power may reach million watts due to short elapse time, and possible damages are caused by large current.

Major features of ESD are related to ambient environment, geometry, charged materials, and so on. Typical output current waveform of an ESD generator defined in IEC 61000-4-2 standard is shown in Fig.23.9 [8].

Rise time of an ESD is on the order of a few nanoseconds, hence high-frequency components above several GHz may exist. The induced voltage level by an ESD is inversely proportional to its rise time. Radia-

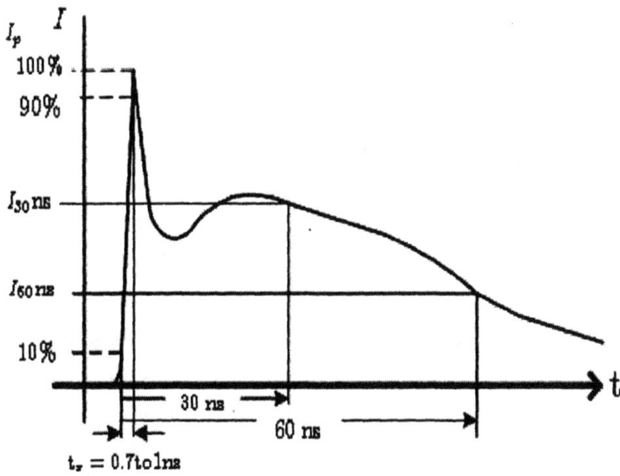

Figure 23.9. Typical current waveform of an ESD generator.

tion caused by an ESD can propagate through openings on an enclosure. Current induced by an ESD can also flow through cables or PCBs to cause radiation.

The coupling mechanisms for transient disturbances are identical to those for steady state signals. Hence, similar reduction techniques as mentioned in the previous Sections can be applied. ESD can be bypassed through a serise *RC* branch or diodes. Use of thin but wide metal foil on the outside surface of an equipment will shield high-frequency current due to skin effect. Filters can be used at I/O interfaces of an equipment which are prone to ESD. Cables should be shielded properly, especially at both ends.

4.2 Lightning

There are about 100 flashes per second happening around the world. Lightning claims 100 to 600 deaths, about 45 million US dollars in building damage, 10,000 forest and brush fires, and about 70% of communication disruptions and power outages in the United States each year. Lightnings can be categorized into two kinds. One is cloud flash which discharges between clouds and may cause damages to aircrafts. The other is ground flash which discharges between clouds and ground, and may cause damages to terrestrial objects.

A lightning process can be decomposed into separation of charges, stepped leader formation, a sequence of upward streamers, channel formation, first return stroke, J- and K- streamers within clouds, dart leader and subsequent return stroke. Electric charges in the clouds are torn apart by convection to store energy which will be released later. Flash is a discharge process, which lasts for about 0.02 to 2 seconds. In each flash, there are 3 to 90 coulombs of charge transferred, there are about 10,000 pulses grouped into 1 to 26 strokes, with time interval between strokes of about 3 to 100 ms. Rise time of a stroke is about 1 to 30 μs, and the duration is about 10 to 100 μs. In each stroke, the current amplitude reaches about 4 to 400 kA at the rate of about 1 to 80 kA/μs, and the peak voltage is higher than 100 MV. Length of return stroke channel ranges from 2 to 14 km. The charge transferred in one stroke is about 0.2 to 20 coulombs, the average energy released in each stroke is 3×10^8 joules with an average power of 10^{13} watts. The ambient temperature can reach around $55,000°$F at which the air is heated adiabatically to generate shock waves(thunders) which propagate at the speed of about 1 mile in 5 seconds.

Important features of a lightning EMP are its frequency contents and current waveform which can be represented by its duration, amplitude, charge and specific energy. The return stroke current modeled by Lin, Uman and Stradler is [39]

$$I_L(z,z',t) = I_0 e^{-z'/\ell}[e^{-\alpha(t-t')} - e^{-\beta(t-t')}], \qquad t \geq t' \text{ and } z' \geq z$$

and $I_L(z,z',t) = 0$ otherwise, where $t' = z'/v$, $\alpha = 10^5$ s^{-1}, $\beta = 3 \times 10^6$ s^{-1}, I_0=10 to 50 kA and ℓ=800 to 3,000 m, assuming that the current path is along the z direction. The double exponential channel base current modeled by Karwowski is [40]

$$I_L(z',t) = I_0 e^{-z'/\ell}\left[e^{-\alpha(t-z'/v)} - e^{-\beta(t-z'/v)}\right]$$

where ℓ=2 km, I_0=10 kA, $\alpha = 3 \times 10^4$ s^{-1}, $\beta = 10^7$ s^{-1} and $v = 1.1 \times 10^8$ m/s. The traveling current source modeled by Heidler is [41]

$$I_L(t) = \frac{I_0}{\eta}\frac{(t/\tau_1)^5}{1+(t/\tau_1)^5}e^{-t/\tau_2}$$

where τ_1=1.26 μs, τ_2=56.3 μs, $\eta \simeq 1$ and I_0=30 kA. The lightning current waveform recommended by the International Electrotechnical Commission(IEC) is a triangular impulse of single polarity as

$$I_L(t) = \frac{I_0}{N}\frac{(t/\tau_1)^{10}}{1+(t/\tau_1)^{10}}e^{-t/\tau_2}$$

where τ_1 is the 10% to 90% rise time which is 19 μs for the first stroke and is 0.454 μs for the subsequent strokes, τ_2 is the 0 to 50% rise time which is 485 μs for the first stroke and is 143 μs for the subsequent strokes, I_0 is peak current, N is a normalization factor.

All major industries should have provisions to deal with lightning. Although the chances are slim, effects of a direct strike on equipments can cause severe thermal, mechanical, structural deformation or electromagnetic injection. Nearby strikes occur more frequently, but their effects on equipments are less severe than a direct strike. Conducted and induced surges appear in equipments due to fast transients of magnetic and electric fields. Although the coupled energy by fields of a lightning is small, the magnitude of induced voltages or currents can be considerable. The magnetic field coupling which induces voltage drop in circuit loops is more hazardous than the electric field coupling. The magnitude of nearby strikes may cause secondary sparking between insulators like isolation transformer, sharp bends in a conductor, and so on [42], [43].

Very often, nearby strikes incur interference to equipments and devices of information, electronics, avionics, communications and power distributions. Polarity-sensitive devices may suffer from nearby strikes. When an interfered object extends over a wide area, the induced currents exhibit attributes of traveling waves.

4.3 Transient Suppression Devices

Transients can be suppressed by using transient suppression devices (TSD) like spark gap devices, MOVs, avalanche diodes, varistors, resistors, inductors and filters. Semiconductor devices like zenor diodes have low impedance and quick response. Large-area avalanche diodes can absorb power at a rate of 1.5 kW/ms. Combination of the above devices can be used to achieve satisfactory suppression.

The following parameters are used to describe nonlinear transient suppression devices. Standoff voltage(V_R) is the highest reverse voltage at which the device stops conducting, which should be higher than the maximum operating voltage of the protected circuits. Minimum breakdown voltage($BV_{\min}$) is the reverse voltage at which the device becomes a low-impedance path for transient and conducts 1 mA. It should be lower than the minimum vulnerable voltage of the circuits. Peak pulse current(I_p) is the maximum allowable pulse current which does not change the performance of protected circuits by more than $\pm10\%$. Maximum clamping voltage($Vc_{\max}$) is the maximum voltage drop across the device when it is exposed to peak pulse current. It should be lower than the minimum vulnerable voltage of the circuit. Usually, the device's voltage overshoot is clamped after about 1 ms of event occurrence. The device must be able to survive the peak pulse power(P_p) which is the product of clamping voltage and peak pulse current. Shunt resistance(R_s) is the resistance of the device while not conducting. Most transient suppression devices(excluding varistors) have R_s greater than 10^{10} Ω. Shunt capacitance(C_s) is the capacitance between electrodes of the device at 1 kHz, which determines the bandwidth, hence the maximum usable frequency of the device.

Typical transient suppression devices can be characterized as $I = GV^a$ where G is the geometry constant of device, a is nonlinearity factor, which is about 2.5 to 60 for varistors, and is 1 for resistors. For bidirectional varistors, a is about 5 for thyristors(silicon carbide) and about 2.5 for zinc oxide. For spark gaps, a is usually very large to stand off large surge current.

Spark gap devices are nonlinear devices typically consist of two to three electrodes encased in a ceramic container filled with inert gas like Ar or Ne. Primary advantages of spark gap devices are low voltage when conducting, high conducting currents(5 to 20 kA in 10 μs), low shunt capacitance(< 2 pF) and negligible leakage current during normal operation. Primary disadvantages of spark gap devices are relatively slow response, varying ignition voltage, prolonged discharge, and not being able to extinguish in dc lines. Spark gap device continues to conduct as long as the gap voltage is higher than the arc sustaining voltage and the current is higher than the arc extinguishing current. It will extinguish only if its gap voltage is lower than the arc sustaining voltage and its current is lower than the arc extinguishing current. In ac lines, spark gap devices usually extinguish at zero voltage crossing.

Metal oxide varistors(MOV) are passive nonlinear resistors with primary advantages of fast response, high energy absorption, conducting wide range of currents(up to 20 kA) and variety of option. Primary disadvantages of MOVs are narrow bandwidth of about 1 MHz, small leakage current during normal operation, performance degradation when exposed to current surges.

Avalanche diodes are passive unipolar silicon diodes with high doping and low clamping levels, characterized by *pn*-junction voltage at forward bias and breakdown voltage at reverse bias. Primary advantages of avalanche diodes are relatively fast response(1 ps), unidirectional or bidirectional operation, wide ranges of standoff voltage(5.5 to 700 V or higher), and wide ranges of maximum clamping voltage(7 to 500 V). Primary disadvantages of avalanche diodes are large parasitic capacitance(50 to 1,000 pF), narrow bandwidth, relatively small transient currents(< 500A), and reduction of circuit's capacitance.

4.4 Transient Suppression Schemes

Transient suppression scheme is designed based on threat standards which define threat type, typical transient waveforms, source impedance and relevant parameters. Very often, actual transient waveforms deviate significantly from those defined in the standards. Valid transient suppression schemes must not affect original circuit functions under normal conditions. The protecting devices clamp circuit voltage, conduct without delay at slight voltage fluctuation around normal, recover to normal conditions after a transient passes, and maintain rated design specifications after operation. In general, when surge voltage is much higher than the operating voltage of the protected circuits, multistage suppression scheme as the one shown in Fig.23.10 is required. The first stage is usually used to dissipate most of the transient energy and peak current, the last stage clamps residual surge to the operating voltage. The isolation series devices must survive the maximum voltage drop between two stages, while not incurring excessive insertion loss [20]. In the selection of protecting devices, one must consider the maximum voltage fluctuation and losses allowed to the designed signal, operating frequency of the signal and characteristic impedance of the lines.

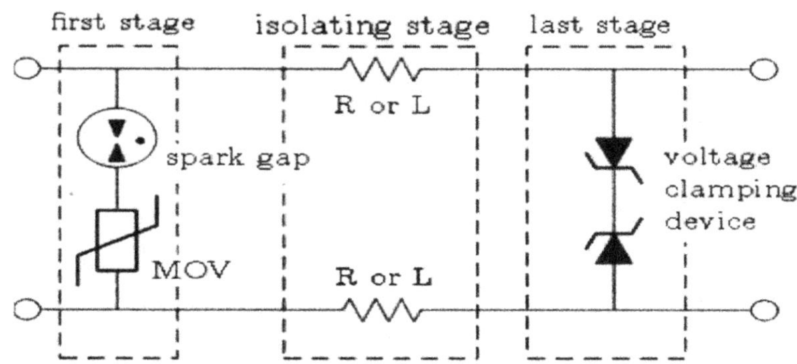

Figure 23.10. Typical multistage transient suppression scheme.

TSDs may be combined for different applications and on different requirements. Line-to-ground protection is used to suppress common mode transients, and line-to-line protection is used to suppress differential mode transients. Current surges can be blocked by high-impedance devices such as series resistors or inductors, or diverted to ground by shunting low-impedance devices such as spark gap devices. Voltage surges can be limited by linear or nonlinear protection devices such as varistors and avalanche diodes.

Major considerations to install TSDs include choosing TSDs with lowest possible inductance, installing TSDs close to the protected circuits, keeping TSDs leads as short as possible, avoiding fuses or circuit breakers in series with TSDs, and so on. Special care must be taken in the installation of devices to ensure sufficiently low ground impedance. Bad grounding of suppression device may significantly degrade its effectiveness. To maintain normal operation of the TSDs, their functions need to be checked on regular basis, and the spark gaps need to be checked for isolation and removal of erosion on electrodes.

5. Reverberation Chamber

Test facilities for EMI and EMS measurements include open area test site, shielded enclosure and others. Open area test sites can be open, enclosed, or buried. Both absorbing and reflecting types of shielded enclosure are in use, screen room and reverberation chamber belong to the reflecting type.

Reverberation chamber emulates an open site environment with incident waves coming in all directions toward the equipment under test(EUT) as shown in Fig.23.11. A mechanical tuner is installed in a shielded room to redirect the waves in all directions. Directional antennas with linear polarization or probes with three axes can be used to illuminate the turner. One may measure the total radiated power from the EUT, immunity of the EUT or shielding effectiveness of its enclosure. Antenna efficiency can also be calibrated in the reverberation chamber.

Field distribution in the chamber is determined by boundary conditions and frequency. The boundary conditions depend on the chamber geometry, configuration of the EUT, instrumentation setup and tuner configuration. The tuner changes the boundary conditions within chamber when it is rotated. Under each new geometrical configuration or new set of boundary conditions, the reverberation chamber supports a new set of complex modes which are different from those in other configurations. Because chamber geometry is complicated, field distribution is hard to compute. Theoretical solutions are provided with modal theory, ray theory or plane wave integral theory. Each theory has its own merits in describing certain chamber characteristics. The modal theory is based on field distribution, mode number and mode amplitude. In the ray theory, multiple rays are launched to the chamber, and multiple reflections determined by chamber geometry are traced to calculate the field strength at any given location. In most cases, the analysis is complicated and its predictions are of little value.

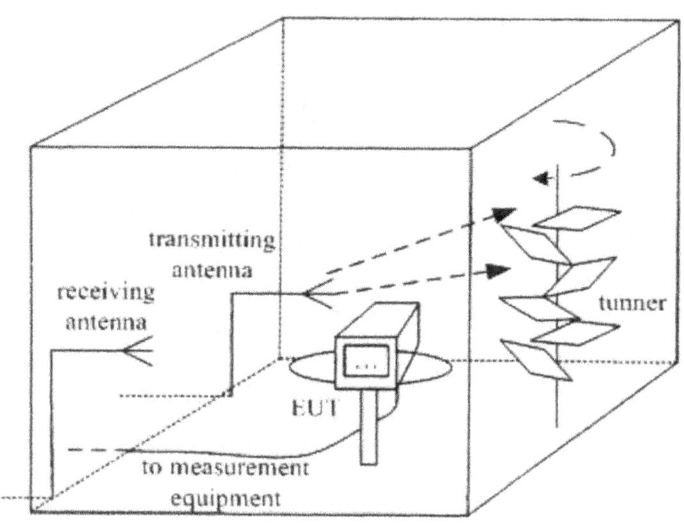

Figure 23.11. Configuration of a reverberation chamber.

To overcome such difficulties, statistical models have been established by treating relevant electromagnetic parameters as random variables, and describing the measurement results by means, standard deviations, extremes, and so on. Once the tuner has been rotated to a sufficient number of postures, the fields at any location become statistically isotropic and uniform. Namely, the waves arrive in all directions and polarizations, and magnitude of the total field is the same within tolerance. Statistical isotropy does not imply that all wave polarizations occur simultaneously. Specific boundary condition renders only one field distribution. With a sufficient number of independent samples in a complicated chamber, all polarizations are equally likely, and the average field magnitude is independent of location, test configuration and chamber geometry. Since measurement in a reverberation chamber is a statistical procedure, large number of independent samples are required.

6. Frequency Management

Fast deployment of wireless communication systems in recent years can be attributed to technology maturity, new service provision, customer expectations, and so on. New wireless services like 3G, DTV, DAB, WLAN and IMT-2000 have been under development or field trials, hence more frequency resources are demanded to have these services implemented. However, frequency spectrum is a scarce resource that can only be used exclusively in the same area and the same period of time for wireless applications. Frequency bands assigned to different applications or services should be properly offset to reduce co-channel and adjacent channel interference. Frequency spectrum is a public resource shared by both public and private sectors. Hence, how to allocate and assign frequency spectrum becomes a critical technical, social and economic issue.

All the developed countries have their own spectrum management policy and authorized agencies to execute the policy. Spectrum management should fulfill the goals of maintaining service order, improving usage efficiency, ensuring quality of communication, satisfying market requirement, promoting industries and economy, and so on.

Domestic spectrum management policy should comply with international regulations, technology feasibility, market requirements and economic development. The ITU(International Telecommunications Union) under the United Nations divides the whole world into three regions (Europe-Africa-Siberia, north America-south America, Asia-Australia) to facilitate interference management. All the wireless applications are categorized into 37 services including fixed, mobile, broadcasting, amateur, aerospace, satellite, navigation, and so on, for the convenience of spectrum allocation and assignment. The ITU also proposes radio regulations as references for all countries to establish their own domestic regulations.

The scope of spectrum management covers allocation, assignment, interference monitoring, charge of usage, planning and trend analysis [44]. The ITU is responsible for global spectrum planning, and the government in each country is responsible for its own domestic spectrum planning and management. In general, domestic allocation should be compliant with the ITU resolution to maintain worldwide compatibility. A given frequency band can only be used exclusively in the same geographical area due to interference consideration. However, it can be reused in other areas with proper separation. Hence, center frequency, bandwidth, power level and interference analysis should be carefully reviewed in assigning spectrum to potential users. Interference monitoring is a necessary measure to ensure normal operation of wireless services. Technological capability and law enforcement authority are the key factors to successfully maintain order of spectrum usage. Since the spectrum is a public asset, part of the revenue generated from its use should be returned to the public. Thus, charge of usage is legitimate, which also serves to motivate the operators to enhance their efficiency of spectrum usage. New wireless applications and services may emerge due to technology advancement, market drive and other factors. The authorized agencies should closely watch the evolution trend and plan a strategy to adapt to new changes.

In the United States, the FCC(Federal Communications Commission) and NTIA(National Telecommunications and Information Administration) are responsible for spectrum management. The FCC is an independent agency reporting to the Congress, which is in charge of all the telecommunication services outside of the federal government. The NTIA represents president of the United States to manage the spectrum used by the federal government. The IRAC(Interdepartment Radio Advisory Committee) under NTIA collects all the requirements from different branches of the federal government and proposes solutions to Director of the NTIA. The FCC also sends representatives to the IRAC for coordination and resolution of possible conflicts.

In France, ART(Telecommunications Regulatory Authority), ANFR (National Frequencies Agency) and DiGITIP(Directorate General of Industry, Information and Technologies) are responsible for spectrum management. The ART reviews applications of telecommunication licenses, assigns spectrum to specialized communication services, arbitrates conflicts between operators, ensures interconnection among networks, facilitates fair competition, and so on. The DiGITIP charters licenses to the operators of public telephone networks, instruments the frameworks of telecommunication regulations, represents the French government to negotiate telecommunication agreements with foreign sovereignties, administrates ANFR, manages French telecommunication companies and postal departments, and so on. The ANFR represents the French government to participate international spectrum conferences like WRC(World Radio Conference), plans and executes spectrum allocation, and so on.

In Japan, the authorized agencies are ICPB(Information and Communications Policy Bureau) and TB(Telecommunications Bureau) under the MPHPT(Ministry of Public Management, Home Affairs, Post and Telecommunications). The ICPB plans general policies to promote the applications of electromagnetic information, plans futuristic development and improvement measures of wireline and wireless broadcasting, conducts research and development on technologies of electromagnetic information and its applications. The TB maintains the order of wireline and wireless equipments usage in transmitting electromagnetic information, plans futuristic development and improvement measures of telecommunication business, allocates spectrum, manages licensing, plans telecommunication and information policies pertaining to international affairs.

In Taiwan, the Republic of China, the MOTC(Ministry of Transportation and Communications) is the governmental organization responsible for spectrum management policy like regulation amendment, fee adjustment, approval of radio station establishment, frequency assignment, and so on. The MOTC authorizes the DGT(Directorate General of Telecommunications) to execute spectrum coordination, station inspection, interference monitoring and remedy, license application/review/issuance, fee collection, and so on.

Global frequency planning for potential telecommunication and wireless applications were discussed through the WARC in 1992, the WRCs in 1995, 1997 and 2000, all sponsored by the ITU. Recently allocated spectrum covers services like high frequency broadcasting, DAB(digital audio broadcasting), DTV(digital television), IMT-2000 and satellite communication. In the WRC-2000, 806-960 MHz, 1,710-1,885 MHz and 2,500-2,690 MHz bands were allocated for the third generation land mobile communication services because the frequencies below 3 GHz are more suitable for such applications. The allocated frequency bands for the first and second generations of land mobile communication might be reallocated to 3G and beyond 3G services. In the same conference, frequency bands have also been allocated for satellite mobile communication as a constituent of the IMT-2000(international mobile telecommunications-2000) system. DTV technologies like European DVB-T system or ATSC system in the U.S., if enhanced by networking and information technologies, may become competitive as compared to CATV and satellite TV. DAB technologies like European Eureka-147 system significantly improve the audio quality of broadcasting. The convergence or coexistence of DAB with conventional AM and FM technologies is an issue to be observed in the near future.

The ITS(intelligent transportation system) is identified as one of the critical infrastructure for national security by the United States. The system is implemented by utilizing advanced computer, information, electronics, communication and sensor technologies to enhance the interactions among users, vehicles and roads, to improve safety, efficiency and comfort of transportation, meanwhile to reduce the impact of transportation systems on environment and ecology. DSRC(dedicated short range communication) and other communication services supporting the ITS also demand new spectrum allocation.

Telecommunication policy in Taiwan was renovated in 1996. Since then, wireless services including paging, trunking radio, mobile data, GSM900, GSM1800, CT2 and low-power digital cordless telephone have been prospering on the market. Among them, the penetration rate of GSM900 and GSM1800 already exceeds 100 %. In the early 2002, five licenses were issued by the MOTC to allocate 170 MHz for 3G services, charging over 50 billion NT dollars which is about 1.5 billion US dollars.

WLL(wireless local loop) is an alternative last-mile solution to the PSTN(public switched telephone network) operators. LMDS(local multipoint distribution service) is a broadband wireless air interface for transmitting voice, data and video. Both technologies require spectrum allocation to provide services. The ISM(industrial, scientific, and medical) bands are unlicensed bands, basically allocated for low-power applications and services. WLAN(wireless local area network) can be viewed as an extension of existing wireline LAN. It has the advantages of easy installation, scalable network expansion, and may find applications in public facilities like hospitals, retail stores, wholesale depots, manufacturing sites, academic campus, stations, shops, and so on. The WLAN complying with IEEE 802.11 series of standards uses the ISM band. Due to its potential market value, spectrum allocation to WLAN has become a focused issue. The competition between WLAN and 3G systems becomes another issue which involves electromagnetic interference, revenue, public interest, licensing, fair competition, and so on.

References

[1] I. Erdin, M. S. Nakhla, and R. Achar, "Circuit analysis of electromagnetic radiation and field coupling effects for networks with embedded full-wave modules," *IEEE Trans. Electromagn. Compat.*, vol.42, no.4, pp.449-460, Nov. 2000.

[2] G. Cerri, R. de Leo, and V. M. Primiani, "Theoretical and experimental evaluation of the electromagnetic radiation from apertures in shielded enclosures," *IEEE Trans. Electromagn. Compat.*, vol.34, no.4, pp.423-432, Aug. 1992.

[3] J. Mehta, D. Bravo, and K. K. O, "Switching noise picked up by a planar dipole antenna mounted near integrated circuits," *IEEE Trans. Electromagn. Compat.*, vol.44, no.2, pp.282-290, May 2002.

[4] P. E. Fornberg, M. Kanda, C. Lasek, M. Piket-May, and S. H. Hall "The impact of a nonideal return path on differential signal integrity," *IEEE Trans. Electromagn. Compat.*, vol.44, no.1, pp.11-15, Feb. 2002.

[5] P. Bernardi, R. Cicchetti, and A. Faraone, "A full-wave characterization of an interconnecting line printed on a dielectric sbla backed by a gridded ground plane," *IEEE Trans. Electromagn. Compat.*, vol.38, no.3, pp.237-243, Aug. 1996.

[6] X. Ye, D. M. Hockanson, M. Li, Y.R, W. Cui, J. L. Drewniak, and R. E. DuBroff, "EMI mitigation with multilayer power-bus stacks and via stitching of reference planes," *IEEE Trans. Electromagn. Compat.* vol.43, no.4, pp.538-548, Nov. 2001.

[7] D. M. Hockanson, X. Ye, J. L. Drewniak, T. H. Hubing, T. P. van Doren, and R. E. DuBroff, "FDTD and experimental investigation of EMI from stacked-card PCB configurations," *IEEE Trans. Electromagn. Compat.*, vol.43, no.1, pp.1-10, Feb. 2001.

[8] T. W. Kang, Y. C. Chung, S. H. Won, and H. T. Kim, "On the uncertainty in the current waveform measurement of an ESD generator," *IEEE Trans. Electromagn. Compat.*, vol.42, no.4, pp.405-413, Nov. 2000.

[9] H. Sasaki, T. Harada, and T. Kuriyama, "A new VLSI decoupling circuit for suppressing radiated emissions from multilayer printed circuit boards," *IEEE Int. Symp. Electromag. Compat.*, pp.157-162, Nov. 2000.

[10] A. Madou and L. Martens, "Electrical behavior of decoupling capacitors embedded in multilayer PCBs," *IEEE Trans. Electromagn. Compat.*, vol.43, no.4, pp.549-556, Nov. 2001.

[11] H. Shi, F. Sha, J. L. Drewniak, T. P. van Doren, and T. H. Hubing, "An experimental procedure for characterizing interconnects to the DC power bus on a multilayer printed circuit board," *IEEE Trans. Electromagn. Compat.*, vol.39, no.4, pp.279-285, Nov. 1997.

[12] H. Shi, F. Sha, J. L. Drewniak, T. P. van Doren, and T. H. Hubing, "An experimental procedure for characterizing interconnections to the DC power bus on a multilayer printed circuit board," *IEEE Trans. Electromagn. Compat.*, vol.39, no.4, pp.279-285, Nov. 1997.

[13] S. van den Berghe, F. Olyslager, D. de Zutter, J. de Moerloose, and W. Temmerman, "Study of the ground bounce caused by power plane resonances," *IEEE Trans. Electromagn. Compat.*, vol.40, no.2, pp.111-119, May 1998.

[14] J. Fan, J. L. Drewniak, J. L. Knighten, N. W. Smith, A. Orlandi, T. P. van Doren, T. H. Hubing, R. E. DuBroff, "Quantifying SMT decoupling capacitor placement in DC power-bus design for multilayer PCBs," *IEEE Trans. Electromagn. Compat.*, vol.43, no.4, pp.558-599, Nov. 2001.

[15] D. M. Hockanson, J. L. Dreniak, T. H. Hubing, T. P. van Doren, F. Sha, C.-W. Lam, and L. Rubin, "Quantifying EMI resulting from finite-impedance reference planes," *IEEE Trans. Electromagn. Compat.*, vol.39, no.4, pp.286-297, Nov. 1997.

[16] M. H. Pong, X. Wu, C. M. Lee, and Z. Qian, "Reduction of crosstalk on printed circuit board using genetic algorithm in switching power supply," *IEEE Trans. Indust. Electron.*, vol.48, no.1, pp.235-238, Feb. 2001.

[17] C. Schuster and W. Fichtner, "Parasitic modes on printed circuit boards and their effects on EMC and signal integrity," *IEEE Trans. Electromagn. Compat.*, vol.43, no.4, pp.416-425, Feb. 2001.

[18] A. Deutsch, G. V. Kopcsay, P. W. Coteus, C. W. Surovic, P. E. Dahlen, D. L. Heckmann, and D. W. Duan "Frequency-dependent losses on high-performance interconnections," *IEEE Trans. Electromagn. Compat.*, vol.43, no.4, pp.446-465, Nov. 2001.

[19] S. Delmas-Bendhia, F. Caignet, E. Sicard, and M. Roca, "On-chip sampling in CMOS integrated circuits," *IEEE Trans. Electromagn. Compat.*, vol.41, no.4, pp.403-406, Nov. 1999.

[20] G. Cerri, R. D. Leo, and V. M. Primiani, "A rigorous model for radiated emission prediction in PCB circuits," *IEEE Trans. Electromagn. Compat.*, vol.35, no.1, pp.102-109, Feb. 1993.

[21] J. P. Hertel, I. D. Flintoft, S. J. Porter, and A. C. Marvin, "Measurement of EMI on network cables due to multiple GSM phones," *IEEE Trans. Electromagn. Compat.*, vol.42, no.4, pp.358-367, Nov. 2000.

[22] C. L. Holloway and E. F. Kuester, "Net and partial inductance of a microstrip ground plane," *IEEE Trans. Electromagn. Compat.*, vol.40, no.1, pp.33-46, Feb. 1998.

[23] J. R. Regué, M. Ribó, J. M. Garrell, and A. Martín, "A genetic algorithm based method for source identification and far-field radiated emissions prediction from near-field measurements for PCB characterization," *IEEE Trans. Electromagn. Compat.*, vol.43, no.4, pp.520-530, Nov. 2001.

[24] X. Ye and J. L. Drewniak, "FDTD modeling incorporating a two-port network for I/O line EMI filtering design," *IEEE Trans. Electromagn. Compat.*, vol.44, no.1, pp.175-181, Feb. 2002.

[25] Q. Yu and T. W. Holmes, "A study on stray capacitance modeling of inductors by using the finite element method," *IEEE Trans. Electromagn. Compat.*, vol.43, no.1, pp.88-93, Feb. 2001.

[26] G. R. Piper and A. Prata, Jr., "Magnetic flux density produced by finite-length twisted-wire pairs," *IEEE Trans. Electromagn. Compat.*, vol.38, no.1, pp.84-92, Feb. 1996.

[27] D. M. Hockanson, J. L. Drewniak, T. H. Hubing, T. P. van Doren, F. Sha, and M. J. Wilhelm, "Investigation of fundamental EMI source mechanisms driving common-mode radiation from printed circuit

boards with attached cables," *IEEE Trans. Electromagn. Compat.*, vol.38, no.4, pp.557-566, Nov. 1996.

[28] D. M. Hockanson, J. L. Drewniak, T. H. Hubing, and T. P. van Doren, "FDTD modeling of common-mode radiation from cables," *IEEE Trans. Electromagn. Compat.*, vol.38, no.3, pp.376-387, Aug. 1996.

[29] D. W. P. Thomas, A. C. Denton, T. Konefal, T. Benson, C. Christopoulos, J. F. Daeson, A. Marvin, S. J. Porter, and P. Sewell, "Model of the electromagnetic fields inside a cuboidal enclosure copulated with conducting planes or printed circuit boards," *IEEE Trans. Electromagn. Compat.*, vol.43, no.2, pp.161-169, May 2001.

[30] G. Caccavo, G. Cerri, V. M. Primiani, L. Pierantoni, and P. Russo, "ESD field penetration into a populated metallic enclosure: A hybird time-domain approach," *IEEE Trans. Electromagn. Compat.*, vol.44, no.1, pp.243-249, Feb. 2002.

[31] W. Wallyn, D. D. Zutter, and H. Rogier, "Prediction of the shielding and resonant behavior of multisection enclosures based on magnetic current modeling," *IEEE Trans. Electromagn. Compat.*, vol.44, no.1, pp.130-138, Feb. 2002.

[32] M. Li, J. Nuebel, J. L. Drewniak, R. E. DuBroff, T. H. Hubing, and T. P. van Doren, "EMI from cavity modes of shielding enclosures-FDTD modeling and measurements," *IEEE Trans. Electromagn. Compat.*, vol.42, no.1, pp.29-38, Feb. 2000.

[33] M. Li, J. Nuebel, J. L. Drewniak, T. H. Hubing, R. E. DuBroff, and T. P. van Doren, "EMI reduction from airflow aperture arrays using dual-perforated screens and loss," *IEEE Trans. Electromagn. Compat.*, vol.42, no.2, pp.135-141, May 2000.

[34] T. Yamane, A. Nishikata, and Y. Shimizu, "Resonance suppression of a spherical electromagnetic shielding enclosure by using conductive dielectrics," *IEEE Trans. Electromagn. Compat.*, vol.42, no.4, pp.441-448, Nov. 2000.

[35] J. V. B. Tejedor, L. Nuno, and M. F. Bataller, "Susceptibility analysis of arbitrarily shaped 2-D slotted screens using a hybird generalized scattering matrix finite-element technique," *IEEE Trans. Electromagn. Compat.*, vol.40, no.1, pp.47-54, Feb. 1998.

[36] M. S. Sarto, S. D. Michele, and P. Leerkamp, "Electromagnetic performance of innovative lightweight shields to reduce radiated emissions from PCBs," *IEEE Trans. Electromagn. Compat.*, vol.44, no.2, pp.53-363, May 2002.

[37] M.-S. Lin and C. H. Chen, "Plane-wave shielding characteristics of anisotropic laminated composites," *IEEE Trans. Electromagn. Compat.*, vol.38, no.3, pp.21-27, Aug. 1996.

[38] P. B. Jana, A. K. Mallick, and S. K. De, "Effects of sample thickness and fiber aspect ratio on EMI shielding effectiveness of carbon fiber filled polychloroprene composites in the X-band frequency range," *IEEE Trans. Electromagn. Compat.*, vol.38, no.3, pp.478 -481, Aug. 1996.

[39] Y. T. Lin, M. A. Uman, and R. B. Standler, "Lightning return stroke models," *J. Geophys. Res.*, vol.85, no.C3, pp.1571-1583, Mar. 1980.

[40] A. Karwowski and A. Zeddam, "Transient currents on lightning protection systems due to the indirect lightning effect science, measurement and technology," *Proc. Inst. Elect. Eng.*, vol.142, no.3, pp.213-222, May 1995.

[41] F. Heidler and C. Hopf, "Lightning current and lightning electromagnetic impulse considering current reflection at the earth's surface," *Proc. Int. Conf. Lightning Protect.*, R4-05, Budapest, Hungary, 1994.

[42] V. A. Rakov and M. A. Uman, "Review and evaluation of lightning return stroke models including some aspects of their application," *IEEE Trans. Electromagn. Compat.*, vol.40, no.4, pp.403-426, Nov. 1998.

[43] P. C. T. van der Laan and A. P. J. van Deursen, "Reliable protection of electronics against lightning: some practical applications," *IEEE Trans. Electromagn. Compat.*, vol.40, no.4, pp.513-520, Nov. 1998.

[44] "Wireless frequency resource planning and management (Chinese)," MOTC-DGT, Taiwan, ROC, May 2002.

Chapter 24

ELECTRICAL PACKAGING FOR MICROWAVE AND MILLIMETER WAVE CIRCUITS

Shyh-Jong Chung[1] and Hao-Hui Chen[2]

[1] *Department of Communication Engineering*
National Chiao-Tung University
Hsin-Chu, Taiwan, ROC

[2] *Department of Electronic Engineering*
Hua-Fan University
Taipei, Taiwan, ROC

Abstract In this Chapter, packaging technology for microwave/millimetre wave chips and circuits is briefly reviewed. Electrical interconnections in chip level and in a multilayered substrate are depicted after a brief introduction to the design concept of advanced packaging. Passive components can be embedded in a multilayered structure which provides more design flexibility. Finally, issues of module housing are discussed.

Keywords: 3-D packaging, chip package, multilayered interconnections, system-on-package.

1. General Description

Microwave and millimeter wave circuits are often enclosed in a package for mechanical and humidity protection, which also provides electrical and thermal connections between circuits and systems. The packages should provide signal pathways from one circuit or MMIC chip to others and prevent parasitic electromagnetic interference. Good package may also improve circuit performance due to better interconnection design, and reduce its physical size by utilizing multilevel packaging configurations. Moreover, by using novel multilayered technologies, the package itself can even become part of the circuits, embedding passive components like resistors, inductors, capacitors, filters and baluns.

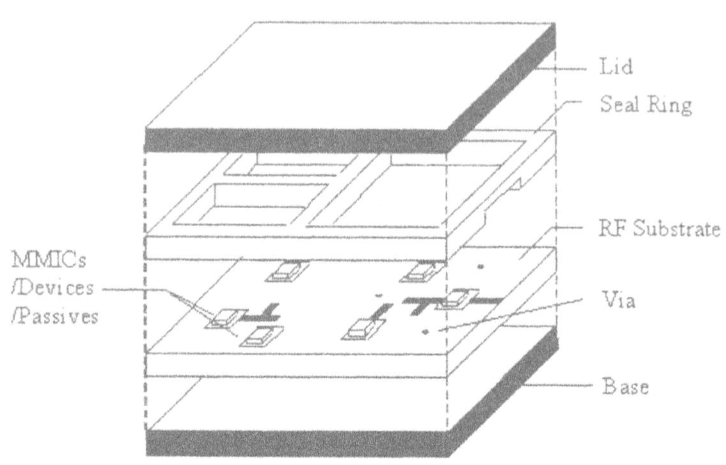

Figure 24.1. Typical packaging structure including metal base, RF substrate, seal ring and top lid.

Fig.24.1 shows a typical structure in microwave or millimeter wave packaging. The structure includes a metal base for heat sinking and mechanical supporting, a ceramic or polymide RF substrate, a metal coated seal ring, and a lid for mechanical, electrical and humidity protection. The RF substrate can be a single-layered or multilayered structure, where MMICs, active/passive components and signal distribution networks are embedded. MMICs or active devices are mounted on the substrate via bonding wires, beam leads, surface mount technology(SMT), flip-chip or ball grid array(BGA). They can also be directly deposited on the metal base by milling a cavity in the substrate. The substrate provides space for the layout of passive circuits and matching networks for active devices. It also provides electrical connection between circuit elements. In most applications, passive circuits and signal distribution networks occupy a large fraction of the total circuit area. To reduce package size, these passive components can be designed and buried in a multilayered substrate using vertical interconnections. The seal ring extends the spacing between RF substrate and the lid so as to reduce proximity coupling. Many parasitic effects can be induced from bonding wires, discontinuities between transmission lines, vertical vias in the substrates and leakage from active devices. These parasitics may cause spurious radiation, increasing undesired coupling between components, and even leading to package resonance. By configuring the seal ring, the package can be properly cabineted to reduce intercomponent coupling, increase isolation and avoid package resonance.

Choice of materials affects the performance of package. Apart from electrical considerations such as dielectric constant and loss tangent, the coefficient of thermal expansion(CTE) and thermal conductivity are two important factors for packaging. The CTE determines thermal match between chips and substrate, as well as between chips and metals when they are assembled together. Thermal mismatch will induce mechanical distortion, crack large-area dies, and sway packaging hermeticity. Thermal conductivity depicts the heat dissipation capability of material, which is often required to be greater than 150-180 W/m/°K for high-power applications [1]. Most of the microwave and millimeter wave chips are made from GaAs which has CTE of around 5.3 to 6.5 ppm/°C. To obtain good thermal match, metal and substrate with similar CTEs should be chosen.

Table 24.1. Thermal properties of some packaging metals.

Metal	Kovar®	CuMo	CuW	Al	Cu	Mo
CTE (ppm/°C)	5.3	7.0	6.5	25.4	17.0	5.1
Thermal conductivity (W/m/°K)	16.7	184.2	209.3	221.0	393.5	159.1

Table 24.1 lists CTEs and thermal conductivities of some commonly used metals [1]. Kovar®(an alloy of iron, nickel and cobalt), copper molybdenum(CuMo), copper tungsten(CuW) and molybdenum(Mo) have similar CTEs as GaAs, and thus can be used as carrier material for MMIC dies. Among them, low-cost Kovar® is widely used for bases and seal rings in packages. However, when mounting MMIC on a Kovar® base, it is better to insert CuMo or CuW shims in between in order to prevent MMIC fracture. These shims also improve heat sinking for high-power chips due to their high thermal conductivities. Table 24.2 shows thermal and electrical properties of some substrate materials, including ceramics, Duroids and polymers [1]-[3]. Commonly used alumina(Al_2O_3) and low-temperature cofired ceramics(LTCC) have CTEs matched to that of GaAs, and are thus suitable for directly mounting low-power

MMICs. For high-power chips, heat spreaders such as molybdenum or diamond should be introduced between chips and substrates with low thermal conductivity. Another way is to mill a chip well in the substrate and mount chip on the thermally conducting carrier. Substrate materials with high thermal conductivity such as aluminium nitride(AlN) can also be used.

Table 24.2. Thermal properties of some substrate materials.

Material	Alumina (99.6%)	ALN	LTCC (DuPont 951AT)	Duroid 3003 6002	RO 4003	Polymide (DuPont 2611)	BCB
CTE (ppm/°C)	7.1	4	4-7	17	13	3	52
Thermal conductivity (W/m/°K)	26	200	3	0.44	0.64	-	-
Dielectric conductivity	9.7	8.5	7.8	3.0	3.38	3.1	2.7
Dielectric loss tangent	0.0006	-	0.0015	0.0013	0.002	0.002	0.08

Besides RF circuitries, dc and low-frequency circuits fabricated in another substrate can also be integrated to the package. RF and dc substrates can be placed on a common motherboard, or be stacked for size reduction. Fig.24.2 illustrates the concept of mixed signal package [4]. RF and dc circuitries are implemented on separate substrates which usually have multilayered configurations that contain vertical interconnections. To provide electrical paths for signals and grounds between two circuitries, a solder-free interconnection(SFI) board is inserted between RF and dc substrates. This SFI board comprises metal contacts(fuzz buttons) and dielectric support. Signals and grounds in the circuitries are connected through these fuzz buttons and vertical interconnections in the RF and dc substrates. The lid for RF circuitry can be attached to the high-power chips for thermal dissipation. To reach a more compact design, RF or dc circuits can be partitioned and implemented on several stack-up substrates. Such a system-on-package(SOP) concept has obtained much attention in recent years, which merges many or all electronic components of a functional system or subsystem into one package [3]-[6]. These components may include RF front-end transceivers, antennas, matching or biasing circuits, DSPs, memories, ASICs, buses and interconnections, control logic circuits(CLC), and so on. Contrary to the system-on-chip(SOC) methodology which includes most of the electronic components in a single chip, the SOP is a heterogeneous integration technology. It is not constrained by a common wafer process. Each IC chip in an SOP can be optimally designed and manufactured [5]. The SOP is thus a quicker, low-cost, efficient and low-risk solution.

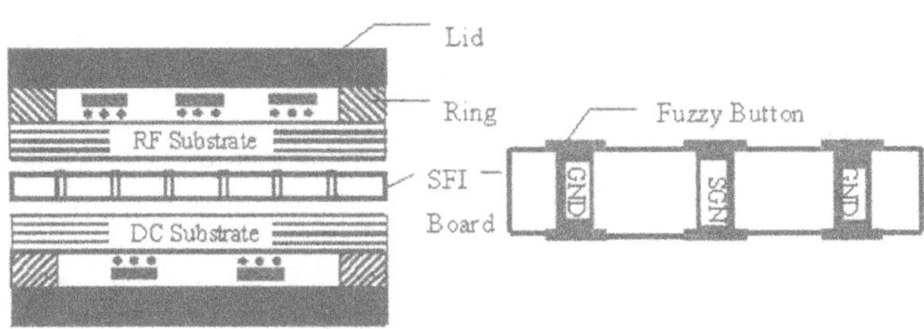

Figure 24.2. Packaging structure demonstrating the concept of system-on-package(SOP) [4].

2. Chip Packaging

2.1 Chip Interconnections

In microwave/millimeter wave circuits, MMIC chips may be packaged in a chip carrier or be directly attached to the RF substrate. Chip interconnections, electrical connections from chip to substrate and lead frame of the chip carrier should be carefully designed to avoid adverse effects such as parasitic reactance, radiation and reflection loss. Principal chip interconnection technologies in use are wire bonding(WB), tape automated bonding(TAB) and flip-chip(FC) bonding as shown in Fig.24.3. Each of them has its own distinct characteristics and preferred applications [7]. Bonding wires are gold, aluminum or copper fine wires used to connect a chip to the lead frame or substrate. This technology is less expensive, flexible and tolerant of chip thermal expansion. Therefore, it is the most commonly used chip interconnection technology. However, parasitic effects like skin effect resistance, radiation loss, mutual coupling and wire inductance may be introduced at the wire bonding interconnections. These parasitic effects become significant with the increase of operating frequency and should be minimized. The skin effect resistance is due to resistivity of the bonding wire and can be reduced by using short, highly conductive wires.

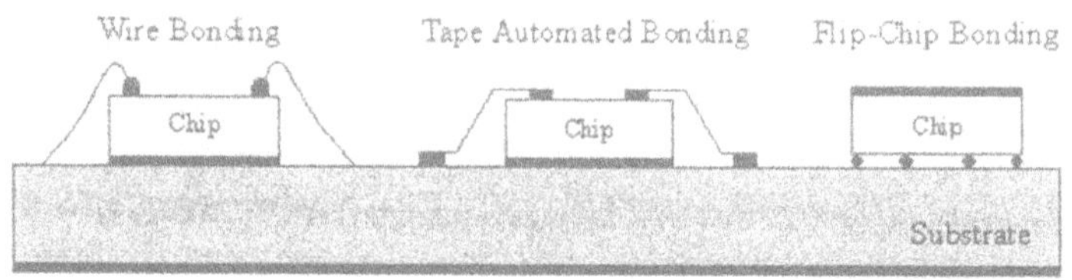

Figure 24.3. **Wire bonding, tape automated bonding and flip-chip bonding for chip interconnections.**

Radiation from wire bonding interconnections results in not only undesired loss but also mutual coupling between bonding wires. Such radiation can be reduced by shortening the wire length from bonding pad or using nonparallel wires, for example, orienting adjacent bonding wires by an angle. The wire inductance has dominant effect, which is in general dependent on the length and shape of the wire. Reducing wire length or using multiple bonding wires can reduce the wire inductance and thus improve the transmission properties of bonding wires at high frequencies.

In addition to minimizing the parasitic effects at interconnections, suitable designs for connecting chip ground and substrate ground should be employed to achieve good ground shielding for the chip circuits. This can be accomplished by directly attaching the chip ground to the substrate ground using molybdenum or copper molybdenum, or by using several grounding vias to connect the grounds as shown in Fig.24.4.

Electrical characteristics of bonding wires can be modeled by an equivalent π-network that consists of a series inductance and two shunt capacitances [8]. The series inductance models the bonding wire while the shunt capacitances account for the discontinuities at wire-substrate(wire-lead) and wire-chip joints. Based on the equivalent π-network, suitable matching networks can be designed to compensate for the parasitic reactance at wire bonding interconnections.

Tape automated bonding uses patterned metal leads to connect chip and substrate. In this technology, a chip is first attached to the inner rim of the patterned leads using gold, aluminum or solder bumps. The

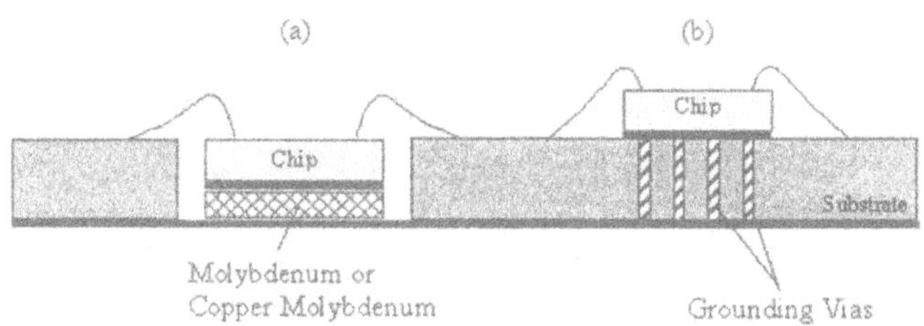

Figure 24.4. Chip ground shielding: (a) direct attachment, (b) grounding vias connection.

bonded chip is then mounted to the substrate. The outer rim of the leads is finally attached to the substrate using bumps. The TAB technology provides a highly automatic, high precision and gang bonding(all leads are bonded simultaneously) solution for high-density interconnections. However, since specific design of patterned leads is required for each chip design, it may not be as flexible as the wire bonding technology. Moreover, electrical characteristics of the TAB interconnection is more complicated because the metal leads have nonuniform width and are closely spaced. The mutual coupling between leads has to be rigorously characterized. Since the TAB technology is usually applied to designs with a large number of interconnections, analysis on TAB interconnections will involve coupled structure with a large number of conductors.

The flip-chip bonding technology uses solder bumps to connect chip and substrate as shown in Fig.24.5. The solder bumps can be gold, gold-tin, lead-tin or lead-indium micropillar. In microwave applications, flip-chip bonding technology is very attractive in comparison with wire bonding or TAB due to its short and stable interconnections. Moreover, the capability of bonding with two-dimensional bond pad array makes this technology a preferred solution for multichip module designs.

However, flip-chip bonding also introduces some parasitic effects that may affect circuit performance. For a flip-chip microwave circuit, parasitic reactance at the bump interconnection, detuning effect on chip circuits, and excitation of parasitic substrate modes are three major issues that should be carefully examined [9]. To reduce these parasitic effects, the bumps should be appropriately designed. The bump diameter and dielectric overlap should be as small as possible to reduce reflection at interconnection. The bump inductance and radiation loss increase with bump height. On the other hand, low-profile bump may lead to detuning effects on the chip, which are caused by electromagnetic inter-

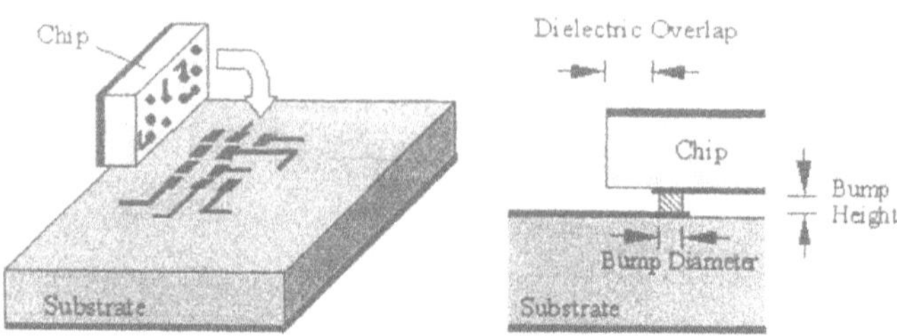

Figure 24.5. Flip-chip technology and its design parameters.

action between chip and substrate. Such interaction not only results in undesirable changes on the electrical characteristics of chip circuits but may also degrade the transition performance of interconnection. The bumps should be made with minimum height to avoid detuning effects and to have low parasitic inductance. Finally, the parasitic substrate modes excited by flip-chip interconnections can induce undesired coupling between interconnections. By increasing the number of grounding bumps, mutual coupling effects can be reduced [10].

Electrical characteristics of flip-chip interconnections are generally described by an equivalent π-network consisting of two shunt capacitances and a series inductance [9]. The capacitance due to dielectric loading by the chip at the transition is determined by the bump height and dielectric overlap. The inductance is caused by changes in direction and distribution of the current density at the bump interconnection, and is therefore dependent on the bump geometry. Since flip-chip interconnection is usually realized using a short bump with small diameter, it is primarily capacitive. An effective shunt capacitance combining both capacitive and inductive effects can thus be used to model the flip-chip interconnection more compactly [9]. When loss is of concern, appropriate resistive elements can be inserted into the equivalent circuit model.

Based on the electrical model, suitable matching networks can be implemented to improve the performance of flip-chip interconnection. Using staggered bump designs can compensate for the capacitive behavior of interconnection and thus reduce reflection [9]. The staggered bumps are also used in resonant flip-chip transition design [11], which improves the transition property of interconnection in a given frequency band. The optimal staggering length is determined by three-dimensional full-wave analysis, the matching network can also be designed using onboard compensation [9]. This approach uses a distributed series inductance on the substrate, which can be implemented by a high-impedance line section, to compensate for the capacitance. An extended design that consists of a series inductance and a shunt capacitance can also be applied. Since the compensation network is applied only to the substrate and can be designed by using a commercial circuit design software, this approach is more flexible and efficient than the staggered bump design.

2.2 Chip Packages

Next, consider the interconnections between a packaged MMIC chip and RF substrate in microwave circuits. A chip is encased in a molded plastic or multilayered ceramic package which is then mounted to substrate using various package-interconnection technologies as shown in Fig. 24.6. Through-hole packages with pins around the perimeter or on the bottom surface, such as dual in line packages(DIPs) and pin grid array(PGA) packages are mounted using pin-through-hole(PTH) technology [7]. The PTH technology requires to drill through holes in the substrate and is, therefore, expensive in manufacturing. Also, the long metal pins may result in significant parasitic reactance and/or loss at high frequencies.

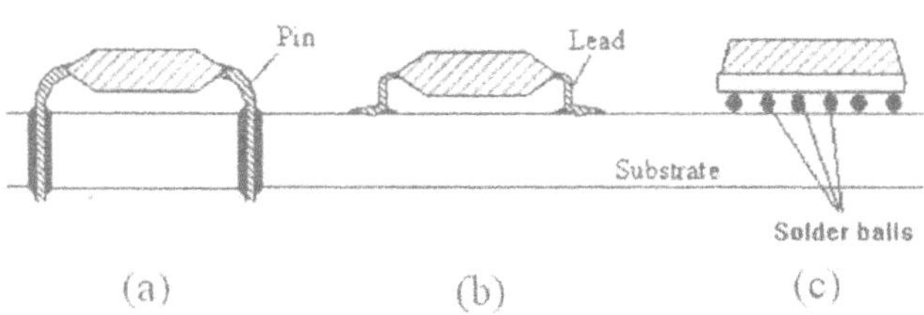

Figure 24.6. Cross sectional views of (a) through-hole package, (b) surface mount package and (c) ball grid array package.

On the other hand, surface mount packages with metallic leads or pads, such as thin quad flat packages(TQFPs) and ceramic leadless chip carriers(CLCCs), are realized using surface mount technology(SMT) [7]. The surface mount packages can be developed with higher I/O density than the through-hole packages, thus reduce the size and cost of system. Also, the surface mount packages offer better electrical performance due to shorter interconnections. Recently, the trend of package shifts from surface mount package to ball grid array(BGA) package. BGA package is a surface mount version of the PGA package, which uses a two-dimensional solder ball array instead of pins to connect the package and substrate [12]. It has the advantage of surface mountability, reduced size and weight, high interconnection density and low cost in assembly. Moreover, the BGA packages have minimum electrical path to the substrate, which leads to low loss and low reflection at interconnections.

Once a chip is encapsulated in a package, not only the package interconnections(pins, leads or solder balls), but also the package structure(package substrate, chip paddle) will affect chip performance. Package parasitic effects such as poor grounding, mutual coupling between leads and parasitic inductance of the path from chip ground to substrate ground may drastically change the chip characteristics. In order to avoid costly design cycles, an accurate electrical model is required to predict parasitic behavior of the chip package. Development of circuit model for chip package based on full-wave simulations or measurements is preferred since high frequency phenomena such as mutual coupling, resonance and frequency dependent reactance and loss can be demonstrated clearly.

Due to complicated configuration of the package and complicated interaction between different parts of the packaged circuits, segmentation techniques like layer-by-layer [12] or step-by-step [13] approaches are usually adopted in the modeling procedure. By these techniques, the whole packaged chip is first segmented into several parts, for example, package interconnection, package substrate and chip paddle. Lumped element circuit model for each part is then developed by conducting full-wave simulations or measurements. The circuit models are next connected by using suitable coupling elements to develop an equivalent circuit of the package. In each connection, tuning procedure can be employed to obtain more accurate results. This technique can reduce modeling cost because a given package needs to be modeled only once. Then, suitable circuit design for optimizing the transition properties of package interconnections can be further developed.

3. Multilayered Packaging

As stated earlier, an RF substrate is used to embed active and passive components, support electronic interconnection between components or circuits, and possibly release heat dissipation. Multilayered configurations can be used to improve circuit performance and reduce package size. When tackling multilayered packaging problems, vertical interconnections between different layers and implementation of buried passive components have been considered.

3.1 Vertical Interconnections

Vertical interconnections of signals or grounds between different layers can be achieved by metal coated vias in the substrate or by coupling apertures on the ground planes. Fig.24.7 shows a schematic of vertical interconnections in a four-layered substrate. Three kinds of vias are used: through vias, blind vias and buried vias. Signal on microstrip line on the top surface is electromagnetically coupled to the strip line through a coupling aperture on the first ground plane, and then linked to microstrip line on the bottom surface using a blind via. The signal is directed to another microstrip line on the top surface with a through via. Two buried vias are built between grounds of the strip line to equalize their potentials. These interconnections create parasitic effects due to discontinuities along transmission lines.

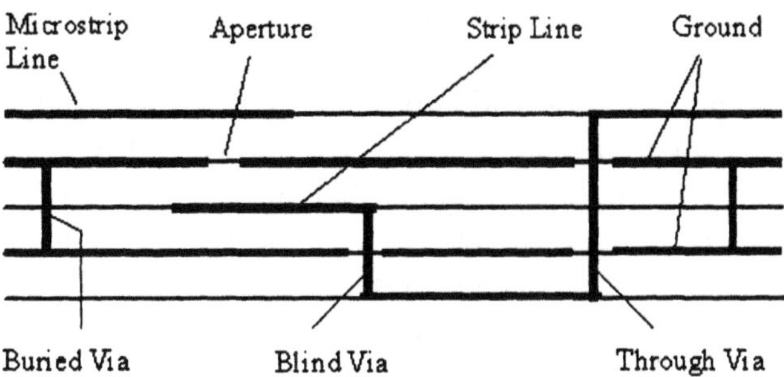

Figure 24.7. Schematic of vertical interconnections in a four-layered substrate.

To model a via interconnection, a series inductance and two shunt capacitances between the via and grounds are introduced at the transitions. These parasitics may cause signal reflection and degrade the circuit performance, especially at high frequencies. Matching networks may be implemented at the transitions to cancel parasitic effects. For an aperture coupled transition, the aperture behaves as a transformer which provides a path for energy coupling between the transmission lines in different layers. The aperture size determines the property of transformer. To each transmission line, the aperture is equivalent to a series load which incorporates the contribution of transformer and other transmission line. To achieve a total power transfer, this load should have a series resistance equal to the characteristic impedance of transmission line. Reactance of the load can be compensated by using an open-end stub.

Loading effect of the transition to the signal line not only creates reflected signal, but also couples signal power to the substrate. The power may be coupled by radiation of the vertical electric current carried along vias or the equivalent horizontal magnetic current on the aperture. The coupled power is radiated into air or is confined between ground planes of the strip line in the form of parallel-plate mode [14], which deteriorates potential balance of ground planes, thus creating noise on the strip line. Also, when guided to another transition in the same substrate, the parallel-plate mode couples power to that transition and degrades the isolation between two transitions. One way to eliminate excitation of such modes is to use vias to connect the grounds of strip lines around the transitions. These vias will keep the ground planes at the same potential and avoid exciting those modes.

3.2 Multilayered Passive Components

Passive components can be integrated and embedded inside a multilayered substrate to save circuit area, increase packaging density, and reduce insertion losses. Fig.24.8 shows the schematics of multilayered three-dimensional inductor and capacitor [3], [15]. As compared to two-dimensional planar inductor, the inductor shown in Fig.24.8(a) occupies smaller substrate area. It also has higher inductance due to higher mutual inductive coupling between turns. Equivalent circuit of the 3-D inductor includes an inductance in series with a resistance to account for conductor loss. The series impedance is then in shunt with a capacitance which accounts for the electrical coupling between the turns and ground plane. The shunt capacitance brings on smaller effective inductance and limits the self-resonant frequency of inductor, which is the transition frequency that the input impedance changes from inductive to capacitive. The larger the shunt capacitance is, the lower the effective inductance and self-resonant frequency are.

To enhance the series inductance, one may increase the total magnetic flux by adding more turns in the inductor. This, however, will increase the electrical coupling between turns and incur larger parasitic capacitance. One may also increase the effective inductance by increasing the separation between turns and ground plane so as to reduce the shunt capacitance. Notice that the image current induced on the ground plane flows in a direction opposite to that of the inductor current, and thus reduces the net magnetic flux flowing through the inductor. Increasing the separation between inductor and ground diminishes the counteracting magnetic flux through the turns and thus increases its effective inductance. An alternative is to pattern the ground plane with slots. This will restrict image current distribution and thus reduce the counteracting magnetic flux [16].

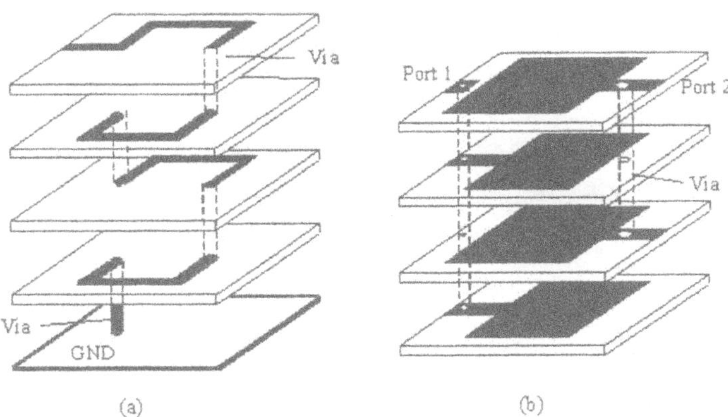

Figure 24.8. (a) Multilayered embedded inductor and (b) multilayered embedded vertical interdigitated capacitor [3].

The vertical interdigitated capacitor(VIC) shown in Fig.24.8(b) is designed with the advantage of three-dimensional packaging [3]. This structure is compatible to conventional MIM capacitors, which is composed of a dielectric layer sandwiched between two metal plates, except with smaller circuit area. The vias equalize potentials of the connected metal plates. Hence, the VIC is basically a parallel combination of MIM capacitors. As compared to planar interdigitated capacitor, the VIC provides larger capacitance because of the higher electrical coupling between electrodes.

Many other passive circuits can also be implemented using three-dimensional configurations. Since the design becomes more flexible, these circuits may have not only smaller sizes, but also better performance. By utilizing broadside coupling between parallel lines, higher coupling coefficient can be obtained. This will benefit the design of some circuits like filters and baluns. For example, filter designed with strongly coupled lines possesses wider bandwidth as well as lower in-band loss. Fig.24.9 shows a schematic of fourth-order multilayered filter, where broadside coupling, aperture coupling and edge coupling are utilized. Broadside coupling is utilized in sections where the strongest couplings are needed, for example, the input/output sections. Aperture coupling can be designed to have coupling coefficient in between those of broadside coupling and edge coupling. By controlling the aperture size, a wide range of coupling levels can be achieved, which is versatile for the design of various filters.

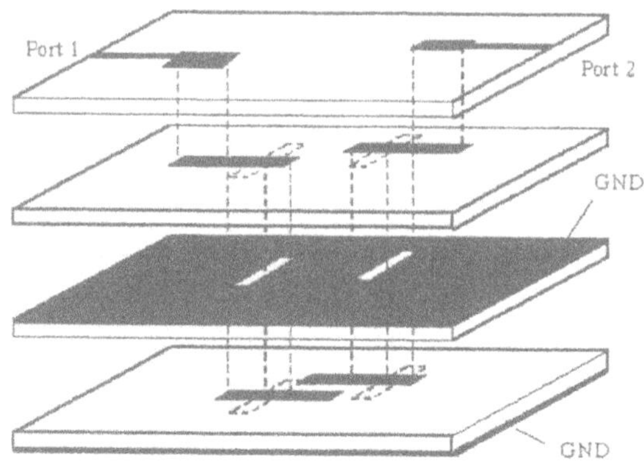

Figure 24.9. Fourth-order multilayered filter using broadside coupling, aperture coupling and edge coupling.

4. Module Housing

As shown in Fig.24.10, a circuit module is hermetically housed with a seal ring and a metal lid to avoid material contamination, mechanical distortion and electromagnetic interferences. However, the housing structure behaves like cavity, and may deteriorate circuit performance due to cavity resonances. When nonlinear active devices or high-Q components are excited, housing resonance will become dominant parasitic effect that needs to be suppressed by suitable housing design. Another important issue that should be considered is parasitic couplings between components and between adjacent ports in the module. These parasitic couplings are due to spurious waveguide modes propagating in the module housing. They need to be choked off especially in designing high-isolation circuits like switches.

Parasitic housing effects can be suppressed by several techniques. A straightforward approach is to design dimensions and/or configurations of the housing structure such that the resonant frequencies of housing and the lowest cutoff frequencies of spurious waveguide modes are raised far above the frequency range of interest. However, as the integration level and operating frequency increase, electrical size of the housing increases, making resonances and spurious modes difficult to avoid.

Moreover, since chips and circuit elements will affect the overall response, it is difficult to design an optimal housing for a given chip. Use of attenuating materials such as absorbers, lossy dielectrics, resistive coat-

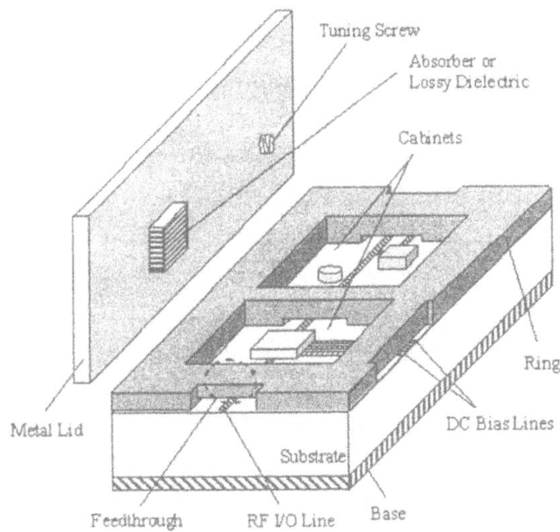

Figure 24.10. Cabinet housing structure of microwave module with tuning screws, attenuating material and feedthroughs.

ings, or tuning elements such as tuning screws, may be more practical. Attenuating materials are used to reduce the electromagnetic fields of resonant or spurious modes. Design of attenuating materials like size and position of absorbers, resistivity and thickness of resistive coatings should be carefully examined to obtain maximum suppression on parasitic housing effects to render negligible influence on the signal [17].

Another approach is to partition the housed circuit into several cabinets by properly configuring the seal ring. Cabinet housing structure can isolate individual chip or component from parasitic coupling and reduce the intensity of spurious modes excited in the cabinet by introducing destructive interferences between modes. Also, the cabinet structure can offer wider frequency band free of resonance [18].

Since housing effects are determined by the housing structure as well as chips and circuits in the module, they can be suppressed by on-board suppressing circuit design. An example of on-board suppressing circuit design is to deposit metal patches on the substrate [19]. The metal patches can block the propagation of parasitic waveguide modes by significantly deforming field distributions of these modes. They can also shift the resonant frequencies of housing outside the operating frequency band. Suppressing circuits can be fabricated on the substrate with other planar circuits in the module. However, guidelines for designing optimal on-board suppressing circuit need to be developed.

In addition to the parasitic housing effects, transition at feedthrough is another important issue. Feedthrough is designed for routing RF and dc lines into and out of the housing. Feedthrough can be implemented by opening hatches on the metal-coated seal ring to offer passageway for the feed lines. On the other hand, when the seal ring is made of dielectric without metallic coating, hatches on the seal ring are unnecessary and the feedthrough can be designed using microstrip-strip-microstrip transition structure.

If multiple RF and dc lines are fabricated in a feedthrough structure, grounding vias can be added between adjacent lines to effectively prevent the lines from undesired mutual coupling [1]. As the use of multilayered substrate increases, feedthrough design using vertical transition has received much attention recently. Vertical transitions can be easily realized by using metal-coated vias. However, parasitic effects of vias at high frequencies may deteriorate the transition performance. To design vertical transitions for microwave feedthrough, electromagnetic coupling structure can be applied [20]. With proper design, good transition can be achieved in the frequency band of interest.

References

[1] D. S. Wein, "Avanced ceramic packaging for microwave and millimetre wave applications," *IEEE Trans. Antennas Propagat.*, vol.43, no.9, pp.940-948, Sept. 1995.

[2] J. J. Wooldridge, "High density interconnect(HDI) packaging for microwave and millimetre wave circuits," *IEEE Aerospace Conf.*, vol.1, pp.369-376, 1998.

[3] A. Sutono, D. Heo, Y.-J. E. Chen, and J. Laskar, "High-Q LTCC-based passive library for wireless system-on-package(SOP) module development," *IEEE Trans. Microwave Theory Tech.*, vol.49, no.10, pp.1715-1724, Oct. 2001.

[4] N. K. Das and H. L. Bertoni, *Directions for the Next Generation of MMIC Devices and Systems*, ed., pp.105-111, New York: Plenum Press, 1997.

[5] C. M. Scanlan and N. Karim, "System-in-package technology, application and trends."

[6] R. R. Tummala and V. K. Madisetti, "System on chip or system on package?" *IEEE Design Test Comput.*, vol.16, no.2, pp.48-56, Apr.-June 1999.

[7] R. R. Tummala, E. J. Rymaszewski, and A. G. Klopfenstein, *Microelectronics Packaging Handbook*, New York: International Thomson Publishing, 1997.

[8] F. Alimenti, P. Mezzanotte, L. Roselli, and R. Sottentino, "Modeling and characterization of the bonding-wire interconnection," *IEEE Trans. Microwave Theory Tech.*, vol.49, pp.142-150, Jan. 2001.

[9] A. Jentzsch and W. Heinrich, "Theory and measurements of filp-chip interconnects for frequencies up to 100 GHz," *IEEE Trans. Microwave Theory Tech.*, vol.49, pp.871-878, May 2001.

[10] T. E. Kazior, H. N. Atkins, A. Fatemi, Y. Chen, F. Y. Colomb, and J. P. Wendler, "DBIT-direct backside interconnect technology:A manufacturable, bond wire free interconnect technology for microwave and millimeter wave MMICs," *IEEE MTT-S Int. Microwave Symp. Dig.*, pp.724-727, 1997.

[11] C.-L. Wang, C.-T. Hwang, R.-B. Wu, and C.-H. Chen, "A resonant flip-chip design with controllable transition band," *IEEE MTT-S Int. Microwave Symp. Dig.*, pp.1423-1426, 1999.

[12] R. Ito, R. W. Jackson, and T. Hongsmatip, "Modeling of interconnections and isolation within a multilayered ball grid array package," *IEEE Trans. Microwave Theory Tech.*, vol.47, pp.1819-1825, Sept. 1999.

[13] T.-S. Horng, S.-M. Wu, H.-H. Huang, C.-T. Chiu, and C.-P. Hung, "Modeling of lead-frame plastic CSPs for accurate prediction of their low-pass filter effects on RFICs," *IEEE Trans. Microwave Theory Tech.*, vol.49, pp.1538-1545, Sept. 2001.

[14] R. Abhari, G. V. Eleftheriades, and E. van Deventer-Perkins, "Physics-based CAD models for the analysis of vias in parallel-plate environments," *IEEE Trans. Microwave Theory Tech.*, vol.49, no.10, pp.1697-1707, Oct. 2001.

[15] A. Sutono, A.-V. H. Pham, J. Laskar, and W. R. Smith, "HRF/ microwave characterization of multiplayer ceramic-based MCM technology," *IEEE Trans. Adv. Packaging*, vol.22, no.3, pp.326-331, Aug. 1999.

[16] C. P. Yue and S. S. Wong, "On-chip spiral inductors with patterned ground shields for Si-based RF IC's," *IEEE J. Solid-State Circuits*, vol.33, no.5, pp.743-752, May 1998.

[17] H.-H. Chen and S.-J. Chung, "Analysis of a partially sealed package for microstrip-line circuits," *IEEE Trans. Microwave Theory Tech.*, vol.46, pp.2124-2130, Dec. 1998.

[18] S.-J. Chung and H.-H. Chen, "Resonant frequencies of a microwave package with vertical diaphragms attached to the top cover," *Microwave Opt. Technol. Lett.*, vol.8, pp.300-302, Apr. 1995.

[19] H.-H. Chen, C.-L. Wang, and S.-J. Chung, "A simple method for blocking parasitic modes in a waveguide-packaged microstrip-line circuit," *IEEE Trans. Microwave Theory Tech.*, vol.46, pp.2156-2159, Dec. 1998.

[20] K. Kitazawa, S. Koriyama, H. Minamiue, and M. Fujii, "77-GHz-band surface mountable ceramic package," *IEEE Trans. Microwave Theory Tech.*, vol.48, pp.1488-1491, Sept. 2000.

Chapter 25

INTRODUCTION TO CELLULAR SYSTEMS

Jean-Fu Kiang and Tsung-Sang Yang
Department of Electrical Engineering and
Graduate Institute of Communication Engineering
National Taiwan University
Taipei, Taiwan, ROC

Abstract In this Chapter, cellular systems in the past, present and future are briefly reviewed to provide basic understanding on their purposes, evolutions, challenges and opportunities.

Keywords: 1G, 2G, 2.5G, 3G, B3G, cordless, wireless, cellular, mobile, standard, network, service, device, capacity, TDMA, FDMA, CDMA, antenna, array, diversity.

1. 1G and 2G Systems

Mobile radio communication systems have been successfully deployed since 1980 in different regions of the world to extend telephone services to mobile users. First generation analog systems like AMPS were developed to provide voice telephony. Due to increasing demand of the mobile users, digital technologies were adopted in the second generation communication systems to provide more capacity and better radio performance than analog systems, and simple data services were also provided.

Mobile systems including paging, cordless telephone and cellular telephone have been widely deployed around the world. Cordless systems like radiophone use radio to connect portable handset to specific base station which is connected to the public switched telephone network(PSTN) by dedicated telephone line. Original cordless system has small coverage and is usually suitable for use inside buildings. More advanced cordless systems increase the coverage range up to a few hundred meters, hence they can be used outdoors with low mobility. In cellular systems, the coverage area is deliberately separated into several subregions, called cells. In each cell, wireless connection is provided between base station(BS) and mobile stations(MS). A call can be forwarded to the PSTN through mobile switching center(MSC) which also deals with handoff if mobile station roams from one cell to another.

Cordless systems have smaller cell coverage, lower transmitting power and lower mobility than cellular systems. Hence, cordless systems are more appropriate for use in office and residential environments, and cellular systems are more suitable for wide-area services.

1.1 AMPS

AMPS was introduced in the United States in 1983. It is an anlog cellular system with handoff and roaming capabilities. The uplink frequency is 824-849 MHz, the downlink frequency is 869-894 MHz. Every logic channel consists of an uplink channel and a downlink channel, separated by 45 MHz. Voice is modulated by frequency modulation(FM) for radio transmission, and frequency division multiple access(FDMA) with channel spacing of 30 kHz is used for multiple access. Cell coverage is up to 32 km, the antenna system is approximately 40 m height. The mobile power level is approximately 600 mW. The cells are connected by mobile telephone switching office(MTSO) to the wireline telephone switching fabric. The MTSO coordinates all switch functions, provides service features implemented by softwares, and manages power control function. Critical processing bottleneck occurs at the MTSO since it makes all the handoff decisions. Choosing larger cell size implies relatively fewer cell crossings, hence lighter load for the MTSO.

NAMPS was proposed in 1991 by Motorola to improve over AMPS in capacity. It divides one 30 kHz AMPS channel into three 10 kHz channels to triplicate the capacity of AMPS.

1.2 USDC(IS-54)

Dual-mode USDC/AMPS system was standardized as IS-54 by the EIA/TIA in 1990. USDC was developed with $\pi/4$ DQPSK digital modulation technique to improve the capacity and performance of AMPS in the United States. It uses the same frequency band as AMPS, uses time division multiple access(TDMA) to support three full-rate and six half-rate users on each AMPS channel, and uses frequency division duplex(FDD) technique for duplexing. The cell size is the same with

AMPS, but the average transmitting power of mobile station is reduced to 200 mW using digital technologies.

1.3 CT2

CT2 is the second generation cordless telephones introduced in Great Britain in 1989. It is used in office environment where user handset can access the PSTN through telepoint. The frequency range is 864.15-868.05 MHz with 40 time division duplex(TDD) channels and 100 kHz channel spacing, and GMSK modulation is used. The maximum cell radius is 100 m and the average transmitting power is only 5 mW.

1.4 DECT

DECT is a cordless system developed by the European Telecommunications Standard Institute(ETSI), which is finalized in 1992 as the first pan-European standard for cordless telephones. It is used to provide local mobility with voice and data services. DECT complies with the open systems interconnection(OSI) which makes it possible to connect with other mobile networks such as ISDN or GSM. The frequency band of DECT is 1,880-1,900 MHz, and frequencies from 1,881.792 MHz to 1,897.344 MHz are divided into ten channels with bandwidth of 1,728 kHz. DECT uses GMSK modulation, multiple carrier/TDMA for multiple access, and TDD for duplexing. The transmitting power is approximately 10 mW, and the effective communication range is up to 500 m. Hence, it is suitable for office and residential environments.

1.5 PACS

PACS system was developed and proposed by Bellcore in the United States in 1992, which is able to support voice, data, image and video services for indoor and microcells within 500 m. It is intended to integrate all forms of wireless local loop(WLL) communication into one system to serve as a last-mile option for local exchange carriers. The PACS system is TDMA-based, it adopts FDD with 1,850-1,910 MHz for uplink and 1,930-1,990 MHz for downlink, it also uses TDD in 1,920-1,930 MHz. $\pi/4$ QPSK modulation is used, and channel spacing is 300 kHz. The mobile unit uses adaptive power control to reduce co-channel interference.

1.6 PHS

PHS system was developed in Japan in 1993, intended for pedestrians in an indoor microcell environment. It uses TDMA, TDD,$\pi/4$ DQPSK modulation. The frequency band in 1,895-1,918.1 MHz is separated into 77 channels with channel spacing of 300 kHz. Forty channels in 1,906.1-1,918.1 MHz are used for public systems, and the other 37 channels in 1,895-1,906.1 MHz are used for home or office.

1.7 GSM

GSM system is a second generation cellular system introduced in 1991 in Europe and standardized by the ETSI. It is intended to solve the fragmentation problems of the first generation cellular systems in Europe. GSM services include emergency call, vediotex, teletex and cell broadcast. Data services in packet switched mode are supported with data rate ranging from 300 bps to 9.6 kbps. Short message service(SMS) is available through integrating with ISDN. GSM operates in the 890-915 MHz band for uplink and 935-960 MHz band for downlink. Channel spacing is 200 kHz, and associated uplink and downlink channels are separated by 45 MHz. Combination of TDMA and FDMA are used in GSM. The data rate of transmission is 270,833 bps, and interleaving scheme is used to reduce the fading effects on the traffic data. The cell size is up to 32 km and the average transmitting power of mobile station is 125 mW.

GSM900 is an early version of GSM system which operates at 900 MHz band. GSM1800 has limited roaming capability between different system providers, hence is not as popular as GSM900. GSM1900 is used only in the United States.

1.8 IS-95

Code division multiple access(CDMA) was first used in military applications, and is introduced to civil communication by Qualcomm in the early 1990s, using the 850 and 1,900 MHz bands. CDMA system uses spread spectrum technique which assigns each user within a cell a specific PN sequence to enable them to share the same channel simultaneously. The IS-95 standard has a chip rate of 1.2288 Mcps and each IS-95 channel occupies 1.25 MHz of spectrum both uplink and downlink, and BPSK modulation is adopted. Sophisticated power control algorithm is used to manage the transmission power of each mobile station to avoid the near-far problem in which the strong signal received from a nearby mobile station masks the weak signal from a far-away mobile station. Power management function ensures that all received signals are at the same level within the cell. IS-95A is the first version of CDMA, which offers a throughput up to 14.4 kbps. In June 1997,

IS-95B specification was completed, which does not change the physical layer of IS-95A. High-speed service is provided through code aggregation by assigning up to seven supplementary codes in addition to the fundamental code for an active high-speed data packet, and data rate up to 64 kbps is available [1].

CDMA system uses vocoder and RAKE receiver techniques. The vocoder can distinguish signal and noise in noisy environment, and the RAKE receiver can reduce intersymbol interference(ISI) due to multipath effects. CDMA system also supports soft handoff because all channels share the same spectrum and any mobile station can communicate with more than one base station at the same time. Usually, the mobile station chooses to communicate with the base station with the strongest received power.

2. 2.5G Systems

GSM system provides connectionless packet service with messages up to 160 characters and data transfer service by circuit switch with data rate up to 14.4 kbps. HSCSD(high speed circuit switched data) is based on circuit switched technology which enables data rates up to 57.6 kbps, but is inherently inefficient for bursty traffic.

GPRS(general packet radio service) is a system with packet switched technology, which uses GSM radio modulation, frequency bands and frame structure. Its transmission data rate is increased to 160 kbps. Voice and circuit switches are supported via base stations and network. Packet switched services and IP connection are provided by adding to the GSM network two network elements: serving GPRS support node(SGSN) and gateway GPRS support node(GGSN). The SGSN keeps track of the location of individual mobile stations and performs security functions and access control. The GGSN encapsulates packets received from external packet networks(IP) and routes them to SGSN.

EDGE(enhanced data rate for global evolution) improves over GPRS by introducing offset 8-ary PKS modulation technique that triplicates the data rate offered by GPRS.

3. 3G Systems

Explosive growth of Internet and increasing demand for all types of wireless services demand larger capacity and higher data rate. Higher data rate enables a broader range of services beyond conventional voice services provided in current cellular systems. Roaring demand for high-speed mobile Internet access will be one of the major challenges for mobile network operators.

There are two major standards using CDMA technology proposed for the third generation wireless systems: cdma2000 by the Telecommunications Industry Association(TIA) and WCDMA by the ETSI, both are recognized as international mobile telecommunication-2000(IMT-2000) standards by the International Telecommunications Union(ITU). Unique standard is desirable to achieve global and seamless roaming, service portability and enhanced services. IMT-2000 is expected to cover the long-term technology requirements for next-generation voice and data services, which is backward compatible to the exising systems.

3.1 CDMA2000

Cdma2000 is designed to be backward compatible with IS-95A/B and retains many attributes of the IS-95 air interface. In IS-95B, higher data rates are provided through code aggregation. In cdma2000, higher data rates are achieved through either reduced spreading or multiple code channels. There are a number of major enhancements in cdma2000 that facilitate advanced data services with higher data rates and capacity [2]. Wider channel bandwidth of 5 MHz and beyond offers higher statistical multiplexing gain among multiple users as well as higher single-user data rate. Multiple supplemental code channels(SCC) can be used to support multiple data streams with different requirements on quality of service(QoS). Fast forward power control of 800 power updates per second provides significant performance improvement in low-mobility environment where most of the high-data-rate applications will occur. Forward link transmit diversity with $N (N = 3, 6, 9, 12)$ carriers of 1.25 MHz each is used to enhance link performance, where the chip rate is $1.2288N$ Mcps with $N = 1, 3, 6, 9, 12$, respectively.

Cdma2000 supports two different deployment configurations. One uses direct sequence(DS) approach for both downlink and uplink, the other enhances the downlink capacity using multicarrier(MC) approach. The peak data rate is 614.4 kbps per Walsh code and code aggregation can be used to achieve 2,048 kbps. IS-95A/B and cdma2000 require that all the base stations broadcast a common pilot training sequence using the same Walsh code in all cells. Each cell of the system uses one of the 511 possible offsets of a 16-bit maximum length PN sequence to enable a mobile station to identify different cells in the system. Mobile station searches for the presence of this sequence to determine if it is in range of an IS-95A/B or cdma2000 base station. On the downlink, a small fraction of users may be assigned large portion of the transmitted power. Hence, capacity of the cell is limited if a high-data-rate user is nearby cell boundary, while large capacity is available if all the high-data-rate users are near the cell center. Once a mobile station is synchronized and link with specific base station, it continues to monitor the pilot to maintain synchronization. Downlink of cdma2000 uses coherent QPSK modulation and requires a coherent reference of the underlying fading channel to demodulate data reliably [3]. If beam-forming technique is applied to cover only portion of a cell, the receiver will require another dedicated pilot for reliable channel estimation.

3.2 WCDMA

WCDMA is a radio interface for UMTS(universal mobile telecommunication system), which is not backward compatible to the GSM air interface. However, handoffs between UMTS and GSM are supported. To facilitate the migration from GSM to UMTS, UMTS network adds WCDMA BSS to connect original GPRS network, which may be common to GPRS and UMTS. Total bandwidth of 215 MHz has been allocated [4]. Within this spectrum, paired bands of 1,920-1,980 MHz(uplink) and 2,110-2,170 MHz(downlink) have been allocated for FDD WCDMA systems, and unpaired bands of 1,900-1,920 MHz and 2,010-2,025 MHz for TDD CDMA systems. The maximum WCDMA chip rate is 3.84 Mcps, each carrier occupies 5 MHz of bandwidth using QPSK modulation technique, and multiple carriers can be deployed in a system [5]. The peak data rate is 2,048 kbps with six Walsh codes. To support widely varying bit rates and maintain quality of service for various services envisaged, the 3GPP(third generation partnership programme) specifies many coding/interleaving schemes, including convolutional codes and turbo codes. Choice of channel coding/interleaving scheme for particular service is unspecified.

WCDMA supports only DS configuration in both downlink and uplink. A mobile station assesses its transmitting power based on its received power, and instructs the associated base station to adjust its power level. WCDMA employs asynchronous operation in which transmission and reception timings of different cells need not be synchronized. The pilots are time multiplexed and broadcasted to all mobile stations. Hence, no additional pilots are necessary.

3.3 3G Networks

Convergence of fixed and mobile services began in 1980s. In the 1980s, mobile communication provides only voice services, and data services are mostly through fixed network between computers. In the 1990s, due to advancement of radio technologies, mobile communication system is able to provide both voice and low-rate data services with low mobility for office use. In the future, mobile terminals with high mobility and high data rate can be connected to fixed networks including PSTN, cable, xDSL, Internet, and so on.

As demand for mobility increases, capabilities are being developed to connect mobile terminals like laptop PCs, PDAs and mobile videophones to heterogeneous networks, and roaming freely across geographical boundaries becomes possible. Mobile all-IP networks of next gereration are expected to provide a substantially wide range of services with convergence and interoperability.

Several access technologies are evolving and emerging. The second-generation GSM is evolving via GPRS, HSCSD and EDGE toward UMTS. In addition, WLAN(wireless local area network) systems like HIPERLAN 2, IEEE 802.11, DAB(digital audio broadcast) and DVB-T(digital video broadcast) are becoming available. For short-range connectivity, systems like Bluetooth and DECT are being developed. Fixed access systems like xDSL(digital subscriber line), ADSL(asymmetric DSL) in particular, increase the user data rate significantly on the last mile [6]. A universal mobile core network is expected to connect a variety of interworking access systems like 2.5G, 3G and WLAN to form a seamless network.

Next generation networks are based on open standards which integrate networks from different operators. Manufacturers are responsible to resolve interoperability issues. The networks are open to all applications, services become versatile, and monopolization becomes impossible.

Next generation mobile networks support global roaming. Apart from physical layer connectivity and radio spectrum allocation, mobility in a hierarchical scheme should be supported. This can be enhanced by the adoption of IP-based wireless/mobile networks.

The 3G systems offer a layered approach for coverage. In small and local areas like offices, homes or stations(picocells), available radio access bandwidth is up to 2 Mbps. Picocell can be covered and connected to a larger microcell which covers a few square kilometers and offers a lower bandwidth of 384 kbps.

3.4 3G Trends and Services

Annual growth rate of mobile communication services went up to about 60% in 1998. Total number of mobile subscribers worldwide is over 400 millions in 2000, and is expected to reach 1,700 millions in 2010 [7]. Such trend has been changing the service models and value chain, as well as modifying the roles of access network operators, core transport network operators, service providers and content providers.

Various mobile technologies have attracted different levels of attention from the industries and public. Expectation on 3G was raised by the successful establishment of 2G systems. Bluetooth proved the feasibility of wireless LAN, and IEEE 802.11 series gain their popularity shortly after.

Mobile systems of the third generation are designed to deliver consistent voice, data, graphical, multimedia and video-based information services like videoconference, regardless of their location in the network(cordless, cellular, satellite, fixed/wireline, and so on) [7]. The third generation systems can support high-speed data and multimedia applications up to 144 kbps for users in the move, and up to 2 Mbps in a local area. When an user moves around, his or her communication link may be handed over to an overlay cell with lower data rate. Wider bandwidth, asymmetrical connections and dynamic bandwidth allocation will offer more system flexibility for both network operators and customers.

New wireless systems offer both real-time and non-real-time modes using common transport mechanisms. Pay load data can be categorized as message, file and data stream [7]. A message is generally a fixed-length packet, which can tolerate time delay and requires low bit error rate. Examples are signaling packets for handshaking and short message service(SMS). A file packet varies in length, which can tolerate time dealy and requires error-free transmission. Examples are Internet page downloading, file transfer, e-mail, and so on. Speech and video applications are connection-oriented data stream which can not tolerate time delay, but allow certain bit error rate. Video broadcasts can bear stream delay if a receiver can be used to record the video stream before playing back.

Mobile phones are substitutes for fixed phones for their convenience and attractive features. Manufacturers, operators and vendors are inspired to provide more services over both fixed and mobile networks. Unified messaging services enable different portables to send and receive various types of messages like voice, data, fax, e-mail, SMS, and so on. Users can use mobile devices to inquire local contents like news, weather, sports, and play game together through networks. Musics and videos(movie preview, sports highlights, videoconferencing) can be fetched through the Internet.

Location services include personal navigation(maps, directions, points of interest), traffic camera for traffic reports, emergency rescue to locate victims, location-based advertising, and so on. Mobile commerce services include ticket purchasing, m-shoping, mobile banking, mobile financial trading, m-wallet, and so on.

3.5 3G Devices

All the CDMA-based systems require wide bandwidth and have an RF carrier with nonuniform signal strength over the band, which imposes high demand on analog frond-end design. CDMA systems generally require fast and accurate power control with a wide dynamic range. Multimode and multi-standard capabilities are preferred for a 3G device to be connectable to 2G, 2.5G and 3G systems. Requirements on RF front-end to operate in multi-standard environment are very different. Hence, only a few components in the transceiver implementation can be shared between 2G, 2.5G and 3G systems.

Instead of trying to specify a new radio interface for 4G systems, one can envisage the integration of existing access networks into a common, flexible and expandable platform. Possible future development for systems beyond 3G is a seamless network including a variety of access systems like GSM, UMTS, DECT, Bluetooth, WLAN and xDSL. All these technologies are optimized for specific purposes, applications and environments. There are many types of mobile devices for different mobile systems. For example, GPRS phone, GPRS PDA, GPRS PC card, UMTS dual-mode phone, UMTS PDA, UMTS PC card and UMTS video notebook. Features like SMS, circuit switched data, WAP/i-mode are optionally supported. Future device designed for operating in a seamless network will have to switch over several air interfaces which are often based on very different requirements.

4. Antenna Systems

Base station links with mobile stations in its service area using antennas. Co-channel interference in cellular system can be reduced by replacing the single omnidirectional antenna at the base station by several directional antennas, each covering a specified sector. For example, a dipole array using corner reflector is designed to have three 120° beams to cover the whole cell [8]. The array geometry can also be adjusted to have three-beam patterns in both 900 MHz and 1,800 MHz bands.

Patch antennas have the merits of simple structure and low cost. Broadband or dual-band designs are also available to achieve sectoral pattern coverage for cellular communication systems. For example, broadband antennas with bandwidths wider than 30% at VSWR ≤ 1.5 near 2 GHz and dual-band antennas operating in 900/1,500 MHz and 900/1,800 MHz bands are presented in [9].

In [10], antenna with sectoral coverage used for local multipoint distribution service(LMDS) is incorporated into base stations to increase system capacity. Sectoral antennas are used to reduce backward radiation to minimize interference from adjacent subcells. Each antenna consists of an eight-element linear array of ASPs [11] and eight back patches which act as reflectors to reduce backward radiation.

Wireless handset antennas are required to be compact, low cost and conformal with handset. Conventional antennas such as monopoles and helices are easy to break off. Dielectric resonator antennas have the advantages of high radiation efficiency, flexible feed structures, simple geometry and compactness [12].

4.1 Antenna Diversity

In mobile communication environments, multipath fading causes signal reception unreliable. Antenna diversity techniques can be used to overcome multipath fading, and are usually implemented by using array configurations. Diversity techniques can be implemented at either base station or mobile stations, with more difficulties to the latter due to space restriction.

Commonly used diversity techniques include space, polarization, frequency and time diversities. Among them, space diversity is the most common one in which elements of an antenna array are separated in location and large mounting space is required.

In the polarization diversity technique, dual polarized antennas are used, assuming that the correlation between two orthogonally polarized field components is small. In [13], a dual polarized array is proposed which consists of 12 identical columns and each column consists of 12 antenna elements to achieve high-gain pattern or to provide tilted beam vertically. Each antenna element is dual polarized aperture-coupled patch antenna. In [14], a dual polarized antenna is used in an indoor cellular system, which is composed of an element with horizontal polarization and another element with vertical polarization. The horizontal element is a circular disk with three notches, feed coaxial line is attached near the center of disk to deliver power to the three feed probes via a three-branch power divider. The vertical element consists of three vertical wire loop antennas.

If space and polarization diversities are combined in an antenna system, multipath fading effects can be equalized, and loss due to polarization mismatch can be compensated for. In [15], a four-branch rod antenna with space and polarization diversities is proposed for street cell environment. Inside the rod cover, patches serve as vertically polarized elements and slot antennas serve as horizontally polarized elements. Two patches and two slots are included in both the upper and lower ends of the rod to provide space diversity.

In [16], discrete lens antenna array is used to achieve both polarization and angle diversities. Each unit of the lens array consists of two patch antennas with orthogonal polarization interconnected by a delay line, and separated by a ground plane. Length of the delay line is varied across the array such that an incident wave can be focused to certain points on the feed side of the array. Plane waves incident from different directions are focused onto different points on the focal surface on which receiving antennas are placed to sample the incoming signal. Multiple receivers correspond to multiple antenna beams, thus enabling beam steering and beam-forming without using phase shifters.

Three processing techniques are commonly used for diversity, they are switch diversity, equal gain and maximum ratio combining(MRC). In the switch diversity, antenna with the strongest signal is selected. In the equal gain diversity, all the antenna elements are weighted by the same gain before combining. In the MRC, signals from all antenna elements are weighted by the instantaneous signal-to-noise ratios(SNR) at antennas and are cophased prior to combining in order to acquire maximum diversity gain. Weighting signals from antenna elements is equivalent to steering the radiation pattern to the direction of desired signal, hence reduces multipath and other interferences.

4.2 Adaptive and Smart Antennas

In the presence of strong interferences, using diversity techniques alone may not be able to achieve required signal quality. Use of adaptive or smart antennas can further improve the communication link by minimizing multipath interferences. Because these antennas are dynamically controlled to steer beams toward desired directions or to avoid receiving interferences from specific directions.

In [17], antenna with switch diversity is implemented using active antenna element accompanied by several parasitic elements. The parasitic elements are constructed to be switchable to three different states of termination by two control inputs. The termination affects current flowing in those parasitic elements, thus changing the radiation pattern. In [18], a switchable four-beam antenna is considered in an indoor wireless local area network(WLAN) to provide pattern diversity.

Antenna beam scanning can also be realized by configuring patch antennas in cylindrical array. The beam direction is determined by delivering power to a subset of elements using RF switching matrix. By incorporating several feed networks into the antenna array, multiple beams are obtained which can be steered independently.

References

[1] J. D. Vriendt, P. Laine, C. Lerouge, and X. Xu, "Mobile network evolution: a revolution on the move," *IEEE Commun. Mag.*, pp.104-111, Apr. 2002.

[2] D. N. Knisely, S. Kumar, S. Laha, and S. Nanda, "Evolution of wireless data services: IS-95 to cdma2000," *IEEE Commun. Mag.*, pp.140-149, Oct. 1998.

[3] R. A. Soni, R. M. Buebrer, and R. D. Benning, "Intelligent antenna system for cdma2000," *IEEE Signal Process. Mag.*, pp.54-67, July 2002.

[4] S. Dehghan, D. Lister, R. Owen, and P. Jones, "WCDMA capacity and planning issues," *IEEE J. Electron. Commun. Eng.* , pp.101-118, June 2000.

[5] M. W. Oliphant, "Radio interfaces make the difference in 3G cellular systems," *IEEE Spectrum*, pp.53-58, Oct. 2000.

[6] W. Mohr and W. Konhauser, "Acess network evolution beyond third generation mobile communications," *IEEE Commun. Mag.*, pp.122-133, Dec. 2000.

[7] W. Johnston, "Europe's future mobile telephony system," *IEEE Spectrum*, pp.49-53, Oct. 1998.

[8] Z. Zaharis, E. Vafiadis, and J. N. Sahalos, "On the design of a dual-band base station wire antenna," *IEEE Antennas Propagat. Mag.*, vol.42, no.6, pp.144-151, Dec. 2000.

[9] F. Tefiku and C. A. Grimes, "Design of broad-band and dual-band antennas comprised of series-fed printed-strip dipole pairs," *IEEE Trans. Antennas Propagat.*, vol.48, no.6, pp.895-900, June 2000.

[10] R. B. Waterhouse, D. Novak, A. Nirmalathas, and C. Lim, "Broad-band orinted sectorized coverage antenna for millimeter-wave wireless applications," *IEEE Trans. Antennas. Propagat.*, vol.50, no.1, pp.12-15, Jan. 2002.

[11] S. D. Targonski, R. B. Waterhouse, and D. M. Pozar, "Design of wideband aperture-stacked patch microstrip antennas," *IEEE Trans. Antennas. Propagat.*, vol.46, pp.1246-1251, 1998.

[12] M. T. K. Tam and R. D. Murch, "Compact circular sector and annular sector dielectric resonator antennas," *IEEE Trans. Antennas Propagat.*, vol.47, no.5, pp.837-842, May 1999.

[13] B. Lindmark, S. Lundgren, J. R. Sanford, and C. Beckman, "Dual-polarized array for signal-processing application in wireless communication," *IEEE Trans. Antennas Propagat.*, vol.46, no.6, pp.758-763, June 1998.

[14] N. Kuga, H. Arai, and N. Goto, "A notch-wire composite antenna for polarization diversity reception," *IEEE Trans. Antennas. Propagat.*, vol.46, no.6, pp.902-906, June 1998.

[15] K. Cho, T. Hori, and K. Kagoshima, "Effectiveness of four-branch height and polarization diversity configuration for street microcell," *IEEE Trans. Antennas Propagat.*, vol.46, no.6, pp.776-781, June 1998.

[16] D. Popovic and Z. Popovic, "Multibeam antennas with polarization and angle diversity," *IEEE Trans. Antennas Propagat.*, vol.50, no.5, pp.651-657, May 2002.

[17] N. L. Scott, M. O. Leonard-Taylor, and R. G. Vaughan, "Diversity gain from a single-port adaptive antenna using switched parasitic elements illustrated with a wire and monopole prototype," *IEEE Trans. Antennas Propagat.*, vol.47, no.6, pp.1066-1070, June 1999.

[18] N. Kuga and H. Arai, "A flat four-beam switched array antenna," *IEEE Trans. Antennas Propagat.*, vol.44, no.9, pp.1227-1230, Sept. 1996.

[19] T. S. Rappaport, *Wireless Communications: Principles and Practice*, Prentice Hall, 1996.

[20] M. Vrdoljak and S. I. Vrdoljak "Fixed-monile convergence strategy: technologies and market opportunities," *IEEE Commun. Mag.*, pp.116-121, Feb. 2000.

[21] A. Springer and R. Weigel, "RF microelectronics for WCDMA mobile communication systems," *IEEE J. Electron. Commun. Eng.* , pp.92-100, June 2002.

Chapter 26

INTRODUCTION TO WIRELESS LOCAL AREA NETWORKS

Jean-Fu Kiang
Department of Electrical Engineering and
Graduate Institute of Communication Engineering
National Taiwan University
Taipei, Taiwan, ROC

Abstract IEEE 802.11 and Bluetooth are proposed for wireless local area network(WLAN) and wireless personal area network(WPAN), respectively. Both comply with the seven-layer model of open systems interconnection(OSI). Both IEEE 802.11b and Bluetooth are assigned the 2.4 GHz frequency band, the IEEE 802.11a is assigned the 5 GHz frequency band. Basic service set(BSS) is the basic IEEE 802.11 network, and multiple BSSs can constitute an extended service set(ESS). Bluetooth network is called piconet, and is established in master/slave mode. Distributed coordination function(DCF) and point coordination function(PCF) are devised in IEEE 802.11 to avoid data collision between multiple users. Service set identifier(SSID), MAC address and wired equivalent protocol(WEP) are three basic security mechanisms in IEEE 802.11 to avoid eavesdropping. Each Bluetooth device has a device address and Bluetooth clock which can be used for link control. Asynchronous connectionless(ACL) and synchronous connection oriented(SCO) are two types of link over piconet.

Keywords: WLAN, WPAN, Bluetooth, FHSS, DSSS, OFDM, WEP, DCF, PCF, BSS, ESS, Piconet.

1. Introduction

Due to rapid growth of wireless communication services in recent years, several wireless data communication solutions are proposed, including IEEE(Institute of Electrical and Electronics Engineers) 802.11 standards, high-performance radio local area network(HIPERLAN), home radio frequency(HomeRF), Bluetooth, general packet radio service(GPRS), third generation(3G) cellular and free space optics(FSO). They are suitable for different environments, at different status of commercial maturity, and have certain common functionality.

LANs were originally invented to cover over several hundred meters or less, and typically link devices within building or campus for data communication. LANs can be implemented with a variety of network topologies(such as token ring and Ethernet), data communication protocols(such as transmission control protocol/Internet protocol(TCP/IP) and advanced program-to-program communication(APPC) of IBM), and transmission media(such as twisted pair, coaxial cable, optical fiber and radio). WLAN simply uses radio or light as transmission medium, and is generally deployed to areas with size similar to the first wired LAN.

IEEE 802 standards provide the framework for organizing IEEE LAN/MAN standardization activities. IEEE 802 standards are based on the open systems interconnection(OSI) model of the International Standards Organization(ISO), which segments data communication protocol into seven distinct functional layers: physical, data link, network, transport, session, presentation and application.

Table 26.1 shows the OSI seven-layer model. The physical layer specifies physical properties of communication media(such as electrical, optical or radio), modulation and channel coding techniques. The data link layer is responsible for transmission, framing and error control(error checking and correction) over particular link. The network layer is responsible for data transfer across network, independent of media and topology of network. The transport layer is responsible for reliability and multiplexing of data transfer across network to the level required by applications. The session layer provides data flow control to ensure proper sequence of data packets. The presentation layer describes the syntax of data being transferred such as the presentation of alphanumeric characters. The application layer describes the protocol for specific applications such as database access or file transfer.

IEEE 802 framework defines the lowest two layers in the OSI reference model: physical and data link layers. The IEEE 802 PHY layer defines the physical characteristics of transmission media and the physical signaling protocol. IEEE 802 segments the OSI data link layer into two sublayers: medium access control(MAC) and logical link control(LLC) layers. The MAC layer defines the protocol for accessing PHY layer, whereas the LLC layer enables multiple user access to MAC. Standardization of low-level network access by IEEE and widespread adoption

Table 26.1. IEEE 802, Bluetooth and Internet protocols compared with the OSI seven-layer protocols of ISO.

IEEE 802/Internet	OSI Layers	Bluetooth
HTTP, FTP, SMTP	Application	Applications
DNS, LDAP	Presentation	RFCOMM/SDP
	Session	L2CAP
UDP, TCP	Transport	Host Controller Interface (HCI)
ICMP, RSVP	Network	Link manager (LM)
Logical Link Control (LLC) Medium Access Control (MAC)	Data link	Link controller (LC)
		Baseband
Physical (PHY)	Physical	Radio

of the IP suite as high-level protocol advocate 802.11 as the dominant WLAN technology. In 1999, two standards, IEEE 802.11a and 802.11b, were proposed as the radio PHY amendments to IEEE 802.11. The IEEE 802.11b supports data rates of 5.5 and 11 Mpbs in the 2.4 GHz industrial, scientific and medical(ISM) band. The ISM bands are license-free bands reserved for use by industrial, scientific and medical wireless equipments. The IEEE 802.11a is assigned the 5 GHz frequency band, utilizes orthogonal frequency division multiplexing(OFDM) modulation technique, and supports data rates up to 54 Mbps.

Due to widespread deployment of IEEE 802.11b, the Wireless Ethernet Compatibility Alliance(WECA) was formed in 1999. The WECA created an interoperability benchmark for IEEE 802.11b, began issuing certifications to equipment vendors in April 2000, and promoted wireless fidelity(Wi-Fi).

2. IEEE 802.11 Network Architecture

2.1 Basic Service Set

IEEE 802.11 defines a hierarchical network architecture for WLAN devices. The end points, also called stations, communicate with one another. An independent basic service set(IBSS) is a self-sufficient network(no connection to wired network) that supports only direct communication between stations that are members of that network. There is no master station or centralized coordination among stations.

BSS may be endowed with centralized distribution, buffering and gateway functionality through an access point(AP). In such case, all stations are connected to the AP and communicate with one another through the AP. The AP may set up connection to wired LAN and provides local relay function for the BSS.

2.2 Extended Service Set

Fig.26.1 shows an extended service set(ESS) with multiple IBSSs connected into a single WLAN. An ESS enables each AP within its network to determine if messages are to be sent to stations within its BSS or another BSS, and to provide gateway access to wired network such as Internet. The APs peform these functions via distribution system(DS) such as Ethernet.

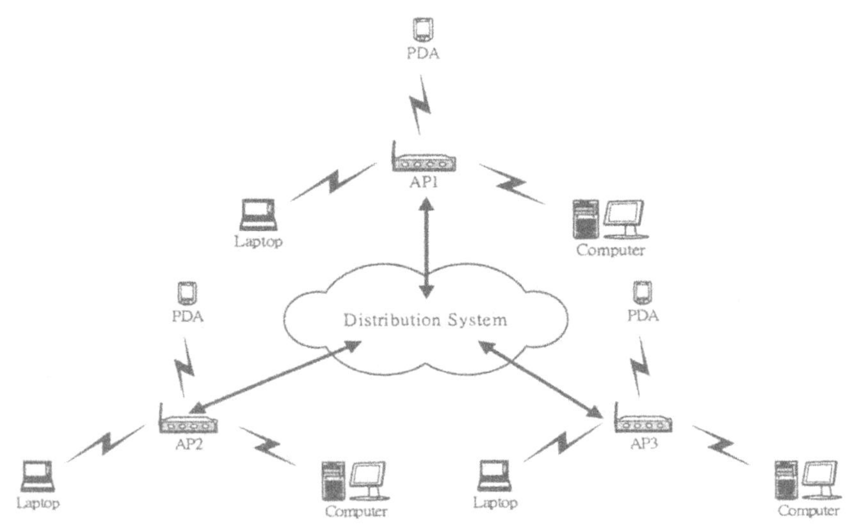

Figure 26.1. Configuration of extended service set(ESS).

3. PHY Layer

3.1 Attributes of IEEE 802.11a

IEEE 802.11a provides faster data transmission and supports various data rates at 6, 9, 12, 18, 24, 36, 48 and 54 Mbps, respectively, with

BPSK, QPSK or QAM modulation techniques. The Federal Communications Commission(FCC) allocates the unlicensed national information infrastructure(UNII) band for this standard, which is comprised of 300 MHz of radio spectrum in 5.15-5.25 GHz, 5.25-5.35 GHz and 5.725-5.825 GHz with maximum transmitting power of 40 mW, 200 mW and 800 mW, respectively.

3.2 Attributes of IEEE 802.11b

IEEE 802.11b standard is designed to be backward compatible with both the IEEE 802.11 frequency hopping and direct sequence systems, and utilizes the same 2.4 GHz frequency band. IEEE 802.11 specifies two types of DPSK modulation: BPSK and QPSK. BPSK supports data rate of 1 Mbps and QPSK doubles the data rate to 2 Mbps. IEEE 802.11b applies complementary code keying(CCK) and packet binary convolutional coding(PBCC) to promote the data rate up to 11 Mbps [1], [2].

3.3 Frequency Hopping Spread Spectrum

In frequency hopping spread spectrum(FHSS), information signals are carried over a fixed sequence of narrow band carriers. Transmission on each narrow band carrier is brief to render the signals less susceptible to eavesdropping or jamming. Synchronization between transmitter and receiver is critical. The hopping sequence of FHSS is generated by PN code generator. In North America, the hopping rate is set at 2.5 hops per second with transmission over each channel lasting for less than 400 ms. FCC regulation specifies 75 hopping channels for FHSS systems in 2.4 to 2.4835 GHz. IEEE 802.11 standard specifies a series of evenly spaced hopping channels with 1 MHz spacing in the 2.4 GHz frequency band. IEEE 802.11 standard defines two versions of GFSK: binary GFSK(1 Mbps) and quaternary GFSK(2 Mbps).

3.4 Direct Sequence Spread Spectrum

In direct sequence spread spectrum(DSSS), PN code having data rate higher than the information bit stream is used to modulate the latter and results in wider bandwidth. In IEEE 802.11b, data stream of 1 Mbps is converted into another bit stream of 11 Mbps using differential binary phase shift keying(DBPSK) modulation. The PN code for 802.11b DSSS is an 11-bit Barker sequence: 10110111000. The DSSS operates in the same 2.4 to 2.4835 GHz ISM band as the FHSS scheme. The 802.11 specifies up to 14 channels for the DSSS scheme. Transmitting power level of 100 mW is used in most of the 802.11 DSSS products.

3.5 Orthogonal Frequency Division Multiplexing

Orthogonal frequency division multiplexing(OFDM) scheme was patented by Bell Laboratories in 1970, which is based on fast Fourier transform(FFT). The OFDM in IEEE 802.11a divides a high-speed data carrier(20 MHz) into 52 low-speed subcarriers(300 kHz each). The FFT enables these 52 channels to partially overlap without losing information. Forty eight of these 52 subcarriers are allocated for data and the other four are allocated as pilots to enable frequency alignment at the receiver.

Symbol duration in each subcarrier is relatively long compared to the maximum delay over the channel, hence multipath interference and delay spread are reduced considerably. Besides, multiple carriers are more immue to channel dispersion than a single carrier. However, multicarrier system has the disadvantage of higher sensitivity to phase noise in local oscillators and carrier frequency offsets. Comparison of different OFDM systems can be found in [3]. In OFDM-TDMA systems, phase noise incurs commom phase error in received symbols and loss of orthogonality of subcarriers, thus giving rise to multiuser interference. In OFDM-CDMA systems, phase noise also provokes multiuser interference and rotates the symbols at decision point. Several researches have been conducted to reduce intercarrier interference [4], and to estimate frequency offset [5], [6]. Recent reserches suggest that OFDM-CDMA is an attractive candidate for wireless communication because it offers robustness to channel impairments, relatively low complexity, and high flexibility in bandwidth assignment. In [7], a programmable configuration of OFDM-CDMA transceiver was developed.

4. MAC Layer

The MAC layer is responsible for managing data transfer between higher-level functions and the physical media. The IEEE 802.11 MAC layer is common to different IEEE 802.11 PHY layers and specifies the functions and protocols required for control and access.

4.1 MAC Services

MAC provides nine logical services: authentication, deauthentication, association, disassociation, reassociation, distribution, integration, privacy and data delivery. An AP can use all these nine services and a station typically uses authentication, deauthentication, privacy and data delivery. Authentication is used to establish the identity of a station and authorize it for possible association. Deauthentication is used to remove an existing authentication. Association is used to link a station to AP and enable the AP to distribute data to and from the station. Disassociation is used to break existing association. Reassociation is used to transfer association between APs. Privacy is used to prevent unauthorized eavesdropping of data through the use of wired equivalent protocol(WEP) algorithm. Distribution is used to provide data transfer between stations via DS. Data delivery is used to provide data transfer between stations. Integration is used to provide data transfer between DS of an IEEE 802.11 LAN and other non-IEEE 802.11 LANs. The station providing such function is called a portal.

4.2 MAC Layer Architecture

Both PHY and MAC layers provide management and data transfer functions. The PHY management function is served by PHY layer management entity(PLME) and the MAC management function is provided by MAC layer management entity(MLME). PLME and MLME exchange information of PHY layer function through management information base(MIB) which is a database of physical characteristics such as possible transmission rates, power levels, antenna types, and so on. These management functions support the main purpose of MAC to transfer data. Data stream is passed to the logical link control(LLC) layer from higher layers. It is packaged into MAC service data units(MSDUs) and then passed to MAC from LLC. The MAC passes MSDUs to the PHY layer through physical layer convergence protocol(PLCP) which is responsible for translating MSDUs into a format that is physical medium dependent(PMD). MAC data transfer is controlled through either distributed coordination function(DCF) or point coordination function(PCF). The latter provides centralized traffic management for data transfers that are sensitive to delay and require contention-free(CF) access.

4.3 DCF

DCF regulates how the medium is shared among stations of the wireless network and how they contend for the medium as peers. Stations with overlapping radio coverage have difficulties in detecting data collision in the wireless media, and the DCF is implemented as a collision avoidance mechanism. To avoid collisions, each station first senses whether the medium is in use before transmitting data. The method of sensing the medium and waiting before transmission is called carrier sense multiple access/collision avoidance(CSMA/CA) [8].

The DCF supports complementary physical and virtual carrier sense mechanism. Physical sensing of the medium is performed by clear channel assessment(CCA). The medium is assumed to be in use under any of the three conditions: any energy above a predefined threshold, any DSSS signal, or a DSSS signal above the predefined threshold. Physical sensing is very efficient, but becomes ineffective to the hidden node problem.

In virtual carrier sensing, information on the use of medium is exchanged through control frames, which significantly reduce the probability of collisions between stations and increase the overall throughput. Because the overhead of using control frames is fixed, smaller data frames imply higher percentage of overhead. Control frames include request to send(RTS) and clear to send(CTS) frames. When a station wants to send data, it broadcasts an RTS frame to the network, which contains the addresses of transmitting and receiving stations and the length of data frame. The AP receving an RTS frame replies with a CTS frame which can be heard by all the stations within the AP's coverage. Receipt of a CTS frame causes all the stations to set their network allocation vector(NAV) which is a timer recording the waiting time before the medium can be contended. The DCF requires an acknowledgement(ACK) frame to be sent upon successful receipt of other frames. Transmission is assumed to be in error if an ACK is not received in a predefined time interval.

Slot times for DSSS and FHSS are 20 μs and 50 μs, respectively, which are supposed to be longer than the sum of signal round-trip time and energy-detecting time. DCF also provides several interframe spaces which are time intervals for a station to wait before contending for medium. Short interframe space(SIFS) is the shortest interval, which is 10 μs for DSSS and 28 μs for FHSS. The other three interframe spaces are distribution interframe space(DIFS), extended interframe space(EIFS) and PCF interframe space(PIFS). All stations need to wait for DIFS before contending for the medium when the last transmission is over. EIFS is used to enable the processing of erroneous frames reported by the PHY layer. PIFS enables a station to have priority access to the medium when operating in the PCF contention-free mode.

Fig.26.2 shows an example of transmitting an MSDU in DCF mode. When a source station senses the channel to be idle, it waits for DIFS plus a backoff interval, and then broadcasts an RTS frame. All the other stations in the BSS receive the RTS frame and set their NAVs accordingly. The destination station responds to the RTS frame with a CTS frame after an SIFS. The other stations receive the CTS frame and update their NAVs. After SIFS of receiving the CTS frame, the source station begins to transmit its data frame. After successful data transmission, the destination station transmits an ACK frame back to the source station after waiting for SIFS, indicating that the transmission is completed.

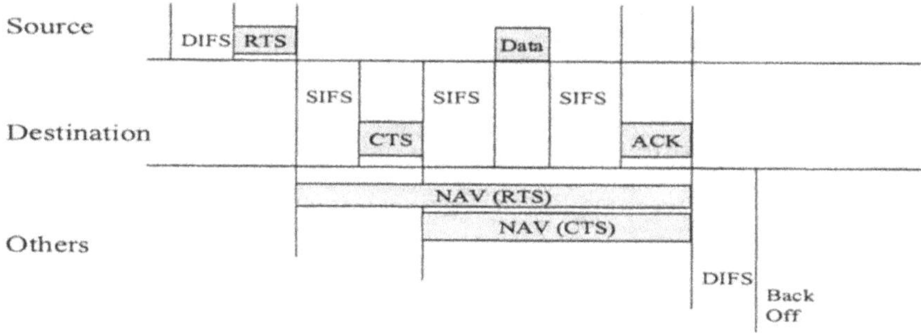

Figure 26.2. Transmission in DCF mode.

A random backoff interval is chosen when a station wants to transmit data frames while sensing busy channel. When the channel becomes free, the station has to wait for DIFS plus the backoff interval. Initial value of backoff interval is determined by multiplying slot time with a random number within a specified window. For example, if the window ranges from one to three and the slot time is 20 μs, the station may have to wait from 20 μs to $(2^3-1)\times 20$ μs. Retry occurs when more than one stations happen to have the same backoff interval. Retries will then occur using windows of $(2^4-1)\times 20$ μs, and then $(2^5-1)\times 20$ μs, up to the maximum value predefined. Due to random nature, two stations contending for the medium usually have different backoff intervals. Hence, any two stations are prevented from repeated collisions.

4.4 PCF

PCF is an alternative to DCF. PCF is coordinated by the AP within each BSS to poll associated stations in contention-free mode. The AP is also called point coordinator(PC) in this occasion. Stations within BSS that are functional during CF period(CFP) are called CF-aware stations. CFP repetition interval determines the repetition rate with which PCF occurs. Within a repetition interval, portion of the time is allotted to contention-free traffic, and the remaining portion is alloted to contention-based traffic, as shown in Fig.26.3.

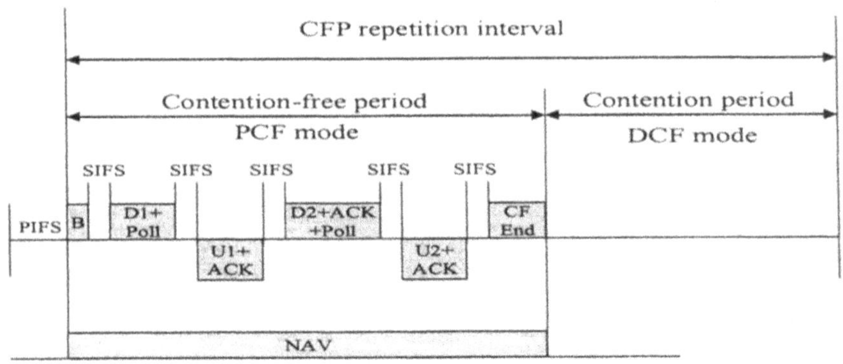

Figure 26.3. Transmission in PCF mode.

At the nominal start of a CFP, PC broadcasts a beacon frame(B) to all the associated stations to initiate the CFP if the medium remains free for a PIFS interval, and all the stations in the BSS update their NAVs to the maximum length of CFP. The PC begins CF transmission an SIFS after the beacon frame is transmitted. It then sends CF-Poll, Data or Data + CF-Poll. If a CF-aware station receives the CF-Poll frame from PC, it will respond to the PC with CF-ACK or Data + CF-ACK frame after an SIFS interval. If the PC receives a Data + CF-ACK frame from a station, the PC sends Data + CF-ACK + CF-Poll frame to all the stations, in which the CF-ACK is used to acknowledge receipt of the previous data frame and Data + CF-Poll is sent to the next station. If the destination station does not have any data frame to transmit, it will send a null function frame back to the PC. If the PC fails to receive ACK for a transmitted data frame, it waits for a PIFS and then continues to send data to the next station in the polling list. When the PC has finished transmission, it terminates the CFP by broadcasting a CF-End frame.

4.5 Power Control and Time Synchronization

The MAC also provides control frames for power control and time synchronization in addition to CSMA/CA control frames(RTS, CTS, ACK, and contention polling). The AP broadcasts time synchronization beacon to the associated stations in the BSS. In an IBSS, algorithm is used to enable each station to reset its timer when it receives a synchronization value greater than its current value.

Station in active mode can receive frames at any time during contention-free period. Stations entering power save mode may inform the AP using control frame. The AP will then buffer transmission data for the station. Station in power save mode will periodically enter active mode to receive beacons, broadcast, multicast and buffered data frames.

4.6 Data Fragmentation

Longer data streams are easier to be corrupted by interference. IEEE 802.11 MAC provides a fragmentation mechanism to break a data stream into smaller units. Frames with length over a specified threshold will be divided into multiple fragments. The frame header contains a sequence control field that indicates the order of fragments. The fragments constituting a frame are transmitted one after another with SIFS sepration to avoid any contention for the medium. Each fragment has its own cyclic redundancy check(CRC), and an ACK is transmitted for each fragment. If an error occurs to a fragment, subsequent fragments are not transmitted until the previous frame is retransmitted and acknowledged. Backoff rule applies to transmission of fragmented frames. Duration for each fragment and ACK frames are used to set the NAV of all the associated stations.

5. Wireless LAN Security

Address of station in a wired LAN is associated with its physical location. On the other hand, station in WLAN is an addressable unit which is a message destination instead of physical location. Wave propagation in WLAN is affected by the floor plan, furnitures, other objects and interference. In WLAN environment, every station is able to hear any other stations. In order to ensure information security, mechanisms are required to grant access only to authorized units so that transmitted data is free of eavesdropping.

There are three basic security mechanisms in the 802.11 standards. The first one is service set identifier(SSID) which is an alphanumeric code assigned to each station and the AP associated with a WLAN. It is not a strong form of security because the SSID is sent in clear text. The second one is a list of MAC addresses that can be managed manually. Only the listed stations are allowed to access the WLAN. The MAC addresses are also transmitted in clear text. The third one is wired equivalent protocol(WEP) encryption algorithm used to encryt data being transmitted to avoid eavesdropping [8].

To begin connecting to WLAN, the intending station broadcasts frames containing MAC address and BSS identifier(BSSID) in all the RF channels. All the APs within range will respond with their own BSSID, MAC address and channel information. The station can then send its data stream through proper channel and begin the authentication process.

The 802.11 standard provides two methods of authentication: open system authentication and shared key authentication. In the open system authentication, any station can conduct authentication as long as it receives the request. The open system authentication is used by default in which all the authentication packets are transmitted without encryption.

In the shared key authentication, WEP is enabled and identical WEP keys are installed in all stations and AP in advance. When the intending station requests a shared key authentication, the AP returns 128 bytes of randomly generated unencrypted challenge text. The intending station then encrypts the text using the shared key and returns the WEP encrypted data. After the AP verifies the integrity of data, the station is authenticated.

Deauthentication is used to terminate existing authentication or disassociate a station. Deauthentication may be initiated by any station or the AP.

6. Introduction to Bluetooth

Bluetooth was developed to link microprocessor-based devices via radio wave. Typical Bluetooth devices are small in size and consume low power. Bluetooth is a wireless network technology suitable for ranges of less than ten meters, and is becoming standard feature of consumer electronics, cellular phones, computer peripherals and mobile computing appliances [9]. Bluetooth special interest group(SIG) was formed in 1998 to promote its commercial penetration. The latest specification of version 1.1 was developed in February 2001, and version 2.0 is targeting on higher data rate. There are two approaches to increase data rate. The first one increases the data rate to 2 Mbps while retaining low cost and backward compatibility with existing systems. The second approach increases the data rate to 10 Mbps or higher, but with higher cost and lack of backward compatibility.

In Bluetooth network, links between devices are established in a master/slave configuration, or called piconet. Bluetooth devices can operate in either master mode or slave mode. A master may maintain a total of 256 connections. Any seven of them can be active concurrently, and the remaining devices are connected with parking technique. A device can be the master of a piconet, while simultaneously be slave to a device in another piconet as shown in Fig.26.4. The set of connected piconets form a scatternet. Device in master mode must set the frequency hopping sequence, and device in slave mode must synchronize to the master in time and frequency by following the hopping sequence. The master also controls how the available bandwidth is shared among slaves by deciding when and how often to communicate with each slave.

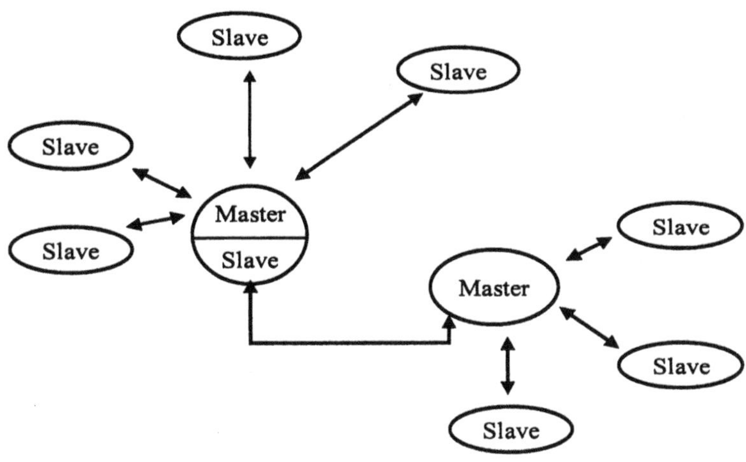

Figure 26.4. Example of Bluetooth scatternet consisting of two piconets.

Protocol layers of Bluetooth are also compatible with the OSI model as shown in Table 26.1. The physical layer includes radio layer, the data link layer is divided into baseband and link controller(LC) layers. Bluetooth uses the 2.4 GHz ISM band with frequency hopping spread spectrum(FHSS) technique. Bluetooth radio operating in asymmetric mode supports maximum data rate of 721 kbps in one direction and 57.6 kbps in the other. The symmetric mode supports 432.6 kbps in both directions.

6.1 Collision in 2.4 GHz

Electromagnetic interference(EMI) problem may occur between IEEE 802.11, 802.11b and Bluetooth products because they operate in the same frequency band. Regulatory agencies, vendors and standard bodies have been addressing such issues. The first notable proposal is an amendment to FCC Part 15.247 which enables adaptive hopping algorithms to change frequency hopping sequence based on previously encountered interference.

6.2 IEEE 802.15

IEEE 802.15 working group is established to examine low-power, short-range wireless network similar to Bluetooth [10]. Task group 802.15.1 is focusing on Bluetooth and task group 802.15.2 is developing recommended practices on how IEEE 802.11 WLAN and IEEE 802.15 WPAN can coexist in the 2.4 GHz band. Task group 802.15.3 is developing IEEE standard for high-speed WPANs to support data rate of 10-55 Mbps over distances less than ten meters. Task group 802.15.4 is preparing standard for simple, low-cost, low-speed WPANs to extend battery life using data rates ranging from 2 to 200 kbps and DSSS modulation. Proposed frequencies are 2.4 GHz and 915 MHz for North America and 868 MHz for Europe.

6.3 Bluetooth Radio

Bluetooth transceiver is an FHSS radio system hopping over channels of 1 MHz width at $2,402+m$ MHz with $m = 0, 1, \cdots, 79$. Bluetooth radio can operate over 79 or 23 channels depending on available frequency spectrum. Bluetooth radio in most countries hopps over all 79 available channels in a pseudo-random manner at 1,600 hops/s to achieve high noise resilience. The maximum transmtting power is 20 dBm(100 mW), and bit error rate of 0.1 % is achieved at the received signal level of -70 dBm.

6.4 Link Controller and Baseband

Link controller decides what data to transmit, how long to wait before transmission, which carrier frequency and transmitting power level to use, in order to execute baseband communication protocol and related processes.

Piconets are formed as needed and remain as long as the participating devices stay in link. Devices in a piconet communicate with one another by sharing the communication channel. Baseband protocol is responsible for piconet and device control functions like connection set up, selection and timing of frequency hopping sequence; modes of operation such as power control and secure operation; medium access functions like polling, packet type, packet processing and link type. Frequency hopping sequences associated with communication channels in different piconets are highly uncorrelated. Thus, multiple piconets may coexist in time and space with minimum interference among themselves.

Any device in a piconet must stay in one of the four states: connected, standby, inquiry and page. In the connected state, the device becomes a member of the piconet. In the standby state, the device is not associated with any piconet or seeking association with any piconet, it is typically idle with only its local clock operating in low-power mode. In the inquiry state, the device learns about the identity of neighboring devices, which must be listening and subsequently respond to the inquiries. In the page state, the device explicitly invites other devices to join it and becomes master of the piconet. The other devices must be listening and subsequently respond to the pages.

To link to a piconet, a Bluetooth device needs to know the frequency hopping sequence associated with that piconet, when and which frequency channel to begin with. The device also needs to know how to formulate, read and write information packets. All these operations require Bluetooth device address and device clock.

6.5 Bluetooth Divece Address and Clock

Each Bluetooth device has a unique IEEE MAC address, called Bluetooth device address. This 48-bit address is partitioned into three parts: nonsignificant address part(NAP) with 16 bits, upper address part(UAP) with 8 bits and lower address part(LAP) with 24 bits. The UAP and NAP constitute a 24-bit organization unique identifier(OUI) which is assigned by numbering authority. The LAP is assigned internally by each organization.

Each Bluetooth device has a free running 28-bit local clock. The clock ticks once every 312.5 μs, which is equal to half of the interval when the radio hops at the rate of 1,600 hops/s [11]. Clock accuracy of ± 20 ppm and ± 250 ppm, respectively, can be achieved by using low-power oscillator in low-power mode.

6.6 Frequency Hopping Sequences

Fig.26.5 shows the frequency selection module(FSM) which decides the frequency hopping sequence. The notation $(x|y)$ means either x or y, $V[m:n]$ indicates bits m through n of word V. The FSM is set by manufacturer to hop over either 23 or 79 channels in different countries. Given the frequency hopping mode and address information, the clock triggers frequency hopping sequence with proper phase.

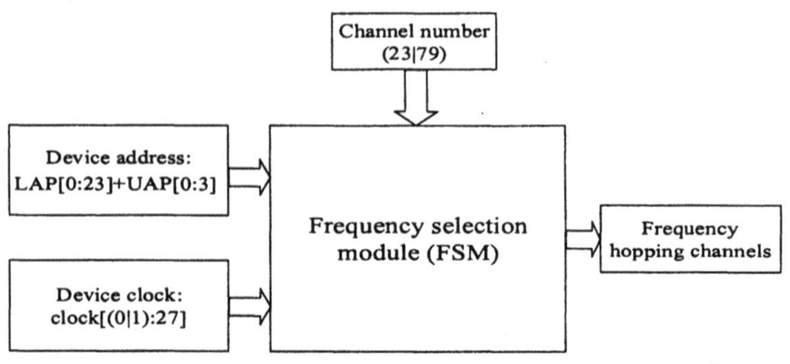

Figure 26.5. Frequency selection module.

6.7 ACL and SCO

Bluetooth baseband protocol supports two types of link over piconet: asynchronous connectionless(ACL) and synchronous connection oriented (SCO). An ACL link exists between master and slave as soon as a connection is established. Since the master does not always transmit to the same slave on regular basis, the ACL link provides a packet switched connection when data is sporadically passed down from higher layers. Pay load data is sent via L2CAP layer and passed to the baseband as L2CAP packets. Broadcast packets are ACL packets that are not addressed to specific slaves. If an ACL transmission from master to its slaves is not acknowledged, it will be retransmitted to assure data integrity and reliability.

SCO link provides symmetric link between master and slave with reserved channel. The SCO link provides a circuit switched connection for time-bounded information such as audio in which data is regularly exchanged. A slave can be allocated up to three SCO links by the same master. Due to the time-bounded nature of SCO data, SCO packets are never retransmitted. The master will transmit SCO packets to slave at regular intervals, defined by the SCO interval which is counted in time slots.

7. Ultrawideband

Ultrawideband(UWB) technique was originally classified for military applications. It is declassified in the middle of 1990s and commercial use is approved by the FCC in February 2002 [12]. UWB uses pulse-based modulation technique rather than harmonic-based modulation, and requires broad spectrum. The pulse width is around hundred picosecond, hence its spectrum can be as wide as several GHz which can support data rate of higher than 100 Mbps. However, broad spectrum of UWB makes it prone to interfere with other devices. To avoid such interference, UWB is limited to short-range applications. It is a potential technology suitable for WLAN or WPAN of the next generation.

References

[1] M. J. E. Golay, "Complementary series," *IRE Trans. Info. Theory*, pp.82-87, Apr. 1961.

[2] R. van Nee, "FDM codes for peak-to-average power reduction and error correction," *IEEE Global Telecommun. Conf.*, pp.740-744, Nov. 18-22, 1996.

[3] F. Cuomo, A. Baiocchi, and R. Cautelier, "A MAC protocol for a wireless LAN based on OFDM-CDMA," *IEEE Commun. Mag.*, vol.38, no.9, pp.159-159, Sept. 2000.

[4] J. Armstrong, "Analysis of new and existing methods of reducing intercarrier interference due to carrier frequency offset in OFDM," *IEEE Trans. Commun.*, vol.47, no.3, pp.365-369, Mar. 1999.

[5] K. Bang, N. Cho, J. Cho, H. Jun, K. Kim, H. Park, and D. Hong, "A coarse frequency offset estimation in an OFDM system using the concept of the coherence phase bandwidth," *IEEE Trans. Commun.*, vol.49, no.8, pp.1320-1324, Aug. 2001.

[6] P. H. Moose, "A technique for orthogonal frequency division multiplexing frequency offset correction," *IEEE Trans. Commun.*, vol.42, no.10, pp.2908-2914, Oct. 1994.

[7] K.-C. Chen and S.-T. Wu, "A programmable architecture for OFDM-CDMA," *IEEE Commun. Mag.*, vol.37, no.11, pp.76-82, Nov. 1999.

[8] "Wireless LAN medium access control (MAC) and physical layer (PHY) specifications," ISO/IEC 8802-11; ANSI/IEEE Std. 802.11, 1999 ed., Part 11.

[9] J. LaRocca and R. LaRocca, *802.11 Demystified*, McGraw-Hill, 2002.

[10] "Wireless medium access control (MAC) and physical layer (PHY)," IEEE Std. 802.15.1-2002, specific requirements, Part 15.1.

[11] C. Bisdikian, "An overview of the Bluetooth wireless technology," *IEEE Commun. Mag.*, vol.39, no.12, pp.86-94, Dec. 2001.

[12] Http://ftp.fcc.gov/Bureaus/Engineering_Technology/News_Releases /2002/nret0203.pdf

[13] B. A. Miller and C. Bisdikian, *Bluetooth Revealed*, Prentice Hall, 2001.

[14] "Wireless LAN medium access control (MAC) and physical layer (PHY)," IEEE Std. 802.11a-1999, Part 11.

[15] "Wireless LAN medium access control (MAC) and physical layer (PHY)," IEEE Std. 802.11b-1999, Part 11.

Index

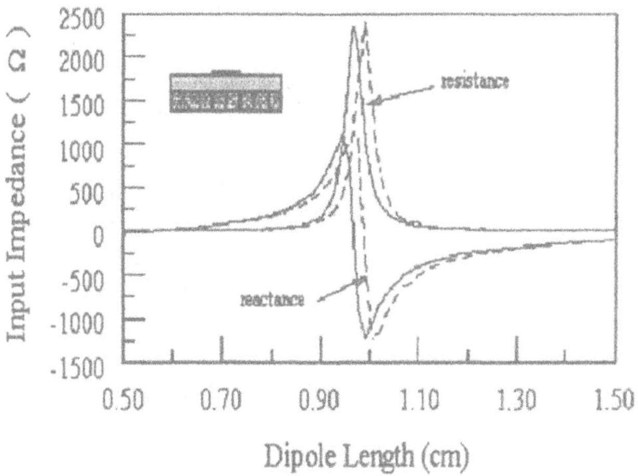

Figure 12.15. **Input impedance of a strip diple on top of a two-layered substrate, —: DOVIE method with complete block filling in the bottom layer, – –: spectral domain method for a two-layered homogeneous substrate.**

array of air blocks and a dipole antenna is located at the interface between the middle and bottom layers. For comparison, the result with a homogeneous top layer is also shown. It is observed that at lower frequencies, the structure with a periodic overlay exhibits higher radiation efficiency because the air blocks reduce the effective dielectric constant of the substrate, hence the surface wave losses.

Surface wave band diagram associated with the periodic structure in Fig.12.16 is shown in Fig.12.17. The band gap in the air block structure is not omnidirectional [10], [11]. The surface wave loss is higher in the band gap. The surface wave power enhances in other directions in the band gap near the broadside(dipole) direction. The surface wave power pattern associated with Fig.12.16 is shown in Fig.12.18. High directivity of the pattern is observed. True efficiency enhancement using EBG structure requires a modeless substrate as describe in Section 2. The analysis of full-sized antennas on modeless structure is similar to that with periodic material blocks, and is currently under development.

4. Microstrip Line on Periodic Substrates

Microstrip line is a basic component in integrated circuits. For microwave EBG applications, it is of interest to study the transmission characteristics of microstrip lines on an artificial periodic substrate. A full-wave spectral domain method based on the DOVIE method de-

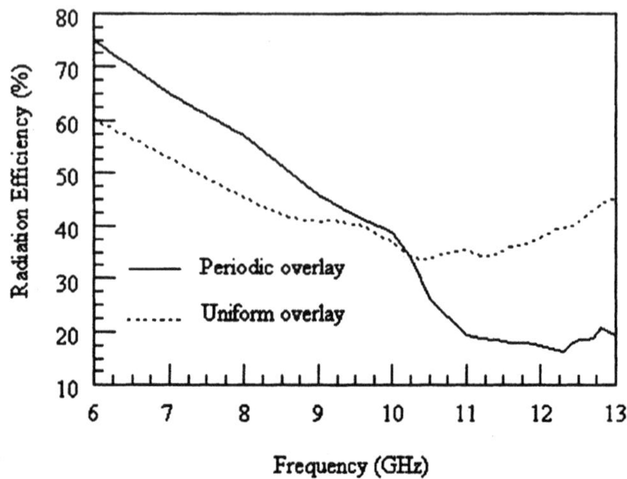

Figure 12.16. Radiation efficiency of a strip dipole on a three-layered substrate, — : top layer with periodic air blocks, – –: homogeneous overlay. All three layers have thickness of 0.635 mm and dielectric constant of 10.2, dipole is 1 mm wide, has resonant length, block period is 8 mm and air blocks are 4 mm × 4 mm.

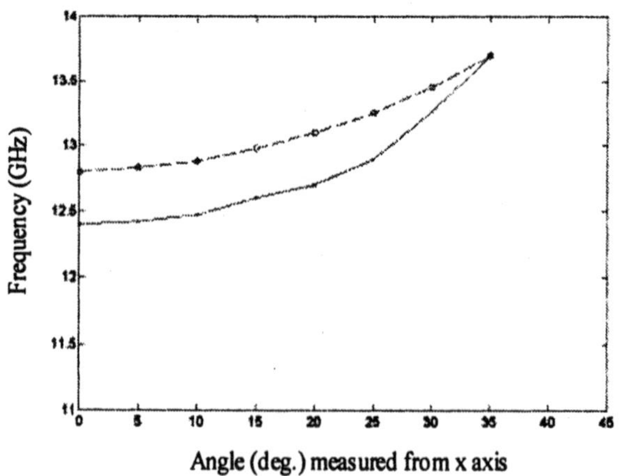

Figure 12.17. Surface wave band gap with a homogeneous substrate and a periodic overlay, parameters the same as in Fig.12.16, —: lower edge of band gap, – –: higher edge of band gap.

scribed in Section 3 has been developed for microstrip line on a substrate with periodic material blocks [41]. The parameters of particular interest

GPSR Compliance
The European Union's (EU) General Product Safety Regulation (GPSR) is a set of rules that requires consumer products to be safe and our obligations to ensure this.

If you have any concerns about our products, you can contact us on

ProductSafety@springernature.com

In case Publisher is established outside the EU, the EU authorized representative is:

Springer Nature Customer Service Center GmbH
Europaplatz 3
69115 Heidelberg, Germany

www.ingramcontent.com/pod-product-compliance
Ingram Content Group UK Ltd.
Pitfield, Milton Keynes, MK11 3LW, UK
UKHW022318190726
13856UKWH00001B/82

* 9 7 8 1 4 7 5 7 4 1 5 7 5 *